NEBRASKA
Pioneer & Heritage Farms

Preserving Our Agricultural Heritage

150th Anniversary Edition

EST. 1895
AKSARBEN
FOUNDATION™

Acclaim Press™
MORLEY, MISSOURI

A farmer and his team of horses in Lancaster, Nebraska, May 1936, Arthur Rothstein, photographer (1915-1985). Library of Congress photo.

Acclaim Press
— Your Next Great Book —

P.O. Box 238
Morley, MO 63767
(573) 472-9800
www.acclaimpress.com

Book Design: Rodney Atchley
Cover Design: M. Frene Melton
Cover Logo Design: M. Frene Melton

For more information about the Nebraska Pioneer and Heritage Farm Awards, contact:

AKSARBEN Foundation
6910 Pacific St, Ste. 102
Omaha, NE 68106
Phone: 402.554.9600
Fax: 402.554.9609
Email: info@aksarben.org

ISBN: 978-1-942613-59-6 | 1-942613-59-8
Library of Congress Control Number: 2016917616

First Printing: 2017
Printed in the United States of America
10 9 8 7 6 5 4 3 2 1

Contents

EST. 1895
AKSARBEN
FOUNDATION

GIVING BACK. MOVING FORWARD.

September 1, 2016

Dear Friends,

Founded in 1895 by Omaha merchants dedicated to the success of the Nebraska State Fair, AKSARBEN Foundation has been a proud advocate for Nebraska Agriculture for 121 consecutive years; funding youth 4-H programs and county fairgrounds, as well as celebrating the cultural heritage of our Heartland through annual events and special recognitions such as the Pioneer and Heritage Farms of Nebraska.

The Homesteading and founding families of our region carved out of the "Great American Desert" a "Great Plains" that feeds our world with its food, family values, hard work and ingenuity. This region, aptly named the "Heartland," has become the economic, cultural and ethical pulse of our great nation, due to those families who have remained steadfast in their commitment to the region through thick and thin.

Like many Nebraskans, my ancestors were Homesteaders, settling in south central Nebraska and co-founding a small community that continues today. Though my family is a few generations removed from the farm, we remain steadfast in our pride and appreciation of our heritage.

On behalf of all current and emeritus members of the Knights of AKSARBEN, and the entire AKSARBEN Foundation family of supporters, we congratulate the Pioneer and Heritage Farm families of Nebraska, and we thank you for serving as the backbone of our great region for generations.

Sincerely,

Jonathan J. Burt
President

6910 Pacific Street, Suite 102 Omaha, NE 68106 | 402-554-9600 | www.aksarben.org

STATE OF NEBRASKA

Pete Ricketts
Governor

OFFICE OF THE GOVERNOR
P.O. Box 94848 • Lincoln, Nebraska 68509-4848
Phone: (402) 471-2244 • pete.ricketts@nebraska.gov

Dear Friends:

"The Good Life" is a phrase beloved by all Nebraskans. It embodies our spirit, our lifestyle and the beautiful surroundings in which we reside.

But in order to make Nebraska such a great place to live, work, and raise a family, it took the strong will, perseverance and a vision of hope from our ancestors who came before us.

These hard-working families toiled in the fields and took great care of their animals that began a great tradition rooted in agriculture. Many of the farms and ranches that were begun over a hundred years ago have been handed down from generation to generation and remain successful operations today.

It is this deep-seated history that has allowed agriculture to flourish and become the state's number one industry. Nebraska is among the top agriculture states in the nation, and our farmers and ranchers have helped us become a leader in several agriculture production categories.

Our past is what is taking us into the future, and we will always cherish and honor our agricultural heritage. It allows us to be optimistic about what lies ahead as is reflected in our updated phrase...."Nebraska: Good Life, Great Opportunity!"

Sincerely,

Pete Ricketts

Pete Ricketts
Governor

An Equal Opportunity/Affirmative Action Employer

STATE OF NEBRASKA

Pete Ricketts
Governor

Department of Agriculture
Greg Ibach
Director
P.O. Box 94947
Lincoln, NE 68509-4947
(402) 471-2341
Fax: (402) 471-6876
www.nda.nebraska.gov

Dear Friends,

Farm and ranch families helped build Nebraska over 150 years ago and remain a vital part of our economy and culture today.

Many of these original farms and ranches have been handed down from generation to generation and continue to be family-owned. While operations may be larger in scope than in the past, the values and ethics that have kept these families together remain strong. These multi-generational family farms today produce food for families around the world.

One thing that has changed is the fact that a farm or ranch doesn't just support one family, it supports several. Whether it's employees who assist in daily operations, or service providers in town helping to keep everything running, all of their families are connected to agriculture directly or indirectly. After all, agriculture is the number one economic driver in Nebraska with one in four jobs derived from the ag industry.

Family farms and ranches have been a cornerstone of the great state of Nebraska and will continue to be part of our foundation for years to come. I commend Nebraska's farm families for their hard work that has made Nebraska the land of the good life and great opportunity.

Sincerely,

DEPARTMENT OF AGRICULUTRE

Greg Ibach
Director

An Equal Opportunity Employer

UNIVERSITY OF Nebraska Lincoln®

EXTENSION
INSTITUTE OF AGRICULTURE AND NATURAL RESOURCES

September 1, 2016

Dear Friends of Nebraska Agriculture:

I am thrilled that the AKSARBEN Foundation in partnership with Acclaim Press are publishing Nebraska's Historic Farms, a recognition of Pioneer and Heritage Farms in Nebraska and celebrates our state's 150th anniversary of statehood. Nebraska's economy is driven by all things agriculture...Nebraska's farmers and ranchers feed the world.

Since Daniel Freeman's first Homestead to today, Nebraska's farmers and ranchers have nurtured their crops and livestock, protected our natural resources and built a strong and vibrant rural economy. These people scratched a living from the soil, dealt with drought, insect pests and weeds and built a food production system second to none on this planet. In the process, the ingenuity and inventiveness of Nebraska's farmers and ranchers created better and more efficient ways of producing food.

Today's farmers and ranchers apply amazing technologies to produce food, protect the environment and care for their families. In the process, their children learn responsibility and caring for crops and livestock in ways that position them for long-term success. And, rural communities benefit from the economic impact of Nebraska's agricultural community.

A hearty congratulations to Nebraska's Pioneer and Heritage farm/ranch families. With grit, determination and ingenuity, these families persevered through bad times and good and built a tradition of commitment to the people, the land, the crops, the livestock and the communities of Nebraska. It is only right that we honor them!!!

The University of Nebraska-Lincoln Cooperative Extension Service salutes the Pioneer and Heritage farm/ranch families of Nebraska!!!

Sincerely,

Charles A. Hibberd

Charles A. Hibberd, Dean and Director
Nebraska Extension

211 Agricultural Hall / P.O. Box 830703 / Lincoln, NE 68583-0703 / (402) 472-2966 / FAX (402) 472-5557
UNIVERSITY OF NEBRASKA-LINCOLN, COOPERATING WITH COUNTIES AND THE U.S. DEPARTMENT OF AGRICULTURE

Nebraska Farm Bureau Federation
5225 S. 16th St., P.O. Box 80299, Lincoln, NE 68501 Phone: (402) 421-4400, FAX# (402) 421-4432

Dear Friends,

This book is a wonderful way to recognize recipients of the Nebraska's Pioneer and Heritage Farm Award. It is quite an achievement for a farm or ranch to be designated a Nebraska Pioneer or Heritage Farm and is a testimony to the importance of family dedication, rural values and hard work. Congratulations to all of the recipients!

It is fitting that Nebraska Farm Bureau partners with the AKSARBEN Foundation in sponsoring the Nebraska Pioneer or Heritage Farm program and is helping bring you this publication. After all, Nebraska Farm Bureau is a grassroots organization led by farmers and ranchers in every county who work to enhance and strengthen the lives of all Nebraskans. It has been 100 years and we continue to be the public face of Nebraska agriculture and the collective voice of farm and ranch families.

Every farmer and rancher has a story. Stories about hard work, honesty, integrity, values and tradition. This publication showcases some of those stories. It also captures Nebraska's beauty and rural ways of life. Farm Bureau works to preserve that way of life. Individually, each Nebraskan's effort is important, but when we put our voices together – we can make amazing things happen for our state.

Thanks to all of you who are a part of agriculture today, who have been involved with agriculture in the past or who just have an interest in learning more about Nebraska farmers and ranchers, who provide us with the best food supply in the world. Lastly, thank you to Acclaim Press for your vision and for publishing this historical document of Nebraska agriculture.

Sincerely,

Steve Nelson, President
Nebraska Farm Bureau

Photos courtesy of Nebraska State Genealogical Society

Nebraska, My Home

Facts and figures related to the 37th state and Nebraska agriculture

The word "Nebraska" roughly translates as "flat water" after the Platte River that flows through the state.

Nickname: Cornhusker State
Motto: Equality before the law
Capital: Lincoln
Largest City: Omaha

Total Area: Ranked 16th among states / 77,358 sq. miles
Population: Ranked 37th among states / 1,896,190 (2015 est.)

Population Density: Ranked 43rd among states / 24 people per square mile

Highest Elevation: Panorama Point - 5,424 feet
Lowest Elevation: Missouri River at Kansas border - 840 feet

Admitted to the Union: March 1, 1867 (37th state)

Boundaries: The state is bordered by South Dakota to the north; Iowa to the east; Missouri to the southeast; Kansas to the south; Colorado to the southwest; and Wyoming to the west.

Counties: 93

Regions: Nebraska is composed of two major land regions: the Dissected Till Plains

Courtesy of Nebraska State Genealogical Society

Cornfields in Southern Nebraska.

Courtesy of Nebraska State Genealogical Society

and the Great Plains. The Dissected Till Plains is a region of gently rolling hills in the easternmost part of the state. The Great Plains occupy most of western Nebraska consisting of several smaller, diverse land regions — the Sandhills, Pine Ridge, Rainwater Basin, High Plains, and Wildcat Hills.

Average annual precipitation: 31.5 inches in the southeast / 13.8 inches in the Panhandle.

Average annual snowfall: 25-35 inches

Highest recorded temperature: 118° F at Minden on July 24, 1936

Lowest recorded temperature: -47° F at Camp Clarke on February 12, 1899

89% of cities in Nebraska have fewer than 3,000 people

National Agriculture Rankings according to the Nebraska Department of Agriculture, USDA, and NASS, February 2016:

1st
- Beef and veal exports
- Cash receipts from meat animals
- Commercial red meat production
- Commercial cattle slaughter
- Great Northern bean production
- Popcorn production
- Irrigated acres of cropland

2nd
- All cattle and calves
- Pinto bean production
- Proso millet production
- Light red kidney bean production
- Bison

3rd
- Corn for grain production
- Cash receipts from all farm commodities
- Corn exports

4th
- Cash receipts from all livestock and products
- Cash receipts from all crops
- All dry edible bean production
- Beef Cows
- Land in farms and ranches
- Alfalfa hay production
- Grain sorghum production
- Soybean production

5th
- Agricultural exports
- Soybean exports
- All Hay, production

6th
- Harvested acres of principal crops
- All hogs and pigs on farms
- Sugar beet production

7th
- Sunflower production
- Commercial hog slaughter
- Oat production
- Winter wheat production

Courtesy of Nebraska State Genealogical Society

An Introduction to the Pioneer Farm

Farmsteader's son with his prize 4-H Club calf, Fairbury Farmsteads, Nebraska, May 1936, by Arthur Rothstein. Library of Congress photo.

When most people think of a family farm, their thoughts turn to a plot of land farmed by a single family where everyone, parents and children alike, works from sunrise to sunset and all have a role to ensure that daily chores are met. However, many family farms across the state are much more than that—the family connection to the farm goes back one hundred years or more, where as one generation moves on to its golden years, the farm is turned over to the next to continue the farming way of life. These farms are special. They are "Pioneer Farms".

Since its inception in 1956, the Pioneer Farm Program has designated nearly 10,000 farms as Nebraska Pioneer Farms. To receive this prestigious award, a farm must have remained within a family line for at least a Century. Since 2014, the Heritage Farm Program has given an award to families who can trace their farm ownership back one hundred fifty years, which is truly something special.

The story of agriculture in Nebraska began thousands of years ago, with people and crops that are gone but not forgotten. This story continues to the 1800s, when the Homestead Act opened the state up for settlement, and the Union Pacific railroad offered up land for sale cheap for anyone hearty enough to move to the prairie and make a fresh start. Settlers came from Eastern states and also from European countries who could sell their small farms in exchange for larger properties in this new state. These people were our first family farmers.

The stories included within these pages span the state, from small farms to big ranches, all providing each family's own unique story of agriculture in rural Nebraska. A timeline provides a touchstone to easily find important agricultural dates, along with a look at the Aksarben Foundation, who oversees the Pioneer and Heritage Farm Awards program and have been so important to the preservation of the Nebraskan way of life.

To the hundreds who have lent their stories and photographs to tell the story of the Pioneer and Heritage Farms and Nebraska agriculture, we offer our deepest gratitude. Your assistance and contributions ensure Nebraska's Pioneer Farms will always be remembered.

We are pleased to present this permanent record of Nebraska farming heritage.

Nebraska Pioneer Farm Award

THE NEBRASKA ASSOCIATION OF FAIR MANAGERS

is Honored to Recognize

Payne Homestead

whose family has owned the same farm in Nebraska for one hundred years or more. By contributing a century of service to Nebraska agriculture, members of this pioneer family have demonstrated qualities of stability and loyalty so essential to fine citizenship.

THE NEBRASKA ASSOCIATION OF FAIR MANAGERS

[illegible]
President

Orville Koch
Secretary

Dated this 1st day of June, 1977

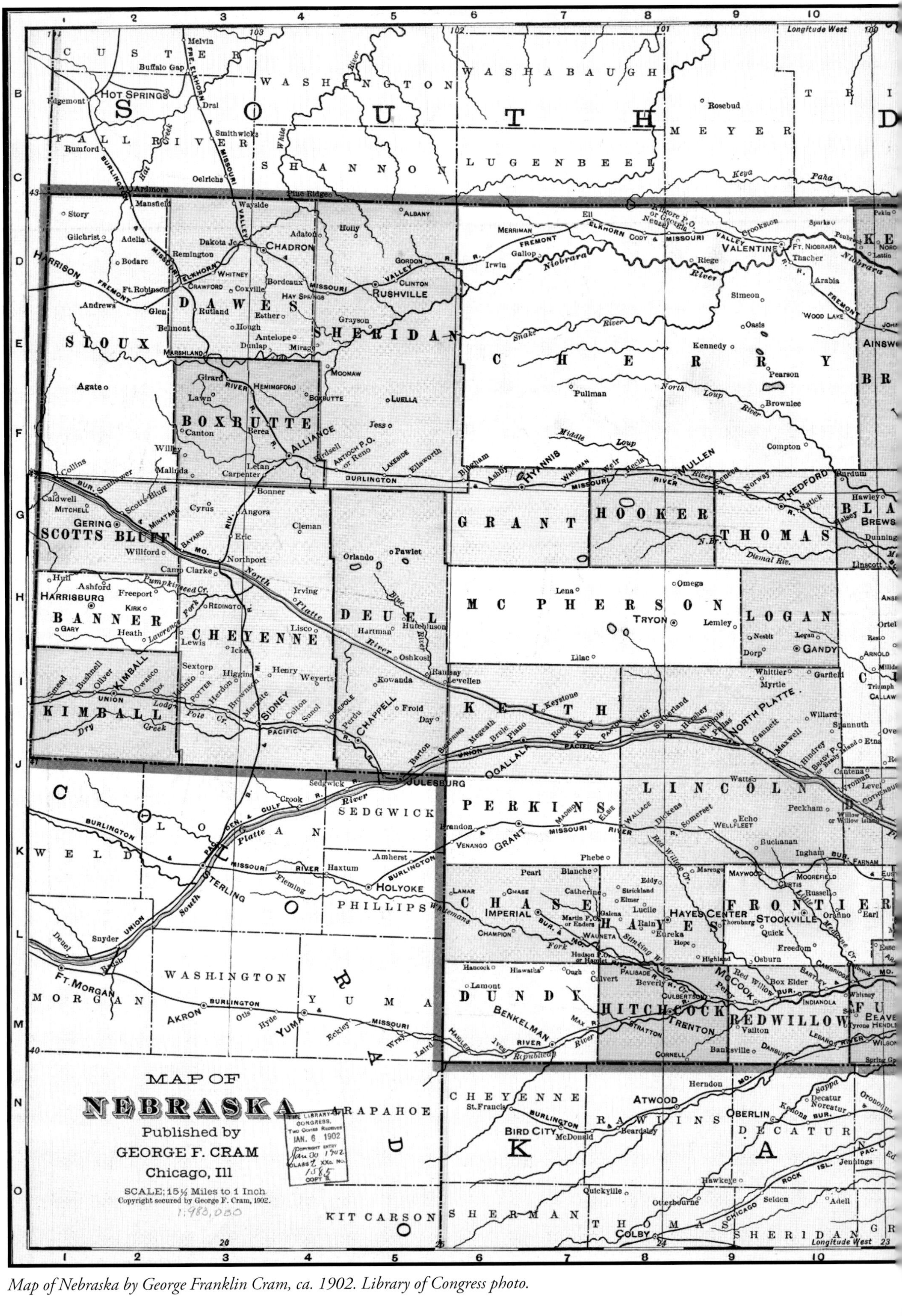

Map of Nebraska by George Franklin Cram, ca. 1902. Library of Congress photo.

13
14
15
16
17
18
19
20
21
22
B
C
D
E
F
G
H
I
J
K
L
M
N
O
BRULE
AURORA
DAVISON
HANSON
MC COOK
MINNEHAHA
M I N N.
DOUGLAS
HUTCHINSON
TURNER
LINCOLN
LYON
OSCEOLA
GREGORY
CHARLES MIX
YANKTON
BON HOMME
CLAY
UNION
SIOUX
O'BRIEN
PLYMOUTH
CHEROKEE
WOODBURY
IDA
MONONA
CRAWFORD
HARRISON
POTTAWATTAMIE
MILLS
A
K
O
T
A
I
O
W
A
BOYD
HOLT
KNOX
CEDAR
DIXON
DAKOTA
THURSTON
ANTELOPE
PIERCE
WAYNE
GARFIELD
WHEELER
MADISON
STANTON
CUMING
BURT
BOONE
PLATTE
COLFAX
DODGE
WASHINGTON
GREELEY
VALLEY
NANCE
BUTLER
SAUNDERS
DOUGLAS
SARPY
HOWARD
SHERMAN
MERRICK
POLK
BUFFALO
HALL
HAMILTON
YORK
SEWARD
LANCASTER
CASS
OTOE
KEARNEY
ADAMS
CLAY
FILLMORE
SALINE
GAGE
JOHNSON
NEMAHA
FRANKLIN
WEBSTER
NUCKOLLS
THAYER
JEFFERSON
PAWNEE
RICHARDSON
PHILLIPS
SMITH
JEWELL
REPUBLIC
WASHINGTON
MARSHALL
NEMAHA
BROWN
MITCHELL
CLOUD
CLAY
RILEY
POTTAWATOMIE
JACKSON
OSBORNE
ROOKS
N
S
A
S
MO.
SIOUX FALLS
ROCK RAPIDS
ARMOUR
OLIVET
PARKER
CANTON
WHEELER
FAIRFAX
TYNDALL
YANKTON
VERMILLION
ELK POINT
LE MARS
CHEROKEE
PRIMGHAR
SIBLEY
SIOUX CITY
IDA GROVE
ONAWA
LOGAN
COUNCIL BLUFFS
GLENWOOD
PLATTSMOUTH
SIDNEY
NEBRASKA CITY
BUTTE
NIOBRARA
O'NEILL
HARTINGTON
PONCA
DAKOTA
NELIGH
PIERCE
WAYNE
PENDER
MADISON
STANTON
WESTPOINT
TEKAMAH
BARTLETT
BURWELL
ALBION
ORD
GREELEY
FULLERTON
COLUMBUS
SCHUYLER
FREMONT
BLAIR
OMAHA
PAPILLION
WAHOO
DAVID CITY
ST. PAUL
LOUP
CENTRAL CITY
OSCEOLA
GRAND ISLAND
AURORA
YORK
SEWARD
LINCOLN
KEARNEY
HASTINGS
CLAY CEN.
GENEVA
WILBER
MINDEN
BEATRICE
TECUMSEH
AUBURN
FALLS CITY
PAWNEE CITY
HEBRON
FAIRBURY
NELSON
RED CLOUD
BLOOMINGTON
ALMA
MARYSVILLE
SENECA
HIAWATHA
BELLEVILLE
WASHINGTON
MANKATO
SOUTH CENTER
CONCORDIA
BELOIT
HOLTON
CLAY CEN.
WESTMORELAND

NEBRASKA FARMERS AND A TIMELINE OF AMERICAN AGRICULTURE

Editor's Note: Items in gray with the state outline are specific to Nebraska agriculture history.

1714

Etienne Veniard de Bourgmont is the first recorded European in Nebraska.

1720

Pedro de Villasur is sent from Santa Fe with forty-five soldiers to scout enemy positions on the plains. As the group nears present day Columbus, Nebraska, they are attacked by Pawnee Indians. Villasur is killed and only thirteen Spaniards make it back to Sante Fe.

1763

Treaty of Paris. All land west of the Mississippi River becomes Spanish territory.

1795

George Washington begins breeding mules after the King of Spain gives him two jack mules.

1800

Treaty of San Ildefonso. The Spanish found it costly to explore this new country and could not see the rewards being worth the investment. They returned the Louisiana to France in 1800.

"Pioneers Crossing the Plains of Nebraska", by Carl Christian Anton Christensen (1831-1912), ca. 19th century.

1803

Louisiana Purchase doubles the size of the United States and includes the region that would become Nebraska.

1804

Lewis and Clark reach the eastern edge of Nebraska.

1815

Frontier farmers ship wheat and meat East.

1819

The U.S. Army establishes Nebraska's first military post at Fort Atkinson.

Will Berg and his working horse team, ca. early 1900s. Courtesy of Vicki L. Kennedy-Weseman.

Covered wagons carried thousands of homesteaders to Nebraska from back east in the 1880s.

Jethro Wood patents iron plow with interchangeable parts.

1819-25

U.S. food canning industry established.

1820

Major Stephen H. Long made an expedition to the Rocky Mountains and back. His opinion of the plains was not favorable. This opinion, shared by many, could explain the reluctance of settlers to make the prairie their home. The plains most certainly offered new challenges to the pioneer.

Agriculture Committee founded in the U.S. House of Representatives.

1820s

New breeds of swine become available (Poland-China and Duroc-Hersey swine developed and Berkshire swine imported.)

1823

Bellevue becomes the first permanent settlement in Nebraska.

1830

About 250-300 labor hours are required to produce 100 bushels (five acres) of wheat with walking plow, brush harrow, and hand broadcast of seed, sickle and flail.

1834

McCormick Reaper patented for wheat.

John Lane manufactures plows faced with steel saw blades.

1718 map from Carte de la Louisiane et du cours du Mississippi by Guillaume de L'Isle, with the approximate borders of Nebraska highlighted. Library of Congress photo courtesy of Bill Whittaker.

1837

John Deere begins making steel plows.

1840s

Factory-made agricultural machinery increases farmers' need for cash and encourages commercial farming.

New breeds of cattle become available (Hereford, Ayrshire, Galloway, Jersey, and Holstein imported and bred.)

1841

Practical grain drill patented.

1842

First grain elevator, Buffalo, New York.

1843

Sir John Lawes founded the commercial fertilizer industry by developing a process for making superphosphate.

1842-44

The word "Nebraska" first began to appear in publications in 1842 when John Freemont explores the Platte Valley and names Nebraska.

1849

Boston hosts first American poultry exhibition.

1850

About 75-90 labor hours are required to produce 100 bushels (2.5 acres) of corn with walking plow, harrow, and hand planting.

1850s

Portable steam engine with drive wheel is adopted for threshers, saws and dairies.

1854

The Kansas-Nebraska Act is passed by U.S. Congress on May 30th. Population in the territory is 2,732.

Threshing wheat with sickel and manpower.

The McCormick Reaper was patented for wheat in 1834.

A family poses with the wagon in which they live and travel during their pursuit of a homestead, 1866. Photographer unknown.

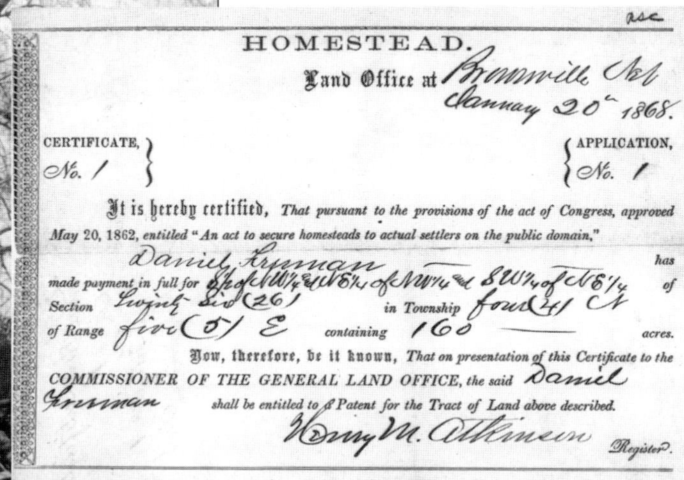

HOMESTEAD.

Land Office at Brownville Neb
January 20th 1868.

CERTIFICATE, No. 1 | APPLICATION, No. 1

It is hereby certified, That pursuant to the provisions of the act of Congress, approved May 20, 1862, entitled "An act to secure homesteads to actual settlers on the public domain," Daniel Freeman has made payment in full for [illegible] of Section Twenty Six (26) in Township four (4) N of Range five (5) E containing 160 acres.

Now, therefore, be it known, That on presentation of this Certificate to the COMMISSIONER OF THE GENERAL LAND OFFICE, the said Daniel Freeman shall be entitled to a Patent for the Tract of Land above described.

Henry M. Atkinson Register.

The Daniel Freeman homestead in Gage County, Nebraska, the first homestead claim under the 1862 Homestead Act. Photo taken in 1904. Library of Congress photo. Top right: Homestead certificate issued to Daniel Freeman, issued January 20, 1868. Note it is application and certificate number one.

1856

Two-horse straddle-row cultivator patented.

1857

Chicago stockyard ship iced pork.

1858

Mason home canning jar invented.

1860

The population in the Nebraska territory rises to 28,841.

Farmers are 58% of the labor force.

Change from hand power to horses characterizes the first American agriculture revolution.

1861

On March 2nd, U.S. Congress creates the Dakota and Nevada Territories out of the Nebraska and Utah territories.

1862

Homestead Act opens Nebraska territory for settlement. Public land is awarded to any head of a family on the condition that the settlers improve the land and live there for five years.

1867

Nebraska admitted to the union as the 37th state on March 1, 1867.

1869

Union Pacific Railroad is completed, leading to a greater influx of settlers to the state.

1874

The Nebraska Relief and Aid Society is formed on September 18th to help farmers whose crops were destroyed by grasshoppers swarming throughout the American West.

Glidden barbed wire patented; fencing of rangeland ends era of unrestricted, open-range grazing.

1881

Hybridized corn produced.

Horse-drawn planters

1888

The first long haul shipment of a refrigerated freight car was made from California to New York.

1889

Carrier of cattle tick fever identified.

1890

Superintendent of U.S. Census declares the American frontier closed. Farmers are 43% of the U.S. population:

- 65% of Americans are rural
- Farming produces 71% of total U.S. exports.

40-50 labor hours required to produce 100 bushels (5 acres) of wheat with gang plow, seeder, harrow, binder, thresher, wagons, and horses; 35-40 labor hours required to produce 100 bushels (2.5 acres) of corn with 2-bottom gang plow, disk and peg-tooth harrow, and 2-row planter.

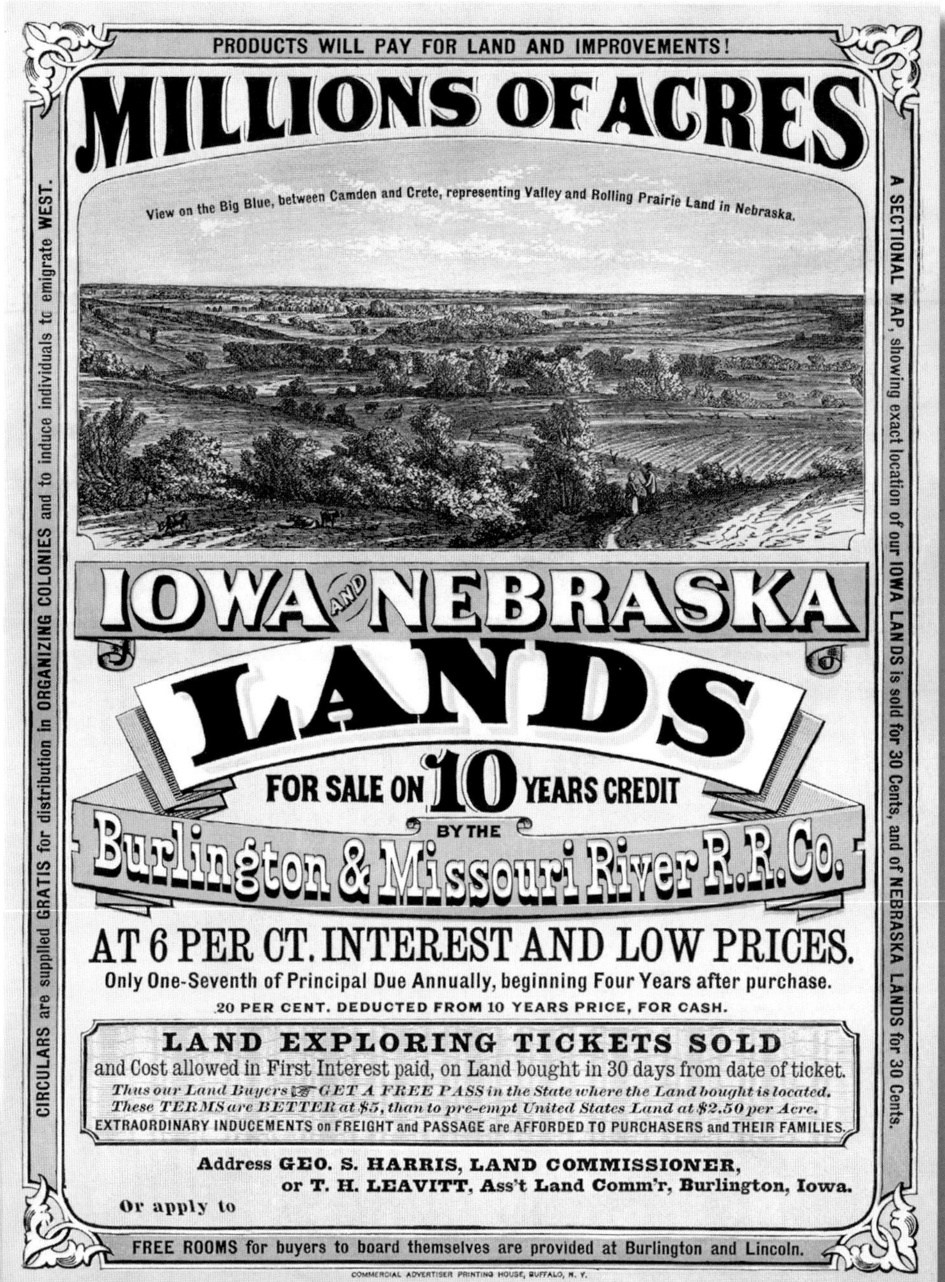

Right: Poster promoting Iowa and Nebraska acreage for homesteading by the Burlington & Missouri River Railroad Co. in the years following the homestead act.

Below: Planting corn in Lancaster County, May 1936. Photo by Arthur Rothstein. Library of Congress photo.

Wheat field on the farm of T.C. Shawver, Dutch flats near Mitchell, Nebraska, July 24, 1910. National Archives photo.

J.M. Clare and Marie Schuler Gillespie, 1913. Courtesy of Betty J. Gillespie Dykman and family

1890-99

Average annual consumption of commercial fertilizer: 1,845,900 tons.

1891-92

First gasoline tractor is built by John Froelich.

1896

Rural-free delivery of the mail.

1900-09

Average annual consumption of commercial fertilizer is 3,738,300 tons.

1902

The Reclamation Act of 1902 earmarks federal aid for irrigation projects.

1905

The first business devoted exclusively to making tractors is established.

1906

Food & Drug Act.

1910-19

Commercial fertilizer use is 6,116,700 tons per year.

Hans and Myrtle Horst with her brother, Walt Stubbs, with the horses and mules used for all farm work, ca. early 1900s. Courtesy of Lynn Horst Sill

Automobiles at the Freemont Tractor Show, August. 9, 1916. Library of Congress photo.

Farmer at sorghum mill, Lancaster County, Nebraska. Photos by John Vachon, October 1938. Library of Congress photo.

Cultivating a potato field, Douglas County, Nebraska. Photo by Arthur Rothstein, May 1936. Library of Congress photo.

John Deere tractor binder

1910

Farmers are 31% of the labor force:

• Farming produces 45% of total U.S. exports.

• One farmer feeds seven people.

1914-1918

World War I expands overseas markets.

1916

Stock Grazing Homestead Act.

1919

Nebraska ratifies Prohibition on January 16th.

1929-1939

The Great Depression

Commercial fertilizer use is 6,599,913 tons per year.

1930s

Frozen food readily available.

All-purpose, rubber-tired tractor with complementary machinery popularized.

1930

15-20 labor hours required to produce 100 bushels (2.5 acres) of corn with 2-bottom gang plow, 7-foot tandem disk, 4-section harrow, 2-row

Feeding the chickens at Fairbury Farmsteads, Nebraska. Photo by Arthur Rothstein, May 1936. Library of Congress photo.

Using a hand tractor for cultivation, Falls City Farmsteads, Nebraska. Photo by Arthur Rothstein, May 1936. Library of Congress photo.

The wife of a homestead farmer, Kearny, Nebraska. Photo by Arthur Rothstein, May 1936. Library of Congress photo.

planters, cultivators, and pickers; 15-20 labor hours required to produce 100 bushels (5 acres) of wheat with 3-bottom gang plow, tractor, 10-foot tandem disk, harrow, 12-foot combine, and trucks.

1933

Agricultural Adjustment Act.

1934

There are approximately 135,000 farms in Nebraska.

1937

Over 90,000 combines working wheat compared with 4,000 in the 1920s.

1940s

One farmer feeds nineteen people.

Commercial fertilizer use is 13,590,466 tons per year.

1941

World War II expands overseas markets.

1944-1945

Congress passes the Pick-Sloan Missouri Basin Project, which authorizes the creation of flood control dams, reservoirs and hydroelectric plants in states drained by the Missouri River.

1945-70

Change from horses to tractors and increasing technological practices characterize the second American agricultural revolution.

Old fashioned pulley driven hay bayler.

Milking a cow in Custer County, Nebraska, May 1936. Photo by Arthur Rothstein. Library of Congress photo.

Corn shocks with Scottsbluff in the background, North Platte River Valley, Nebraska, September 1941. Photo by Marion Wolcott. Library of Congress photo.

1945

10-14 labor hours required to produce 100 bushes (2 acres) of corn with tractor, 3-bottom plow, 10-foot tandem disk, 4-section harrow, 4-row planters and cultivators, and 2-row picker.

1948

Farmers shipping by truck.

1950

One acre produces thirty-eight bushels of corn.

Commercial fertilizer use is 22,340,666 tons per year.

One farmer supplies 15.5 persons (est.)

1954

More tractors than draft animals on the farm.

1955

6.5 labor hours required to produce 100 bushes (4 acres) of wheat with tractor, 10-foot plow, 12-foot row weeder, harrow, 14-foot drill, self-propelled combine and trucks.

1960

Farmers are 8.3% of the labor force:

- One farmer feeds forty-six people.
- Farming produces 23% of total U.S. exports.

Hybrid corn seed planted in 96% of corn acreage.

Farmer on tractor in the 1940s.

Frank E. Hagemeister beside his Farmall tractor, September 1941. Photo by Marion Wolcott. Library of Congress photo.

Filling a homemade silo with corn, Waterloo, Nebraska, September 1941. Photo by Marion Wolcott. Library of Congress photo.

Threshing beans in the North Platte River Valley, September 1941. Photo by Marion Wolcott. Library of Congress photo.

1965

There are approximately 82,000 farms in Nebraska. The average size of each farm is about 588 acres.

5 labor hours required to produce 100 bushes (3 acres) of wheat with tractor, 12-foot plow, 14-foot drill, 14-foot self-propelled combine, and trucks; 99% of sugar beets harvested mechanically.

1966

Fair Labor Standards Act extended to farm workers.

1970s

Centre-pivot sprinkler systems introduce a fundamental change in the history of Nebraska agriculture, making it possible to cultivate land that previously could not be irrigated.

Commercial fertilizer use is 43,643,700 tons per year.

No-tillage agriculture popularized.

1970

Chicago Stockyards close.

1974

Gerald Ford of Omaha became President of the United States.

1950s-style open-cab combine.

Preparing to plant seeds in the 1950s.

The Ericson farm in 1960, owned by Knowlen and Harriette. Courtesy of Kevin Ericson.

Nebraska grain bins and elevator, 2015. Photo by D. Ramey Logan.

1975

3.75 labor hours required to produce 100 bushels (3 acres) of wheat with tractor, 30-foot sweek disk, 27-foot drill, 22-foot self-propelled combine, and trucks; 3.33 labor hours required to produce 100 bushels of corn with tractor, 5-bottom plow, 20-foot tandem disk, planter, 20-foot herbicide applicator, 12-foot self-propelled combine, and trucks.

1980s

Commercial fertilizer use is 47,411,166 tons per year.

More farmers use no-till or low-till methods to curb erosion.

1980

One farmer supplies 75 people.

1981

USDA founds Agriculture in the Classroom for school children.

1982

Initiative 300 prohibits individual farmers from selling their land to corporations.

1983

Americans eat 25% more red meat than in 1945.

1989

After several slow years, the sale of farm equipment rebounds; more farmers begin to use low-input sustainable agriculture (LISA) techniques to reduce chemical applications.

1990

Farmers are 2.6% of the labor force.

One farmer supplies 100 people.

1994

Farmers begin using satellite data to manage crops:

- 29% of Americans are rural.
- One farmer feeds 129 people.
- Farming produces almost 19% of total U.S. exports.

Farm Bureau celebrates its 75th anniversary.

1997

First weed and insect resistant crops appear (soybeans & cotton).

2000

By the early 21st century, there are fewer than 50,000 farms in Nebraska, with the average size of each farm about 932 acres.

One acre produces 137 bushels of corn.

2000s

One farmer feeds 155 people.

2002

In Nebraska, a total of 7.5 million acres are irrigated, a 21% increase from the year 1992.

2004

In Nebraska, there are 48,300 farms covering 45.9 million acres.

Corn accounts for 22% of state farm receipts.

The average value of an acre of cropland is an estimated $1,750.00.

2005

Nebraska ranks 4th among the 50 states with $11.2 billion in farm marketings. Of this, about $7.3 billion came from livestock production and $3.9 billion from cash crops.

Nebraska ranks 3rd behind Texas and Kansas in the total number of cattle on farms, including 61,000 milk cows.

2006

Cuba buys $30 million in food from Nebraska, including corn, wheat, soybeans, and other products.

Sod House Living

Home for Nebraska's Early Pioneers

The William R. Dowse Sod House, located in Comstock, Custer County, which was built in 1900 and is listed on the National Register of Historic Places. It is now open as a museum.

With approval of The Homestead Act in 1862, families from the eastern U.S. and even across Europe came to settle the Great Plains, including Nebraska. Under the provisions of the act, settlers could obtain title to a quarter-section (160 acres) of land for a nominal fee, provided that they built a house, made certain improvements, farmed the land, and occupied the site for at least five years.

The eastern quarter of Nebraska had an ample supply of trees with which to build log homes. However, as settlers moved into the central and western parts of the state, trees became scarce as prairie took over. Many homesteaders, if not most, spent everything they owned purchasing the land, farming equipment and transportation to their new homes, so importing materials for building a home was out of the question.

The answer was to use the materials readily available; sod, cut into thick rectangular bricks, formed suitable walls when stacked upon each other, could be cheaply built and provided excellent insulation against the summer heat and winter chill.

Prairie grass with strong and dense root networks were preferred, such as buffalo grass, slough grass and big bluestem. At the house site, the soil was usually removed up to two feet deep, leaving bare earth that would serve as the dirt floor of the home. The actual sod would first be cut short or burned off, then scored or plowed in workable blocks that would be stacked as bricks.

Toward the end of the 1800s, a new plow was invented to create uniform sod cuttings. Known as the

Homestead family in front of their sod house with windmill on the roof on an adjoining building, ca. 1884-85. Photo by Solomon D. Butcher. Library of Congress photo.

Three adults and a child sitting along the roadway in front of their sod house and windmill, Custer County, Nebraska, ca. 1880-90. Photo by Solomon D. Butcher. Library of Congress photo.

The George O. Waters homestead near Comstock, Nebraska, 1887. Photo by Solomon D. Butcher. Library of Congress photo.

Pictured l-r: Lillie Semier holding daughter, Daisy, son George standing next to chair, and J.D. Semier standing with donkey, with their sod house in the background, near Woods Park, Custer County, Nebraska, ca. 1880-90. Photo by Solomon D. Butcher. Library of Congress photo.

grasshopper plow, the devise could cut a uniform strip of sod three to six inches deep and twelve to eighteen inches wide. The strip would then be cut into consistent blocks.

Stacked double-wide, the house walls could easily be three feet thick. Oftentimes the house would be topped with a sod roof and later, as materials were more readily available, the roofs were shingled or covered with tarpaper, boards or metal.

Although cheap and relatively easy to produce on the prairie, sod house living had many disadvantages, including dirty floors, leaking ceilings, and low light due to the expense of windows. However, just as the log homes used in the east, sod houses served their purpose and were even built through the Great Depression when many Nebraska families suffered financially.

Many examples of sod homes still stand today including the following sites listed on the National Historic Register:

Dowse Sod House near Comstock, listed on the NRHP and operated as a museum

Jackson-Einspahr Sod House, Holstein, listed on the NRHP

Gustav Rohrich Sod House, Bellwood, listed on the NRHP

Wallace W. Waterman Sod House, Big Springs, listed on the NRHP

AKSARBEN Historical Timeline

1895-2015

The current AKSARBEN Foundation, which began in 1895 as the Knights of Ak-Sar-Ben for the purpose of attracting visitors to the city of Omaha for the Nebraska State Fair, has evolved to become a leading Nebraska non-profit organization whose mission is to influence change for the betterment of youth, the economy, and Heartland communities. Along with the Nebraska Farm Bureau and the Nebraska Association of Fair Managers, AKSARBEN Foundation presents the Pioneer and Heritage Farm Awards each year. An overview of the Foundation's notable history follows over the next several pages.

1895

Twelve men went down to Mardi Gras to bring back the floats and save the State Fair from moving from Omaha.

First Parade, "Feast of Mondamin", was held September 19.

Meetings were held at Old Fellows Hall, 14 & Dodge.

Total membership was 554 business leaders who paid $10.

The official proclamation of the Knights of Ak-Sar-Ben, 1895

1910

Total membership was 1,590

Theodore Roosevelt, former U.S. president, at the Ak-Sar-Ben Den, 1910

Ak-Sar-Ben Parade marches west on Farnam Street, 1916

"Buy A Bond"

1919

Ak-Sar-Ben's purpose was directed to agriculture, the economic backbone of the region.

200 acres at 63rd & Shirley was purchased to build a permanent community center.

In its 25th year, total membership was 3,841.

Floats from Ak-Sar-Ben Electrical Pageant, 1918. "The Allies Answer"

"Lafayette, We've Come"

An Ak-Sar-Ben street carnival (King's Highway), 1919

Harness Races initiated at Ak-Sar-Ben Field, 1919

Racing fans view the Great Western Circuit Horse Racing Meet at Ak-Sar-Ben Field, 1921

The Imported Egyptian Ballet, 1923. Top row, l-r: G.N. Childs, R.N. Prawitz, C.W. Detrick, Ralph Kastner, A.D. Anderson. Bottom row, l-r: L.T. Peterson, Billie McKeague, John Ward, George Walker, H.L. Berkshire, L.E. Fitch, and Arlo McLarnan.

1921

Memberships were 4,280.

Racing was introduced on a 1-mile track.

Ak-Sar-Ben Den, located at 20th and Burdette Streets, as it appeared in August 1926

1927

The original Ak-Sar-Ben Den was destroyed by fire along with most of the organizations historical records.

Ak-Sar-Ben Den fire, June 27, 1927

Laying the cornerstone for the new Ak-Sar-Ben Coliseum, 1928

1928

The first Stock Show opened November 3.

The Coliseum was dedicated on May 25, 1928. Adjoining barns were added as well.

1929

Racing was banned in Nebraska.

Ak-Sar-Ben Coliseum, completed with cattle and horse barns, April 1929

1935
Racing resumes.

Spectators rise for a better view at Ak-Sar-Ben Track, 1936. Front row, l-r: Mrs. John Latenser Sr., Mrs. J.J. Gillin, Ms. Eileen Keliher-Jeffers, Mrs. Lawrence M'Cague, Mrs. Donald K. Howe, and Mrs. James E. Davidson. Standing in back: Mrs. Ballard Dunn (with polka dot tie), Mrs. Joseph M. Baldridge, Mr. Baldridge, Ms. Betty Paxton, and Mrs. Amos Thomas.

1938
Ak-Sar-Ben installs an ice rink in the Coliseum.

Ak-Sar-Ben purchases the Douglas Street Bridge for $2.3 million and promises to make it free to the public to typify the bond of friendship between Nebraska and Iowa.

1939
First year of the Ak-Sar-Ben Knights hockey team.

Races attendance reached 150,000.

The Ak-Sar-Ben Bridge between Iowa and Nebraska, Omaha, Nov. 1938. Photo by John Vachon (1914-1975) courtesy Library of Congress.

Ak-Sar-Ben Knights hockey team

1940
Ambassadors become a part of Ak-Sar-Ben.

1941
Women are allowed to become members.

The B-26 Martin Marauder, "The Ak-Sar-Ben Knight", with members of the crew, ca. 1940s. Pictured front row, l-r: Sgt. D.B. Reed, Ft. Smith, Ark.; Sgt. J.T. Ridge, Morristown, NJ; Sgt. D.L. Ohlsen, Los Angeles, Cal. Back row: Lt. Martin V. Frank, New York City; Lt. W.J. Amsberry, Anacortes, Wash.; and Lt. John D. Ashford, Omaha, Neb., Pilot.

Ak-Sar-Ben Coliseum served as a motor repair shop during World War II.

1942
US Army Transport set up on the Ak-Sar-Ben Grounds. No Hockey, Coronation or Races were held until 1945-1946.

"Food for Freedom" and "Smiles for Soldiers" created to meet wartime challenges of the nation.

Ak-Sar-Ben Field was used as a motor depot during World War II.

1943-45

4-H Beef Show temporarily moved to Omaha Stockyards.

Ground breaking for Ak-Sar-Ben 4-H building, March 16, 1945. Pictured l-r: George Garrison, Walter White, K.D. Fouts, L.I. Frisbie, G.A. Lineweaver, R.A. Turner, and Ed Janike.

1945

Horse races resumed.

4-H Building honored the 4-H youth who sacrificed their lives in WWII.

Attending the Ak-Sar-Ben Rodeo in 1948 are (l-r) Susan Murray, Queen of the Cheyenne Rodeo; Mayor Glenn Cunningham; Emily Reynolds, Ak-Sar-Ben Queen; and Norma Jean Bell, Cheyenne Queen's Lady-in-Waiting.

1944

Good Neighbor Program was started.

Ak-Sar-Ben 50 Year Celebration held at Creighton.

1947

First Ak-Sar-Ben Rodeo held, which replaced the horse show.

Ak-Sar-Ben Bridge paid off becoming the first free-toll bridge across the Missouri River in Nebraska.

Dedication of Ak-Sar-Ben 4-H Club building, November 2, 1946

An aerial view of Ak-Sar-Ben Racetrack and other facilities, early 1960s.

1954
Ak-Sar-Ben memberships topped 20,208.

Electrical parades resume for one year to commemorate the Omaha Centennial.

1956
The Pioneer Farm award program (100 years) begins.

1962
Ak-Sar-Ben Western Hall of Fame started to honor persons who helped build the West.

"They're off!" at the Ak-Sar-Ben Racetrack, early 1960s.

J.J. Isaacson, known as "Mr. Ak-Sar-Ben" in the 1960s

1966

Ak-Sar-Ben renovates the Coliseum and Barns covering the brick with metal and updating the interior at a cost of $2 million.

1963

To honor the memory of the original pioneer trek across the plains, Claire Wolf started the Ak-Sar-Ben Trail Ride from Lincoln to Omaha in time to participate in the grand entry at the rodeo.

Financial assistance for students of Omaha University and Creighton University, 1960. Pictured l-r: Dr. Milo Bail, President of University of Omaha; V.J. Skutt, Ak-Sar-Ben governor; and Very Reverend Carl Seinert S.J., President of Creighton University.

Nine hospitals receive a hospital gift from Ak-Sar-Ben, 1962. Pictured front row, l-r: J.D. Anderson, Lester Wehmer, James B. Moore. Second row: Francis Bath, Arthur Coad. Third row: Sister Ingeborg Blomborg, Mrs. Phyllis Schmidt. Fourth row: Miss Kathleen Regan and V.J. Skutt. Top row: John Hurley, Bing Grunwald, and Hal Perrin.

Royalty of Ak-Sar-Ben, 1962, with A.F. Jacobson and Kathleen Regan (Mrs. John Hughes)

Royalty of Ak-Sar-Ben, 1963, with J.L. Thurman and Tegwin Compton (Mrs. Timothy Westover)

1967

"The Ak-Sar-Ben Story" was released in April.

The Ak-Sar-Ben property was annexed into the City of Omaha.

1970

Membership dues increased to $12. Total membership was 53,622.

1972

Ak-Sar-Ben provides more than 500 scholarships annually.

To save the Orpheum Theater from possible demolition, Ak-Sar-Ben purchased the theater and donated it to the city of Omaha.

1976

Ak-Sar-Ben contributed $1,205,004 to 62 programs across the region.

Ice Show performance at Ak-Sar-Ben Coliseum, April 1966

Aerial view of the Ak-Sar-Ben Racetrack

1982

All-time record in memberships, 71,635.

The first River City Roundup was held as a celebration of Western Heritage and as a salute to agriculture.

1979

Two Russian/U.S. hockey games are played at Ak-Sar-Ben Coliseum.

Ak-Sar-Ben donates $300,000 to Schramm State Park Aquarium and $250,000 to the Henry Doorly Zoo Courtyard and Cat Complex.

1985

Ak-Sar-Ben provides more than 500 scholarships annually.

1979 Racing Program

Coronation & Scholarship Ball

1986

Ak-Sar-Ben racing monopoly becomes a thing of the past with dog track in Iowa and other tracks in Minnesota and Missouri.

The Coronation Ball is suspended for restructuring.

1988

Ak-Sar-Ben Coronation returns as a fund-raising event for the Ak-Sar-Ben Scholarships.

1992

Ak-Sar-Ben transfers the racetrack to Douglas County and gifts away $17 million in property assets to the community.

River City Roundup is voted "Omaha's Best Annual Event."

1996
Ak-Sar-Ben restructures.

Last season of racing on the Ak-Sar-Ben Grounds.

1998
Ak-Sar-Ben supports development of Qwest Center (now CenturyLink Center).

CenturyLink Center, formerly known as Quest Center

Horatio Alger Career Scholarship recipients

2002

AKSARBEN property redeveloped.

Last year of AKSARBEN events on the AKSARBEN property.

2005

AKSARBEN partners with Horatio Alger to provide $250,000 in need-based scholarships annually.

2003

Wrangler ProRodeo Tour Playoffs, televised on CBS, moves to Omaha.

AKSARBEN events moved to Qwest Center.

2008

AKSARBEN supports the AKSARBEN Village Project.

2010

AKSARBEN 4-H Stock Show adds North Dakota as an eligible state to compete.

AKSARBEN awarded 50 individual $6,000 scholarships, bringing the total awarded since 1945 to more than $7 million.

The Obelisk in Stinson Park, part of the Aksarben Village development in Omaha.

The Bruce family of Gage County proudly display their Pioneer Farm Award received in 2015.

Stock Show & Rodeo

2014

AKSARBEN launches the 4-H stock show AG Leaders Scholarship program, providing $100k in annual Scholarship awards to 4-H stock show exhibitors.

The Heritage Farm Awards Program (150 years) begins

2013

AKSARBEN annual events and fundraising efforts net over $1 million for AKSARBEN Scholars.

2015

AKSARBEN launches its largest scholarship program, the AKSARBEN Horatio Alger Program, providing over $1 million in annual scholarship awards to Heartland youth.

Photo courtesy of Nebraska Extension

Farm Histories

Editor's Note: The following Pioneer Farms submitted the stories about their individual farms. Some are categorized according to the founding family surname, some by present owners, and some according to the farm name. Please refer to the index as needed.

ABBENHAUS – William Abbenhaus acquired the original quarter section of land in Knox County, Township 30, Section 25 in 1885. The deed to this land was a Timber Culture Certificate or "Tree Claim" during the Grover Cleveland administration and was signed on April 4, 1885 and filed in Niobrara, NE.

In 1903, he purchased 160 acres to the west from the State of Nebraska as a Sale of Indemnity School Lands and in 1904, another 160 acres of "University Land" was purchased to the north of the "Tree Claim." William passed away in 1930. The original and north quarter were then owned by one of his sons, Frank, and then later by grandsons, Leonard and Gerald. The west quarter was sold over the years, but was repurchased in 1969 by Gerald Abbenhaus, Frank's son, William's grandson.

The sun peaks out of the horizon at the Abbenhaus farm.

On March 1, 1996, the farm was purchased from Gerald by his great-nephew, Joseph Abbenhaus and wife, Andrea. They farm this along with other land that they have acquired. Some of the original cottonwood trees from the "tree claim" are alive and standing. The original frame home, built in the 1890s, remains but is inhabitable.

Joe and Andrea have three children: Tyler, Kylie and Eric. They, as well as Joe's parents, Jim and Kay, raise corn, soybeans, alfalfa and beef cattle southwest of Bloomfield in Knox County. Tyler is the sixth generation to be involved in the Abbenhaus farming operations. Kylie and Eric are currently pursuing agricultural degrees. *Submitted by Joseph Abbenhaus.*

AHRENS – Charles S. and Karoline Coney purchased 160 acres in Pilger Township, Stanton County, NE in 1891.

With the use of horsepower they raised the traditional crops of small grains, corn, hay and pasture for the livestock. Cattle and hogs were raised along with poultry and fowl. Coming from Michigan there was also an orchard with a variety of fruits.

Charles S. Coney

Charles served as a teacher in Pilger Township for many years before being elected to serve as Stanton County Superintendent of Schools. Charles was re-elected for Superintendent four consecutive terms.

Charles and Karoline raised five children: Wilfred, Olga, Albert, Elsa and Edward. The farm, named Springbrook Stock and Grain Farm was purchased by Edward and his wife, Tillie, as the other four family members continued the westward movement of the time. Their quest for opportunity and success took them to California and Washington.

Ed and Tillie Coney family. Pictured l-r: Rolland, Tillie, Donna, Stanley, Larry, Edward, Charles, Karoline, Carol, Marilyn, and Nancy.

Edward and Tillie raised eight children: Marilyn, Stanley, Nancy, Charles, Carol, Donna, Larry and Rolland. Their conservative values, rich culture, perseverance and heritage led them to pass Springbrook on to their eight children in equal part. It was purchased by Marilyn, Charles and Larry and remains that way today. *Submitted by Marilyn Ahrens.*

ALBERT – In 1906, John H. and Emma (Haflinger) Albert purchased 40 acres east of Louisville, NE, which later expanded to 160. My father, Ervin, was born on the farm and had a brother, Earl, and a sister, Esther (Stohlmann). On August 13, 1939, he married Dorris (Parsell). To this marriage, I (Ray) was born in 1941 and my sister Margaret Kay (Snyder) in 1943.

My parents purchased the farm from my grandparents in 1945. In 1955, an adjoining 40 acres was added. Through inheritance and the purchase of additional land, the farm was expanded to 600 acres.

The Albert family, ca. 2015.

In 1966, I married my wife, Donna (Tlustos). We had two children, Kevin, born in 1968 and Renee (Yardley), born in 1971. Kevin married Krissy (Loos) in June 2000 and they have two children, Kole (11) and Kanon (9). Later that same year, in December, Renee married Brady Yardley. They also have two children, Braydon (14) and Emma (10).

In 1977, my father, Ervin, passed away and my mother, Doris, moved to town. Donna and I then purchased the home place and moved there in 1979, where we still reside. Kevin is currently farming this land, producing corn, soybeans and alfalfa hay. I, however, am semi-retired but still help out in the fall during harvest and supply seed and fertilizer in the spring. While also raising 40 head of Saler stock cows, I enjoy keeping up the family farm for generations to come.

ALTHOUSE – The north 1/2 of the northwest 1/4, Section 18, Township 11, Range 9E of the Sixth Principal Meridian in the County of Cass and State of Nebraska was transferred from The US to John Morrison on Feb. 26, 1873, according to the *Abstract of Record Title* issued by the Thomas Walling Company of Plattsmouth, NE. One day short of 115 years later, on Feb. 25, 1988, the property transferred to Steven and Tamara Althouse. The farm had passed through several ownerships in the years between Mr. Morrison and the Althouses, but since Aug. 9, 1905, the date it was purchased by Benjamin H. Landis, it has been continuously owned by members of the same family.

Benjamin Heistand Landis was Steve's great-great-grandfather. One of his daughters was Edith (Landis) Cooley, her daughter was Marian (Cooley) Althouse, her son is Clarence Althouse Jr. and his son is Steven Althouse. The farm passed into another branch of Benjamin Landis' family through his daughter Gladys (Landis) Daft in 1944 but Steve was given the opportunity to purchase the farm from the Daft heirs in 1988, thereby keeping the land in the family.

Steve and Tamara live in the house built by Benjamin and Sallie Landis in 1906 and they have raised three daughters on this farm. It continues to be a place where family is the priority and farming is our proud profession. It is our hope that the land will remain within Benjamin's family for generations yet to come. *Submitted by Steve and Tamara Althouse.*

Landis House, circa 1906. Benjamin Heistand Landis (second from right), Sallie Landis (far left).

Anderson farm in 1951

ANDERSON – Andrew Peterson Anderson of Ceresco, NE bought the northeast 1/4 of 12-14-6 in Merrick County from T.B. Hord Land and Cattle Co. in August 1911 for $10,400. In 1914 his son David and new bride, Dora (Breyer), moved to the farm. Upon arrival they found their new home infested with mice and bed bugs. In 1916 David purchased their first Fort touring car. David and Dora built a new house in 1918.

David's main occupation was farming. Corn, oats, wheat and alfalfa were grown along with feeding cattle, milking a few cows and raising hogs. David and a good friend, Rynold Butt, owned and operated the threshing rig with Rynold in charge of the engine and David the separator. It took from 18 to 20 men to keep the operation running smoothly. The women were in charge of cooking dinner and lunch to feed all those hungry men. The rig would move from one farm to the next and next year would start at the other end and do in reverse. David enjoyed hunting, fishing and many fun times in the cow pasture when the neighbor men gathered to play ball. David was school board treasurer for Dist. #21 school for many years. Dora raised chickens, ducks and geese along with tending a large garden and flowers. Neighbors would come to help, when needed as, in pitting cherries or picking the down from live geese for their pillows and feather ticks.

Travelers drive the road just west of the Anderson farm looking back to the east, July 20, 1924. At the time this road was the highway, known as the "white way" or the "great white way". It has since been moved one mile south and known as Hwy. 92.

Men walk on top of chopped corn stalks to make silage, 1935. To the right is David Anderson beside the tractor, the first on the farm with steel wheels. Myrtle Anderson is standing halfway up on the left side.

David took ownership of the farm after his father's death in 1940. Several other parcels of land were added to the original farm thru the years.

Dora continued living on the farm for seven years after David's death in 1965 and the next year deeded the farm to their only child, Myrtle (Anderson) Holtorf and husband Hans and they took over the operation along with their own farming operation which was only three miles away. In 1976 and 1977 the property was deeded to their daughter, Marilyn (Holtorf) Kyes and husband Edward when they took over the farming duties. They still live on and farm the property in 2015. A daughter and three sons were raised in this Kyes family and two sons have their own farming operations close by. We have a photo of the farm as it looked at the time it was bought in 1911 and also one in 1951 showing the house David and Dora built in 1918 and the growth of the building site. The house still stands after being remodeled and added onto four times since Edward and Marilyn took ownership.

In June 2014, Edward and Marilyn celebrated 50 years of marriage and that same year her parents, Hans and Myrtle Holtorf, would have celebrated 75 years on Feb. 18 and Myrtle's parents would have celebrated 100 years of marriage also on Feb. 18. *Submitted by Marilyn (Holtorf) Kyes.*

ANDERSON – In 1909, Emil and Charley Anderson purchased this farm ground from Joseph and Hatti Nelson. In 1913 Andrew P. Anderson, father of Emil and Charley, inherited Charley's share of the farm from his estate. In 1918 Joseph Anderson, son of Andrew P. and Clara M. Anderson, bought that share from them for $1.00 "with love and affection" as stated on the deed. In 1926, Joseph purchased Emil's share of the farm for $8,750. The current house on the farmstead was built by Joe in 1927 where he and his wife, Jennie lived after they married in 1928. They had two sons, Leonard and Lloyd, who farmed with Joe for several years until Leonard left the family farm. Joe and Jennie moved to town in 1962 and Lloyd and his wife Dorothy moved to the farm. Lloyd farmed with Joe until his death in 1967. Lloyd with help from his sons: Steve, Larry and Greg continued to farm. The farm is owned by Lloyd and Dorothy Anderson and farmed by their son Greg with part-time help from their grandson Joe.

Anderson farm taken in 1954.

ANDERSON – Due to the 1880 flood of the Elkhorn River, my great -grandfather Johan and my great-grandmother Anna Stina would move from the land in the Elkhorn Precinct that John homesteaded under the Homestead Act of 1869. They would move approximately 28 miles, six miles south of Oakland, Burt County NE. The move took place Feb. 25, 1881. The average snow fall for that February was 17.30 inches, rain .36 inches with an average temperature of 17 degrees. On the 25th there was fog with a temperature of 37 degrees, and a "fresh wind." Besides Johan and family, Anna Stina's cousin Daniel Anderson and family, brother Jonas Johnson and family, and Johan's father, my great-great grandfather, Anders and his second family would make the move. The families moved in one day in covered wagons, taking cattle as well as all of their possessions. Fanny, a cow, was pulling a wagon and slipped on an icy bridge over the Elkhorn River. She kept the furniture, etc. from falling into the icy waters. The account from the Fremont Historical Society refers to them "as the Swedish Colony" and gives the reason for the move as being for religious reasons.

Johan Anderson

John and Anna Stina would remain on the farm until September 1906. They then retired and moved to Oakland, leaving the operations of the farm to Gustavus, their ninth son. After my great-grandfather died, April 8, 1926, my grandfather Gus inherited the northeast quarter (NE1/4) of the southeast quarter (SE1/4) of section 36, in township 21, North, Range 8. Gus would then buy, from his brothers and sisters, their inherited land. Thus returning the homestead to its original size. The land bordered a popular trail used by Native Americans.

Gus and Jenny would have five children, my mother Zonia being the fourth. Gus and Jenny farmed the homestead until they retired and turned over the farming to my uncles Vincent and Bertel. My grandfather, Gus, died Aug. 24, 1960 leaving the farm to his twin sons.

Bertel was the surviving heir, who expanded the farm into its current 321 acres. Bertel died June of 2007, at the age of 93. Since then, I have been proud owner, caretaker and descendant of our family's homestead marking 134 years on this land.

ANDERSON – In 1876, when he was 28 years old, Nels Anderson came to Phelps County from Sweden. He purchased a half section of railroad land, took a timber claim, and preempted an additional quarter, making in all 640 acres, located approximately four miles north of the present village of Funk. On Sept. 1, 1894, Nels purchased an additional quarter, the Southeast 1/4 of Section 29, Township 7, North Range 17 west of the sixth principal meridian. This brought his total land holdings to 800 acres in Phelps County. Nels was the first settler in Anderson Township, which bears his name. He built a 16 by 22 foot frame house, bought a wagon and team and began farming, raising a little wheat the first few years.

Nels and Anna Anderson family. Pictured l-r: Nels, Hannah, Carl, Melvin, and Anna

In 1878 Nels married Anna Almen. Nels and Anna had three children: Hannah (1880), Melvin (1883) and Carl (1891). As Nels continued to improve and add to the farm, he grew wheat and corn and raised livestock. Hannah, Melvin and Carl remained in Phelps County all their lives. Mel and Carl worked the farm with their father. They each married, but had no children. Mel and his wife, Selma, built a home on part of the Anderson property. Carl and his wife, Freda, moved into the home place after Nels and Anna retired and moved into Funk. Hannah married Fred Youngquist and moved a few miles away to a farm just west of Holdrege. Fred and Hannah had one daughter, Evelyn, born in 1915. Evelyn married Harry Dahlstrom - they lived in Holdrege and had one daughter, Jane, in 1947.

When Nels Anderson died in 1938, the farm was divided among his three children. The property received by Melvin and Carl has since been transferred to other owners. Hannah received title to the SE 1/4 of 29-7-17. This portion of the original Anderson farm passed in 1974 from Hannah to Evelyn Youngquist Dahlstrom, only granddaughter of Nels and Anna. At the time of Evelyn's death in 2006, the farm passed to Jane Dahlstrom-Quinn, the only great granddaughter of Nels and Anna Anderson. There have never been any structures on this quarter. At the present time the farm is leased - it is pivot-irrigated and corn and soybeans are raised. The family is proud of their heritage in Phelps County and they feel honored that this land has passed down in the family for 121 years. *Submitted by Jane Dahlstrom-Quinn, Nels Anderson's great-granddaughter.*

ANDERSON – On May 11, 1893 Hamilton G. and Electa Swiger purchased 168 acres from Valentine and Sarah Newman for $4,775. Legal description – NW quarter, Sec. 5, Township 20N, Range 10E in Burt County, NE. Also a strip of land eight rods wide north and south by 160 acres long, east and west along the north side of property, containing eight acres. On Sept. 15, 1893 they sold it to their daughter, Mary E. and husband, Otto W. Anderson.

On June 9, 1917 Otto and Mary purchased 40 acres adjoining on the south from John D. and Belle Jensen – north half of the north half of southwest quarter of Section 5, Township 20N, Range 10.

Otto and Mary had six children born on the farm: Anne, who died in infancy; Nannie; Nettie; Waitmun; Carson and Nellie. Upon the deaths of Otto and Mary, the farm passed to the five children.

In 1946 Waitmun and wife, Merle (Crannell) Anderson, purchased the farm from his brother and sisters.

Waitmun and Merle had two sons, Robert H. and James W. Waitmun died in 1957 and Merle in 1959. The farm passed to their sons. Robert married Joan (Bromm) in 1954 and moved on to the farm in 1957. Robert and James farmed it together for several years. When they dissolved their partnership, Robert retained ownership of the farm. Robert and Joan have two sons, Gary L. and Michael J. and five grandchildren. Robert and Joan remain living on the farm. The farmland is rented to son Gary.

Through the years crops raised have been corn, soybeans, alfalfa and oats. Chickens, hogs and cattle were raised and a few cows milked. Today corn, soybeans and alfalfa are raised. *Submitted by Robert H. Anderson.*

ARMSTRONG – William Armstrong now owns the century farm located in Webster County, Pottsdam Township, Section 2, Range 4, Township 10.

In 1872 Lorenzo Armstrong purchased 160 acres where wheat, corn, barley, oats and alfalfa were raised. Livestock consisted of hogs, cattle, horses and chickens. The farm passed from Lorenzo to his son Clarence and then to William Armstrong, son of Clarence and grandson of Lorenzo.

Over the years, another 800 acres have been added to the original 160 acres. Corn, wheat, soybeans, alfalfa and milo are the crops now raised on this farm.

The original house still remains on the farm.

William and Margaret Armstrong have no children. *Submitted by William Armstrong.*

ARMSTRONG – On March 4, 1903 William M. Armstrong, a Scots-Irish immigrant, purchased 160 acres in Madison County, BE in the southeast quarter, Section 18 of Township 24, North Range 2, from Julius F. Redman and his wife M.J. Redman. William moved there from Butler County in 1903. His wife Hannah, of Scottish heritage, and their two young sons, Robert and Willi, followed. Prior to their final purchase of the land, the US Government had granted it to Julius Redman via a patent on Feb. 1, 1875.

William M. Armstrong and grandson Bill., 1928.

Crops grown on the Madison County farm originally included the following: corn, oats, pasture, hay. In those early days, the Armstrong family participated in Threshing bees as the need arose to process grain both for their family farm as well as for other farmers in the community. According to Edna Armstrong, the men threshed the grain while the women prepared the food to feed the hungry men and farm workers. Currently, crops grown on the farm today include corn, soybeans and pasture; 155 acres remain.

William's son, Willi, married Edna Barritt and to them was born their son, William D. "Bill" Armstrong. Willi and Edna, at that time, lived at an adjoining farm just to the north of William Armstrong. Owning land, as well as passing it on within the family, was a value important to William M. and Hannah Armstrong; on June 8, 1952 their land passed on to their grandson Bill.

Bill Armstrong married Helen Brandt in December 1950. Four children were born to them: Randall L. Armstrong, James W. Armstrong, Janet H. Armstrong, Steven D. Armstrong. Nine grandchildren followed: Rhiannon J. Armstrong; Shauna L. Beckman, nee, Armstrong; Landon D. Armstrong; Andria K. Armstrong; Jamie K. Dittrich, nee, Armstrong; Alexander J. Armstrong; Andrew W. Armstrong; Azlin K. Armstrong; Kaitlin H. Armstrong.

William M. and Hannah Armstrong's grandson, Bill, and their descendant, along with Helen, and Bill's wife, Mary, and several great grandchildren of Bill continue to pursue the American dream while continuing to take a moment to appreciate our ancestors' sacrifices, their accomplishments and their dreams. *Written by Janet Armstrong, daughter of Bill.*

ASCHOFF – In 1891 Ernst Aschoff, a German immigrant, purchased land west of West Point, from two men who had homesteaded adjoining lots. There had been a brickyard on part of the land. Many of the older buildings in West Point were made with these bricks, including the original Cuming County Courthouse. Ernst became a US citizen in 1905. Ernst married Bernadina Gentrup, who died during childbirth, when their son Henry was born. Ernst later married Suzanne Burger and had a daughter, Mary (Feldhaus). Ernst died in 1928.

Henry went to Europe for WW I. Henry married Mary Guenther in 1928 and moved into the original house. They farmed and had livestock, poultry and a garden. They had two children, Anthony and Bernadine.

Anthony married JoAnn Koke, from Gretna, in 1957. A new house was built for Henry, Mary and Bernadine. Anthony and JoAnn moved into the original house and raised a family, including Cathy Dobias, Carl, Frances Belling, Mark, Paul Kenefick-Aschoff, John, Cletus, Alan and Betty English. Anthony became owner of the land in 1984, after Henry, Mary and Bernadine's deaths. A dairy herd was added to the farm. Anthony was active in St. Mary's Catholic Church, Knights of Columbus, Farmers Union, League of Rural Voters, Farm Crisis Hotline and other organizations that helped people in need. Anthony and JoAnn have 20 grandchildren and seven great grandchildren.

Members of the Aschoff family

In 1994 Carl married Jan Glodowski from North Bend. They moved onto the farm when Anthony and JoAnn moved into town. Carl and Jan established their Ash-Glow dairy herd. They are active in the dairy industry including hosting the Cream of the Crop dairy show.

After Anthony's death in 2005, ownership of the farm went to JoAnn. To continue family ownership, a grandson is interested in purchasing the farm.

ATKINSON – John C. and Polly Peckham Atkinson came to Nebraska in 1866 to homestead near Steinauer. John hauled logs from the Missouri River to build a cabin where they lived until 1872. That year they moved to the farm located northwest of Pawnee City. Their house burned to the ground on the next Easter Sunday so they rebuilt on the same spot. In about 1885-6, they added on to their house, completing the structure as it stands today. There, they raised five daughters and two sons, having lost a son to blood poisoning just after moving to Pawnee. Their second son, Charles, came home from western Nebraska where he had a homestead, to take over the farm when they retired and moved to town sometime between 1893 and 1896. Charles and Mabel had three children. The second son, John, was killed in a hunting accident when he was 21. Their eldest son, Ernest, returned home about 1926 after working as Gage County Agent, to help his Dad with the farm. Charley died of a heart attack in 1933. Ernest and his wife Ruth had two daughters, Patricia and Betty. He kept the farm going through the Depression, WW II, and into the '50s. When he died of a heart attack in 1959, his daughter Betty and her husband Dewey Clark

Atkinson family ca. 1886.

took over the farm. They raised four daughters: Connie, Penny, Polly and Debbie, on this farm. Dewey died 2014 from cancer, but Betty and Polly still live on the farm, which is in a trust for Polly. The farm ground has been rented out to a distant cousin of Dewey's since the mid '90s when Dewey decided to slow down a bit. There are five grandkids and nine great-grandkids, making a total of seven generations to spend time on the farm. *Submitted by Betty Clark.*

BABCOCK – Harry Babcock moved from Wyoming to Nebraska in 1925. He married a young lady from southern Brown County in 1930. They first rented 1,520 acres of grassland near her girl-hood home for $50 a year.

Harry A. Babcock Sr. and Margaret Babcock (front) with their children, Harry A. Babcock Jr. (Bill), Lorene, and Harley, 1954.

All of the land that they purchased was grassland in Brown County. The first parcel was 340 acres (see 21-26-21) for $800. In 1931 they moved onto a building site in the northwest 1/4, 3-26-21, purchasing the 940 connecting acres. Margaret's father told Harry "Son, you'll never make it." Harry's dad, visiting from Wyoming, reminded him that he had never owned any land, and he encouraged the young couple. This is where they raised their two sons and one daughter, and lived until 1980, when they purchased a large, old house in Ainsworth.

In 1946 Babcocks bought the 640 acres known as "the Johnson Homestead" (see Otto and Alma Johnson family). In 1955 they bought 480 acres near the Calamus River. The final acquisition was in 1988, when they bought 740 acres adjoining their home place.

By this time they were spending a lot of time at their home in Ainsworth. Their oldest son and his family lived five miles east of Harry and Margaret's building site, so they frequently did some of the ranch work for them.

The main income for the Babcocks was from the cattle. The Sandhills are easily eroded, not suitable for farming, but they produce good quality grass. This was cut, stacked or baled and fed to the cattle in the winter.

The only ground that was broken was 120 acres in the NW 1/4, 13-26-21 for corn, alfalfa and small amounts of oats and rye. This was flat land and suitable for farming. They moved irrigation pipe by hand for several years, and in 1957 they purchased a pivot irrigation system. The corn was ground and the insilage was fed to the cattle in the winter, along with the hay.

Harry began buying cattle in 1929. He bought a few head of Shorthorn cows and Holsteins. Since he did not have any land yet, he had a neighbor keep them for the milk until the spring of 1930. Then he borrowed $1,000 to buy 13 head of cattle, six horses and a few pieces of old machinery from Margaret's brother.

In the 1940s and 1950s the Babcocks raised mostly Hereford cattle, then integrated some Charlois, and ended with all Black Angus. In the early 1990s the total number of their herd had reached 999. Margaret had always worked right beside Harry, and at this time she asked if they could buy one more cow or calf - to reach 1,000. Harry modestly declined.

A small portion of the Hereford cows owned by Harry and Margaret Babcock in the 1950s with farm buildings in the background.

The last of the cattle (33 head of bred Black Angus cows and one bull) were sold on Feb. 4, 2005 at the Burwell Livestock Auction. Harry received recognition from the auctioneer for selling cattle there for many years. Margaret had passed away in 1999, but seated beside Harry were his daughter, Lorene, and his oldest son, Harry Jr. "Bill." The market was good that day. Harry was pleased. He was 98 years old at that time.

At the Brown County Fair in September 2006 Harry accepted an award from Ak Sar Ben for the family's ownership of the 640 acres that had been settled by Margaret's parents in 1906. The float in the parade held 26 family members.

Harry passed away on Sept. 13, 2008 at the age of 100 years and 11 months. The two sons, Harry Jr. and Harley, inherited equal portions of the land. Lorene received other assets, being glad that the ranch is still in the family. *Submitted by Lorene Babcock Mercier.*

BALDERSON – Sinclair and Mary Balderson left their home at Marian, IA in the spring of 1886 for their homestead five miles north of Benkelman, Dundy County, NE. The homestead in those days was 160 acres of land given to anyone by the US Government who would live on it for five years

Those were hard years for Sinclair and Mary. Their diet consisted of potatoes, beans, prairie chicken jack rabbit and wild berries and plums picked along the creeks.

Balderson farm, Spring 1931

About this time it was determined by the Government that families could not make a living on 160 acres in the semi-arid region of southwest Nebraska. So another 160 acres was offered if the homesteader would plant trees on it. This Sinclair and Mary took advantage of by filing on the quarter next to their homestead quarter. This gave them the whole south 1/22 of Section 18, township 2, range 37, west of the 6th Principle Meridian, containing 320 acres more or less.

In 1906 Sinclair and Mary moved to a farm just north of the city limits of Benkelman. This farm was purchased for back taxes that the previous owner had not paid because of the drought and financial panic of the 1890s.

In 1888 Sinclair and Mary's only child, Lee Ray, was born. When Sinclair passed away in 1929 Lee became the owner and operator along with his mother Mary. Lee married Edith Delaney in 1926 and to this union two sons were born, Ray Lee in 1930 and Frank Lee in 1931.

Lee passed away in 1974 and Edith in 1983, at which time Ray and his wife, Maxine (Herrick) Balderson, purchased Frank's share of the farm and they became the owners and operators of the then 1600 acre farm where it stands today. *Submitted by Ray Balderson.*

BALLAGH – Robert A. Ballagh, Sr. was born in Ireland on September 30, 1844, and at the age of 29, came to America. After working five years, he went back to Ireland to marry Sarah Jane McLeod on September 11, 1878. On their wedding day, the newlyweds returned to America and made their home at Lynchburg, Virginia, opening a mercantile store. When a fire destroyed their business, the Ballaghs moved to Jackson, Virginia, where they began a dairy.

Robert moved his family to Spalding, Nebraska in 1882, residing there for two years. In 1884, he claimed a homestead of 160 acres, in the northern part of Garfield County. In the spring of 1885, together with his wife and three children, Anna, Hamiliton, and Robert A. Ballagh, Jr. (Robbie), Robert moved into a sod house he had built on his claim. Mr. Ballagh realized that he knew very little about the new life to which he had brought his family. To help him learn, he subscribed to booklets and other material furnished by the State University. Robert and Sarah were instrumental in establishing the first Church and school in the community, and with a post office, 'Ballagh' was on Nebraska maps from 1890 until 1945.

Following the death of Robert A. Ballagh, Sr. in 1934, his son, Robbie continued managing the ranching operation, which consisted of the original homestead acres, along with additional acres he and his father had purchased. Robbie had married Harriet Emlin Gumb on April 22, 1914. They lived in the house he built using cement blocks he made from the native sandy soil. To this union, four children were born: Robert Glen, Gordon Adair, Edith (Mrs. Kenneth Weller), and Esther (Mrs. Francis Knight).

Gordon A. Ballagh returned to the ranch after serving in WWII, and once again the land was passed on to a son. It was while spending 18 months in a German prison camp, that Gordon came to know for certain that he wanted to return to the ranching lifestyle. Gordon married Marjorie E. Rowse in 1945, and they eventually built a new house next to the sandstone one that Gordon's father had built. Gordon and Marjorie had five children: Rowan, Janell, Lynn, Ronda, and Eric. Gordon and his family initially raised Hereford cattle, but transitioned to an Angus herd in the 1960's.

By the 1980's, Lynn A. Ballagh had married Amy C. Ballagh, a school teacher from Kansas, and established his family on the 'home place'. Together they raised their four sons: Aaron, Ryan, Devyn, and Kellen. Lynn's family continued to develop a registered Angus cow herd, established both a fall and spring calving program, and added acreage to the ranching operation.

Currently, in 2015, all four of Lynn and Amy's sons are involved in the ranching business with them, which still includes the original 160 acres. And, with the 6th generation of 4 boys and 1 girl now learning to ranch, it looks like Robert A. Ballagh, Sr. chose a good place in Nebraska to call 'home'!

160 acres homestead: SW ¼, Section 12, Township 24 N, Range 15 W, Garfield County.

BARR – James A. Barr purchased 160 acres of land, northeast 1/4, Section 22, Township 11, Range 3 in York County in the fall of 1877 from a homesteader. During his farming venture, he owned over 1,000 acres in York County, plus land in Howard and Merrick Counties. There's two farms still owned by Barr relatives. The rest have been sold.

James, who was from Canada, returned to Canada and married Mary Ann Miller in 1880. They had 12 children, but three died.

James raised wheat, barley, corn, shorthorn cattle, registered Clydesdale, Belgian and Percheron horses for work, stud service and to sell.

Joy Mervin Barr, the youngest son, got the mortgaged farm when his parents moved to York in 1926. He paid the mortgage in later years. Joy raised horses, cattle, hogs, corn, wheat and oats.

Joy married Elora Gelvin on Sept. 4, 1924. They had five children: Nina, Adena, Robert, Glen and Joyce. They had five grandchildren. Glen got kicked by a mare and died at age 7.

When Joy died in 1966 the four children inherited the farm. Robert, who married Mildred Kopke in February 1953, purchased each of the sister's 40 acres, so by 1977 they owned the original homestead. Robert raised wheat, soybeans, but mainly milo.

Robert and Mildred did some remodeling themselves on the original house and still live on the farm and raise some Angus cattle.

Robert and Mildred had two children, Joy Marie and Timothy. When Robert retired from farming, their son Tim started farming the land. He raised corn and soybeans.

Their married daughter, Joy Marie Sherman, passed away in 2006. *Submitted by Mrs. Robert Barr.*

BEHRNS – Joachim (James) F.E. Behrns was born Aug. 4, 1817 and Anna Sophia Goljohn (Godejohann) was born on Jan. 28, 1814 in Mecklenburg, Schwerin, Germany. Joachim's German occupation was a brick-maker. Four children were born to this union: Henry, 1845; Sophia, 1848; John, 1852 and Minnie, 1857.

Behrns family – from left, Henry and Mary, children: Sophia and Josef. The house was built in 1882.

In 1862, the Behrns' family sailed to America, landing at New York and proceeded to the Stoll homestead near Nehawka, NE. Joachim purchased 85 acres of land that fall and then on Feb. 17, 1866 James purchased 75 acres (northeast 1/4, Section 22) Avoca precinct in Cass County from Peter and Lena Peters for $1125.00. The southern boundary of this "75 acres" was a rock wall.

Three years later, Feb. 8, 1869 Joachim sold the 75 acres to his eldest son, Henry, who had married Mary Sturm on Oct. 29, 1868. Henry and Mary raised five children: Henry, Mary, Minnie, Josef and Sophia.

The present home was built in 1882 when Josef was 2 years old. A four-level barn was built in 1886. Henry named the 75 acres "the Oakwood Stock Farm."

Henry and Mary moved to Nehawka in 1907 and at that time Josef "Joe," who had married Lena Jorgensen, took over the farming operation. To this union three children were born: Florence, Helen and Leslie. In May 1909 Joe and Lena purchased this farm.

Leslie married Anna-Marie Straub on April 24, 1941 - to this union four children were born: Edward, Barbara, Roger and Maxine.

In 1949, Joe and Lena moved to Avoca, as Leslie and his family moved onto the home place and continued farming.

Leslie and Anna Marie acquired the 75 acres in 1968. The ownership is now in the fifth generation of the Behrns family: Edward Behrns, Barbara Vaughn, Roger Behrns and Maxine Schatz.

Our farm is eligible for the Heritage (150 Years) Award in 2016. *Submitted by Maxine (Behrns) Schatz.*

BELL – William Bell and Eliza (Wessidone) were married Jan. 22, 1865. They arrived from England with daughter Harriet Ann in America, set-

Frame house built in 1885

Sod house and sod barn built in 1877

ting in up-state New York in 1870. In 1877 they traveled to Nebraska, homesteading 160 acres located on the southwest 1/4, Section 8, North Range 15, Macon Township 3 of Franklin County, NE. They built a sod house and sod barn, living in the sod house until May 23, 1885 when they moved into a new frame house.

Daughter Harriet Ann (second generation) married William J. Knapp in 1896. (William came from England to America in 1888.) They were the parents of five children.

Son David C. Knapp (third generation) and wife, Alwina (Ruhs) Knapp, continued on the farmstead after purchasing it from siblings. Their daughters, Shirley Ann (Knapp) Nelson and LaDonna Knapp (fourth generation) are now owners. LaDonna Knapp continues to reside on the homestead.

Farming the land as of 2015 are Stephen D. Nelson (fifth generation) and Scott D. Nelson (sixth generation). *Submitted by Shirley Ann Nelson.*

BENDIXEN – In 1890 Michael and Bridget Leahy purchased 160 acres located in the southwest quarter of Section 20, Township 21 North, Range 4 East of 6th Principle Meridian in Cuming County, NE.

The farm passed from grandparents Michael and Bridget Leahy to their daughter Margaret Eucks to their granddaughter Peg Bendixen.

The 160 acres now belongs Peg and her husband, Carl A. Bendixen. They raise corn and hay on the farm. *Submitted by Peg Bendixen.*

BENTZINGER – Jacob Bentzinger Jr., Neal Bentzinger's grandfather, acquired the 160 acre farm from Charles Wilhelm on Feb. 28, 1908. The farm is in Osage Precinct No. 27-7-11, just north of the Otoe, Johnson County line.

A new house was built in 1918. In 1939, the farm passed to William Bentzinger and his sisters, Edna Bentzinger and Leona Holscher. William and Helen Bentzinger, Neal Bentzinger's parents, bought the sisters shares of the farm in 1950.

In 1972 William Bentzinger died. In 1973 Helen Bentzinger moved to Cook and at that time Neal Joyce Bentzinger bought the farm and moved there. We still live here on the farm. We raise corn and beans and have a large pasture. We have a daughter, Gail Stanberry and a son, Russ Bentzinger and two grandchildren. *Submitted by Neal Bentzinger.*

Bentzinger farmstead in September 2009.

BERG – On March 24, 1913, William and Ida Berg purchased an 80 acre farm in Adams County located just one mile east of Kenesaw, NE. This farm, (east half of the southwest quarter of Section 26, Township 8, Range 12) was located in the adjacent section to his home farm he purchased in 1889.

The William Burg family, 1941

Fischer - Upon William's death in June of 1956, his oldest daughter, Erna Berg-Fischer, acquired ownership. At that time she lived on an adjoining 40 acres to this 80 acres.

Augustin - In 1960, William's second eldest daughter, Elva Berg-Augustin's husband unexpectedly passed away and she decided to move into town, so Erna's family moved on Elva's farm place nearby. Therefore, Erna traded Elva this 80 acres for Elva's home farm.

Kennedy - Elva passed away on Dec. 25, 1980 and her youngest sister, my mother, Vera Berg-Kennedy accepted ownership. In 1994, over a span of the next four years, Vera began gifting this 80 acres over to her two children, Randy Kennedy and Vicki Kennedy-Weseman.

Will Berg and his working horse team.

Weseman - Vera Berg-Kennedy passed away on May 19, 2013. In settling her estate and exchanging ownership of property, Vicki Kennedy-Weseman became owner of the entire 80 acres. Vicki has one son, Jason Kennedy Weseman, which is the great grandson of William and Ida Berg.

In June of 1967, the town of Kenesaw flooded. Soon after, work began on plans for a ditch to surround the village to prevent future flooding. My aunt, Elva Berg-Augustin sold easement rights for the ditch to cut through the middle of this 80 acres. Today, it is a dry land 80 growing wheat, corn and soybeans as crop rotations. This 80 is a high producing farm with rich soil. In 2013, it was recognized as a 100-Year Family Farm. *Submitted by Vicki L. Kennedy-Weseman.*

BIERE – Henry and Emma Biere purchased this farm near Lorton on Nov. 19, 1901 from Herman, Anna and Froche Redelfs. The farm is located 4-1/2 miles south of Dunbar on Nebraska Highway 67.

On Sept. 28, 1857 the ownership of the NE 1/42-8-12 in Otoe County, NE had been transferred from the US Government to Fred. W. Hattendorf. Over the next 44 years, the ownership was transferred to several individuals.

The Biere family immigrated from Haustenbeck, Germany to Gasconade County, MO in the 1870s. Grandson, Henry moved to the Brock, NE area in 1891. Emma Nowack's family moved to the Brock,

The steam-threshing rig owned by Henry Biere and sons.

Boeckenhauer family farm, fall 2015

NE area from Woolham, MO having immigrated from Pominerland, Germany in 1863

On Dec. 7, 1902, Dean's father, George H. Biere, was one of four children born to Henry and Emma at their home. Over the years, Henry Biere and sons George and William did custom threshing for area farmers with a steam-threshing rig made in Port Huron, MI. The Biere's raised corn, wheat, alfalfa and prairie hay. Livestock included: milk cows, beef cows, hogs and chickens.

Upon Henry and Emma's retirement they transferred the ownership of the farm to their son, William. When William and Bertha retired, the farm and home were rented and Biere's retained ownership. Bertha continued ownership after William's death on Feb. 20, 1989. Upon Bertha's death their nephew, Dean Biere, purchased the farm on July 1, 1994 from her estate. Henry, Emma, William and Bertha had made improvements to the house over the years. In 1995-1996 the home was remodelled extensively and is now lovingly cared for by the tenants. At the present time, Dean continues to farm the land raising corn, soybeans and prairie hay. *Submitted by Dean A. Biere.*

BOECKENHAUER – The farm was originally purchased by Alfred Haglund and his wife, Charlotte (Peterson) Haglund, who were natives of Sweden and came to Wayne County from Mercer County, Illinois in 1891. They settled on their farm four miles south and two miles west of Wakefield just east of the ghost town of LaPorte. Alfred had been the foreman for the owner of a large farm in Illinois and brought along some purebred Shire horses and Shorthorn cattle that he had acquired from his employer. He became a successful farmer and livestock raiser.

Alfred and Charlotte Haglund

One of Mr. Haglund's goals was purchasing 80 acres of land for each of his seven children. This he did accomplish in his lifetime. The 80 acres in this farm were purchased in two 40-acre sections, one on Nov. 2, 1898 and the second 40 on April 23, 1907. It is located in Wayne County, the north half of the northeast quarter of Section 25 Township 26 North, Range 4. It is located immediately to the south of the family's homestead.

Esther Haglund

This farm was the land that was given to Alfred and Charlotte's daughter Esther in 1939. There was no homestead on the land, as Esther never married or lived there. Instead her brother-in-law Ezra Boeckenhauer (husband of Esther's sister Lydia) and her nephew Robert Boeckenhauer rented the land from her and farmed it until her death.

At that time according to Alfred's will the land was to go to Esther's nieces and nephews (8 in all). Robert Boeckenhauer purchased his cousin's and sister's shares in 1962 and was the owner until his death in 1992.

Robert & Inez Boeckenhauer & family, 1965

According to Robert's will the land was awarded as an undivided 80 to his four children: Juanita Boeckenhauer Reed-Boniface, Karen Boeckenhauer Weeks, Lauren R. Boeckenhauer and Maurice E. Boeckenhauer. When Maurice died in 1993, his share went to his widow, Delores Meyer Boeckenhauer. Juanita, Karen, Lauren and Delores are the current owners.

Brian Nelson of Wakefield has rented the farm since the early 1990's. The land consists of 40 acres of dry land cropland and 40 acres of pasture. Ezra and Robert Boeckenhauer were early adopters of terracing and worked with the Soil Conservation Service in the early 1950's to terrace the hill on the west end of the land, which has been cropped. Another unique feature of the land is that the proposed railroad that ran to LaPorte came across this parcel of land and still today, one can see evidence of the railroad bed in the topography of the land.

There is a small metal tank on the northeast corner of the land set into a poured cement frame easily accessed from the road. This was a way that Alfred showed his hospitality to anyone passing by with horses or cattle that needed to be watered.

The large grove provided plentyof entertainment for the Boeckenhauer grandkids, hiking in the grove, climbing the tree with the seven trunks and often picking the bounty of mulberries.

Our family's roots run deep in Northeast Nebraska, in our heritage is the gift of a wonderful ancestry, the men and women who came to this state and were willing to take the risks that must be faced in a new land. In 1880 our great-grandfather used oxen and a walking plow to prepare for his sod corn. This century farm is a tangible reminder of the vision of our Great Grandfather to purchase land and the wisdom and sacrifices made by our grandparents and parents to "keep and care for the land" during the many changing times in agriculture—weather, prices, conservation practices, technology, scientific discoveries in animal husbandry and plant growth, and world events the Great Depression, and World Wars. As current owners, we take pride in our agriculture heritage and work to pass these values onto the next generation.

BOESIGER – The Lancaster County land that is now Pine Grove Stock Farm (SE 1/4, Section 19, South Pass Precinct, SE 1/4 19-7-7) was purchased by my great grandfather, Frederich Jacob Boesiger (Oct. 18, 1845 – May 4, 1928) from Wilhelmina Rausch Cramer on Feb. 5, 1906 for $12,000.00. It was a wedding present to my grandparents, Frederick Jacob Boesiger (Oct. 14, 1886 – May 16, 1964) and Maude Sargent Boesiger (Oct. 8, 1887 – Sept. 29, 1978) on March 13, 1913. It passed to my parents, Donald Fredrick Boesiger (Dec. 31, 1915- Nov. 27, 1999) and Alice-Leone Ambler Boesiger (April 23, 1915 – Nov. 10, 2014) on

Grandmother's death. After Mother's death, it passed to her children: Pamela (May 2, 1943), Donna Mankin (May 24, 1945), Fredrick (June 28, 1947) and Susan Landau (June 25, 1951), who established the LLC in 2015. My Great Grandfather also purchased the NW 1/4 of the adjacent section to the east, which contained the family orchard lost in the 1930s drought. This section eventually went to Madalyne Boesiger Latshaw (Oct. 29, 1921 – Nov. 24, 2008) and then to her three children; her son Earl Latshaw (Oct. 3, 1961) currently farms both quarters.

My grandparents raised corn, wheat, oats, hay (red clover, alfalfa, prairie, brome), cattle, hogs, chickens, guinea hens with horses and mules. We assume they continued what Great Grandfather had raised. The orchard provided apples, mulberries, pears and cherries. Milo replaced wheat and oats in the 1960s and was replaced by soybeans; alfalfa is the sole hay crop. There is currently no livestock.

Mason A.T. Cooper and carpenter Chris Dohe created an unusual tile-block and wood barn in 1918. Pilasters reinforce and support a completely freestanding roof. Doubled rafters make up the double-gabled profile and are supported by wood buttresses anchored over the tile walls and floor joists embedded in the tile walls. Neighbors were skeptical and thought it would blow over in the first windstorm. *Submitted by Pamela Boesiger.*

BOETTCHER – Adolph and Emilie Boettcher immigrated to America in 1884 from Zericke, Germany. He was employed as a cooper, which was a wood barrel maker in a brewery at West Point, NE. Adolph and Elsie Boettcher were the parents of nine children: Robert, Bruno, Gustav, Bernhard, Elsie, Marie, Hulda, Ernest and Ella.

Richard and Dessi Boettcher with children Cody, Tony and Shauna.

Adolph's sons, hearing of the homestead availability on the Fort Randall Military Road, commonly known as the Three-mile-strip, persuaded their parents, in 1893, to move to Boyd County and homestead five miles northwest of Spencer. Here they built a stone-frame house for their abode. At that time, the area had not been surveyed and no decision had been made as to whether the land was in South Dakota or Nebraska. In 1895, the survey was completed and the Boettchers found themselves in Boyd County, NE. The homestead is located in the Southwest quarter of Section 5 of Township 34, Range 12 in Boyd County, NE. Taxes for the acreage in 1894 was $9.18 and in 1895 was $2.94.

Early photo of the Boettcher stone-frame house.

Richard and his John Deere tractor

In 1996, Adolph's great-grandson Richard (Dessi) Boettcher purchased the property. Richard married Dessi Weakly in 1978, to this union three children were born: Shauna who lives in Yankton, SD; Cody (Ashley) who lives in O'Neill, NE and Tony (Kari) of rural Spencer, NE. Tony now lives on the home farm and assists Richard with planting and everyday farm chores. They have seven grandchildren.

All that remains of the original buildings is a corner of the stone house and the barn, which is still in use today.

Richard also owns a tract of land homesteaded by his great-grandfather's daughter, Hulda Boettcher. Both tracts of land have been in the Boettcher family and name for over 122 years.

Richard plants corn, beans, hay and alfalfa. He always has had a herd of Angus cattle. Richard continues to farm this tract of land along with other land they have acquired over the years. *Submitted by Richard and Dessi Boettcher.*

BOHLING – On Aug. 28, 1872 George Ottens purchased 160 acres in Nemaha County. Crops grown on the farm consisted of corn, beans, wheat and hay. Livestock was cattle, chickens and hogs.

Ownership of this farm passed from George Ottens to his daughter, Mary and her husband. She had married William Bohling who bought this land. The next owners were William's children, Henry and Anna Bohling. Henry's son, Wilbur, and wife Lois became the next owners. Larry Bohling, son of Lois and Wilbur, inherited the land. On July 3, 2014 Ryan (son of Larry) and his wife Denise bought the farm.

The farm still retains the original 160 acres. Corn and beans are the crops grown now and cattle are on the farm. A portion of this farm that has never been tilled is native prairie with the Oregon Trail wagon tracks visible.

Ryan and Denise have two children, Emily and Johnathan.

BONDERSON – Ben Bonderson was born in Skane, Sweden. He immigrated to the USA with his parents, Andrew and Kate, in 1863 when Ben was 11 months old. They journeyed to Utah on the Mormon Trail and returned to Omaha in 1867. Andrew and Kate Bonderson bought a farm near Emerson in 1887.

Ben & Hilda Bonderson farm house

Nels and Johanna Peterson and their daughter Hilda, the future wife of Ben, immigrated to Omaha from Helsingborg, Sweden in 1887. They were hired to manage the Bonderson farm until Ben's brother John and his new bride, Jenny, moved to John's parents'

The Bonderson brothers, l-r: Jules, John and Ben

farm in 1898. Ben spent time at his parent's farm near Emerson and he bought himself a near-by farm (Section 29-27-7) for $12 per acre in 1891. The farm he bought was four miles southeast of his parents' farm and was owned by Jesse and Nancy Wigle. Jesse Wigle was the first white man to settle in Dakota County, NE and his daughter Emma was the first white child born in Dakota County. The Wigles lived on several farms before moving to this farm in 1883 where he died. While Wigle built the original small, three room house, he and his family lived in a building that later became the present day chicken coop. When Ben married Hilda Peterson in 1895 they added on a new, bigger kitchen. Ben and Hilda, like Ben's brother John, added a great deal to their house in 1910 as their families grew. 1910 was also the year their father Andrew died.

Ben passed away in 1943 and Hilda in 1968. They had five children: Adolph, Alma, William, Helen and Virgil. Helen lived in the house until her death in 2005 at 99 years old. She and her sister Alma had many family gatherings in their home and lovingly took care of the beautiful house and farmstead. The great niece of Ben and Hilda, Jenny Susan Nelson Rudberg and her husband Ted along with their children, Eric and Anna as well as Anna's husband Pete and their daughter Inga have owned the old Bonderson house and farm since Helen's death in 2005. Jenny is the granddaughter of John and Jenny Bonderson and daughter of Ruth Bonderson Nelson (Neal). They have painted, repaired and restored the beautiful farmhouse and the 1880s barn, 1900s hay barn, 1930s corn crib, and various animal buildings. It is on the National Register of Historic Places, and it also has been designated a 100 year Nebraska Pioneer Award Farm. It is the Rudberg family's honor and pleasure to keep alive the memory of Ben and Hilda Bonderson and their children. The fields are still farmed by Ben and Hilda's grandson Rodney, great grandson Kevin and great, great grandson Ben Bonderson. *Submitted by Jenny Rudberg.*

BONDERSON – John Bonderson's parents, Andrew and Kate, immigrated from Skane, Sweden in 1863. They traveled to Utah on the Mormon Trail and then returned and settled in Omaha in 1867. Their son John was born there in 1871. In 1887 Andrew and Kate bought a farm 4-1/2 miles northeast of Emerson at a price of $9.00 per acre. Andrew died on this farm in 1910 and Kate died there in 1913. In Omaha in 1898 their son John married Jenny Anderson, who had immigrated from Smaland, Sweden. They then moved to John's parents' farm. John and Jenny raised their nine children and farmed there until John's death in 1950 and Jenny's death in 1951. All of their children: Arthur, Dell, Harvey, Elmer, Rena, Fern, Elaine, Merrill and Ruth, have passed away. Family members have continuously owned the land. It is presently owned by John E. Nelson, the grandson of John and Jenny and the son of Ruth Bonderson Nelson (Neal).

John and Jenny Bonderson farm house

The small original farmhouse from 1887 was added onto in 1910. This large, beautiful home stands proudly today. The farmstead still has the cattle barn, a 1930s corncrib and various outbuildings. Ben, the older brother of John, also has a 100 year Nebraska Pioneer Award Farm, located approximately four miles southeast of John's farm. *Submitted by Jenny S. Nelson Rudberg.*

BORNEMEIER – On Oct. 27, 1903 Simon F. and Katherine E. (Rueter) Bornemeier purchased 160 acres in the north 1/2 northwest 1/4 and west 1/2 northeast 1/4 3-10-10 Cass County.

The barn was constructed in 1915 and the current dwelling was erected in 1929.

Aerial view of the Bornemeier farm

In 1922 the land was transferred to Simon and Katherine's son, Edward F Bornemeier. In 1927 Edward and his wife Anna M. (Schlueter) purchased an additional 80 acres adjoining the original tract. Chester E. and Valda M. (Gerhard) Bornemeier purchased the three 80s from Chester's parents. In 1952 Chester received the Cass County Soil Conservation Award for his conservation practices, primarily terraces.

In 1886 a strip of land 100 feet wide had been deeded to the Missouri Pacific Railroad Company. This became part of the 26 mile Mo-Pac Trail from Lincoln to Wabash.

Since childhood, Chester and Valda's son, Joel, has been active in the farm operation and continues farming. In addition, Joel is also a pilot, so there was a runway installed near the home site in the late 60s. Joel continues as an aerial crop sprayer and has lived in the home on Fletcher Avenue since 1985.

BORUCH – The Lambert Boruch Trust in Polk County, NE was homesteaded by Blazy Boruch in 1885, with 80 acres. Blazy then purchased an additional 340 acres bringing the total acres to 420. Blazy's son,

The Boruch farm

George Boruch, farmed the homestead until his death. At that time one of George's sons, Henry Boruch, started farming. Henry farmed the homestead until 1963, Henry then brought another farm and George's other son, Lambert Boruch, started farming the homestead. Today one of Lambert's son's, Greg Boruch, is currently farming and lives on the original Homestead place. *Submitted by Greg Boruch.*

The Boruch family homestead residence.

BREDEHOFT – Born in Ahlerstadt Hanover, Germany in 1857, John Bredehoft immigrated to Wisconsin in 1868. He eventually settled in Orange, CA, where he worked as a teamster, freighter and vineyard culturist. He married Maria Hauschild of Orange, CA, and established orange groves there with Maria's brother.

In 1890, John and Maria turned to farming in Polk County, NE, and in 1903 purchased 320 acres in Stewart Township in York County. A year later, 160 acres were purchased in Waco Township. There they began to build up a grain and livestock farm where short horn cattle were raised.

The Bredehoft family, ca. 1941. Pictured front row, l-r: John Christian, Roland, and Bertha (Wilksen). Back row: Juanita (Bredehoft) Buck and Pauline (Bredehoft) Gray.

One of their sons, John Christian Bredehoft and his wife, Bertha (Wilksen), bought the Waco Township property, where they continued to raise shorthorn cattle and Duroc hogs, along with grain farming. John and his brother, George, also did custom threshing for their friends and relatives. Bertha kept busy raising her family and had a butter, cream and egg route for a number of years. John and Bertha raised three children on the farm: Roland, Pauline (Les Gray), and Juanita (Wilbur Buck).

Roland joined the operation and worked with his father, taking time to serve four years during World War II as a radio operator mechanic, and was promoted to Staff Sergeant. While stationed in the Marianas, he received a medal for "skill, courage and devotion to duty" while participating in the longest over-water flights ever made over enemy controlled areas in the central and western Pacific islands.

After service, Roland rejoined the family farm and also had his own business leveling land for irrigation. He married Gertrude (Richert), who had been a telephone operator before they married. After his father's death, Roland and Gertrude continued to farm the land. Later, another 80 acres was added, brining the farm to 240 acres in all.

Currently, Gertrude Bredehoft and daughter, Deb Huffman, are the owners of the farm. After Gertrude moved off the farm to York in 1955, she sold the acreage to Kathy and Rich Hankel. Kathy Hankel (Buck) is the great-granddaughter of John and Maria Bredehoft.

BREEZY POINT FARM – In 1910, George C. Bruning purchased 160 acres in Thayer County, NE, the northeast 1/4 10-2-4. At that time, he raised cattle, hogs, chickens and grains. George and his wife, Margaretha (Reichert), lived on the home place until they built a home in Bruning in the 1920s. George's son, Paul G. Bruning, acquired the farm from his father and continued to raise grains and livestock. Paul and his wife, Alberta (Schweitzer), lived on the home place until 1971 when they moved to Bruning, to live in the house that George and Margaretha had built.

Breezy Point Farm after the 1984 tornado

Above and below: aerial views of Breezy Point Farm.

In 1975, Paul and Alberta's son, Douglas P. Bruning, and his wife Janelle (Houck) moved to the home place. In 1978, they replaced the original home with a new one.

In 1984, a tornado destroyed the original farm buildings and their home. Doug and Janelle rebuilt their home and replaced the farm buildings, but did not continue to raise livestock due to the destruction of the barn, fences, etc.

In 1996, Douglas acquired the farm and continues to raise corn and soybeans. Douglas and Janelle have two daughters, Michelle (Jason) Paulk and Maggie (Justin) Jiskra, and four granddaughters: Emma Jiskra, Ruby Jiskra, Caitlin Paulk and Skylar Paulk.

BRIDGES – Patricia A. Bridges, surviving widow of James A. Bridges, is the current owner of approximately 5,000 acres located in Township 27, North Range 32, west of the 6th P.M., 20 miles north of Mullen in Cherry County, NE.

The original settlers were; Rollin G. Roth and Winnie Roth, Catherine Doerr Rotg and Edward H. Roth. In April 1908 each settler acquired 640 acres per the Kinkaid Homestead Act for a total of 1,920 acres.

Originally only cattle, horses and prairie hay were raised on the farm as no grains grew in the barren sandy soil.

There have been no other owners but the original family members. After the deaths of Dr. and Great Grandma Catherine, Mildred

Roth McDuffee and Elsie Roth were the only children of Dr. and Mrs. Roth so Elsie Roth wanted to stay and ran the ranch and cares for her Mother and her Uncle Ed. She ran the ranch from circa 1940 until she sold it to her niece, Patricia A. (McDuffee) Bridges and her husband, James A. Bridges in April 1984. They have lived on the ranch since then until Jim's death, Dec. 31, 2010. Pat continues to reside at their ranch home and maintains the ranch and the guest ranch. She leases the ranch pasture lands and hay meadows to Jed and Jerry Connely where they run some of their cow/calf, purebred Angus cattle and put up the hay crops.

Pat and Jim Bridges (deceased Dec. 30, 2010) at the Double R Ranch

Pastures are leased and are raising cow/calf pairs most of the year. The meadows produce many bales of prairie hay for winter feed. Pat and Jim have been utilizing the empty buildings and some of the original homestead to start and maintain successfully, a guest ranch with three cabins for lodgers. The guests can hunt, fish, hike, or just relax and enjoy the quiet and peaceful seclusion. The guest ranch was begun in 2000 with two cabins and now has three cabins for guests. In addition to the fishing and hunting, the guest ranch offers birding-hiking-star gazing-ranch tours and some seasonal ranch activities. They have a good White-tail and Mule Deer population and quite a lot of Pheasant, grouse, prairie chicken and dove for excellent hunting.

The original three room sod house is open for viewing. The sod house was originally built from sod taken from the meadow west of the cabin area. It had a shake shingle roof and was finished on the outside with about a six inch layer of cement. This made the house cool in the summer and warm in the winter. It has three separate entrances, all on the front side of the house. Each room had large double hung windows with accompanying window seats. These windows had to be replaced to help with preservation of the sod house in its entirety. There was good ventilation and circulation with the original windows. There is an attic that was in use as Dr. Roth's pharmacy and library and a place for general storage of seasonal items. Dr. Roth's office building is now a little Gift Shop for our guests to buy souvenirs. Both buildings are available for our guests to view. There are quite a number of items of interest that belonged to the Roth family to view.

The original School District 66 schoolhouse was gifted to the Bridges. It has been restored and is open for visitation. It contains a lot of the family library and school books that the district didn't want. It also has a circa 1885 Upright Grand Piano that has been tuned most years and was used by Pat to give piano lessons to kids in the area until last year.

The original barn is still standing. It was built in part with lumber brought on the train from Grandin, MO by the Roth family. They brought enough lumber with them to build the structure for the sod house and to build the small frame building that was their original dwelling, and later became Dr. Roth's office and lumber for another frame building which became the plant house and wash house.

James A. and Patricia (McDuffee) Bridges were awarded the NEBRASKA Pioneer Farm Award by the Knights of AK -SAR -BEN on Aug. 16, 2008 at the Cherry County Rodeo at the Cherry County Fairgrounds in Valentine, Nebraska. *Submitted by Patricia A. (McDuffee) Bridges.*

BRITTENHAM – On Dec. 31, 1901, Lorenzo Brittenham and his brother Ellis traveled west from Davenport, IA. They homesteaded adjoining quarters. Lorenzo's the northwest quarter of 18-10-27 and Ellis' the southwest quarter of 18-10-27. Lorenzo raised cattle, chickens and some corn and improved with a sod house dug into the bank and a well. On Christmas Eve 1905, Ellis was killed in an accidental discharge of a gun. Lorenzo bought the quarter from Ellis' widow, Esther (Flanigan) Brittenham, giving him 320 acres. Lorenzo and his wife Nellie (Surber) moved out of their soddy into Ellis' while they built a house. They moved into their new house in 1919. Lorenzo and Nellie had seven children: Hazel, Floyd, Clara, Jacob, Dorothy, Oliver and Clarence.

Their son Jacob took over the farm around 1940 with his wife, Mildred (Heath). They had four children: Edith, Floyd, Wayne and Eldon.

A painting of Lorenzo and Nellie's sod house given to them by their neighbor, Jessie Rose, around 1915.

In 1973 Eldon moved back to the farm with his wife Shirley (Johnson) with their three boys: Dale (Vivian), Rick (Cathy) and Lynn. They continued to run cattle on the land and raising some chickens, alfalfa and prairie hay. In 2001 Eldon and Shirley received the Aksarben Homesteaders award, celebrating 100 years. Later on in 2001 Lynn and his wife Melissa (Davis) took over the management of the farm, also continuing to raise cattle, chickens and hay. Lynn and Melissa with their three children: Austin, Holly and Jacob still live on and work the land. In 2013 Lynn and Melissa took possession of the farm, bringing with them another 60 acres purchased by Lynn and Melissa. The Brittenhams still live in the original house built by Lorenzo and Nellie. They love the land and the heritage it brings and look forward to continuing their legacy along with Jacob, Holly, Austin and his wife Lonna and their children. *Submitted by Lynn and Melissa Brittenham.*

BROWN – On May 18, 1893 Charles T. Brown homesteaded 160 acres located in the west quarter of Section 32, Township 2, Range 32. Wheat and corn were the crops grown and cattle was raised on this farm.

The farm passed from Charles T. to his son Morris Brown then in 2004 became known as Brown's Valley Farm when Charles' grandson Cermit and his wife, Marjorie, acquired the farm.

Charlie and Belle Brown, sod house, 1893

Today crops grown are wheat, corn and milo. At one time there was a barn with a drive-in mow but it no longer remains. The hand dug well which was 100 feet deep still remains.

The farm was awarded the Nebraska Pioneer Farm Award.

Cermit and Marjorie have three children: Sondra, Nathan and Justus. Their grandchildren are Hannah Brown, Ian Brown, Zeke Brown and Isabelle Brown. In 2006 Cermit retired from teaching school but still farms 100 acres. *Submitted by Cermit Brown.*

BRUCE – A 320 acre farm located in Gage County, Elm Township, north half of Section 13, T-2-N, R-5-E was sold to Tamme and Katharine

Cropland on the Bruce farm

Zimmerman in 1893. Tamme was said to have been the first permanent white settler in Gage County.

On Oct. 2, 1925 this farm passed to Eilert in 1911, then to Fannie (Zimmerman) Bauman. Fannie was Tamme's daughter and Eilert Zimmerman's sister. Fannie had married Edd Bauman. On June 21, 1962 the farm passed to Tom (Fannie and Edd's son) and Lena Bauman. Tom and Lena's daughter, Angeline (Bauman) Bruce and husband Bill acquired the farm in 1975. On May 22, 2007 ownership passed to the sons of Bill and Angeline, Tom and Clare Bruce and Jon and Suzanna Bruce.

Originally the farm probably grew wheat and corn and raised cattle, as did most farms. The farm still contains 320 acres and corn and soybeans are grown there now. Unfortunately the farm house, windmill, root cellar and garage were removed sometime in the 1960s.

Angeline Bruce receives the pioneer farm award

Angeline grew up near Pickrell, NE with her brother and parents. Angeline finished nursing college in Nebraska. She married Bill Bruce and moved to El Granada, CA then retired to Brookings, OR. Bill and Angeline were always advised to sell the farm but they did not as Angeline did not want to lose her last connection, land wise, to Nebraska. Sometime in the 1960s Marvin Spiller, nephew of Tom and Lena Bauman, started farming the land. Marvin also farmed the land for Bill and Angeline Bruce. When Marvin retired, his son Greg farmed the land for Bill and Angeline and farms it today for Tom and Clare and Jon and Suzanna Bruce.

In addition to their sons, Bill and Angeline have five grandchildren: Nicole, Allan, Adriann, Evelyn, Eileen and one great-grandson, Aiden. *Submitted by Jon Bruce., researched by Ross Bauman.*

BUCKLEY – When Mathias Buckley immigrated from Goppingen, Wurtemburg, Germany in 1852 and came to America, he probably had no idea what he was setting into motion: a ranching tradition that would span 107 years and span five generations.

Rudolph and Emily Buckley with children Carl, George, Jessie, Bill, Roy, and Emerson.

In the mid-1890s, Mathias's oldest son, Rudolph, along with his wife, Emily and four young children: George and Jessie, twins, Roy and Bill moved to western Nebraska. Traveling by covered wagon they arrived near the community of Belmot with 25 cents and a dozen chickens.

In 1908, Rudolph and Emily moved their family with six children: twins George and Jessie, Roy, Bill, Carl and Emerson, west to Sioux

County, making a homestead application at the Alliance Land Office on 640 acres that is known today as the Buckley family's Rafter Z Ranch.

Rudolph's youngest son, Emerson and his wife, Lillie, raised their family of six on the ranch while continuing to breed Hereford cattle.

Today Emerson and Lillie's two sons, Jr. and Jack, continue the family legacy with Jack's son, Will and two grandsons, Wyatt and Colton Buckley waiting to carry the ranch on into the future.

On June 15, 2008 nearly 450 guests were on hand to celebrate the 100 years and five generations of homesteading with the Buckley family.

After 107 years in the northwestern corner of the Nebraska panhandle, with fluctuating cattle prices and drought conditions, the Buckley family feel very fortunate to continue making their home where their ancestors put down roots. *Submitted by Jack F. Buckley.*

BUCKMAN – Fredrich (Fred) D. Buckman, son of Heinrich and Katharine (Strockman) Buckman was born August 16, 1853, in Siedenburg near Mellinghausen, Hanover, Germany. In 1871, he moved to America and hired out to farmers near Platteville, Wisconsin. On December 20, 1878, he married Wilhelmina M. Bunsen, daughter of Hans C. and Christina E. (Kahl) Bunsen. She was born in Neudorf near Lutzendorf, Holstein, Germany, on June 29, 1857, and moved to America with her parents in 1872. Their first child, Magdalena (Lena) K. was born September 26, 1879 at Platteville.

Descendants of Fred D. Buckman (Steinhoff), 1988. Front row: Shirley Steinhoff Stohlmann, Clara Kehlbeck Steinhoff, Arthur Steinhoff, and Cindy Steinhoff Drake. Back row: Danny Stohlmann, Kim Stohlmann Roecker, Craig Stohlmann, and Kirby R. Drake.

After the death of a son in early infancy, the Buckman's moved to Nebraska. They lived about six months on the old Mohr place (west of Avoca in Cass County), and also spent some time in the Straub settlement (east of Avoca). On December 29, 1888, they purchased the NE ¼ Section 28, Range 10, Township 11 (Weeping Water Precinct, Cass County) from Addison P. Weston. Previous owners had been George W. Wiswell, Albert Hayman, and Selden N. Merriam.

They lived on the farm until about 1900 when they moved to Avoca where Fred built a new elevator with B. C. Marquardt. During this time the William Coatman, Henry Brockhoff, and George Sudman rented the farm.

The Buckman's were members of the First Lutheran Church of Avoca (North Branch) located across the Cass and Otoe County line in Otoe County on Highway 50. Their great-granddaughter and her family still attend this church.

On April 28, 1909, Lena Buckman married William H. Steinhoff (b. January 26, 1880, Platteville, WI, to August and Bertha (Genthe) Steinhoff, Syracuse). Lena and William moved to the farm where their first son Ervin A. F., was born on December 21, 1910. Ten years later their second son Arthur F.E., was born on March 7, 1920. On March 21, 1922 Fred and Wilhelmina Buckman deeded the farm to their daughter Lena.

Wilhelmina Buckman died in Avoca on November 11, 1928. Lena K. Steinhoff died June 28, 1934. After the death of his wife and the untimely death of his daughter Fred Buckman died March 3, 1937. Ervin Steinhoff served in World War II. After his discharge from the army he settled in Nebraska City and died on March 5, 1963.

Arthur Steinhoff and Clara M.M. Kehlbeck (b. November 4, 1918, rural Avoca, daughter of William and Minnie (Leefers) Kehlbeck were married January 24, 1940. (See Henry Kehlbeck Family). Prior to her death, Lena Steinhoff had deeded the farm to her son Arthur. Arthur and Clara moved to the Buckman/Steinhoff farmstead and William Steinhoff continued to live with them until his death on February 29, 1944. He missed the birth of his first grandchild by four months. Shirley A. Steinhoff was born June 4, 1944, and their second daughter Cindy was born December 23, 1956. Shirley married Danny Stohlmann (b. January 14, 1943, d. March 13, 1994, son of Ellsworth & Norma (Luby) Stohlmann, Louisville-see Frederick Stohlmann Family) on November 15, 1963. Shirley continues to reside on the Frederick Stohlmann farmstead south of Louisville. Her two children Kim Roecker and Craig Stohlmann (along with their families) reside near her.

Cindy married Kirby R. Drake (b. January 10, 1954, son of Arthur & Gwineth Drake, Elmwood) on October 7, 1978. Cindy and her husband Kirby moved to the Henry Kehlbeck farmstead (located in the opposite corner of the Buckman/Steinhoff farmstead), and her mother's birthplace (see Henry Kehlbeck family). They have one son, Quentin A. Drake.

Arthur and Clara continued to live and operate the Buckman/Steinhoff Farmstead until the mid-1990s when they retired and lived in Lincoln, Syracuse, and finally a nursing home in Ashland. Arthur passed away June 21, 2003 and Clara passed away July 31, 2006. All of the Buckman's and Steinhoff's are buried in the North Branch Cemetery (First Lutheran Church).

Shirley Stohlmann inherited the Buckman/Steinhoff farm from her parents and has rented the home and farm ground to local residents. The farm has now been in the family for almost 128 years. Portions of the two story wood home are over 140 years old. A prominent feature on the farmstead is a large oak tree grown from an acorn planted by Lena Steinhoff in 1896. The rest of the older buildings on the farm were built for/by Fred Buckman and William Steinhoff. A building that served as a granary, machine shed and garage was built in 1970, and a new grain bin in 1981 was erected by Arthur Steinhoff.

BUESCHER – The initial 80 acres that are part of the Buescher farm were purchased by John Gerhardt and Catherine Buescher in 1882. There oldest son, Henry developed the farm and in 1886 he and Mary Rempe married. Their first house is still part of one of the current buildings (as is the original barn). In 1887 Henry purchased adjoining property and in 1888 acquired the original 80 acres from John. This comprised the quarter of Henry's farm, which is the southwest quarter of Section 35, Township 4 North, Range 8, West of the 6th principle meridian in Nuckolls County. In 1902 they built the 3 story house pictured. Their 5 children were born on the farm, John, Elizabeth, Frank, George, and Cornelius. George became the next owner; he and Margaret Ross married in 1932 and raised their four daughters, Ellen, Bonnie, Charlyn, and Colleen there, residing on the farm 44 years. George lived there from birth to retirement at age 77. The next to live on the farm was Colleen, who currently owns the original quarter. It is being farmed by her nephew Gene Himmelberg, the fifth generation to work the land. A portion of the original 80 acres will soon be transferred to yet another generation, Michelle Janda. She and husband Phil are refurbishing the 1902 house with their children, Samantha, Blake, and Connor, who represent the 7th generation to call the farm home. The "Buescher Homestead" is a recipient of the Knights of AK-SAR-BEN Nebraska Pioneer Farm Award.

The Buescher family home

George Buescher and his team of horses

Since the initial horse drawn equipment used in 1882, crops planted on the farm through the years included corn, wheat, milo, soybeans, oats, and hay. Livestock was a mainstay, primarily cattle, hogs, and sheep. In addition, milk cows, turkeys, ducks, and chickens were raised, along with vegetable gardens and fruit trees.

BURKE – Michael "Mike" William Burke, son of Johan and Ella (Conway) Burke, was born in March of 1865. Mike and Ella had eight children: two in Ireland, two in Iowa and four in Kansas. Mike was the fourth child and born in Iowa.

Mike came to Oshkosh on foot, stopping during a storm at the only house between Ogallala and Oshkosh. He homesteaded in 1891 on the south side of the river. His older brother, Hugh Burke, homesteaded a farm four miles west of Oshkosh on the north side of the river in 1887.

Mike married Maggie Kennedy in Marshall County, KS. They had met when she came to visit her sister, Nara (Kennedy) Duffin. Mike and Maggie had six children. Mike was present when the town of Oshkosh was founded and given its present name. He was also an original member of St. Elizabeth Catholic Church. Previous to the founding of Oshkosh, he traded at Hartman post office.

Edward "Ed" William Burke was born Aug. 10, 1892. He was Mike and Maggie (Kennedy) Burke's second child and was born in a sod house near "Penn" school, located four miles west of Oshkosh, on the north side of the river. This was originally homesteaded by Ed's uncle, Hugh Burke, in 1887. Ed had an older sister, Frances, and the family later included Mary, Cathryn, Esther and Joseph.

Ed attended school at "Penn" school through the 8th grade. His mother (Maggie) passed away in 1909, and in 1910 the family moved to the south side of the river, which Michael had homesteaded in 1891.

Ed and his siblings worked hard gardening, butchering, meat curing, churning butter, raising chickens, cattle and hogs and other daily tasks. Entertainment on Sundays was provided by rodeos held at the river's edge with buggies and wagons forming an arena on the south side of the river.

Just before entering the Army on May 2, 1918, Ed homesteaded a place one mile west of his father's and when his father died (in 1948) he took over both homesteads. By Sept. 12, 1918, Ed was a corporal, and one of his tasks was to train and break horses for the cavalry. By the time Ed was on a ship, ready to go overseas, peace was declared and WW I was over. He was discharged on Feb. 18, 1919. He met his future wife, Marvel Elva Hall, at a "welcome home" dance given for the returning servicemen. Edward married Marvel (Hall) Burke on Dec. 28, 1921, and together they had three children - Janice Ruth, Keith Edward and Jeanne Elizabeth. Ed was born, raised and died in Garden County. He never lived anywhere else.

Janice (Burke) married Jonathan "Jack" Seibert on June 1, 1943. They had five children together - Carolyn, Robert "Bob," Cheryl "Sherri," John, and Janice "Gaylene." In 1971, Janice and Jack took over the two homesteads of Mike and Ed Burke, which they later inherited in 1995. On May 25, 1998, Jack passed away. Janice (Burke) Seibert still resides at the two homesteads, which her son, Bob, helps manage and will later inherit. *Submitted by Janice Burke Seibert.*

BYGLAND – Iver S. Bygland Sr. immigrated from Norway in 1889. After making a few stops along the way, he eventually settled in Boone County, NE in the area known as "Bradish". Iver Bygland Sr. acquired land, the northwest 1/4 section 8-21-5 from Levi Gutru for $3,000.00 on March 17, 1891. Over the next few years he was successful enough to acquire another four quarters of land. He married another Norwegian immigrant and to this union seven children were born. He was known for being kind and generous, helping other immigrants to become established in this area. In 1909 he ran and successfully won a seat in the legislature. He served for six years and then ran for a seat in the House serving one term. He was very popular and was well liked by his constituents.

Senator Iver S. Bygland

Stephen Bygland, their first born son, acquired the home quarter in the 1950s for a sum of $4,000.00. In December of 1957 he married Edna Luebke. They raised two children, Frank and Mary Bygland Yrkoski. Frank acquired the farm in 2010.

In the history of the farm there have been cattle, hogs and chickens. Crops included corn, wheat, oats, soybeans and alfalfa. In addition to the row crops, Iver planted an orchard near the house.

Historic buildings still standing on the farm include the original house which is now a work shop. The barn, with the additions of a milking parlor on the east side and holding pens for the horses, was built in the teens and twenties. The existing house was built in 1912 with indoor plumbing, 1-1/2 baths, five bedrooms, and central heating with a coal burner, plus its own electrical plant. In August of 1991 the family was presented the Aksarben Pioneer Farm Family Award for 100 years of ownership. *Submitted by Frank Bygland.*

CARNES FARM #1 – Cecelia Carnes and Richard Carnes farm, 40 acres located in the Northwest Quarter of the Northwest Quarter of Section 25, Township 13, North Range 6 in Bluff Precinct in Hamilton County, Marquette, NE.

Richard and Cecelia Carnes standing by their pasture gate, August 2001.

Cecelia's Great-Grandfather Frank Johnson was born in Smaland, Sweden in 1872, spending several years at Osco in Henry County, IL.

In 1879 he came to Nebraska and settled in Hamilton County in 1887. He purchased the 40 acres of land from John and Ellen A. Lock on March 10, 1893 for $800.00.

Frank Johnson died Feb. 28, 1920 and the land passed to his wife, Annetta Johnson and in 1933 to their twin granddaughters, Mildred Randahl Holtorf (Cecelia's mother) and Myrtle Randahl Clarke.

In 1955 Myrtle Randahl Clarke's share was purchased by Cecelia's mother and father, Mildred and Carl Holtorf. In 1979 the farm was purchased by Cecelia and Richard Carnes.

The land was leveled in 1970 and an irrigation well was put in.

Cecelia and her husband, Richard, live one-half mile south of this farm, where they have a cow, calf and grain farming operations.

They have two daughters, Tammy Hecox and son-in-law Don Hecox, Billy Jo Tomasek and son-in-law Steven Tomasek, four grandchildren and one great-grandchild.

Cecelia and Richard Carnes received the Nebraska Pioneer Farm Award for this farm in August 1993.

CARNES FARM #2 – Cecelia Carnes and Richard Carnes farm contains 120 acres. Eighty acres is located in North 1/2, NE 1/4 in Section 35, Township 13, Range 6, Bluff Precinct and 40 acres is located in Northeast 1/4, Northwest 1/4 in Section 35, Township 13, Range 6, South Platte Precinct in Hamilton County, Marquette, NE.

Cecelia's Great-Grandfather Frank Johnson was born in Smaland, Sweden on April 20, 1850. He immigrated to America in 1872, spending several years at Osco in Henry County, IL.

In 1879 he came to Nebraska and settled in Hamilton County. Frank Johnson purchased first 40 acres from the Union Pacific Railway Company on Feb. 27, 1894 for $100.00. The second 40 acres also was purchased from the railroad on Sept. 21, 1899 for $200.00. The third 40 acres was purchased from Delbert E. and Ellen F. Danhauer on March 23, 1901 for $500.00.

When Frank Johnson died Feb. 28, 1920 the land passed to his wife, Annetta Johnson and in 1933 to their twin granddaughters, Mildred Randahl Holtorf (Cecelia's mother) and Myrtle Randahl Clarke.

In 1955 Myrtle Randahl Clarke's share was purchased by Cecelia's mother and father, Mildred and Carl Holtorf. In 1979 the farm was purchased by Cecelia and Richard Carnes. They live one-fourth mile north of the farm, where they have a cow, calf and grain farming operations.

The Carnes pasture land is historic in an even bigger way as the Oxbow Trail runs through it, with wagon ruts still visible today.

The Carnes have two daughters, Tammy Hecox and son-in-law Don Hecox, Billy Jo Tomasek and son-in-law Steven Tomasek, four grandchildren and one great-grandchild.

Cecelia and Richard Carnes received the Nebraska Pioneer Farm Award for this farm in August 2001.

CAROTHERS-BURGE – In 1885 my Great-Grandfather Robert Leeper and Sarah Carothers homesteaded the Northwest Quarter, Section 19, Township 10, North Range 36 West of the 6th P M in Perkins County, NE. Oct. 25, 1905 their son, my grandfather, Robert Savage and wife Nellie Carothers inherited this land, 160 acres. He also bought 80 acres from his Aunt Virginia whom they called Jennie, the West 1/2 of the Northeast Quarter, Section 19, Township 10, Range 36 in Perkins County, NE.

Robert S. Carothers departed this life Sept. 5, 1925 leaving the land to his wife and children. Children were: Roy, Glenn, Rex, Robert, Fern Miles and May McCormick. On Sept. 18, 1929, Fern Miles inherited the land described above, containing 240 acres. Upon her death she deeded it to her daughter Lucille Bastian and son Robert Miles. Lucille and Dick Bastian bought Robert's 1/2. April 23, 1999 Lucille (1st cousin) and Dick Bastian sold the 240 acres to Sharon and Dale Burge. Sharon Burge was the daughter of May McCormick, daughter of Robert Savage Carothers and a sister to Fern Miles. *Submitted by Sharon Burge.*

CERVENY – The 160-acre Donald Cerveny farm, which has been in the Cerveny family for 125 years, is Section 17-5-3 in the south half of the southwest quarter and the south half of the southeast quarter. They received the Nebraska Pioneer Farm Award in 2004.

In was originally owned by Vaclav Cerveny, who immigrated to the United States from Czechoslovakia with his wife, Barbora, and five children – four boys and one girl. After being in Illinois for a few years, they moved to Saline County in 1891 and purchased the farm near Swanton on March 6, 1891. He was one of the early settlers of the Swanton vicinity. They moved to Swanton when they left the farm.

In 1896, Vaclav sold the farm to his son Joseph and wife Emma Cerveny for $4,000. Joseph was born 1869 in Chicago, Illinois. He came to

The Cerveny farm, May 2008.

William and Donald Cerveny harvesting wheat.

Nebraska as a young boy with his parents, and attended school when he could, through the sixth grade. Children in those days attended school only when they were not needed at home for helping with the work. Emma was born in 1871 near Tobias. Joseph and Emma had four children: William, Emma, James and a baby boy died in infancy and was buried in a tiny country cemetery still in existence. The home consisted of one room with a cave underneath. They farmed until 1918, when Joseph sold it to his son, William and wife, Alice Cerveny.

William farmed mostly with mules and Vaclac and Joseph farmed with only horses and mules. William did buy a Fordson tractor later in farming, and used it for a short time before purchasing a McCormick-Deering 10-20 built by International Harvester, which ran on kerosene. He farmed until 1946 when his son, Donald and wife, Irene began farming.

Donald's dad William helped with the farming in the early years, and Irene helped out after that. She worked in the fields with the tractor, hauled grain and various other farm chores. Besides this farm, they farmed other land acquired over the years. They raised corn, wheat, soybeans, milo, oats, alfalfa, and hay over the years and also had cattle, hogs and chickens. Water was added to the house in 1950, and wood burning stoves were replaced with an oil heater and later it was replaced with a furnace. Also added was a pantry, bathroom, kitchen, and the porch was enclosed. The house was remodeled again in 1979, when a larger living room and basement were added. They had three children—Cheryl Weishahn, Lincoln; Patricia Wine, Phoenix, Arizona; and son Jack, who is deceased. They also have four grandchildren, three great-grandsons and four great-granddaughters. They retired in 1990 and continued to live on the farm until 2015, when they moved to Wilber. They enjoyed doing more traveling after they were retired. Currently the farmed acres are crop shared and alternate corn and soybeans. The pastures are rented, and there are CRP (Conservation Reserve Program) acres.

William Cerveny

The work on the farm was demanding, and hours were long a lot of the time; but, it was a life they enjoyed and are proud of their accomplishments and farming history.

Donald and Irene received the Pioneer Farm in 2004.

CHISM – In 1894, William T. Chism bought 80 acres in Richardson County, Nebraska, and later purchased two additional 80-acre parcels across the road, making a total of 240 acres in the West Muddy and Porter Precincts, Township 3, and Liberty. William raised pigs, chickens and cattle. In 2011 the farm passed down to grandchildren Patricia Hampel McCloud, Debi Marts, and Yolanda Marts and the land produces soybeans and corn.

CHRISTENSEN – The land purchased by George and Effie Baker is located in 20-1-18, four miles south of Alma, Harlan County, NE on Highway 183. They purchased 160 acres April 17, 1906, which was homesteaded by William S. Campbell June 11, 1881, purchased by C.T. Slawson in 1884 and sold to George and Effie for $6,400.

Christensen farm, December 9, 1975

George and Effie purchased another 160 acres March 11, 1907, which had been homesteaded by Herbert B. and Rhoda E. Dow Nov. 18, 1890; purchased by Owen and Mamie Carroll on Jan. 23, 1906 and sold to George and Effie. William and Alta Baker Walker purchased all of it in 1942 and after the death of Alta in 1989, the land passed to the seven children: Floyd, Ruth, Margaret, Robert, Donald, June and Jean Walker. Greg Christensen (son of Jean) purchased the shares of the six older siblings and currently farms the land.

The barn pictured was moved from a farmstead further south, which was located on part of the land they purchased, and re-located to the present farmstead. Several picture frames have been made from the siding after it was destroyed. The house from the south farmstead served temporarily as Prairie Dog School until a new one was built. Elsie, Alta and Loren Baker attended school there; all seven of the Walker children and two of the Walker grandchildren also attended, as well as children and grandchildren of Alta's siblings. The Bakers had six children: Guy, Ethel, Everett, Elsie, Alta and Loren and 21 grandchildren.

Corn, wheat, oats and alfalfa were the main crops; cattle, hogs and chickens were raised. Crops now are wheat and corn and Greg pastures and feeds cattle there.

The farm was recognized as a Pioneer Farm on July 29, 2006 at the County Fair. *Submitted by Jean (Walker) Christensen and Greg Christensen.*

CIHLAR – Vincent Cihlar was born in 1868 in Bohemia and emigrated to America in 1886. He married Kristina Pavlik and they were the parents of 11 children. The family settled near Walnut, NE in 1904. Vincent passed in 1952.

His youngest son, Edward, married in 1934 to Marie Cernousek, they had three children and farmed the land.

The oldest son of Edward, Wallace, farmed with his father. Wallace and Patricia (Nelson) have no children. Wallace spent two years in the Army, returning in 1959 to continue farming with his father and is still actively farming in 2015.

Edward passed away in 2007. *Submitted by Patricia A. Cihlar.*

CLARK-LANG – David Clark (1815 - 1901) visited section 26, in northwest Washington County in 1867. He moved from Lorrain County, OH in 1868 with his wife, Lucy Foster and their son, Hiram. They claimed 160 acres on Bell Creek using a Military Bounty Land Warrant they had purchased in Ohio from the family of a veteran of the War of 1812.

The Clarks were pioneers in the Alder Grove Community. They raised horses, cattle, corn, wheat and prairie hay. David purchased more land from the Union Pacific and eventually owned about 2,000 acres He retired in 1886.

David's son, Hiram, married Elizabeth Whisinnand in 1871. They had seven children: Anna, Duke, Bingta, Art, David, Libby and Vola The eldest daughter, Anna, married Charles T. Lang, a native of Davis County, IA. He was a graduate of Iowa Wesleyan and teaching at Fontanelle. Anna was teaching at Hewitt School. They married in 1893. After several years of teaching, banking and business, Charles returned to the freer life of farming in 1904, first on a farm near Tekamah and in 1910 the family moved to the original land warrant land near Alder Grove.

Anna and Charles had three sons: Roscoe, Francis and Richard. Roscoe died in a farm accident. Francis and Richard attended the University in Lincoln and returned to farm their father's land. Richard married Elizabeth Skinner in 1930. She had attended Wayne Normal and was teaching at Alder Grove. Elizabeth was the granddaughter of Thomas Riley Wilson, an early pioneer and the first surveyor of Washington County.

Richard and Elizabeth had five children: Joan (Price), Russell, Wayne, Charles R. and Kathy. Richard farmed all his life at Alder Grove. He raised corn, soybeans, oats and alfalfa, fed about 25 steers each year, milked a few cows, raised about 75 hogs and some chickens. Elizabeth had a large garden.

Russell served as an agricultural missionary of the Methodist Church in Chile from 1960 to 1975. He was an extension agent in Dodge County from 1976 to 1996. He married Nancy Hinton of Knoxville, TN in 1959. They have five children: Cristina, a food scientist; Rowan, a food service manager; Wesley, a statistician; Jon, a physicist and Robert, a miller. They have 12 grandchildren. Nancy passed away in 2008. In retirement Russ lives on the land warrant farm and has written several books of local and state history.

CLOPTON – In 1884, David A. Clopton came from Jackson County, Iowa, to sign Homestead papers for 160 acres of land located at Section I, Township 33, Range 21 in Keya Paha County, just north of Springview.

In 1887, following marriage to Kate E. Blakey, the young couple added an adjoining 160-acre Tree Claim Site, making a total of 320 acres. The Tree Claim required the planting of 10 acres of trees and maintenance thereof for eight years.

David and Kate planted over 4,000 windbreak, fruit, nut and flowering trees, as well as numerous bushes and vines to prove-up their acquired land. As with many other pioneers, they ranched and farmed to live and improve their way of lives.

Along the way, David and Kate raised a family of four children: Myrtle, Ira, Ray, and Wilda. In 1920, David and Kate sold the ranch to their youngest son, Ray, who had returned from the Army in WWI and married Manila Bay Lambert. David and Kate retired.

Ray and Manila operated the ranch together for 45 years. They had three children: Dorothy (who died in infancy), Dean, and David. Manila died in 1965. Ray stayed on alone for another seven years before transferring ownership to Dean and David, and retiring in 1972. He died in 1976.

Although neither Dean nor David remained on the ranch, they remained extremely conscious of their heritage and remained active in its maintenance and operation. Before he died in 2008, Dan transferred his ownership share to David...and the legacy continues. *Courtesy of David R. Clopton.*

COADY – Matthew Micheal Coady purchased the southeast 1/2, 32-16-7 March 2, 1906 from George and Rosa Prange. Matt married Elizabeth McGinn Sept. 8, 1909. They had two children, Leo and a daughter dying in infancy. Leo married Catherine Gorman Feb. 14, 1946. They had four children: Micheal, Richard, Patricia and a son died in infancy. Micheal married Melvina Ruhe June 1989; they had two children, Patrick and Catherine.

Matthaew and Elizabeth (McGinn) Coady, ca. 1909

Matt and Leo farmed the land with horses into the 30s, raising cattle, hogs, corn, chickens, eggs, milk and cream. Matt and a neighbor purchased the first pull type combine in the county after going to tractors. As a young man, Leo sold the *Omaha Bee Newspaper*, also delivering it sometimes on horseback. Also sold the *World Herald* throughout Saunders County. Matt and Elizabeth lived on the farm most of their married life, moving to Fremont October 1949. Leo and his family moved to the farm, living there until August 1959 when Sand Creek flooded, getting four feet of water in the house. The creek had never gotten in the house before. A native hay meadow was kept until in the 1960s.

Matt unexpectedly died December 1959 of a heart attack while visiting the farm. Leo, unable to support his family, had to quit farming due to years of flooding. He got a job but continued to raise cattle.

The farm was rented until 1972 when Richard Coady began farming. He farmed the land until his death in 2010. Then Micheal and Patricia started farming with the help of good neighbors, raising corn and soybeans.

Hopefully the next generation will continue to love the land. Realizing how Matt struggled to keep the farm in the great depression of the 1930s. Along with each generation struggles. *Submitted by Patricia J. Coady.*

COLLETT – William J.F. Collett acquired 80 acres on June 24, 1878 located in York County, Township 8, Section 11, Range 4.

This 80 acre farm passed from William J.F. Collett to his son, also William J.F. Collett and then to grandson William W. Collett and finally to great-grandson William D. Collett.

Today the farm produces corn and soybeans. The house built in 1906 remains on the farm.

In 1978, the Aksarben Pioneer Farm Award was presented for this farm remaining in the same family for 100 years.

William and Sandra Collett are still farming and have two daughters, Krystal Collett and Brittany Collett.

CONNELL – The original purchaser was John Connell in 1889. He was a great-great uncle who purchased 168.19 acres for $350.00 in Township 32, Range 50 in what is now known as Whitney, NE.

In 1891 John went into a partnership with his brother, Wm. D. Connell and wife Eva (great great grandparents). From their union three children were born: Clarence F. Connell (great uncle), two great aunts, Lillie (Connell) Paterson and Gertrude (Connell) Harris.

Upon the death of William, the children gifted their shares to their mother Eva in 1922. In 1924 Eva (great great grandmother) sold the land to her son, Clarence F. Connell and wife Ethiel. From this union, seven children were born: three aunts: Gwendolyn, Mildred and Betty Jean; four uncles: Clarence, Robert, Charles, Warren (father).

When Clarence F. and Eva passed away, the land was signed over to Charles Connell (uncle) in 1967. In 1970 Warren (father) and wife Frances (mother) purchased the land and operated it until their deaths. From this union three children were born: Alan P. Connell, Cheryl K. (Connell) Reeves, Terry L. Connell.

Farm signs (both photos) made from original logs from the Homestead House of 1889

In 2012 Alan, Cheryl and Terry jointly received ownership of the original acreage and will pass it to their five children. The land is now known as The C.A.T. Ranch L.L.C.

On the original homestead was a three room log house. Later in years there was an addition of a kitchen and bathroom added with a dug out cellar under it with an outside entrance, through the years two more additions were added. This home is no longer standing. In 2009 one of Warren and Frances' grandsons removed the old house to put up a new one. Some of the logs from that original house were made into signs when you enter the property.

Alan P. and Barb Connell have three daughters: Shawna, Kristie and Wendy and three granddaughters: Blake, Emma and Janessa. Terry L. and Carol Connell have a son Tanner and daughter Charlee; three grandsons: Ryan, Hudson and Brock and one granddaughter Harper. *Submitted by Terry Connell.*

CONNELLY – Edward Frank Connelly was born on April 29, 1863 in Lindsay, Canada. In October 1871 Edward's parents, William and Sarah Jane Connelly, moved from Lindsay, Canada to St. Bernard Township, Platte County, NE. In August 1880 William Connelly received a land patent (document #2951) for 160 acres of farm land in St. Bernard Township.

Edward Frank Connelly was born on April 29, 1863 in Lindsay, Canada. In October 1871 Edward's parents, William and Sarah Jane Connelly, moved from Lindsay, Canada to St. Bernard Township, Platte County, NE. In August 1880 William Connelly received a land patent (Document #2951) for 160 acres of farm land in St. Bernard Township.

On Sept. 8, 1883, Edward Connelly was the highest bidder at the Platte County Sheriff's sale of 80 acres in Section 8, Joliet Township (T019N R3W of the 6th Prime Meridian). Nine years later (March 1892) Edward and wife, Margaret, purchased the 80 acres to the west of their farm to get 160 acres, NE1/4 S8. Edward initially built a small home on the farm. He used horses to till the land. He also had milk cows, feeder cattle, pigs, chickens plus a large garden. He planted and harvested oats, corn, wheat and alfalfa.

In the 1890s Edward quadrupled the size of floor space by adding two large rooms to the north side of the existing structure. He also created a basement under the new additions plus adding a vegetable cellar. He also built five bedrooms on the second story, which spanned the entire north-south span of the lower floor.

In the mid 1920s Edward installed a Delco 32 volt DC Light plant, which included 16 2-volt glass batteries. Insulated copper wires and light fixtures were installed in all rooms. During this time frame a bathroom, a pantry and a porch were added to the house.

Prior to Edward's death in 1926, he and Margaret transferred the farmstead to their youngest son, James E. Connelly. In the mid 1940s, James purchased a tractor and a few implements. After a couple of years, horses were phased out of the farming operation. After James E.'s death, his wife, Catherine, became the owner of the farm.

In April 1968, a "Warranty Deed - Joint Tenant" was filed. The land title is now in the name of Catherine Connelly and/or James D. Connelly (oldest son of James and Catherine).

In June 1984 James D. Connelly, as the surviving Joint Tenant, has sole ownership of the farm. Today corn, soybeans and alfalfa are the crops on this irrigated farm. The first irrigation well was dug in 1956. *Submitted by James D. Connelly.*

COON – Jason Bloodgood purchased land from the Union Pacific Railroad in 1881 (NE quarter of the NE quarter and the south half of the NE quarter of section 7 in Kearney County (120 acres) for $300 and in 1884 purchased the south half of the NW quarter of Section 7 in Kearney County (79 and 38/100 acres) for $198.45. His children inherited the land, and in 1900 his daughter, Margaret (Maggie) Bloodgood Bliss and her husband John H. Bliss purchased it from her siblings.

The yard at the Coon farm in the early days.

John H. Bliss had a house built on the property in 1918 in anticipation of son, Frank, returning home from World War I. However, when Frank arrived home he decided to homestead his own land in Wyoming.

John and Maggie asked their daughter Fern (twin sister to Frank) and her husband, Glenn Coon, to move into the house. In March 1919, Glenn, Fern, and their three children, Margaret, Loretta, and Bob moved from Shelton into the new home. Two other children were born on the farm, Amelia (Midge) in 1924 and Glenn, Jr. in 1926. Glenn Sr. and Fern farmed the land and eventually bought Fern's siblings' share of the land. During the late 1930's the original route for the tri-county irrigation canal cut through the property, causing a portion of the farm ground to be changed to pasture ground.

Glenn Sr. did the farming with a team of horses. He was uncomfortable with using a tractor, but bought a Farmall for his sons to use. They raised wheat, milo, corn, and alfalfa. Fern passed away in 1944.

In 1949, after Glenn Jr. married, he and his wife Joyce lived on the farm. Glenn Jr. continued to do the farming, renting the land from his father and making several improvements to the house and yard. Even though he purchased newer tractors over the years, he kept the last Farmall H his father had bought him and had it in running condition.

Following his father's death in 1972, Glenn Jr. purchased his siblings' share of the land. He continued to farm until his retirement from farming, then rented the farm ground to an area farmer. Although he no longer actively farmed the ground; Glenn Jr. continued to live in the house he was born in and worked to keep the farm maintained; spraying weeds, chopping musk thistle, and checking the cattle and fences in the pasture. He also participated during harvest season by trucking.

Glenn Jr. spent his entire life on the property passed down from his great-grandfather. He passed away in 2001, after an accident on the farm with his H tractor. Glenn Jr.'s four children now own the farm under the name Yellow Rose Glen. The main crops grown are wheat, soybeans, and corn. None of the land is irrigated.

CRAYS – The 80 acre Crays Family Farm in Saline County is now owned by Don and Sheila Elznic Crays with Roland and Norma McKellips Crays. It was owned by John and Mary Jane Tosh Crays from 1889 to 1905, Charles and Minnie Davies Crays from 1905 to 1945, Donald and Anne Herynk Crays from 1945 to 1998 and Don, Sheila, Roland and Norma Crays from 1998 to the present.

On Dec. 21, 1889, John Crays purchased the S1/2 NW1/4 of 22-6-1 from his son-in-law and daughter, William and Belle Nevins. The following spring they moved there with three of their children. They lived on the farm for 15 years before selling it to their son, Charles and wife, Minnie.

Charles and Minnie owned and lived on the farm until they retired in 1938 and moved to Tobias, NE. In 1945 they sold the farm to their son, Donald and his wife, Anne.

Donald and Anne moved to the home place in 1938 where Donald had lived all of his life except for about two years. Their farm was a cow/calf and grain operation. Following the death of Anne, Don and Roland's mother, Don and Sheila purchased the 80 acre farm from the heirs of Anne Crays in 1998. They then sold an acreage with the house and buildings to Roland and Norma. Roland and Norma now live on the farm.

Don farmed the land raising corn, soybeans and alfalfa on it along with having a Polled Hereford cow/calf operation. Don and Sheila's daughters are Stephanie (Kirk) Jensen, Linda (Grant) McKay and Clara (Tim) Grace and their grandchildren are: Cody, Casey and Jessica Jensen, Bailey McKay, Kane, Weston and Xavier Grace and Seth McKay (deceased).

This is the third farm Don and Sheila own that is a Century Farm. *Submitted by Don R. Crays.*

CRAYS – The Don R. Crays Family Farm located in Saline County was purchased by Joseph Herynk, grandfather of Don Crays. Joseph purchased 80 acres, the N1/2 NE1/4 of 6-51, from George and Lena Struckman in 1902 and another 80 acres, the S1/2 NE1/4 of 6-5-1, from the Mary Hothan heirs in 1908.

Aerial view of the Don Crays family farm

Joseph Herynk and Anna Dlouhy were married in 1909 and lived on the farm the rest of their lives. Joe passed away in 1959 leaving the south 80 acres to his daughter, Anne (Donald) Crays. Anna Herynk passed away in 1965 having lived in the same section all of her life. In December of 1965, Don and Sheila Crays purchased the north 80 acres from the Anna Herynk Estate. Following the death of Don's mother, Anne Crays, Don and Sheila purchased the south 80 acres from the heirs of Anne Crays in 1998.

According to a newspaper article in the July 7, 1916, *Tobias Express* - Joseph Herynk is putting up a fine new house. It is rather a large house, full two stories and from the road looks like it might have eight or nine rooms. It is now nearing completion and will soon be ready for occupancy.

Don Crays and Sheila Elznic were married in 1965 and have lived on this farm of 160 acres all of their married life. Don has farmed for over 50 years raising corn, soybeans and alfalfa on this farm. They also have a Polled Hereford cow/calf operation. Their daughters are Stephanie (Kirk) Jensen, Linda (Grant) McKay and Clara (Tim) Grace and their grandchildren are: Cody, Casey and Jessica Jensen, Bailey McKay, Kane, Weston and Xavier Grace and Seth McKay (deceased).

This is one of three farms Don and Sheila own that is a Century Farm. *Submitted by Don R. Crays.*

CUNNINGHAM – Nicholus Edward Cunningham Sr. acquired 360 acres located in Jefferson County, Lincoln Precinct, Section 31, Range 2, Township 1 in May 1886. Corn, wheat, oats and hay were grown to support livestock consisting of cattle, hogs and horses.

The farm was passed from Nicholus Edward Sr. to Nicholus Edward Jr. and to James Edward Cunningham in May 1964.

This 360 acre farm now grows corn and soybeans. The Century Farm is now in the possession of the third generation.

James Edward and Patricia R. Cunningham have three children: Allan James, Paul Edward and Roger Wade Cunningham. James Edward retired in July 1997.

DAUEL – Henry H. and Laura (Schultz) Dauel purchased 80 acres near Malmo, NE in July of 1908 from Wilhelm and Albertina Keeker. It is located in the east half of the northeast quarter of Section 33 in Township 16 (Douglas), range 6, east of the Sixth principle meridian in Saunders County, NE.

After their passing, it went to their children: Henry F., Arthur and his wife, Catherine (Wittmersehouse) and Leona and her husband, Ervin Nelson. Art and Catherine took total possession in 1975. Art and Catherine's son, Robert "Bob" and his wife Loretta "Lori" (Codr) acquired the farm in 2005. Through the years corn, beans, alfalfa, oats and wheat have been raised. There is 15 acres of original prairie. At one time, ruts from a wagon trail could be seen in it. A cow/calf operation, fat cattle and feeder pigs were livestock that were raised on the farm.

Lori, Chloe and Bob Dauel

In 2007, a new house was built on the original home place, where Bob and Lori presently reside. In 2002, they incorporated their farming operation, known as "Duck Creek Farms." The Dauel's trucking company is also based here.

They continue to farm this tract along with other land they have acquired through the years. *Submitted by the Dauel's.*

DEALEY – Herman Czenkusch purchased this farm from William F. Bohlander and Rosa C. Bohlander, his wife, and Edward Bohlander, single, on Feb. 22, 1909. Emma Czenkusch Hemmann inherited the farm upon the death of her father, Herman, on April 22, 1918. On Feb. 20, 1974 Emma Hemmann, single, deeded to her son, Loren E. Hemmann, Trustee. Upon the death of Loren E. Hemmann on May 22, 1980 the farm passed to his daughter, Jeannette L. Hemmann Dealey. She subsequently deeded to herself as Trustee and to her husband, Donald D. Dealey, Trustee and ownership remains the same to date.

The 160 acre farm is located in Buffalo County, northeast 1/4 of 2-10N, 17W. Corn, wheat, oats and hay were the grain crops grown and cattle was raised. The farm is grassland only now.

The original house that was hand built by G.P. Hemmann, husband of Emma was moved to a neighboring farm in approximately 1952. No other buildings remain today.

Jeannette L. and Donald D. Dealey have three children: Karen Dealey, Tim Dealey and Rachel Vonderfecht. Grandchildren are: Ashlyn Dealey, Keegan and Kyler Vonderfecht. Jeannette retired Dec. 31, 2007 but is still doing some title searching and working on the farm, controlling mask thistles and fencing. *Submitted by Jeannette Hemmann Dealey.*

Jeanette and Donald Deadley with their children and grandchildren.

DEAN – Omar J. Dean "OJ" and Mary L. Judy were married in 1892. On October 28, 1909, Omar and Mary Dean applied for the water rights and began the purchasing process for eighty acres of land in Morrill County. This land was located at the Southwest quarter of Section 29 in Township 20, Range 49 West of the Sixth principle meridian in what was nicknamed the "Rocky Ridge" community. Omar and Judy had five children: Faye, Eunice, George Wayne "Spud", Wyatt, and Olivia. The Dean children attended public school at District No. 159 with approximately forty-five other students. Their teacher was Nelson J. Mitchell. "OJ" and his son, "Spud" helped to dig the waterway now known as the Northport Irrigation District. They used horses, shovels, slips, and a fresno scraper. This canal now supplies irrigation water for 16,180 acres of farm ground in Morrill County. Omar and Judy entered their eternal place of rest in 1946 and 1953, respectively.

Barney Dean (3rd generation) and his great-grandson Landon Dean (6th generation) venture through the pasture to look for horses.

George Wayne Dean "Spud" - In the early 1940's, the responsibility of the farm was transferred to their son George Wayne "Spud" who married Milo Brown in 1922. To this union, two children were born – Helen Louise and Harry Wayne "Barney" Dean. Spud and Milo had a hog and cattle operation while farming alfalfa and corn. When Milo wasn't taking care of the children, she was milking twenty- three head of milk cows. She sold the milk to the local creamery until the law was passed requiring Grade A. The family continued to milk cows until the 1950's. Also on the farm, Spud managed thirty-two hound dogs and six saddle horses. The dogs and horses were used for hunting coyotes. Feed for the dogs consisted of ground corn that had been cooked over an outdoor fire. Helen and Barney also used the horses for getting to school where there was a shed for keeping the horses during the school day. Many square dances were held in the "big barn" located next to the family home with much of the music furnished by Milo's father, Charlie Brown. In the mid 1960's, Spud suffered from a heart attack; he and Milo retired from farming and moved to the town of Bridgeport and their son, BARNEY resumed the farm and cattle operation. Spud was also a World War I Army veteran. He died in 1980 and Milo passed in 1999.

Harry Wayne Dean "Barney" served in the US Navy from 1944 – 1946. On July 4th, 1949 he married Eleanor M. Decker, and the two of them had three sons – William, Michal, and Monte. When Barney took over the family business in the mid 1960's, he purchased an additional 320 acres of adjoining farm land where he continued to raise corn and alfalfa. There are many memories of family, friend, and neighborhood events on the homestead: ice skating parties at the lake on the property; horse riding adventures in the pasture; and bonfires with hot chocolate. Barney was a Bridgeport School Board member for thirteen years, and also participated in the Morrill County Sheriff's Posse and the Bridgeport Rifle Club. Their children also attended school at the "Rocky Ridge" country school until it was closed, and they began attendance at the school in Bridgeport. Eleanor kept busy with her involvement in Happy Hour Extension Club, 4-H, and Sunday School. She also was a bookkeeper for NAPA for over thirty years. In 1994, while lying on the floorboard of his pickup, Barney survived a tornado that completely destroyed the original home where three generations of Dean's (OJ, Spud and Barney) had raised their families. Instead of rebuilding, Barney and Eleanor purchased a home in Bridgeport, and Barney leased the farming to a neighbor until his son Monte took over the family business in 1996.

Monte W. Dean wed Tammy K. Faessler on July 28, 1979. Together they have raised four sons – Jerod, Jacob, Joshua and Jase while farming the family homestead and the farm that they reside on northeast of Broadwater. Their sons all attended Bridgeport Schools where they graduated from high school before pursuing their own careers and beginning their own families. Presently, Monte and Tammy farm the homestead that has grown to 212 acres of irrigated farm ground and 748 acres of pasture. They also farm 640 irrigated crops of corn, wheat, alfalfa, and sugar beets in addition to 640 acres of pasture land at their rural Broadwater residence. They operate a cow/calf operation of 250 head. Tammy is a registered nurse and has served as the school nurse for Bridgeport Public Schools for twenty-one years. Monte and Tammy have made plans to build a home back on the original homestead sometime soon. Two of their sons have expressed an interest in eventually helping with the family farm/ranch operation as Four J Land and Cattle LLC. The Dean family hopes to carry on their legacy in agriculture as Monte and Tammy have been blessed with one granddaughter, five grandsons, and two grandchildren on the way. They are grateful for the blessings of living and raising a family on the farm. *Courtesy of Tammy Dean.*

The original home that three generations lived in prior to 1994 when a tornado completely destroyed the residence and surrounding buildings.

DEERTZ – Grandpa and Grandma, Dad and Mom, didn't come to Nebraska in covered wagons. They were born here. But when I hear the word "pioneer," I can't help but think of them.

When you look through the old financial records you can tell that life was always a struggle. And the stories they told of the "dirty thirties;" the years the crops dried up, the day the sky was filled with dark clouds of grasshoppers.

Mom had her chickens and garden. Grandma made rag rugs from clothes that were beyond mending. Besides farming, Grandpa and Dad were blacksmiths and mechanics. Grandpa was "old school" and used a coal fueled forge. Dad was more modern and wanted an electric welder. He didn't have money to buy one so he made one from the generator of a 32 volt light plant. They repaired the neighbors' broken machinery. Sometimes they were paid with cash, sometimes with promises of proceeds from next year's crops. It didn't matter, you didn't say no to a man who was down on his luck, at least back then you didn't.

If you couldn't raise it, grow it or make it, you might have to do without it. Life was hard. But you didn't have to look far to find someone who was worse off than you. Like the man who had used up his credit at the store in town and came to them and asked if they would

Third, fourth and fifth generation of born in Nebraska pioneers.

grind some wheat into flour so his wife could bake some bread. Well, you didn't say no to a man who had hungry children to feed, at least back then you didn't. *Submitted by Carl L. Deertz.*

DIDIER – "The Farm That Time Forgot" – This farm nestled in the Barada Hills of southern Nebraska was acquired from the US Government in 1855 as a land grant to Great Grandmother Mary Peneaux (a Lakota Sioux Indian) and also her husband, John Didier – one of the earliest white settlers in southeastern Nebraska. Their home is on display as a historic landmark – relocated to the nearby town of Brownville, NE.

The logs of their old cabin were built from trees that were growing during George Washington's presidency – according to the record of the "tree rings."

These photos show the 1854 Didier log cabin before and after its restoration in 2012.

The descendants of John and Mary have resided at the farm continuously until this day – a 160 year span. The farm is a historic treasure that dates back further than most farms in the region – and has continued to produce crops for well over a century and a half – benefitting generations of descendants. It has changed hands many times in the ensuing years – yet always remained in the family.

Today, Wilbur Last and his family continue to live on the same land where he was born and raised over 80 years ago. He is proud of his heritage and the land of his inheritance, which bears the footprints of his forefathers. Their hard work is indelibly etched in every acre – form the old leaning fences to the hand dug wells.

The farm has been written about numerous times in a local newspaper, *Your Country Neighbor* by columnist Vicki O'Neal – granddaughter.

The farm is a place of celebration where descendants gather to share good times and memories – They are grateful for all the hard work invest in this farm and are determined to keep the acreage in the family for many generations to come. *Submitted by Wilbur Last.*

DILL – Andrew Dill married Susan Franklin in 1857 in Coletta, Whiteside County, IL. The family came to Cass County, NE Feb. 10, 1876 and settled on a homestead southwest of the village of Murray that Andrew had purchased the previous November in 1875.

Andrew had purchased 160 acres in section 21, in West Rock Bluff precinct in Cass County, NE from a nephew, Rozzel Morrow. Rozzel's parents, Andrew Morrow and wife Sarah Dill Morrow, had purchased original property in October of 1868.

Andrew Dill's son, Benjamin Dill, inherited the farm upon the death of Andrew in 1917. Benjamin married Emma Jane Adams. Emma Jane and Benjamin's son, Lester, farmed the homestead until Lester's death in 1958. The east 80 acres sold to Lester's sisters, Ethel and Grace. Grace sold her undivided half to Ethel.

The west 80 acres was purchased by Lester in 1956 for $13,000. He and brother, Franklin Dill, owned the farm until Lester's death. This was purchased by Harold Rice, son of Ethel Dill and William Rice, and their grandson, Paul Rice, in 1968 from the daughters of Franklin Dill for $40,000. This reunited the original homestead that had belonged to Andrew and Susan Dill.

Over the years the farm produced corn, wheat, prairie hay, alfalfa, red clover, milk and beef cattle, horses, hogs, chickens and bountiful gardens to feed the large family.

The four square two-story farm house built in the early 1900s currently is occupied by the sixth and seventh generations. The land continues to be farmed by the descendants.

The Rice family received the Ak-Sar-Ben 100-year Farm Family award in August 1976 at the Cass County Fair. Attending that award ceremony were Will, Harold, Paul and Michael Rice. (Grandson-in-law, great-grandson, great-great-grandson, and great-great-great-grandson of Andrew Dill.) *Submitted by Paul M. Rice.*

DIRKS – Martin H. Dirks and his wife, Foelkelina Folken Dirks, came from Germany to Platte County in 1882, the year they were married in Germany. They had seven children: Thalha, Diedrich, Mina, Anna, George, Ricka and Lena.

George and Anna Louisa Dirks with children, l-r: Lois, Milton and Lucille.

On Feb. 10, 1883, Martin H. Dirks, purchased 80 acres in Platte County, located in the east half of the southeast quarter of Section 29 in Township 19 North, Range 1 East of the 6th P.M. from the Union Pacific Railway Company.

On Dec. 22, 1887, Martin H. Dirks purchased 160 acres located in the east half of the southwest quarter and the west half of the southeast quarter of Section 29 from the Union Pacific Railway Company.

On May 11, 1920, Martin H. Dirks and Foelkelina Dirks, husband and wife, transferred 80 acres (the home place) in the southeast quarter of Section 29 to their son, George Dirks.

Martin H. Dirks died Oct. 13, 1921. George married Anna Louisa Schutte on Oct. 21, 1924 and they had three children: Lois Dankert, Milton Dirks and Lucille White. Foelkelina died in 1939 after moving to Columbus and living with her daughters, Ricka and Lena.

On Feb. 25, 1939, George Dirks purchased 40 acres in Platte County, located in the east half of the south half of the northwest quarter of Section 29 from John and Julia Kopetzky.

On Jan. 12, 1953, George Dirks and Anna Louisa Dirks transferred the home place, the west half of the southeast quarter of Section 29 to George Dirks and Milton H. Dirks. Milton was away serving in the Korean War, returning to the farm in 1953 and Milton and Esther E. Janssen were married July 26, 1953. The two families lived together for a short time in the brick house built in 1947 until George and Louisa retired to Columbus in 1953. Milton and Esther have four children: Kathleen Siefken, Keith Dirks, Sherilyn Raders and Joyce Bennett.

On Nov. 21,1963, Milton H. Dirks and Esther E. Dirks, purchased 80 acres of native grass pasture in the northeast quarter of Section 29 from the Friedrich a/k/a Fred Goedeken Estate heirs.

George Dirks died June 16, 1967.

On Nov. 15, 1968, Anna Louisa Dirks, widow, of George Dirks, transferred her interest in the farm to Milton H. Dirks and Esther E. Dirks.

The Dirks family received the Aksarben Pioneer Award in 1983.

Louisa died on April 21, 1991.

Milton H. Dirks died April 30, 1997.

Esther E. Dirks continues to reside on the now 132 year-old family farm and daughter, Kathleen and son-in-law, Fred farm along with their son, Ryan Siefken. Milton remembers the pasture having tracks and ruts in the hillsides that came from wagons travelling to Christ Lutheran Church from the now St. John's Church area. The ruts were eroding the ground so Milton closed the ruts to prevent further erosion.

The original 75 acres of pasture from UPRR and additional 80 pasture acres purchased from the Goedeken family in 1963 remain native prairie grass.

DIXON – In 1910 Ed Reynolds purchased 160 acres located in Buffalo County, Section 7, Township 11, Range 15. Corn and wheat were grown to support hogs, cattle and chickens.

The farm was passed from Grandfather Ed Reynolds to Lloyd Dixon, father of Phillip Dixon. Phillip acquired the farm in 1967.

Corn and soybeans are the crops now raised on this 200 acre farm. The only original buildings are a chicken house and wash house.

During the 1930s drought, the state husking contest was held here.

The Aksarben Award was presented for this 100 year old farm.

Phillip and Corliss Dixon have two children, Denise Zweiner and Chad Dixon and five grandchildren: Kaden Zweiner, Shelena Zweiner, Marrina Zweiner, Westin Dixon and Tanner Dixon. Phillip retired in 1998. *Submitted by Phillip Dixon.*

DOBIASH (DOBIAS) – Vencil and Anna Dobiash (the H was dropped from the name around 1910) grandparents of Jim, were married in Bohemia in 1861 and came to Cleveland, OH in 1866. The family moved to Otoe County in 1868 and on to Green Valley, Holt County, NE in 1885, Son Joe married Bertha Jirska (born in Prague, Czechoslovakia in 1889). They left Green Valley in 1910 because the wet seasons caused their horses to become sick. Joe and Bertha bought two quarters of land five miles north of Atkinson, NE (NW quarter 5-30--14 and SW quarter 32-31--14). They paid $7000.00 for the land, which they purchased from Albert and Rosa Morell. An old cement block house and no granaries--no trees--on the property.

Ray and Ruby's last family picture taken Christmas 2013, shortly before Ray's death.

After Joe died, Grandma Bertha, called Betty, stayed on the farm with bachelor son Charley until she and Charlie moved into town because of deteriorating health, in the mid 1950s. Grandson James Jr. farmed the land until Jim's son, Raymond and wife, Ruby Meyer Dobias, bought the land in 1959 and moved a house from town to replace the "little house on the prairie" which was torn down and re-built west of the new house as a garage. (Interesting note: Grandma had a wild prairie grass called Prairie Sand Reed in one corner of her yard. It lay dormant about 50 years until about six years ago, when it re-surfaced and was identified by the University of Nebraska. It still flourishes in a small plot.) Ray and Ruby raised seven children here. Raymond passed away Jan. 22, 2014. Ruby still resides on the farm. Children are Nancy Sich, Deb Leininger, Marsha Meyer, Barbara Buller, Becky Lindgren, Michael and Calvin. There are 24 grandchildren and 24 great grandchildren.

DOUBLE R FARMS (THE RACEK FARMS), MORSE BLUFF, NE – On July 22, 2005, the Lorin Racek family, of Morse Bluff, NE, received the 2005 Knights of Ak-Sar-Ben Nebraska Pioneer Farm Award for maintaining possession of their family farm ("Double R Farms") for at least 100 years – to be exact, 123 years (at that time) of consistent ownership by five generations of the Racek family. Their commitment to their family's history led to the Racek family being named as one of the 2005 recipients. Lorin passed away in June of 2000, leaving his wife, Nadine, as owner of the farm until she passed December 28, 2014. Their children include Larry Racek (a fifth generation Racek), who operated and managed the farm through 2006 until his death at age 57 in January 2007; Jerrine Racek formerly of Denver, CO, sole survivor, who now owns and manages the family farm; and Lorna Racek Pelan, who passed away in June of 2002 at age 47. Lorna's five grown children (Chris, Shawn, Jennifer, Candice and Brittani) live in Denver, CO.

Libbie & Joe Racek on their 50th anniversary 6-16-64

The house and outbuildings that are still standing on the property were built in 1914 by Lorin's parents, Joe and Libbie Racek. Looking back over the family's years on the farm, Larry said he remembered a number of very difficult times. *"I remember Grandpa Joe Racek talking about the tough times and also remember all the older guys around here talking about almost losing their farms in the '30's."* Great-grandfather, Mathias Racek ("Mike, Jr."), purchased the 80 acres from Francis and Annie C.S. Galton of Galesburg, Illinois on July 10, 1882. Mike Jr. sold the acreage to his son, Joe Racek (Larry's grandfather), on January 16, 1914 — five months before Joe married Libbie Pabian on June 16, 1914 at Wahoo, NE. While the house and outbuildings were being built, Joe and Libbie temporarily lived with Joe's parents, Mike Jr. and Anna, on the "home place" farm (just around the section) which was homesteaded in 1869 by the Raceks and where Joe was born in 1886. (NOTE: Joe's brother, Louis Racek and wife Emma eventually inherited the homesteaded farm from Mike Jr. and received the *Ak-Sar-Ben's Nebraska Pioneer Farm Award* in the 70's. Larry Racek purchased the "home place" farmstead from Louis and Emma Racek in 1977.)

Nadine & Lorin Racek on their 50th anniversary 8-12-98

Upon completion of their newly-built home in September of 1914, Joe and Libbie moved across the section to a new life (the same farm house where Jerrine Racek lives today). Several years later, they had one son, Lorin, who was born on the Racek farm on November 8, 1923.

They had a full-fledged working farm, raised crops, cattle, hogs, chickens, geese, ducks, milked cows, butchered and smoked their meats, kept an orchard and had a big garden, canned, churned butter, quilted and made soap.

Joe and Libbie Racek retired in 1965 and moved to a home they purchased in Morse Bluff (230 2nd Street next to the Methodist Church), whereupon their son, Lorin, and his wife, Nadine, and three children, Larry, Jerrine and Lorna, moved to the farm.

Joe died in 1967 and Lorin and his son, Larry, continued farming together, naming the farm *"Double R Farms."* When Lorin's mother, Libbie, passed away in 1986, Lorin inherited the farm and continued farming with Larry. They raised crops, livestock, did custom work, and sold used equipment and parts. At one point, Nadine raised several hundred chickens, ducks and geese and milked a couple of cows.

"Double R Farms" (2¾ mi. south of Morse Bluff off Hwy. 79)

Lorna Racek Pelan with Parents, Nadine & Lorin Racek (1995)

Jerrine Racek, sole survivor and now owner/manager of Double R Farms – with Mother, Nadine Racek (2003)

"The Pelan Family" at Lorin & Nadine's 50th Anniversary (1998) L-R: Brittani, Candice, Jennifer, Shawn, Chris, Lorna & Arnie

Nadine Racek (Mrs. Lorin Racek); son, Larry Racek; and daughter, Jerrine Racek accept the farm award at the July 22, 2005 Saunders County Fair in Wahoo, NE

Barn renovation before and after

Their children, Larry, Jerrine and Lorna, helped with the chores and were actively involved in 4-H clubs, participating in county and state fairs as members of the *Purple Ribbon Livestock Club* and the *Clover Cuties*.

Lorin, who graduated from North Bend H.S. in 1942, farmed in the Morse Bluff area all his life, but he was also a musician and played the saxophone/clarinet in his own band, as well as with numerous territory big bands. He was a fireman for the Morse Bluff Volunteer Fire Department, the Fraternal Order of Eagles #200 of Fremont, and the Saunders County Livestock Feeders Association. He enjoyed driving around to look at the crops (as did his mother, Libbie), going to farm sales all over the country with his son, Larry, and working in the machine shop. He especially loved his Case farm equipment.

Nadine, who graduated from Prague H.S. in 1944, was an elementary education teacher for seven years at Districts 54, 91 and 92. But, she was best known as the *Kolache Queen*, having received numerous awards for her tasty pastries. In 2012, NET featured Nadine in *Nebraska Stories*, in a segment entitled, *Kolache Queen*. NET continues broadcasting this segment during *The Best of Nebraska Stories* each year during fundraising. Until her death, Nadine continued her lifelong passion of baking for shows, farmers' markets on Saturday mornings in Fremont, as well as special orders. As a member of St. George's Catholic Church, Nadine sang several years in the choir and was past President of the Altar Society. She collected Heritage Village pieces and enjoyed gardening, tending to her flowers, baking for benefits and charities, and helping others.

Larry Racek Large Allis Chalmers Collection & Late Model Farm Machinery Estate Auction March 30, 2007 (Stock Auction)

Larry, who graduated from North Bend H.S. in 1967, managed and operated *Double R Farms*, as well as farmed other land he owned and rented. He also raised cattle. For several years, Larry served as a volunteer fireman with the Morse Bluff Volunteer Fire Department and was Treasurer of the District 54 school board until it closed in the mid-90's. He was a member of the Saunders County Livestock Feeders Association and attended St. George's Catholic Church in Morse Bluff, NE. Attending county and state fairs, auctions and sales topped his list of fun things to do. He collected antique Allis-Chalmers tractors and equipment, as well as having a sizable miniature farm equipment toy collection. He enjoyed participating in Outlaw, Bush Pullers, and the National Tractor Pulling Associations, and also loved designing and building projects in the machine shop. In the 60's, Larry played drums in *The Racek Combo*, as well as with other notable local bands.

Larry Racek (was a 5th generation Racek operating the Racek farm, Double R Farms)

Jerrine, who graduated from North Bend Central in 1970, attended one year at Wayne State College in 1970-71, and received a B.S. in Music Engineering from the University of Colorado in 1994, Summa Cum Laude. You may have heard Jerrine belting 'em out as a featured vocalist with the *Dorsey Brothers Orchestra*, or performing for President Gerald Ford and First Lady Betty Ford at the annual *Gerald Ford Ski Classic* in Vail. Perhaps you caught her "riding the wave" at the premiere jazz club, *Hofbraukeller*, in Munich, Germany. Then again, you may have heard her performing in Denver at *Dazzle's*, the *Brown Palace Hotel*, a *Colorado Rockies'* game, the annual *Glenn Miller Festival*, or in concert with the *Colorado Symphony* or the *43rd* and *101st Army Bands*. More than likely you'll remember Jerrine when she sang with the *Paul Moorhead Orchestra* from 1973-1980, singing in ballrooms throughout the Midwest, and the *Dean Bushnell* and *Glenn Shull Orchestras* singing in the Denver area (from 1982-Present).

Lorna attended North Bend Central and Fremont Bergen high schools, as well as the Grand Island School of Business where she received her certificate. She married Arnold Pelan on September 23, 1972. They had five children: Chris, Shawn, Jennifer, Candice and Brittani (who all live in Denver, CO). Lorna was talented at interior decorating, flower arranging, arts & crafts, baking, catering and party planning. She took pride in her home and was very passionate with her flowers and yard work. She had the world's most beautiful flower beds and landscaping. She enjoyed her family, her dogs and loved people. She had a warm heart, a love for God and a passion for life. Lorna's first grandchild, Blake, was born June 12, 2005 to her youngest daughter, Brittani, and husband, Ryan Burke. A second grandchild, Brooklyn, was born May 21, 2008. Sadly, Lorna never met them because of her untimely death. They are the only great-grandchildren of Lorin and Nadine and are the seventh generation.

After Larry's death in 2007, Nadine and Jerrine had a record-breaking farm equipment sale (2,500 attendees), which included 42 antique Allis Chalmers tractors, capturing the attention of the *Wall Street Journal* who reported the auction as the largest of its kind.

In 2009, Nadine and Jerrine honored Larry's ardent wish to preserve the *Double R Farms* barn by renovating it using Larry's last corn crop.

Five generations of the Raceks have owned, lived, raised families, and worked on this farm for 143 years because of the courage and foresight of their ancestors.

The first Racek to own what is now known as *Double R Farms* was Mike Jr. At the age of 14 (in 1861), he came to America with his parents (Mathias and Maria Suchy Racek) and siblings (James, Joe, Mary and Anna) to an Illinois town where his father worked for a cattle feeder for a year, outfitting the family with an old team of horses, a covered wagon, and a horned cow. With this equipment and 25 cents in their pocket, they set out for Nebraska. But, halfway across Iowa, near Cedar Rapids, one of the horses died. They stayed in the Cedar Rapids area for more than a year and in 1868 once again started out to homestead in Nebraska with a horse and a cow. (NOTE: the horned cow that the family originally brought from Illinois to Nebraska killed Mike Jr.'s mother, Maria Suchy Racek. She is buried on the hillside east of Morse Bluff in an abandoned cemetery with unmarked graves. Mathias Racek is buried in the Killian Cemetery west of Morse Bluff).

The family followed the Platte River on the north side until they came to North Bend, NE. The river was too deep to cross at this point so they were stopped once again. Fortunately, they met an Indian who could speak a little English who guided them to a crossing further west towards the town of Linwood. The family intended to homestead on land southwest of Linwood where the Tichacek farm is presently, but a Pawnee Indian camp on Skull Creek was located nearby, so they traveled east and camped on the John and Fred Homer land west of Morse Bluff. From there, they started looking for four 80's to homestead. They went to the Colon area but could only get one 80. So they returned back to where they came in contact with a land agent. The land agent told them there were three 80's south of Morse Bluff (at one time owned by John and Julia Vachal, Louis Racek, Emma Hunt and Ed Virka). This land is still owned by Racek descendants.

Two of the brothers (Mike Jr. and James) built a sod house on the fence line and just about lost their homestead rights for having one home for two homesteads. They wanted to live together and still have half of their home on their land, because the Homestead Act read: "One home for each homestead." Louis Racek once said that when he and the Virka's were making a fence, they came upon some stove lids and other articles, so Louis figured that was where the sod house once was.

After the brothers got their homestead claims (Mike Jr. homesteaded a farm three miles south of Morse Bluff, having walked to Lincoln, NE twice for homestead rights -- May 20, 1872 and August 15, 1874), they worked in Omaha three days a week leveling hills at 25 cents a day. They walked to Omaha, which took one day, and the walk back another, leaving two days per week in which to farm. Joe, one of the brothers of Mike Jr., was not of age to homestead (age 21), so he went to Omaha to work, and, when he came of age, the Homestead Act was discontinued.

Mike Jr. married Anna Kavan; James married Veronica Beznoska; Joe married Barbara Kohout; Mary married Taylor Slack; and Anna married John Mensik. Mike Jr. and Anna's children were: Joe, John, Louis, and Anna. Mike Jr. lived on the "home place" for 71 years (one of the original 80's located just across the section from the *Double R Farms*) – the same site where Joe Racek's brother, Louis Racek, and Louis' wife Emma lived, farmed and received the Knights of Ak-Sar-Ben Nebraska Pioneer Farm Award.

It is this type of history that the Raceks sought to preserve. They loved the land that was their home for generations. Having lost her parents and siblings, it is now Jerrine's responsibility to carry the farm into the next generation. After brother Larry's death in 2007, Jerrine stayed on the farm with her Mother for eight and a half years. Nadine passed away on December 28, 2014 leaving Jerrine as the sole owner and manager of *Double R Farms*. Jerrine remains on the farm. In those eight and a half years, Mother and daughter worked together renovating the farm, renting out the land for cash rent, as well as renting the Astro shed and grain bins. In 2013, Nadine and Jerrine worked with their Tenant to bring irrigation to the farm for the first time.

Jerrine Racek and Bob Harris on their wedding day, June 4, 2016.

On June 4, 2016, Jerrine Racek married Robert Lee "Bob" Harris at St. George Catholic Church in Morse Bluff, NE. Having reconnected with one another after 37 years after they first planned to wed, they now stand together overseeing the future of Double R Farms. Not everyone gets a second chance! One of the first projects they plan to accomplish is salvaging the farm Aermotor windmill and renovating it to stand in its former spot on the farm.

DOUGLASS – On August 14, 1901, Volney G. and Florence Douglass purchased "Green Valley Stock Farm", 80 acres in Pt. SW1/4NE1/4; NW1/4SE1/4 5-1-15 in Franklin County for $3,000. They had two daughters, Marie and Ruth. Due to Florence's failing health they headed to Montana, hoping the higher altitude would help Florenc. While there, Marie passed away in August of 1906; she is buried at the north entrance of Yellowstone National Park by Gardiner, Montana. The family returned home to Bloomington, and Florence passed away in January 1907.

Volney married Sara Jones on June 10, 1914, and to this union another daughter, Emily, was born. Volney and Sara raised Holstein cattle, Poland China hogs, and corn. In 1926, Volney and Sara built a new house. Volney and Sara both lived on the farm until their deaths in 1946 (Volney) and 1969 (Sara).

Emily married Kenneth O. Kahrs on August 1, 1934 at the family farm. Kenneth and Emily moved in with her parents and remained on the farm until 1981. They had six children: Douglass, Ruth, Marie, Charles (Bill), Judith, James, and Sara, all born on the same farm as their mother. Kenneth, Emily and the boys continued farming and raising stock cattle and hogs.

In April 1981, Doug and Betty (Kennedy) Kahrs moved to the farm; Kenneth and Emily had moved into their new home they built in Franklin. Doug and Betty lived there until 2000, when they moved to Franklin and their youngest son Dennis and his family moved to the farm.

January 4, 1998 Valley View Farm, Inc. (Doug, Betty and their three boys) bought the family farm where they raised corn, hogs and cattle. They sold the hogs in the late 1990s and bought more cattle. Today they still farm the original 80 acres and raise cattle.

The original buildings on the farm survived the 1935 Republican River Flood; however, the makeup of the land was changed forever. January 17, 2006, the house that Volney and Sara built burned; they replaced the house and Dennis and his family continue to live there.

Doug and Betty will celebrate 60 years of marriage in 2016 and are still actively involved in the family farming/ranching operation, along with their four children: Vicki Splattstoesser, Franklin, NE; Bruce Kahrs, Naponne, NE; Donald Kahrs, Bloomington, NE; and Dennis Kahrs of Bloomington. They have 13 grandchildren and 14 great-grandchildren.

DRUEKE – The 160-acre farm east of West Point, NE, was originally homesteaded in the early 1880s by two siblings, Margaret Eberline and Richard Schumph. Each built their own home on their own 80

Henry Drueke farm, June 7, 1987.

The Duerfeldt farm, possibly taken before 1892

acres. After Margaret's death in 1890, a neighbor, Charles Grunke, purchased both plots from Richard as an investment. Then, Frank Drueke purchased their entire 160 acres from Charles in 1896. Frank deeded the land to his son Anton when Anton married Teresa Ortmeir in 1901. Anton and Teresa moved the house from the south 80 acres and combined it with the home on the north 80 acres to create one home. A large horse barn was added to the farm in the early 1900s. Anton and Teresa raised three children: Frank, Katherine and Elizabeth. Upon Anton and Teresa's death in 1913, the farm was deeded to Frank and his sisters. Frank was too young to farm, so the land was initially rented out.

On August 23, 1926, Frank married Agnes Johnson and the newly married couple took over the farm. Frank and Agnes milked a few cows, raised hogs and chickens, and started their family. Frank and Agnes also had three children: Henry, Frances Jean and Richard. The family built a newer, modern house on the homestead in 1960, and their 'few' milk cows blossomed into an award-winning dairy herd. After Frank's untimely death in 1963, Henry took over the farming responsibilities.

In October 1970, Henry married Bernadette Kosuth and the couple purchased the farm from Richard and his mother. Henry and Bernie had one daughter, Mary Fran. Together this generation of the Drukes expanded the farming operation, adding several buildings and growing the successful Brown Swiss dairy herd. The family not only milked the cows, but also showed the prize-winning herd in County, State and National shows at both 4-H and Open Class levels. Henry and Bernie sold the milk cows in 1999 and started a small stock cow herd. Although they are now 'retired' from livestock, they still reside in the home and farm the land, raising corn, soybeans and alfalfa.

DUDEK – William Dudek purchased the first 160 acres of this farm located along the Frenchman River in a warranty deed dated Oct. 18, 1897 from Lemuel and Margaret Allison. The Allisons had acquired the land as a timber claim, signed by President William McKinley. Between 1901 and 1907 other adjoining parcels were purchased, and the farm was named Frenchman Valley Ranch.

William Dudek and his first wife, Eva Ady, had seven children: Fred, Anna (Follet), Eve (Keech), Mary (Byrket), Bill, Nellie (Ele) and Joe. Eva died in 1910 at the age of 26. In 1911, William married Hattie, who had worked for him tending to the house and children. To this marriage one child was born, Leonard, in 1916.

When Leonard was eight, his father deeded him the pasture land which remains in the Dudek family today. William died in May 1942 while Leonard was serving overseas in WWII. William's land was auctioned off to settle his estate and Hattie purchased the cropland adjoining Leonard's pasture, which he inherited upon her death in 1966.

Upon Leonard's death in 1968, the land was transferred to his wife, Dorothy and their daughter, Mary Lisa. Mary Lisa Dudek acquired this farm in August of 2000, and passed away in May 2013.

DUERFELDT – Gustav Christian Duerfeldt migrated to the USA from East Prussia with his father, Johan Duerfeldt, (mother deceased in 1841), three brothers and two sisters by sailing vessel, *The Bart Caroline*. After sailing 116 days, they arrived and settled in Buffalo, NY, on Sept. 25, 1846. The depression of 1857 encouraged a group of Buffalo colonists to move west. As treasurer of the group, Gustav C. came to Nebraska in 1858 with $4,200 in gold for a down payment on selected acres for the group in Richardson County near Arago. He himself chose a farm of 80 acres in the Barada Precinct Section 30-3-17. He went back to Buffalo and in 1859, returned to Nebraska with $14,000 in gold to make the final payment on all the selected acres.

In September 1861, Gustav C. and Louise (Parchen), two sons and one daughter came to Nebraska to live. In October 1861 they moved into a log cabin. In 1869, Gustav C. bought 40 adjoining acres and in 1882 he bought 50 more adjoining acres, making the farm 170 acres total. Gustav C. and Louise had 12 children. Of them, my grandfather, Gustav William "Gus" was the first son born in Nebraska. Gustav was one of the first to plant fruit trees and was instrumental in advanced farming practices.

Gus married Elizabeth (Adami) in 1892. They had three daughters and one son (Clifford). Gus took over the farming operation in 1894, which included raising livestock and annually feeding several carloads of cattle and hogs. Fat cattle were driven by hoof about 10 miles to a rail siting at Straussville. Hogs were hauled in a box wagon by a team of horses. On May 6, 1907 Gustav C. transferred the 170 acres to Gus. Gus planted approximately eight acres of catalpa trees near the farmstead known as "Catalpa Lane Farm." For years Gus was involved in public affairs, a stockholder and branch manager of the Barada Exchange of Southeast Nebraska Telephone Company, stockholder and president of the Barada Bank, precinct assessor and Woodmen of the World insurance agent.

When Clifford married Edna (Ruegge) in 1919, he took over the farming operation and Gus and Elizabeth moved to Falls City. Clifford and Edna had three children: Lester, Bernice and Marvin. On Feb. 6, 1929, Gus sold the farm to Clifford. Clifford farmed with horses until 1941 when he bought his first tractor. Corn was shucked by hand until 1953 when he bought a mounted corn picker. A threshing machine was used to thresh oats, wheat and rye until the late 1940s. He continued feeding cattle and hogs and milking cows. He always had a hired man to help with the farming until 1945 when Marvin graduated from high school. In 1959 Clifford bought 190 adjoining acres, making a 360 acre tract. Clifford and his two sons farmed together until 1984. They farmed 700 acres, mostly corn to feed 400 feeder steers bought at 750 pounds and fattened to 1300 pounds. Several hundred hogs were both farrowed and sold as fat hogs. The boys both milked 12 to 15 cows. Corn, clover and alfalfa were raised. Soybeans weren't raised until the 1970s.

Marvin married Norma Lee (Constantine) in May, 1951. So they could live in the original house built in 1893, Clifford and Edna built a home over the hill from them. Marvin and Norma Lee had five children: Barbara Elliott, Sharon Bennett, Glenda Kermoade, Richard Duerfeldt and Cathy Scholl. They have also been blessed with 12 grandchildren and 13 great-grandchildren. On Oct. 30, 1970, they bought the original 80 acres from Clifford plus another 80 adjoining acres. Marvin and Norma Lee bought Clifford's remaining 200 acres when he passed away in 1989 and bought another 160 acres in Section 8-3-16.

In 1961 Clifford, Edna, Marvin and Norma Lee received Aksarben's Pioneer Farm Award. In 1979 they received the Richardson County Conservation Award for having 160 acres totally conserved with terraces and waterways. Today all but two waterways have been converted to tile. Marvin continues to raise corn and soybeans. He raised feeder steers and steers on grass until 1995. Norma Lee, down through the years, has maintained a large garden producing hundreds of jars of canned food and 50 laying hens and 250 broiler chickens which stopped in 1995 due to health. Marvin's view of farmsteads has changed immensely. There's not as much livestock in pens and feeder cattle raised has dwindled. Barns have become obsolete, corn cribs have been replaced by grain bins, and machine sheds are now a must have. Although he's seen these changes, he wouldn't imagine living anywhere else. *Submitted by Marvin Duerfeldt.*

DUNKLAU – Shortly after his wife's death in 1887, Herman Dunklau, moved his family from Arlington, NE to Wayne County, NE. They came to Wayne because they heard the water was better. After homesteading a quarter section for 16 years (SE ¼, Section 9, Township 27, Range 3), Herman Dunklau expanded and bought the neighbor's farm across the road in 1904 for $6,480, or $40.50 an acre. Upon his death, Herman left this new farm to his son John Dunklau, John and his wife, Mina (Kruse) raised seven children in the newly expanded 4-square house – Ruby, Edward, Lloyd, Kenneth, Marvin, Alden, and Wilma. After John, then Mina's deaths, the farm was divided between their seven children. Alden Dunklau & his wife, Darleen (Roberts), lived in the house with their four children, Kathy, Lou Ann, Michael, and Randal. When Alden died in 1976 the farm still had several owners. In 1983, Darleen bought the remaining family shares to make the quarter section complete again. Mike Dunklau and his wife Kim (Chase) lived on the farm from 1978 until 1985. All three of their children, Ryan, Leah, and Alissa were born while they lived there.

The house on the Dunklau farm

Randy Dunklau, the great grandson of Herman Dunklau, has farmed the land and lived in the house since 1986. Together with his wife, Shari, they have raised four children, Jesse, Samantha, Jorge, and Joseph. The home has hosted many gatherings and parties over the years and still remains a beautiful home and productive farm today.

DUSATKO – My great grandfather, Friedrich Bruning, emigrated from Zetel, Germany to Eureka, IL in 1857. In 1862 he enlisted in the Army of the North in the Civil War. He was honorably discharged in 1865.

In 1898 he moved to Thayer County and in July 1900 he purchased the land in Thayer County, NE located at NW 1/4 of section 7, township 4 and range 2. On Nov. 30, 1909 his daughter, my Grandmother Margaretha Bruning Reichert, purchased the land.

The house, taken in the 1900s.

In 1964 the land was passed down to Loyd Reichert. In 2002 when Loyd Reichert (my father) passed away, the land was passed down to me, Maxine Dusatko and my sister, Joyce Bank. We are the current owners of the land.

In 2000 when my Dad was no longer able to live at the farm, the buildings were sold at auction to be moved. The house that was built in 1910 was purchased by a distant relative and moved into Bruning.

The land is currently rented by a local farmer. *Submitted by Maxine Dusatko.*

DVORAK – On this the 150th anniversary of the statehood of Nebraska, we are proud and humbled to share that the Dvorak family has owned and lived on the southeast quarter of Section 23, Township 21, Range 4 in Cuming County since 1898 when Anton Dvorak purchased the land.

The Dvorak farm

The following year he married Anna Mestl and they were blessed with six children: Herman, Joseph, Rose (Jonas), Rudolph, Bohumila (Kampschnieder) and Ann (Brodecky). In 1932 Herman married Adelia S. Hajek. They raised four children: Herman E., Maxine (Sr. Irene), Fr. Franklin and Martha (Minchow) on this farm. Herman E. married Clara Reisdorff in 1966 and they too raised their children: Teresa (Arens), Paul and Roger here. When Herman E. and Clara moved to Dodge in September of 1997, their son, Paul, remained on the farm. He married Nancy Bousquet in 1999. They are currently raising their four children, the fifth generation: Grace, Charles, Sophia and Henry on the same farm.

In 1939 Herman was instrumental in bringing rural electricity to the area, having one of the first transformers installed on his farm.

Over these 118 years, there have been many changes in the appearance of the farmstead. Buildings were added as needed. A large garden, fruit trees and various farm animals allowed for self sufficient survival during the depression and war times.

Farming practices have also seen many advances. First was the introduction of steam engines. Herman and a team of neighbors shared a steam engine that was used to shell the grain. He also saw the introduction of the tractor. For many years those tractors rode on steel lug wheels. It was during these 'dirty thirties' that Herman developed a hay sweep for his F20 Farmall. This was very helpful in gathering the dry thistles and other tumbleweeds, which needed to be burned in the fall.

In 1942, without bus service and students too young to drive, Herman E. rode six miles to St. Wenceslaus School in Dodge in an open, two-wheeled, horse drawn wagon with fellow neighbors. They had the luxury of an enclosed wagon during the winter. By the time Herman E. was involved in the farming operation, he saw the introduction of the corn picker and the combine, which greatly altered the harvest. Self-tying balers decreased the need for farm laborers.

Corn, beans, alfalfa and prairie hay continue to be the crops produced and sheep grown for mutton and wool.

Due to the changes in the farming operation, many of the farm buildings have been removed. The family home remains. Paul, a carpenter, has brought the interior and exterior of the house into the 21st century by doing extensive remodeling.

The Ak-Sar-Ben Pioneer Farm Family Award was presented to our family on Aug. 15, 1999, at the Cuming County Fair in West Point, NE. *Submitted by Herman E. Dvorak Sr., Irene Dvorak, Fr. Franklin Dvorak and Martha Minchow.*

DYKMAN – The James Milton Clare Gillespie family settled in western Nebraska in the early 1900s. Clare, as he was known, came to Cheyenne County from Iowa in 1907 to homestead northeast of Dalton on the very south edge of Morrill County. His brother, Bert, and several cousins preceded him. Clare filed on his homestead claim six miles east and three miles north of Dalton, NE on July 2, 1907. He and his brother Bert batched and proved up on their homesteads, which were one mile apart. On Aug. 11, 1913 at Bridgeport, NE, Clare filed notice of intention to make final five year proof to establish claim on the 8th day of October 1913. Witnesses were William Blumenthal, William Ravert, William Schuler and G. Frank Babb. Clare paid the sum of $14.00, the amount of Register and Receiver for the entry. Clare married Anna Marie Schuler on Dec. 12, 1912. They moved to their homestead and lived for two years in the sod house Clare built. Six

Cleve and Helen Stibor Gillespie, the second owners

J.M. Clare and Marie Schuler Gillespie, 1913

The Gillespie family at the presentation of their 100-year award. Betty Dykman, the present owner, holds the plaque.

children were born to Clare and Marie, three sons and three daughters. They purchased additional land to the south of the homesteaded land. Clare was a master farmer who had a love for the land. He first farmed with horses and mules. Marie was a wonderful helpmate to Clare and grew a large garden, canning or selling the extra produce. In March of 1947, Clare and Marie purchased the Cedar Creek Fish Hatchery, five miles south of Broadwater, NE and moved there. In 1957, the year following Clare's death, Marie moved into Dalton and lived there until her death on July 16, 1975. Cleve Gillespie, oldest son of Marie and Clare, purchased the Homestead after Marie's death and owned it until his death in July of 1983. Betty J. Gillespie Dykman, daughter of Cleve and Helen, purchased the 480 acres after Cleve's death and still owns it at this time. She and her brother, L. C. "Bud" Gillespie, still reside in the Dalton-Broadwater Community. Bud and wife, Shirley, are living two miles northwest of the homestead. Betty's grandson Cody Dykman, great great grandson of Clare, lives 1-1/2 miles west of the Homestead. The family still grazes cattle on the entire 480 acres.

DYSON – In 1892 John W. Dyson acquired 160 acres located in Burt Logan County, Township 29, Range 9E, NE. Crops grown were corn and oats. Livestock was cows and hogs which was the main source of income. J.W. had a 22 pen farrowing house and farrowed twice a year. At one time, J.W. farmed a total of 280 acres. J.W. Dyson retired in 1923 at which time the farm passed to his son, John Lester Dyson. John Lester farmed until 1965.

In August 1987 Donald L. Dyson, grandson of J.W., acquired this 196 acre farm. Donald raised Hereford cattle and hogs until 1996. Corn and soybeans are the crops now grown. Donald lives in the five bedroom house that was built in 1906. The 1896 house has central heating, bathroom, hot and cold running water – all for $4,500.

Donald and Wanda have been married 30 years. They have two children, Sara Donaldson and John S. Dyson and five grandchildren: Alex, Andrea, Rachael, Rehanna and Alicia.

Donald retired in January 2004 and the farm land is now leased to a neighbor. *Submitted by Donald L. Dyson.*

ECKERT – John Sr. and Albertina Damschneider Eckert, German emigrants came to the United States in 1872. The John Eckert, Sr. family

John Eckert Sr. family, 1899. Standing: Otto Eckert, Charlie Eckert, Albert Eckert, Martha Eckert (Wusk), Bertha Eckert (Krie), Frederike Eckert (Lehman). Seated: John Eckert Sr., Albertina Eckert, Reka Charles, Henry Nolte, Pauline Eckert (Nolte) holding Edith Nolte, John Eckert Jr., Helen Eckert, Bertha Smedieke (Eckert) holding Ella Eckert. Front Row: Herman Lehman, Clara Lehman (Nelson)

moved to Dixon County in 1882. They purchased 160 acres on May 5, 1883, located in Dixon County, Concord Township southeast quarter of section 4, range 4. Their first home on this farm was a dugout on the southwest side. Later, a building site was built on the northeast side of the farm, which consisted of a house and various barns. There is not a building site today. The Eckert family raised corn, oats, alfalfa, cattle, chickens, hogs, and horses. Today the farm raises corn and soybeans on 160 acres of irrigated land.

When John Sr. and Albertina moved to Oklahoma in 1903, they sold the farm to their daughter Frederike and her husband Rudolph Lehman. Earl and Wilma Eckert (great-grandson of John Sr. Eckert) purchased the farm on March 4, 1963. They raised corn, soybeans, oats, cattle, chickens, and hogs over the years. The farm is currently rented by Earl and Wilma's granddaughter Julie and her husband, Nick Kvols. Today they enjoy their retirement years at the Brookdale retirement community in Wayne, Nebraska.

Earl Eckert and Wilma Albrektson were married on July 21, 1948. They have two daughters, Marilyn and Anita. They are the proud grandparents of Angela, Eric, Julie, Jacob, and Joshua along with great grandchildren Carter, Carlie, Claire, Ross, and Chayce.

ECKHOUT – The Loup River Freighter Hotel was located on the land that would become the John and Eva Eckhout farm. The owner, Jerome Lalone, lost the land in 1893. The man who held most of Lalone mortgages bought it for $325.00.

Five years later in 1898, Felix and Flavia Eckhout became owners of the 440 acres of this property, including the house. They had ten children, one of whom, George, took over the farm. George married Lulu Pfeiffer, and along with his other acres, they homesteaded 80 acres in Section 7 of Sartoria Township in 1907. They had two children, Elizabeth and Bernard.

Later in life, Bernard took over the farm. He married Mary Nickman and they had seven children. John, the youngest, took over the

The Eckhout farm

farm. He married Eva Linden, and they had four children, Byron, Rita, Tara, and Nathan. John passed away June 9, 2009, and Eva remains the farm today.

Several Eckhouts lived in the Freighter house. John and Eva lived in it for three years, then it was donated to the Buffalo County Historical Society. It was moved to the Trails and Rails Museum in Kearney in January 1983, restored, and is now open for display.

EDWARDS – John M. Edwards, son of Charles S. and Mary Dent Edwards, was born near Varna, IL, on July 22, 1839. He became a schoolteacher and married Frances " Fanny" Thompson on Feb. 1, 1871 in Chicago. She was a native of England and came to the US by way of Canada. He was not able to hear from birth and she had lost her hearing from disease when she was a child, but they did not let their handicaps keep them from living normal lives. Fanny was an accomplished seamstress. John left teaching and worked at a surface coal mine in Kansas for two years, then moved back to a 94-acre farm near what is now Toluca, IL, which was owned by his father. Their son, Harry, and daughter, Adelaide, were born there. They farmed there about 10 years but the weather was wet and damp and they decided to move to Nebraska to a drier climate.

John had two brothers, Robert and Charley, who had land in Richardson Township. In the summer of 1883, John came out to look at this land but it had no water or wood or buildings. He heard about the Willis Richardson farm, which was for sale. He looked at it, liked it, went back home and told his father about it. On Sept. 15, 1883 Charles purchased the 160 acres for $2.900.

John and Fanny arrived at their new home on March 4, 1884, having come by rail with their two children and belongings to Germantown, now Garland, NE. They worked hard and they enjoyed entertaining their friends. Fanny died of consumption and sometime later, John married Louisa Dixon from Seward County, also a native of England.

Harry Edwards married Laura Burge, daughter of Alfred and Alice Burge of Dwight, on Jan. 1, 1900. They had two sons, one dying at birth and Floyd who lived on the home place, and one daughter, Alyce Thelma (Mrs. Lawrence Schmidt) of Lincoln.

Floyd Edwards married Ada Fouts, daughter of Samuel and Minnie Fouts of David City, formerly of Octavia, on June 1, 1926. Their daughter, Ermal, married Richard Collister of Paxton. They moved to the home place near Dwight in 1979.

EDZARDS – Gerd Leenerts purchased the SW quarter of Section 29-5-2 in Fillmore County, Nebraska for $6500 in December of 1899, becoming the first in the family to own this ground. Upon his death in 1915, his wife Gretje inherited the estate which then transferred to her son Tom in 1923 at her death.

Tom and his wife Tena farmed the ground until his death in 1934. During this time, some acres were sold to the State of Nebraska for the improvement of Highway 81. Tena owned the estate until her death in 1946, at which time it was bequeathed to their 5 children, George, William, Theodore, Frank Leenerts, with their daughter Minnie (Leenerts) Harms receiving the quarter of land.

Herman and Minnie Harms farmed the ground until their retirement when their daughter Ardean and her husband Alton continued with the farming. After Minnie's death in 1964 and Herman's death in 1974, the estate was willed to their 3 daughters, Ardean Edzards,

Above and bottom left: aerial views of the Edzards family farm.

Roene Sloey, and Ramona Maske who then sold the ground at public auction. At that time, Ardean's son, Larry Edzards (great, great grandson of Gerd Leenerts) and his wife, Betty, purchased the land in 1975. They have continued to farm to this day.

In 2005, the Edzards Family received the Nebraska Farm Family Award from the Knights of Ak-sar-ben with Ardean (Harms-Edzards) Cogger, Larry and Betty Edzards, and their children Micky and Katina and grandchildren attending the ceremony.

EICHMEIER – August John Eichmeier and Margretha (Hennings) were married in Germany in 1880. On March 3, 1883 they landed in New York and went to Illinois with her parents. In 1884 they came to Saunders County in Nebraska, where August worked as a farm hand. In 1890, he was able to purchase 160 acres in Green Precinct-Sl/2NW 4-13-8 and S1/2NE5-13-8 from homesteader Fred Eberhardt, and later 140 more acres.

August was a county commissioner from 1906-1910. They raised a family of five children: August Henry, Anna (Claus) Frahm, Rosa (Max) Hass, Lillie (William) Kinyon and Ernest.

Their son, August H., remained on the farm after the parents' deaths. He married Ida (Witte) in 1922 and they had a son, Milton D. After Ida passed away, August later married Martha (Cornell) in 1933.

Milton took over the farming operation upon high school graduation in 1940 and continued farming until he retired. He spent his 90

Eichmeier farm, 2014

The Eichmeier family in 1891

years on the farm. Milton married Alice Quick, who continues to live on the farm, in 1948. Four girls were born to them and grew up on the farm. They are Shirley (Ron Bottorff) Eichmeier, Nancy (William Ham) Burke, Judy Eichmeier and Mary (John) Bergan. Over the years, the farm has grown to 440 acres, which will continue on in the Eichmeier family through their grandchildren - Jill (Chaput) and Chris Kaufman and children, Cash and Carter; Paul and Danielle Eichmeier and children Easton, Hastings, and August; Casey and Megan Bryan and children Wesley, Henry, and Grant; and grandson Luke Bryan. *Submitted by Alice Eichmeier.*

ELMS LAWN FARM – John T. Swisegood purchased 80 acres (later increased to 160 acres) on Nov. 26, 1877. The farm is located in Richardson County, Township 2N, Section 8, Range 15. Ownership of the farm has passed through the generations as follows:

1. John T. Swisegood.
2. Cora Swisegood (daughter), Bert Swisegood (son), Fred Swisegood (son), Roy Swisegood (Lee Roy Swisegood Sr.) (son), and Minnie Swisegood (daughter).
3. Lee R. Swisegood Sr. and Orra A. Swisegood (wife)
4. Orra A. Swisegood.
5. Lee R. Swisegood Jr. (son), Isabelle Stukenholtz (daughter), Ruth Spencer (daughter), Arta Stephenson (daughter), and Vashti Becker (daughter).
6. Lee Roy Swisegood Jr. and Thelma M. Swisegood (wife)
7. Thelma M. Swisegood.
8. Pegeen Swisegood Fast (daughter) and Gayle A. Swisegood (son).

Pegeen Swisegood Fast and Frederick S. Fast have three children and three grandchildren. Gayle A. Swisegood married Linda K. Swisegood, and they have two children and two grandchildren.

ELZNIC – Owners of the Elznic Family Farm located in Fillmore County are Evelyn Elznic and Sheila Crays. Leona Goeury purchased 160 acres, the SE 1/4 of 4-7-1, in 1895 from John and Bridget Ryan. In 1899 she purchased 80 acres, the S1/2SW1/4, of 4-7-1 from Calvin and Catherine Songster.

In June of 1902, Ida Dehner arrived in Exeter, NE on the Orphan Train. Ida was 6 years old at the time and was placed in the home of Miss Leona Goeury. Although Leona wanted to adopt Ida legally, Ida preferred to keep her own name. Leona rented out the farm but had her own hogs and cattle. Ida married Frank Elznic in 1919. Leona lived with them on the farm until her death in 1925. Following her death the farm was owned and farmed by Ida and Frank.

After their son, Clarence, married Evelyn Madison, they lived on the farm where Clarence raised wheat, milo, soybeans and corn besides having cattle and hogs. The land was owned by Clarence after the death of his mother in 1974 until his death in 1975. Then the land was deeded to Evelyn Elznic and Sheila Crays, Clarence and Evelyn's daughter. Evelyn lived on the farm until she moved to Exeter in 1978. She now resides in Fairmont, NE.

Farm photo taken by Wilson & Wilson Studio, at the time the largest producers of aerial farm photos in the world, ca. Fall 1949.

Don and Sheila Elznic Crays live on a farm by Tobias, NE, where they have a Polled Hereford cow/calf operation. They raised corn, milo and soybeans on the Elznic farm for a number of years.

They have three daughters: Stephanie (Kirk) Jensen, Linda (Grant) McKay and Clara (Tim) Grace and grandchildren: Cody, Casey and Jessica Jensen, Bailey McKay, Kane, Weston and Xavier Grace and Seth McKay (deceased).

Don and Sheila now own three farms that are Century Farms. *Submitted by Don R. Crays.*

ERICSON – What is still the home quarter of ground was homesteaded in 1892 by John Godfrey Ericson. Born in Smaland, Sweden in 1863, his parents, Eric and Greta, moved to New Sweden, IA when John was just 3 years old. In 1881 the family again moved to the Axtell area.

In 1887 Godfrey, as he was known, married Mary Broman, also born in Sweden and were blessed with six children. Emil was the only child to carry on with farming. Seven years before Godfrey homesteaded here, Emil's future bride's family, the Swanson's, homesteaded the quarter just south of the Ericson claim. Emil and Verna brought three children into this world, two of which continued farming while the third, Gwendolyn, did own some land in the area, but chose to move to Chicago and was employed by the IRS. She now resides in Phoenix. Gerald and his wife, Theone Ericson, tragically lost their lives in June of 1993 as a result of a car accident a half mile from their home. Knowlen, the youngest, farmed the homestead until he retired, so to speak, in 1991. He lost his battle with cancer on his 77th birthday, Sept. 2, 2005. His wife, my mother, Harriette (Bloom) Ericson, still lives in her home town of Axtell.

Knowlen and Harriette had two sons, Bradley, who teaches at the University of Nebraska Kearney and Kevin who farms and lives on the original homestead.

Currently the farm consists of close to 1,400 acres including the Swanson homestead. All but a few pivot corners are irrigated with conservation tillage and no-till the preferred management practice. GPS, Auto-steer and moisture probes seem to have taken over the pitch fork and the pump jack. With the exception of crop dusting, all facets of the operation are done by me and Michael Sorensen, the grandson of Gerald and Theone. Though technically not an Ericson it's

The farm in November 1918

The farm in 1960, owned by Knowlen and Harriette

up to him to carry on the legacy. He has been my "turn to man" since he was 15. I have three wonderful children: Gavan Ericson, Teresa Morgan and Brent Bishop, none of which have any interest in agriculture. *Submitted by Kevin Ericson.*

FAGAN – John Teahon acquired 160 acres located in Sec. 17-18-22. In 1895 he acquired a tree claim of 80 acres located in Sec. 8-18-22. Livestock on the farm was hogs, cattle, chickens and horses. This farm passed to William Teahon and then to F.C. Fagan and is now in the F.C. Fagan Family Trust.

The farm still retains the original 160 acres plus the 80 acre tree claim. Now corn, wheat, alfalfa and soybeans are crops grown.

F.C. (deceased) and Ruth have seven children: Patricia, Cecelia, Ellen, Anna, James, Mary and Elizabeth. Their grandchildren are: Heather, John, Agusta, Annie, Christine, Amanda, Andrew, Paige and Jessie. Kaeli is their great grandchild. *Submitted by Ruth Fagan.*

FERGUSON FAMILY RANCH – Hugh Robertson Ferguson was born near Webster, Dodge County, Nebraska October 5, 1879 to John and Jean Ferguson who had immigrated from Beith, Scotland in 1868. He moved with his parents and siblings from Webster, Dodge County to Blaine County in 1891 to the Pleasant Valley community east of Brewster and lived just south of the North Loup River. Jessie May Cox was born in Blaine County on August 10, 1886 to Dr. C. B. W. and Virginia King Colvard Cox near Buffalo Flats, now known as Purdum, in Blaine County. Her parents moved from Nathan's Creek, North Carolina after the Civil War. Hugh and Jessie were married in Comstock, Nebraska, on March 21, 1908 and lived on Hugh's Kincaid south of his parent's home. Here their son Colvard Robertson was born on May 29, 1909. On March 10th of 1909 they purchased the SE ¼ of Section 22, Township 22, Range 22 located 6 miles south of Brewster, Blaine County, Nebraska, from William H. Connor for $1600, the home place. This adjoined Jessie's homestead, which they later sold to her brother Hubert Cox, and was near her parents who now lived a couple miles south of Brewster. Colvard recalled riding in the house as it was moved across the hills from the Kincaid to the new property in Section 22. The family moved and the house was in place for the birth of their second son, Harold McClellan, on March 19, 1911. Hugh, a good carpenter like his father, added onto the house and built a barn, shop, granary, salt house, chicken house and hog shed and this became the home place where now five generations have lived. On October 12, 1926, Jessie passed away. Colvard and Harold lived in Broken Bow with their Grandmother Cox and Uncle Bob and graduated in 1927 and 1929. Hugh lived on the ranch where he met a school teacher, Mollie Radil, who was teaching at the school south of the ranch. They were married on June 10, 1930. Hugh added to his land holdings 160, 440, 480, and 640 acres at a time, and swapping with other landholders as necessary to consolidate holdings. During the Dust Bowl many sold their Kincaids and between 1931 and 1943 the ranch acquired seven sections of ground bringing the total for the ranch to 8 ½ sections or 5440 acres. Names of the pastures hold the names of those early Kincaiding neighbors: Charlie Mitchell Place, Blind Man Place, Jim Way Place, Barton Place, Bodenhamer Place, Otto Diecken Place, Yost Place. Hugh raised Hereford cattle, Percheron working horses and Quarter Horses, sometimes having as many as 60 to 100 horses on the place. He sold one team of driving horses, Tony and Jake, to a man from New York City who said they would be well taken care of as he had a heated stable. Sons, Colvard, and eventually Harold joined their father in the ranching and farming business, married, built homes and raised their families on this original quarter section. Colvard Ferguson married Jessie Pickering on April 30, 1939 and moved a house from the Barton place for his family, which soon had three daughters, Judy Eacker, Patricia Keeney, and June Winberg. Harold married Maebell Pinnell October 13, 1940 and had one son Mark Allan. After 13 years in Broken Bow and a short stay in Denver after the death of Maebell, Harold and Mark moved to the ranch in February 1944 and he, too, moved a home onto the home place. Harold married Winifred "Peggy" Maloy on November 9, 1945 and they had four daughters: Linda Kyle, Debra Ferguson, Joanne Ferguson Cavanaugh, and Elizabeth Ferguson. When Hugh was approaching retirement, Colvard and Harold purchased the cattle and operated under the name of "Ferguson Brothers". Colvard liked to farm so did that and had a plane and did aerial spraying for other farmers.

Hugh and Jessie Ferguson, 1914

Harold and Colvard, 1914

In 1950 Colvard died at age 41. In 1951, Hugh deeded to Harold and Colvard's widow, Jessie, the parts of the home place, section 22/22/22, where their homes were located and the remainder of the original home place to his second wife Mollie. By March 1952, Hugh had deeded the remainder of the ranch property to Harold and to Colvard's three girls. Hugh died in 1953 and Harold continued to operate the ranch. He usually had a summer hand to help with haying until his children were old enough to help. The girls all loved mowing and raking prairie hay in their swimsuits, not so much picking up bales. They also helped their mother, who was a wonderful cook, with meal preparation-an important component of any ranch operation, as well as canning. In about 1965 when Colvard's girls all became adults they took possession of the southwestern most property and Harold the property around the home place and continued to operate. In 1970 Harold and Peggy were awarded the Blaine Thomas Soil and Water Conservation District Conservation Award for planting nearly a mile of shelterbelts on the property during the 1950s. Upon the death of Harold in 1981, Peggy took on the responsibility of the operation for the next 20 years. In 1989 upon Mollie's death, the remainder of the original ¼ section was inherited by Mark Ferguson. Upon Peggy's death in 2013, Harold's children came into possession and daughter Elizabeth has leased and operated this portion of the ranch as a summer operation. She is assisted by her daughters who live nearby and their families and occasionally her siblings. Improvements such as underground water lines and cross fencing have been installed. Colvard's girls still own and operate their portion of the ranch and upon the death of their mother in 2009, own her portion of the original home place. March 2011 marked the 102nd year that the original home place owned by Hugh and Jessie Cox Ferguson had been in family operation. On August 28, 2011, the family gathered at the Brewster United Church of Christ to receive the Nebraska Pioneer Farm Award. John & Jean Ferguson were charter members of the church in 1899, son Hugh joining later in the year & helped build the original church in 1909. We are thankful that we were able to grow up on, and after more than 100 years, work and play together as a family on what we've come to love and know as the "Never Sweat Ranch" in the Nebraska Sandhills.

FISCUS – The first Fiscus arrived in Philadelphia on September 2, 1743 with his family from Germany. Over the years, the family moved

Adam Wilson and Emma Fiscus at their homestead northeast of Bayard, Nebraska. Courtesy of Joyce Russell

to Maryland, North Carolina, Indiana, Iowa, Nebraska and on to other states. They left Germany, as did many others, for a new life and land of their own. One ancestor, John Adam Fiscus, was a soldier in the Revolutionary War. He fought in the Battle of Guilford Courthouse, which was the turning point of the war and led to victory for the patriots. The Fiscus descendants are very proud of their heritage and the faith and endurance of those who came before.

Adam Wilson Fiscus was born in Owen County, Indiana in 1847. He married Emma Dean of Owen County in 1868. They were successful farmers in Indiana, taught by Wilson's father, John A. Fiscus. Wilson was the Assessor and the 1900 census taker in Audubon County. Five children were born to this couple, three died as children and were buried in Indiana. The family later moved to South Dakota, then to Kansas. They had dried out in Kansas the summer of 1906.

Stories were being told of land sales further west, and in February 1907, Wilson and his son, Eugene, took the train to Bayard, Nebraska. They liked the area, and Wilson bought not one, but two places East of Bayard - one was 80 acres, the other 40. No improvements were on these places. In September of 1907, he bought the relinquishment on Henry Wells homestead, with a "Sears" house on the place, and 80 acres, ten miles northeast of Bayard. The legals are: S1/2NE1/4, SE1/4NW1/4, Section 4, Township 21, Range 51, West of the 6th P.M., Morrill County, NE. Five years after purchasing the homestead, he sold the other two places.

In 2009, The Nebraska Pioneer Farm Award by the Knights of Ak-Sar-Ben Foundation and the Nebraska Farm Bureau presented a plaque and an outdoor sign to the Wilson Fiscus family for 100 years of continuous ownership of the farm. These were presented to Joyce Russell and Jeannie Tyree, Wilson and Emma's great grand daughters. Jeannie and Gary Tyree still live on this farm and it has been in the family for 108 years.

FISCUS – Eugene and Florence Fiscus homesteaded 160 acres in the Haynes Precinct in Morrill County, Nebraska 12 miles northeast of Bayard, on March 18, 1909. The legals are: E1/2NW1/4. W1/2NE1/4, Section 34, Township 22, Range 51, W of the 6th P.M., Morrill County, Nebraska. They moved from Woodruff, KS, and prior to that lived in Cedar County, near Hartington, NE. Gene bought a threshing machine and custom harvested grain each fall. He sold this thresher before moving to Morrill County.

While the homestead house was being built, Gene and Florence lived east of Bayard, where their daughter, Bessie, was born. In the fall of 1909, they, and their 3 daughters - Dema, 8, Elva, 5, and Bessie, 1 year, moved to their new home. The girls were raised there - over the hill half a mile from the Roy and Bessie Fiscus family. Roy and Gene homesteaded on the same day. Gene and Bessie were brother and sister, and Roy and Florence were brother and sister, thus their children were "double" cousins!

Gene and Florence's home is above the irrigation ditch, and is dry land. Gene raised cattle and hogs. He bought another thresher and custom harvested grain each fall to keep his family fed. He was gone during the fall harvesting, except on weekends. Florence and the girls did the chores, and learned their duties well. The most difficult task was hauling water from a well a quarter mile "down" the hill. Bessie was most thankful for a "new" well in the yard, with the best water in the county! After Gene and Florence passed away, Ivor "Red" and Bessie McRae lived on the homestead where their two daughters, Joyce and Jeannie were raised. After Red died in 1963, and Bessie in 2006, the place was passed on to Joyce and Lew Russell and Jeannie and Gary Tyree. At present, Lew and Joyce, and their daughter, Sherri and Lee Eads and family live on the place. The Eads children are the fifth generation to live on the homestead and the Eads live in the original homestead house.

Eugene Fiscus is on the engine for this threshing machine. Courtesy of Joyce Russell

At the Fiscus Family Reunion in 2010, the Nebraska Pioneer Farm Award from the Knights of Ak-Sar-Ben Foundation and the Nebraska Farm Bureau was presented to Joyce Russell and Jeannie Tyree. The plaque stated *"For long and meritorious service to agriculture as exemplified by continued ownership within the family of the same Nebraska farm for 100 years or more. Nebraska has been enriched by the courageous pioneer spirit and loyalty to the land exhibited by members of this family, down through the years."*

FLATUNG – On July 20, 1912 Andrew J. Flatung purchased 195 acres from George and Ana Higgins for $12,000.00. On Aug. 28, 1912 he bought the SE l/4 of section 4, T-1N, R-2 (160 acres) for $12,000.00 from McDonnell, Young and Higgins. The family moved into the farm home that year. Andrew farmed the land until he and his wife, Emma, moved to Fairbury in 1942. At that time his son, E.C. Flatung, and his wife, Emma, farmed the land until 1953. In 1953 his youngest son, Leo, and his wife, Sophia, took over the farming operation. He purchased the farm in 1961 and farmed there until his death in 1971. His widow continued to live there until her death in 2002. Upon her death the farm passed to their two daughters, Sandy Moss and Sharon Kettelhut. They formed a partnership, Flatung Family Farm. Presently Mark DeBoer rents the farm ground. Leo's grandson extensively remodeled the house and in 2013 Mark and his wife, Dani, moved into the house. The farm now consists of 572 acres. *Submitted by Sharon Kettelhut and Sandy Moss.*

Andrew and Emma Flatung homeplace

FOLLSTAD – Louis T. Miller, the original owner, immigrated from Ontario, Canada with his parents, Jacob and Catherine Miller, and eight siblings in 1881. In 1886 he acquired 160 acres located in Burt County, Bell Creek Township, N2NW4 29-22-9 and S2SW4 20-22-9. Additional land was purchased in 1889, 1895 and 1913.

Louis married Emilie E. Steckelberg in 1900 several years after she immigrated from Germany. Crops were corn, oats, barley, alfalfa, prairie hay and timothy hay; livestock was beef cattle, dairy cattle, hogs, horses, chickens and vegetables were also raised.

Their son, Richard F. Miller, married Hilma M. Hedquist in 1928. Richard F. Miller died in 1993 when Hilma M. Miller became the owner. Following her death in 1998 three of their children inherited the farm and are the current owners: Virginia P. Miller Follstad, Stanley R. Miller and Norman R. Miller. The farm now is 240 acres that produce corn and soybeans.

The original dwelling built in 1898 is still on the farm where it is the home of the current tenant Jay and Deanna Johnson and their family. It has all the distinct design characteristics of Queen Anne style architecture, which has been recently enhanced with authentic paint and maintenance.

The long-standing horse and dairy barn, the hog barn, and a garage are all in good condition and are still in use for various purposes. The concrete block corn crib and grain storage building was constructed in 1935 but is no longer used.

The Miller farm, April 2015

Discovery of arrowheads on the east ridge in the S2SW4 overlooking the natural springs creek is evidence of the presence of Native Americans. Most likely this was a lookout for animals who came to drink which then became game food.

A large grove of native trees is at the north border of N2NW4. Louis Miller added many trees to this area.

The Ak-Sar-Ben Foundation Nebraska Pioneer Farm Award was presented in 1990. *Submitted by Virginia Follstad.*

FRANK – The farm that Larry and Marilyn Frank own and reside on is the homestead of Larry's great-grandparents Michael and Mary Cernousek. Michael died in 1884 at the age of 37 while digging a well. Mary continued to live there and raise their four children. She later married Frank Novak. After living on the land the required number of years for the homestead, Mary Novak received the original abstract of title on June 23, 1886, recorded in Book 2, page 487 for the SW 1/4 of Section 17, Township 30, Range 7, Knox County, NE.

Michael and Mary's daughter, Emma, married James Frank (Emma and James were Larry's grandparents). In 1908, Mary Novak deeded the farm to her son, Joseph Cernousek. Joseph deeded it to his son and wife, Frank and Velma Cernousek, in 1965.

Larry and Marilyn Frank purchased and moved onto the farm on March 1, 1974 where they continue to farm and reside. They have purchased additional land through the years and developed pivot irrigation. They raised their three children on this farm. It was a family operation with everyone working hard either in the dairy barn, hog facilities, with beef cattle or in the fields. Their children are Julie (Chris) Heinrich of Lincoln, NE, Mark (Tara) Frank of Yankton, SO, and Garron (Shelly) Frank of O'Neill, NE. They have nine grandchildren who all love to visit the farm and as they grow older are helping on the farm. Their sons Mark and Garron maintain a working interest in the farm and are both employed in agriculture related careers.

As of March 1, 2016, the Franks have owned and resided on this farm for 42 years. They received the Nebraska Pioneer Farm Award in 1999. *Submitted by Larry L. and Marilyn K. Frank.*

FREDENBURG – On Jan. 13, 1888 Benjamin Fredenburg purchased 80 acres in Nemaha County, located in the West half of the Southeast quarter of Section 32 in Township 6 North of Range 13 East from the Silas Frazier family, who settled the property under the Homestead Act, Certificate Number 1213.

Benjamin Fredenburg sold the 80 acres to his daughter and son-in-law, Hattie and George Christy, in 1892. In 1912 George Christy purchased the East half of the same section to bring the farm to 160

At the Nebraska State Fair in Lincoln. George Christy would go all around Nemaha County to pick up fruit from various orchards, then take it to Lincoln, and finally would make a display of all the fruit. George exhibited fruit at the State Fair for 41 years.

At the St. Louis World's Fair in ·1904, where some of the fruit from the Christy Fruit Farm was displayed. George Christy also helped create this display of fruit.

acres. George and Hattie lived on the farm from 1889-1937. According to a 1920 Census, they planted 300 apple trees, 40 peach trees, 300 pear trees and 400 cherry trees. Other fruits that they planted were grapes, strawberries and raspberries. They also had working horses, hogs, dairy cattle, chickens and bees. Grain sorghum, prairie gross, corn, wheat and oats were also raised on the farm. The Christy's Fruit Farm was known for their fine fruit across the nation from the early 1900s through the 1930s, and until the bad freeze in 1941.

After George and Hattie retired, their son Chan and his wife, Evalyn and family and later, his daughter Gladys and husband Roland Evans lived on the farm and moved the farm ground in to a general row-crop and wheat operation.

When the George and Hattie Christy estate was settled, it was purchased by another son, Floyd Christy and his wife Ann in 1954. After Floyd's death, it was left to the children of his sister, Zora Horn and brother, Chan Christy.

Currently the farm is owned by 14 great great grandchildren of Benjamin Fredenburg, making them the 5th generation to own the farm.

FREEMAN – On July 31, 1909 William and Agnes Schlichting acquired 80 acres located in Saunders County, NE.

The farm passed to their daughter, Emma Wittmersehouse in 1927. In 1941 it passed to her son, Merlin Wittmersehouse. His son and daughter-in-law, Merlin and Bernice Wittmersehouse gained the farm in 1962. In 1993, the farm passed to their daughter, Rita Freeman.

Rita and Ronald V. Freeman have two children, Amy Ellis and Laura Stempa. Grandchildren are Ryland Lehman, Brinly Lehman, Harper Ellis and Carson Stempa. *Submitted by Rita Freeman.*

FRITZ – Jacob Herman Fritz was born in 1874 in Germany. He, his father, step-mother, four siblings and four half-brothers and half-sisters sailed into Ellis Island in 18880. Five more children were born in the US. Herman married Amanda Audiss in 1900. On Feb. 25, 1904 a homestead certificate was signed by Theodore Roosevelt, granting Herman 160 acres in the SE 1/4 of Section 23 in Township 35N, Range 14W in Boyd County, NE. Herman and Amanda's children were Venus, Fern and Roy.

Roy Herman Fritz was born Feb. 23, 1910. Roy was married to Hazel Mills on March 5, 1945. Roy took over the original land and added ad-

Herman and Amanda Fritz, first generation

Roy and Hazel Fritz, second generation

Jim and Gladys Fritz, third generation

ditional land to the operation of row crops, alfalfa and cattle. Roy and Hazel's children were Sharyn, Marlene, Jim and Gail.

James Roy Fritz was born April 3, 1949. Jim was married on Oct. 28, 1983 to Gladys Goethals. Jim and Gladys currently live on the original home place. They have added more land to the operation and raise Black Angus, Hereford and Charolais cattle, as well as corn and alfalfa for feed.

The Fritz family was proud to receive the Aksarben Pioneer Award in 2004. *Submitted by Gail Spencer.*

GAREY – In the 1870s, after returning from serving in the Civil War, Andrew Jackson Garey operated his blacksmith shop at Jenningsville, PA. His son, Robert S. Garey, who had married Katharine (Farr) Garey, was working in the nearby timber and quarries. The economics of the area were not good at that time and both families, plus other relatives and friends, answered the call in 1878 to migrate to Nebraska to claim their homesteads in Furnas County, NE.

The Garey family homestead. Note the sod house and windmill.

The original 160 acre homestead was taken in two adjoining 80-acre tracts, making it a mile in length (E2 of the SE4, the E2 of the NE4 of Section 10-2-23). More land was soon accompanied by the purchase of SW4 of Section 11-2-23, which was purchased from a neighbor who sold it for $20.00 so he could return to Pennsylvania. This was the first site of a new two-room Sod House with a storage basement and 160 feet deep, hand dug well.

Wheat, corn, potatoes, garden vegetables, cane and wild plums were raised and stored. Grains were later ground at the area mill and the cane taken to the sorghum mill to make molasses. Quail and prairie chickens helped supply the family meat.

Three months of the year the children walked a mile south to the Sod School House, encountering an occasional rattler. The children all carried heavy walking sticks to protect themselves.

The continuous ownership of these homesteads have been held by the following Garey Families - Andrew Jackson Garey and wife Katharine Garey (1878-1898), (son) Harry and Lena Garey (1898-1942), (son) Rubert and Margaret Garey (1942-1967), (son) Douglas and Betty Garey (1967 to 2015).

We were honored by AK-SAR-BEN in 1978 with the presentation of our 100 Year Family Farm Award. *Submitted by Douglas B. Garey.*

GEISERT – Wilhelm August Geisert immigrated to America in 1886 from Eichstetten, Germany with his siblings and their widowed mother. His oldest brother Reinhard had settled a claim near Ogallala, Nebraska in Keith County in 1884 and August soon purchased his own home and farm nearby from George R. Smith.

Five generations of the August Geisert Family have kept the legacy alive on the homestead including August and wife, Nettie (Siffring) & children, William, Albert, Edward, Riga & George; Edward and wife, Louise (Behrends) & children Eldon, Maxine, Gerald & Connie; Gerald and wife, Rosemary (Castner) & children, Elizabeth, Kristin, Timothy & Nathan; Nathan and wife, Valarie (Johnson) & children Jakeb, Jordan & Jensen.

Keith County lies east of the Rockies and at the southern edge of the Nebraska Sandhills. The region was referred to as The Great American Desert, unfit for civilization with its expansive, mostly treeless, windswept prairie. It is remarkable that the settlers and their successors have turned this area into productive farm and ranch ground through dedication, conservation and innovation.

August Geisert, middle, with two of his sons Albert and Bill, loading stacked hay from the threshing machine onto the hayracks around 1915.

Horses were used on the farm until Ed Geisert purchased a new John Deere GP tractor in 1929. A progression of tractors and equipment followed. Contrasting the efficiency and productivity of August's horses, Nathan now runs a John Deere 8310 RT equipped with GPS guidance and on board computer.

Nathan and Gerald's interest in wind generation lead to putting up two 80 ft. wind towers, supplying power needs to the homesteads. The family raised cattle and some sheep. Nathan eventually sold the cattle and takes in cows to graze the stalks instead. Dryland wheat and corn was a staple until irrigation was utilized in the late 1960's. In Nathan's lifetime, the farm has produced both irrigated and dryland corn & wheat, popcorn, edible beans and sunflowers.

GEORGESEN – In 1888, George Georgesen purchased a 125 acre tract of land in Dakota County for $10 per acre. The land was located FIVE miles southwest of Homer and contained a timber pasture of burr oak, red elm and walnut trees which was well suited for a livestock farm. George married Georgiana Sorensen and they later purchased adjoining parcels of land to enlarge the farm to 240 acres.

In 1927, their youngest sons, Henry and Elmer, enlarged the farm to 320 acres with the purchase of the adjoining William Rockwell farm. This became the residence of Henry and his bride, Lola Buckner. Henry and Elmer farmed together and owned the local threshing machine, a "Red River Special" powered by a "Hart-Parr 18-36." In 1947, the land was divided with Henry keeping the east half and Elmer the west half which is currently owned by his son, Boyd.

Henry's son, Nolan, married Margaret Andersen in 1951 and began farming with his father until being drafted into the army during the Korean War. Upon his return from Korea, Nolan moved back to the farm. Oliver equipment was a common sight at the farm in the 1950s as Nolan's brother, Vernon, owned the Oliver dealership in Walthill, NE at the time. Nolan purchased the farm from Henry in 1961-1963. Land from the adjoining Alfred Harris farm was purchased in 1988 and 1997. Today, Nolan is retired and rents the farm to his son, Eric.

Eric purchased an adjoining farm in 1994 from Charles Pedersen that has remained in the family since 1883. Eric married Sharlene Drefs in 2000 and they took up residence at that farm. They have a daughter, Erin and son, Matthew. Today the Georgesen farm produces corn, soybeans and Angus cattle. *Submitted by Eric Georgesen.*

Georgesen farm, 1914

GERLACH – The family of Conrad Gerlach (1852-1928) who came to Harvard shortly before Christmas 1886 from Norka, Russia had their roots in Hesse, Germany where Ernest Gerlach was born in 1727 at Isenberg. Today the last of this lineage of Volga-River Germans in Harvard are Mrs. Gary (Sandra) Samuelson and Dr. Don R. Gerlach, whose grandfathers, George and Conrad H., were sons of Conrad. Conrad's father Heinrich (1822-1907) joined his son's family in 1890, having left his first wife Catherina (1824-1862) nee Lofink buried in Norka and bringing his second wife Maria Katherina nee Schneider and their son John with him.

Henry Gerlach and second wife Maria Schneider.

Heinrich's forebearers were part of the great migration of Germans to the Volga River in the 1760s-70s where the Russian Empress Catherine the Great attracted fellow Germans to settle with the promise of religious toleration, land grants and freedom from military service. It was Ernest Gerlach who is recorded in the 1775 Norka census with his son Johann George (1752-1820) who left Isenberg, Hesse and settled in Norka on the Volga; Johann George was the father of Peter (1779-1847) and grandfather of Heinrich (1822-1907) and great-grandfather of Conrad (1852-1928).

When Heinrich Gerlach died in 1907 his family included his second wife, their son John and the children by his first wife: Katherina Elizabeth (Mrs. John Aschenbrenvec) still in Norka, Conrad, Katherina Marie (Mrs. Adam Hamburger) and Eva Katherina (Mrs. John Krieger). These were the only survivors of eight children by Heinrich's first wife and six by his second wife.

When Conrad Gerlach and his wife Christina (nee Hamburger) arrived in Harvard in 1886, their family included Conrad H, George, Elizabeth (Mrs. Henry Hamburger) and Amelia (Mrs. William Poppe). His wife died in 1892 and he made his home with both sons.

The long tradition of Gerlach's living by farming was broken when Heinrich's son John and his great-great-grandson Don R. turned to other pursuits, the latter earning bachelor's (1954), Master's (1956), Doctoral (1961) degrees from the University of Nebraska and becoming a professor of history at the University of Akron, Ohio 1962-94. Don received several student scholarships, a Regent's Fellowship, and a Fulbright grant, which enabled him to study at the University of London, England (1956-57). In 1958-60 he served in the US Army, which included a 13 month tour in Korea with the First Calvary Division at an education center. There he administered the University of Maryland's overseas program and taught American History. He was discharged with the rank of Sp-5 (sergeant). The author of four books and numerous articles, he retired in 1994 and in September 2010 returned to his native town. Heir to the farm northwest of Harvard where he and his father were born (1932 and 1904), he resides at 508 N Fillmore Ave. in the house, which his grandparents acquired in 1929; built in 1919 by Mr. and Mrs. George Fishback, the house was sold by the Fishback heirs for $4,250. Don grew up in the house after his mother died in 1936.

Conrad Gerlach with his children: rear L – George, R – Conrad H. and daughters Amelia and Elizabeth.

Ralph R. Gerlach (1904-67) continued the family tradition of farming, married Evelyn Ida Bishoff (1906-36) whose parents, Otto W. (1876-1953) and Anna (nee Hachtel, 1880-1924) were also of German ancestry, so all of Don's family were of German stock. Evelyn's grandfather, John A. Bishoff, was a Civil War veteran (Confederate), as was her great-grandfather, John Hergenrother (Union); both men were Harvard homesteaders in 1871. Evelyn was a school teacher in Clay County schools and Red Cloud before her marriage in 1929. She died at the Mayo Clinic of accidental electrocution during treatments for a skin disorder. Ralph Gerlach remarried in 1947 to Rose Marie nee Wiegand of LaCrosse, KS and a descendant of Heinrich Gerlach's second wife.

Ralph Gerlach was a respected member of the Harvard Community, serving in various offices: as township clerk (1944-1951), member of the Farmers Union Elevator Board of Directors (1959-63), of the Harvard School Board for two terms (1946-1952, member of the Rural Fire Department and the Community Club and treasurer of Zion Evangelical United Brethren Church, which became part of the United Methodist Church. Ralph's parents, Conrad H. (1873-1957) and Marguerite, nee Pfenning (1874-1952) were also longtime members of Zion Church, Conrad having served as Class leader and Trustee, member of the Administrative Council, Sunday School Superintendent and teacher of the adult class (in German). His wife Marguerite was also born in Norka, Russia and came to America in 1893. Her children included Helen (Mrs. Reuben Schultz), Gladys (Mrs. Adam Schafer), Esther (Mrs. Henry Hohnstein) and Ralph. Conrad H. was a boy just past 3 when he got caught in the blizzard of January 1888, coming home from Country school east of Harvard, almost missing his parents' home on N. Adams St. One of his ears was frostbitten. As a hired farmhand he acquired his first pocket watch, in hunting case, which he plowed up during his first field work for a Mr. Nelson. In the 1893 depression he worked for the railroad while living in Inland. On Nov. 8, 1904 he and the family moved from Lynn Township to the farm northwest of Harvard, which was given the Nebraska Pioneer Farm Award by the Knights of Ak-Sar-Ben for continued ownership of the land by the same family for over 100 years.

Dr. Don R. Gerlach

Like his grandfather Gerlach, Don was active in church affairs. Confirmed in 1959 by the Anglican Bishop in Korea, he found the Episcopal Church had turned Apostate in 1976. He then helped form the Angelican Catholic Church to maintain traditional faith, morality and doctrine by the organization of a new parish in Akron, OH, the Diocese of Midwest and the Provincial Church which he served in various capacities; Creator of the Archives at the University of Akron, teaching at Holywood Seminary, member of its Board of Governors, and holding various Diocesan office; regular lay delegate of annual Diocesan synods and 20 Provincial Synods. He served on his parish vestry as clerk and senior warden, was active as a lay reader and acolyte until he moved to Nebraska in 2010. *Submitted by Dr. Don R. Gerlach.*

GIVENS – Samuel Givens purchased this farm March 1, 1901 for $2,625.00 located at E half of NE quarter section 4 range 6 township 7. This land plus his homesteaded property was tilled by various tenants from the community.

After Samuel and his wife Louise passed away, their estate was sold at public auction the spring of 1945. At this auction Elsie, their daugh-

Samuel and Louisa Givens, 50th wedding anniversary

Ralph and Elsie Mulder

ter, and husband Ralph Mulder, purchased this 80 acres for $30.50 per acre My parents owned and farmed this property for the remainder of their lives.

Upon the death of my parents, I, Ray, inherited this farm in 1957. We, Ray and Marian, farmed it until I retired in 1995. Currently our daughter, Diane, is life estate owner and our nephew, Roy Mulder, is the present tenant. *Submitted by Ray Mulder.*

GIVENS – My great grandfather, Samuel Givens, homesteaded the N 1/2 of the NW 1/4 section 4, township 6 range 7 in 1878. It remained in the Givens family until the estate was settled with a public auction in April 1945. At that time my grandparents, Ralph and Elsie Mulder, purchased it for $30.50 per acre.

William and Dorothy Mulder

Upon the death of my grandparents, the Estate was settled and this property was given to the oldest son, Howard, who in turn sold it to my father, William in the late '50s.

Having the desire to farm we, Sharon and I, moved to the residence after both my parents were deceased and became the owners in 1993.

Samuel and Louisa Givens, 50th Anniversary

Roy, Sharon and Adam Mulder, June 2007

Pictured l-r: John Fisher, Lucy, Elsie (mom), Grandma Givens, and Grandpa Givens, ca. August/September 1902.

During the late 1950s many conservation practices were begun and put in place over the years to make this a productive and impressive residence. Mostly, corn and soybeans are raised with some hay to supply the livestock. *Submitted by Roy and Sharon Mulder.*

GOLDFUSS – Frank Goldfuss emigrated to the US from Germany in 1880. He arrived in 1882 in Holt County, NE, taking land on a tree claim, 160 acres six miles north of Atkinson, NE. Agnes Hanel was born in Heinzendorf, Germany. Migrated to the US in 1885. They were married on Aug. 17, 1885 and made their home on Frank's tree claim. They had 11 children born on the homestead, with George being the oldest. He was born April 12, 1886 and grew up on the farm. George married Josephine Dobias on July 14, 1914 at Creighton, NE. Their six children were Elmer, Evelyn, Charles, Harold, Edith and Leonard. George lived on several farms. They moved back to the homestead and purchased the place in 1939. George and Josie moved into Atkinson in 1945 and he passed away on Aug. 6, 1968. Josephine passed away April 19, 1985. Harold was born Oct. 21, 1926 on this farm and spent his entire life on the homestead until his death July 6, 2011.

Grandma Agnes churned enough butter to buy the material to build this 10-room house in 1909. The north part was removed and became a chicken coop.

Harold married Bethene Anderson (born June 25, 1934). Our daughter Beth was born Aug. 25, 1959 and lives in Powell, WY. We have spent our entire married life here on the homestead. Harold and Bethene bought the NE 1/4 from Harold's folks. They bought 80 acres west of the homestead from Harold's mother. Total land owned is 236.96 acres. Harold passed away on July 6, 2011. Bethene still lives on the place.

GORDON – Thomas Carlyle Jackson arrived in Blaine County in 1883. He acquired 160 acres from Reuben Lowell, the original owner of the homestead in 1887, located in the SW 1/4 sec. 5, T 24, R. 25. T.C. was granted a land patient in 1907 for a tree claim of 160 acres located in the NW 1/4 sec. 8, T. 24, R. 25. The farm passed to Mary Jackson, daughter of Reuben Lowell, wife of T.C. Jackson. The farm then passed to Mary (Jackson) and husband, Ed Arnold and to Lois Arnold Gordon in 1988.

T.C. Jackson lived in a dugout in the banks of a hill from 1883. Still on the farm are the house built in 1918, a barn and windmill. The family has been volunteer weather observers for the National Weather Service, starting with T.C. in 1908, then daughter Mary took over. Mary and Lois both received recognition for this service.

The present home was a school house that Ed Arnold brought to the farm to use as a building to extract honey. Lois and husband turned this building into living quarters and call it the honey house. Ed was a beekeeper for many years and sold to the Sue Bee Company plus neighbors.

The land is now used for cattle pasture, sandhill hay, alfalfa and a few goats and chickens are raised by Lois and husband.

Lois and husband, Keith E. Jahn are both in their 90s and have been married just nine years. There are not children, but Lois' sister, Margaret, has five daughters and sister Rachel has two sons. *Submitted by Lois J. Gordon.*

GRANTHAM – Thomas Young (T.Y.) Grantham and his brothers, Clint and Frank Grantham, homesteaded along the Deadhorse Creek southwest of Chadron in 1885. The ranch was purchased for $1.00 to $2.00 per acre, and extended east past Chadron Creek. They ran 3000-4000 head of sheep, and kept 500 lanterns going at night to keep coyotes away. T.Y. and his wife, Nancy, had a son, Thomas Clay Grantham (T.C.). Thomas Clay married Sarah (Sadie) Pearson. Clay and Sadie had five children: Elam Grantham, Mabel (Mrs. Ray Haden), Hazel (Mrs. Herman Claussen), Wesley, a sister, Helen, who died at the age of 4 and a brother who died at birth. They continued to live in the same house that

Clay, Elam, Raymond, and Mark Grantham, Easter Sunday, April 17, 1953.

Ty and Nancy Grantham

Elam and Ida Elsie Grantham

Clay, Codie, John, Kim, Carey, Justin, and Cole Madsen and great-grandpa Ray Grantham on the 4-wheeler in 2010.

was built by T.Y. and Clay Grantham at the time of the homestead in 1885.

T.C. and Sadie homesteaded 959 acres in 1885 located in Sections 17 and 18, Township 32, Range 49 in Dawes County, NE. Corn for silage and hay was grown and livestock was hogs, sheep, Hereford cattle, horses and chickens.

Thomas, Clay, Sadie, Hazel, Elam and Mabel

Elam Grantham married Ida Elise Schmechel on Sept. 4, 1924. They lived in a house that Elam built on the corner just 1/2 mile west of Elam's birthplace and the homestead for the Grantham's. On Oct. 20, 1925, a son, Frederick Raymond, was born to Elam and Elsie. He was their only child. Raymond, attended Flag Butte and Rose Hill schools located in the Deadhorse Community. He graduated from Chadron High School before enlisting in the US Army. He was inducted on Sept. 14, 1944 and served in Europe.

After his discharge in 1946, Raymond returned to the family farm and attended Chadron State College before marrying Phyllis Ann Burkhiser on Aug. 20, 1947. Elam and Elsie moved to the home place in 1944 when Clay and Sadie Grantham moved into Chadron. Elam and Elsie took over the ownership of the Grantham homestead, keeping it in the family and continuing the pioneer heritage. They raised a lot of horses, Hereford cattle, sheep, wheat, corn, oats, alfalfa, chickens and hogs.

Elam and Elsie continued to farm and ranch the Grantham homestead until Elam passed away in 1968, at the age of 66 years. Elsie Grantham continued to own part of the homestead along with Raymond until her death in February 1996. The homestead was then owned and operated by Raymond and Phyllis Grantham and their children: Connie, Kim and Mark. Phyllis Grantham passed away in October 2000.

Kim married John Robert Madsen Feb. 20, 1970 and moved into the original homestead house where Elam and Elsie Grantham and others before them lived. They raised three children: Justin, Jennifer and Julie, who helped them and Raymond continue the farming and ranching operation before purchasing 160 acres of the original homestead and other land adjoining the original homestead. John and Kim Madsen continue to live along the Deadhorse Creek on the original homestead while Justin and his wife, Carey, along with their three sons: Clay, Cody and Cole, live just south along the Deadhorse Creek continuing to be a part of the farming and ranching operation today raising cattle, wheat, alfalfa, chickens and spring crops. Jennifer and Julie and their children continue to help with the farming and ranching operation.

Ray, Phyllis, Connie (Moore), Kim (Madsen) and Mark

Raymond Grantham at the age of 90 years continues to own and operate the remaining 800 acres of the original homestead, which now produces wheat, alfalfa, grass hay and cattle.

Over the years, the Grantham family has endured many hardships, (floods, hard winters, depression of the '30s, death, drought, dust storms, low cattle, sheep and hog prices), yet survived and continue to believe they live in the best part of the world, continuing the family heritage and pioneer spirit passed over seven generations. *Submitted by Kim Grantham Madsen.*

GRIMM-VANKLEEK-LeMASTERS – Brothers Henry, Jacob and Nathaniel Grimm immigrated to America from Germany settling in Walker, IA. In 1872 Henry and Jacob came to Knox County, NE staking a tree claim on the Middle Branch Creek in Walnut Grove Township, which was recorded in Niobrara, NE. They built a dugout in a hill with a fort around as protection from the Indians.

In 1873 Henry brought his wife, Catharine, and 11 children with a team and wagon, leading a cow and crossing the frozen Missouri River at St. Helena, NE. Food was provided by trapping, hunting and gardening. Several settlers soon came, living near Henry's Fort while building their homes. A school, cemetery and town, Grimton, soon sprang up being named after Henry Grimm.

The settlers endured harsh winters, prairie fires, storms, drought, disease and grasshoppers. The Sioux Indians weren't a threat but stole chickens and livestock for food.

Henry's daughter, Millie, married George Edward VanKleek Dec. 5, 1895,12 days before Henry passed away at age 71. They lived with Catharine, building a two story home and large red barn with a hay mow featuring a white star in front in 1898. Catharine passed away in 1920, age 92.

Dollie was George and Millie's only child. She attended Grand Island Business College where she met and married Howard Jay LeMasters. They cared for Millie, building a farm-ranch operation. Their son, Max, married Vivienne Boelter in 1949, living as a three generation family until Millie died, age 82.

H.J. LeMasters passed away in 1977 and Dollie in 1990.

Max and Vivienne built a herd of 100 head of Hereford cattle, irrigating corn and alfalfa from the Middle Branch Creek. Hereford/Angus cross soon evolved.

Max passed away, age 49. Daughter, Patty, married Ed Sukup and bought into the ranch through FHA in 1981, living on a nearby farm and managing a Machine Shop. Vivienne continues living on the Homestead, now being 143 years since the Tree Claim was recorded.

Many arrowheads and Indian artifacts were found throughout the years. *Submitted by Vivienne LeMasters.*

GROUND – William H. Ground (first generation) and wife, Sarah, living in Illinois, filled all of their worldly possessions in a covered wagon with other family, also including a cottonwood tree (still living today) and traveled to Juniata, NE in 1883, Verona Township 16-18-11 purchasing 240 acres. Eleven children were born at the farm.

William H. Ground and brothers ran various businesses in Adams County, including a funeral parlor in Kenesaw. A horse drawn hearse built by the brothers can be seen today at the Hasting's Museum.

Lester L. Ground (second generation) married Ann Bourg and took over farming in the early 1900s and raised milk cows (which Lester milked until age 94), chickens, hogs, along with wheat, oats, corn and hay using horse drawn equipment. It was a big day when tractors were brought to the farm.

Ann loved to garden, raising fruits and vegetables along with beautiful flowers. They had two children, Doris and William L. Ground, who took over farming in the 1960s.

William L. Ground (third generation) married Betty Shay and to this union two boys were born, William P. Ground and Dan L. Ground, who lives with his family just west on the original farm.

William P. Ground (fourth generation) purchased the farm in 1990 and farms there today. When he retires he will pass it to his son, J.R. Ground (fifth generation) and his son, Brendon (sixth generation).

The original house, built in 1883, is still being lived in by William P. Ground. Three additions and a complete remodeling have been added over the years.

One of the awards received was for the farm being the family for over 100 years, from the "Nebraska Pioneer Farm Award" by Ark-Sar-Ben. The family is proud of their legacy and hope it continues for generations to come. *Submitted by William "Bill" Ground.*

GRUBER – Emil Gruber was born in Germany in the late 1800s and emigrated to the US as a young man. It was Emil's desire to farm and raise a family. He began his dream by marrying Maria Friedrieka Marquardt and petitioning for naturalization in 1913.

The Gruber family after the death of Emil. From left to right are Maria, George, Leonard, Oscar and Elsie Gruber.

Emil and Maria Gruber were the first of the Gruber family to live on the family farm west of Elwood. They raised four children on the farm. They were George, Leonard, Oscar and Elsie.

George, the eldest of the Gruber children, was born in the same house that sits on the family farm today, as was George's daughter, Coleen (Simmons).

George attended rural school outside Elwood and assisted his father Emil in farming from the time he was a small boy until Emil's death in 1933.

On Oct. 29, 1933, George married Anna Marie Maahs, the daughter of Carl and Kate (Kalthoff) Maahs, at Immanuel Lutheran Church. George and Anna had two children, Coleen and Douglas.

George and Anna were good stewards of the land. They took great pride in planting trees for a weather barrier, fruit trees and a large garden. George kept abreast of the latest conservation methods in order to preserve the land from wind and water erosion.

The Gruber farm has been in the family since the spring of 1908. Before his death in 2009, George lived on the land for a total of 94 years.

George's granddaughter Gail and her husband Ray Ball now live on and operate the Gruber family farm.

George Gruber was presented the Knights of Ak-Sar-Ben Pioneer Farm award Saturday, Aug. 2, 2008 by the Gosper County Fair Board.

The Pioneer Farm award is given to families that have maintained ownership of a farm for 100 years or more.

GUENTHER – August Martin Albert Guenther and Alwina (Alvina) Agusta Beccard Guenther purchased the farm two miles north and one mile east of Dunbar in the NW1/4 of Section 6, Township 8, Range 13, AC Belmont, Otoe County, on May 18, 1881, for $3,300. Prior to the purchase, August and his father rented a farm west of Nebraska City in 1875 but the grasshopper plague hit Nebraska that year and they lost all of their crops. Two cows and a few chickens provided a meager sustenance through the first winter but the following years were prosperous, and by 1881 August and Alvina were able to buy this farm. They grew corn, wheat, and soybeans, and raised hogs, chickens and milk cows.

The first home was a very small house with two rooms downstairs and one or two upstairs. A new home was built in 1889 and the old house converted into a granary with a buggy shed built on the south side. Also, with the help of their family, they built a horse barn and a corncrib in 1900, and a cattle barn in 1915, most of which are still in reasonable condition in 2015. Both barns had oak dance floors and were the typical 'peg' type construction.

August and Alvina had five children; Frederick, Emma, Gustav, Heinrich and Eda; over time, they were able to purchase five 160 acre farms in Otoe County, one for each child. When son Heinrich (Henry) married and made his home on this farm in 1916, August and Alvina retired and moved to Nebraska City. Henry and his wife, Nellie, built a new home in 1925; that home, as well as the above mentioned out buildings are depicted in the accompanying photos.

In 1950 Henry and Nellie's only child, Leta, and her husband, Glenn Bremer, moved to this farm, carrying out the farming operations until they retired in 1971, moving from the farm to Nebraska City in 1982.

Painting of the farm by Lincoln artist Dorothy Ballweg, given to Glenn and Leta Bremer by their three children and spouses – Nancy and Lanny Goering, Lorin and Dorothy Bremer, and Patricia and Wayne Heather – in 1982 when they moved from the farm to Nebraska City.

Picture of the Guenther farm in 1952.

Their three children helped in the daily operations while still living at home; Henry drove out to the farm nearly every day to work on a variety of projects until his death in 1975. They grew corn, milo, wheat, oats and hay, and raised chickens, cattle and hogs, also milking a small herd of Shorthorns. A new granary and garage were built in 1957. There was always a large garden and beautiful trees, flowers and shrubs. Glenn and Leta continued to participate in and oversee the farming for many years after moving to town. They received the Pioneer Farm Award in 1991.

HABERMAN – Our farm has been in the Haberman name for 150 years, passed down from Anton Haberman, who emigrated from Austria. He came to this country in 1850 at the age of 15 and first settled in Wisconsin. Then prior to the Civil War, he homesteaded in Nebraska Territory, which was occupied by peaceful Native Americans. Anton's land was recorded at the Dakota City Territorial Capital, but registered in Cedar County as the first homestead in Precinct 2.

Anton, the homesteader and some of his grandchildren.

His wife, Frances, also emigrated from Austria at the age of 12, arriving in New York in 1852 on the ship, *Jason*. She and her family traveled by covered wagon pulled by oxen to Wisconsin. She became known as one of Cedar County's earliest pioneer women. Anton came here alone and lived in his wagon until a sod house could be built. His first task was to dig a well, 80 feet deep and eight feet in diameter, lined with homemade bricks. These bricks were crafted from chalk rock that was found in the area. Anton ground the chalk rock, mixed it with water and baked it. He dug this well six feet off the water vein and dug a line to connect them. As of the year 2000, the well was still used and still remains a special landmark on the farm.

Down through the years, the Haberman family has been involved with St. Boniface Church in Menominee. Anton was instrumental in its construction. The bell in the bell tower was donated by Anton and his family.

The farm was later passed to his son Joseph and wife Nettie, then to Anthony and wife Leona and now Wilfred and wife Joyce. Their son Chad and his wife Toni live on the farm also. They have a daughter and three sons. Thus, their children are the 6th generation to live here. The farm manages row crops and maintains a herd of stock cows. For more than a half century, the dairy operation has been the primary way of life on the Haberman farm. *Submitted by Wilfred and Joyce Haberman.*

HACKBARTH – Fred and Ruby Hackbarth were married in 1906 in Custer County. In March 1907 they moved to the land they'd bought from homesteaders near Hershey (SW 1/4 Section 12 for $1,560 and NW 1/4 Section 12 for $3,500). They moved with two teams of horses and wagons to a sod house where their four children were born – Bertha, Walter, Olive and Frieda.

Farming was done with horse drawn equipment. In 1914 they built a two story house. Fred and Ruby later acquired an additional 640 acres.

Fred and son Walter farmed together until 1943 when Walter died following a farm accident.

The farm house in 1914 with horse drawn farm equipment.

The Hackbarth home in 2014 at 100 years old.

Fred and Ruby's daughter, Bertha and husband, Ed Kuenle, and their family moved from a rented farm to the home farm where Fred and Ed farmed together. Ed and Bertha's four children: Marian, Loretta, Carol and Phil grew up on the farm and attended a one-room country school through 8th grade. Marian helped her dad with field work starting at age 10, driving a 9N Ford tractor. She continued to help until she graduated from high school.

When Fred and Ruby passed away, their three daughters each inherited 320 acres.

Ed and Bertha farmed all their lives, raising corn, wheat and cattle – Bertha died in 1983 and Ed in 1984. Their daughter, Marian, and husband, Phil Merrill, purchased Marian's siblings share of the farm. They did a major remodeling on the old farm house, installed pivots and sub drip irrigation. They rented out the crop land, but Phil continued to put up hay and custom feed calves.

Phil and Marian's three children – Mike and wife Laura, Martin and wife Sara and Karen and husband John Binkley and their families visit the farm often. *Submitted by Marian Merrill.*

HAEFFNER – Christian Haeffner purchased the land Jan. 1, 1883 for $500.00. The farm, which was common school land, is two miles east of Burr, NE. It is in South Branch precinct No. 36-7-10 in Otoe County. Christian Haeffner was known for his horticulture ability and planted the first ceder trees in the area; many are still standing today.

When Christian died in 1926, the farm went to his son, Henry. He passed away a short time later, so Minnie Haeffner, Christian's wife, remained there until her death in 1946. Raymond and Leta Haeffner, Henry's son, purchased the land to settle their estate.

Haeffner farm stead, early 1980s

Raymond had farmed the land since he was 16. The house was built in 1900. Raymond died in 1992 and Leta died in 2007, when the land was purchased by their daughter and son-in-law, Neal and Joyce Bentzinger. Russ Bentzinger, their son, now lives in the house and farms the land. He is the great great grandson of Christian Haeffner. He raises corn and soybeans and has a big pasture for cattle. *Submitted by Joyce (Haeffner) Bentzinger.*

HAJEK – The Hajek homestead, located four miles northeast of Odell, NE in Paddock Township, started in 1889 by Frank A. and Mary Hajek. The land, which was originally part of the Otoe Indian Reservation, was purchased from Frank Husa for $3 an acre. The legal description is SE 1/4 of Section 5-1-6.

Theodore and Donnetta Hajek along with their three daughters, Alyssa, Brianna and Kelsey standing in front of the granary.

Frank and Mary were blessed with 12 children. The family attended St. Mary's Catholic Church in Odell. The children attended Plainview School District 78 of Gage County, located within walking distance of their home. Tragedy struck Aug. 18, 1909, when the home burned to the ground. The family was only able to save very few items, which included their treasured organ and a large kitchen range. With the help of neighbors and others, they began building a new home the following week. The family slept in the granary and hay loft while a new home was built. The granary is still standing and is the oldest building on the farm.

After Frank's death in January 1945, Frank's son, William A. Hajek, returned from service in WW II and acquired the land. William married Henrietta (Skoch) in 1946. They were blessed with nine children. William continued to farm the land, had a small dairy operation and was also a skilled carpenter in and around the Odell area. William was also known for his musical talents and had a small polka band, the Bill Hajek Band, in which a few of his children also played.

William passed away in January 1990. His son, Theodore F. Hajek, purchased the farm later that year. In 1991 Theodore and his wife, Donnetta (Oestreich), accepted the Ak-Sar-Ben Pioneer Farm Family award for the farm remaining in the same family for 100 years. Theodore continues the tradition of farming, raising beef cattle and operating Hajek Enterprises. *Submitted by Theodore Hajek.*

HAJEK-SINDELAR – On August 26, 1878 Frank & Anna Hajek purchased 155 acres from Joseph & Anna and Frank & Mary Krajek located in Colfax County. The South 1/2 of SE Quarter, South 1/2 of SW Quarter and the NW Quarter in Section 33. Township 20. The amount paid was $2,500.00. Also purchased was the East half of the East 1/2 section in Section 33. This portion of land was 35 acres more or less and was purchased for $250.00.

The land was passed onto William & Tillie Hajek (son of Frank & Anna Hajek). William & Tillie Hajek passed it onto their children, Anna Hajek Sindelar (Husband Lumir Sindelar), Mildred Hajek Krepel (Husband Edwin Krepel) who lived on the home place until 1989. Mary Ann Hajek (Husband Clarence Najmon). The farm and tract of land was purchased by daughter Anna Hajek Sindelar and she passed it onto her children. Lois Sindelar Holloway (Husband John Holloway), William Sindelar (Wife Pat), Joseph Sindelar (Wife Marilyn), Mary Sindelar Gall (Husband Larry Gall).

The original land was pasture but was later turned into productive crops of corn and beans. Son William and wife Tillie took up residence in the home and had 3 daughters. Anna, Mildred and Mary Ann. Daughter Mildred and husband Edwin Krepel lived there after William and Tillie retired. After Mildred and Edwin Krepel retired in 1989, the

Members of the Joe Sindelar family, l-r. Front row: Logan Sindelar and Caleb Grigsby. Middle row: Brittin Sidelar, McKenzie Sindelar, Marilyn Sindelar, Joe Sindelar, Mya Grigsby, and Blake Sindelar. Back row: Jared Sindelar, Rachel Sindelar, Adam Sindelar, Sara Sindelar, Shane Sindelar, Justin Sindelar, Steve Grigsby Jr., and Brenda Sindelar Grigsby.

house and buildings sat idle. In 2000, the farm buildings were cleared and the barn was moved to the Sindelar family feedlot at Cedar Hill Cattle Co. The windbreak and trees planted by William Hajek were kept and in 2003, Joseph and Marilyn Sindelar built a new home on the original farm where they now live currently.

The crop land is now farmed by Anna Hajek Sindelar's sons Bill, Joseph (wife-Marilyn) & sons. Jared, (wife Rachel), Adam and Justin (wife Sara). Joe and sons also own and operate the family feedlot (Cedar Hill Cattle Co.) with his brother (Bill) and wife Pat. The feedlot was started by their parents, Lumir and Anna (Hajek) Sindelar. The 18,000 head facility is located 2 miles from the Hajek Farmstead.

HAMMANS – John L. Hammans homesteaded in Custer County, NE near Callaway in 1884. This 160 acres is located on Tallin Table, the SW quarter of 11-15-25. John and some of his neighbors from Drakesville, IA came to Custer County by covered wagon. One of these covered wagons was pulled by a yoke of oxen. John built a sod house on that land and his sister, Cynthia Hammans, "kept house" for him. In 1886 he returned to Drakesville, IA, his boyhood home, for his bride, Isadora Ena Taylor and they resided in the sod house. Their two sons, Francis and Earl, were born there. A third child, Edna, was born shortly after the family moved from the sod house to the new four room frame house on the adjoining property which John and "Dora" later purchased. A 400 foot well was dug on the homestead and they were unable to get a well, so water had to be hauled. This homestead received the century farm designation in 1984, being owned by son Early Hammans of Kearney, NE, who was 92 years old at that time. Following Earl's death in 1993, the homestead was purchased by R & M Acres of Gibbon, NE, of which Edna Hammans Randall was a part. Edna died in 2005 and the homestead continues to be owned by R & M Acres, of which granddaughter, Janice Randall Martin, and great-granddaughter, Sina Martin Lehn a part.

John L. Hammans, wife Dora and son Francis. Sod house, Custer County, NE.

HAMMANS – John and Dora Hammans moved to Buffalo County near Gibbon, NE in 1906 from the Homestead in Custer County, which John claimed in 1884. At that time the family consisted of Francis, who remained in Custer County for a time, farming the homestead and adjoining land which had been acquired; Earl and Edna, who were in High School. Dora died in 1938 and John in 1949.

John L. Hammans centennial farm as it appeared in 1917. In front are John Hamans' REO and Earl Hammans' Ford.

Family members lived on the farm continuously until 1974, as granddaughter Rhea Hammans Solomon and husband, Rollo and their children - Melvin, Ardyce, Sheryl and Aleda, occupied the farm. John and Dora's daughter became the owner after John's death. As the years progressed the land was placed in a family cooperation, R and M Acres, Inc. of which Edna was a part.

A milestone for this particular farm was that the first Rural Electric Administration (REA) pole in Buffalo County was set on this farm in 1938. The electric lines were energized in December of 1938 for the first phase of rural electric power from Elm Creek to Gibbon. NE

HANSEN HOMESTEAD FARMS – On March 13, 1882 Gertrude Sorgenson Hansen, one of the early settlers of Webster County, made Homestead Application for 160 acres near Red Cloud, NE in the Northeast Quarter, Section 24, in Township Three North, Range 12 West. She was born in Skjelbred, Sonforeder, Norway in 1852 and migrated to the US in 1869.

She moved onto the claim in June of 1882 and her Homestead Application was approved in June 1887. She married Ole Hansen on March 18, 1885. Their sod home was built with a board roof, board floors and board partitions which made it habitable for all seasons. It was valued at $150 and the land was valued at $10/acre. In 1882, she raised 30 acres of corn and owned four horses, 17 head of cattle and 20 hogs.

When Gertrude died in August of 1935, her son Andrew Hansen became the executor of her estate and the original Homestead was assigned to daughters Martha Hansen and Sarah Hansen Bergman and when Sarah died in 1945 her portion of the Homestead was assigned to Andrew Hansen and Martha Hansen.

In February 1960 the Homestead was purchased by Andrew's son, Richard Hansen, and wife Irene who on Dec. 14, 2012 transferred ownership to Hansen Homestead Farms which is an LLC operated by their son Dennis, wife Sally and their sons, Matthew and Cole. Richard and Dennis continue to farm this land. *Submitted by Patti Hansen Yaussi.*

HANSEN – Mr. and Mrs. Knud Hansen and their two children emigrated to the USA from Denmark. Knud, 35 years old, walked from Omaha to Homer. He took a homestead on Wigle Creek four miles southwest of Homer in June 1868. His wife, daughter Carrie and a stepson George arrived later.

The Hansen's lived in a dugout near Wigle Creek, which is a short distance from the present house. A daughter, Lizzie, was born there in 1872. She was 9 years old when the family moved into their two-story wooden-stone house. Two more children were born to the Hansen's, Christian and Louise.

Hans Christian Rasmussen came from Denmark and married Louise Hansen of Homer. They had seven children: Harry, Lawrence, Francis, Grace, Ada, Earl and Donald. In 1906 H.C. Rasmussen added four more rooms to the house to accommodate the large family.

On Dec. 22, 1940 Donald Rasmussen married Margaret Gotch of Emerson, NE. Margaret was the teacher at the one-room school house known as Wigle Creek School District #16. Donald and Margaret had four children – Russell, Kent, David and Rita.

David Rasmussen married Della Lindahl of Allen, NE in 1973. They had three children – Linda, Lisa and Dave. They have lived on the farm since 1974 and are present owners. A new home was built by Margaret in 2002. Margaret passed away in December 2012 and Donald 1992. In 2014 a tornado took the barn, which had been built in 1921. *Submitted by Della Rasmussen.*

HANSEN – Rasmus Hansen emigrated from Denmark in 1882 at the age of 19 to Illinois to work out his indenture on a dairy farm. Rasmus earned $10.00 a month and in two years he paid off his debt and moved to Omaha where he raised cabbages and other produce. While living in Omaha he met and married Serena Bartelsen on March 31, 1888. In the spring of 1892, they traveled by train to Stanton County to work for the Wisherd Ranch. They rented an entire box car into which they packed their meager belongings, a team of horses and several milk cows. Rasmus managed the ranch for several years until purchasing their original farmstead of 240 acres on March 1, 1895. Rasmus and Serena had a family of seven sons and two daughters. Due to Rasmus and Serena's hard work, they built a family estate of 2,560 acres in Stanton and Boone County.

Katherine and Norman Hansen

The farm ground that Rasmus had purchased from Joseph Rozmarin on March 1, 1907 was transferred to his son Otto Hansen on Nov. 13, 1934 for "One Dollar, Love and Affection." Otto Hansen and his wife, Nina (Taylor), raised two sons, Raymond and Norman, and two daughters, Arlene and Myrna, on the family farm. Otto spent his life farming the land until 1978 when he retired. Otto and Nina passed away in the spring of 1980 and passed the farm on to their two sons, Raymond and Norman Hansen. Norman and Raymond divided up the farm and Norman and his wife Katherine (VanDieken) bought out Raymond's share in the farmstead a few years later.

Norman maintained the farmstead and crop shared the ground out for several years before planting it to CRP acreage. It has remained in the CRP program since and continues as a wildlife area as well as a home to a large bee colony. Norman passed away on Oct. 30, 2014 and is now owned by Katherine Hansen, who is maintaining it as Norman would have wanted. Katherine is planning on eventually passing the land down to her two daughters, Fauneil Meir and Brenda Hansen-Styskal and their families. *Submitted by Brenda Hansen-Styskal.*

Pictured l-r: Stephanie Meier, Joshua Meier, Alexander Meier, Scott Meier, Katherine Hanson, James Meier, Norman Hanson, Todd Meier, Brenda Hansen-Styskal, Candice Meier holding Maisy Meier, Fauneil Meier holding Miley Meier with Ryan Meier in front helping great-grandpa. Photo taken when Norman received the Nebraska Pioneer Farm Award. The original barn and house are in the background.

HANSMIRE – John H. Hansmire Sr. (1817 – 1875) was the first homesteader and came to this country from Germany (Prussia) to get away from being drafted into the army. Prior to coming to homesteading in Nebraska, he was located in Muscatine, Iowa. With the California Goldrush of 1849, he went overland to California to seek his fortune. He survived but did not become wealthy. It took John six weeks to get home working his way home on a freighter going around South America. He had four children, John, Millie, Henry J., and Etta. John H. Hansmire became a citizen of the United States in 1864, at which time his last name was changed from the Prussian spelling Huntzmeyer to an Americanized spelling of Hansmire.

Henry Harrison Hansmire and Emily Louise Priefert's wedding photo, June 17, 1913.

In 1869 John H. Hansmire and three children came to Nebraska. The fourth child, Millie, married and stayed to live in Muscatine, Iowa. They came in a wagon pulled by a large team of mules and arrived in the spring.

The farm was homesteaded in 1869. The 160 acre homestead is southwest of Fairbury, Nebraska, near Reynolds, Nebraska (location of Farm: NE ¼ Section 11 Township 1N Range 1E Jefferson County, Nebraska). It is located on Buckley Creek, an ideal spot for a homestead, because it contained water for the family and their animals, wood to build a home and to use as fuel, and land that could be farmed. When the homesteaders came they all took homes along streams where there was water and fuel. They did not want out on the open prairie. This left a lot of land for free grazing. A log cabin was built and it took two weeks to get the finish lumber they needed from Nebraska City, located on the Missouri River, which was the nearest trading post at that time. Later on Beatrice, Nebraska, about 25 miles distant, was their trading post.

Henry J. was not old enough to get 160 acres and could get only 40 acres to homestead in 1881. (Location: Part of NE ¼ SE ¼ Section 2 Township 1N Range 1E). He had a little dugout on the northeast corner, which is just north of the home place. The original 200 acre homesteads remain with the family but acres have been increased since then.

The family of Henry J. Hansmire and Susie Elizabeth Riddle, l-r: Henry Harrison Hansmire, Henry J. Hansmire (father), Jessie Frank Hansmire, Henrietta Hansmire (on floor), Susie Elizabeth Riddle (mother), and Arthur Ferdinand Hansmire. Photo ca. 1896.

Continuation of the farm ownership passed from John H. Hansmire Sr. to his son Henry J. Hansmire, and subsequently to Henry Harrison Hansmire (1890 – 1963) and his wife Emily Priefert Hansmire (1890 – 1972). Their children were William (b. 1914, died shortly after birth), Leo (1917 – 1943), Mildred (1915 – 1933), and Floyd Walter Hansmire (b. 1920). With the passing of Henry Harrison Hansmire, the farm passed to Floyd Hansmire. Floyd married Ruth J. Harvey in 1942 and they had 3 children, William, Susan, and Julie. Floyd and Ruth Hansmire and their three children, William, Susan and Julie, have continued to care for the land since the passing of their grandfather and grandmother, Henry Harrison and Emily Louise Hansmire. The land became Hansmire, Incorporated in 1976. It was owned for many years by William, Susan and Julie Hansmire and is presently owned by William and Julie Hansmire.

HANSON-THIEMANN – In the year of 1856 my great-grandparents, William and Sarah Jane McClintock, left Galesburg, IL and traveled by wagon across Iowa, ending their journey in Pawnee County, NE. On the way they stopped in Ft. Des Moines, IA where my great-grandfather became a citizen of the US.

Pictured of the house built in 1876. Photo taken ca. late 1920s with Grandmother Mary Jane, Mother Florence and Uncle Art.

He had come to the US from Ireland when he was 17 and was still a subject of Queen Victoria, so he pledged his allegiance to the US and became a lawful citizen.

It was winter when they finally settled in a carved out place in the side of a bank on the original farm that was a land grant issued by James Buchanan. My grandmother, Mary Jane, who was born Jan. 19, 1856 was a small baby that first winter. They had 160 acres from the land grant and when the Homestead Act was enacted in 1862 by Ulysses Grant, they proved up on 160 more acres. The land belonged to my great-grandparents. My grandmother grew up on this land. What farming was with oxen or horses. In 1876 they built a two story house that was home to four generations. The lumber was hauled from Brownville, NE.

My grandmother married Frank Monagon and after living in Kansas, came back to the farm. My mother, Florence Alice, was 11 years old when they traveled back by covered wagon, which took two weeks.

She grew up there and later began teaching n country schools. She later taught in town schools and after marrying my father, William Hanson, they moved to the farm. Her parents moved to Pawnee City.

My father did farming and used horses, sometimes four head, to pull the machinery to plant what corn, oats, milo and alfalfa for hay. My mother raised chickens and had a big garden. They lived through the '30s of the depression years when crops failed due to drought and grasshoppers. My father had an old Fordson tractor by then.

I was born in 1928 and grew up remembering the depression years and the hardships that followed. We had some very bad storms that also added to the problems. I attended Country School District 16, which was the school my grandmother and mother attended. In 1956 my grandmother was 100 years old and that year the farm received a 100-year plaque given by Aksarben.

After teaching country schools a few years, I married my husband, Ross Thiemann. My parents retired and moved to town. We kept the

The Thiemann family on May 29, 2000 on original homestead land.

farm going but started a cattle herd using the land mostly for pasture and hay. In 2006 the farm was again honored with a plaque for being a Pioneer Farm Family for 150 years.

Four generations on homestead land: Mary A. Thiemann (4th), Joyce Winter (5th), Amanda Crouch (6th), and Michael Crouch III "Trey" (7th).

In the year of 2000 in May we celebrated our 50th wedding anniversary by gathering our children and grandchildren with a picture of us on a hay rack on the original homestead land. Our children – Roger, Joyce Winter, Rodney and David are the fifth generation to walk on the land. Our grandchildren – Amanda, Jacob, Jason, Amy, Sarah, Seth, Megan, Matthew and Alex are the sixth generation. Recently our great-grandson, Michael Dean Crouch III "Trey," walked on the land with his parents, Amanda and Michael Crouch II. "Trey" is the seventh generation to run and play where his ancestors lived over 160 years ago. This land had Indians roaming over the trails to seven generations of family living and making memories of Pioneer days to the present. *Written with love by Mary Alice (Hanson) Thiemann, great-granddaughter of Wm. and Sarah McClintock.*

HARDESSEN – Otto and Sophie Hardessen were both immigrants from Germany. In March of 1915, they moved from Columbis, Nebraska to Brule, in Keith County, where they had purchased a 160-acre farm in January of that year. They went from having electricity to kerosene lamps and carrying water from the windmill. With a crop failure the first year, they found their income from chickens and cream a life saver. In the early 1920s, Sophie's interest in chickens grew from a 240 egg, lamp heated incubator to a 24,000 all modern electric incubator with a modern ventilation system hatchery, and a 6,000 baby chick brooder. Otto raised corn and hay and had the first irrigation well in the county in 1928. He also bought cattle to fatten that were shipped out on rail to market in Omaha.

Otto and Sophie Hardessen in their chicken hatchery, 1930s.

Otto and Sophie moved to Ogallala in 1944, at which time they turned the farm over to their son Carroll and wife Mary, who are still the present owners. They raised corn, alfalfa, sugar beets, hogs and cattle, but gradually changed to hay, corn and a 1,000 head feedlot. In the mid-1950s, neighbors helped to move one large chicken house to a new foundation, creating a barn for storing hay, machinery, and a place for livestock. Three trench silos were built for corn silage to feed the cattle. The original four-room house that was on the land when purchased in 1915, and where Carroll was born and lived for 95 years, was added on to and remodeled several times. He and Mary lived there for seventy years before moving to Brookdale Senior Living in Kearney, where they now reside.

Carroll and Mary Hardessen, November 2012.

Carroll was always progressive in hi farming and feedlot operation, being one of the first farmers in the county to practice minimum tillage. He often manufactured machinery that he needed, including an automated loader for sugar beets after they were topped by hand, a front loader tractor with a scale for measuring feed, and was one of the first to use overhead sprinklers to keep the cattle cook in summer. The Hardessens received numerous conservation awards over the years.

Cattle in the Hardessen feed lot, 1960s.

In the early 1980s, they acquired an additional 160 irrigated acres and their son John farmed with them until recent years when the land was rented out to neighbors. Carroll and Mary celebrated their 71st wedding anniversary in August of 2015. In addition to son John, they also have two daughters, Jane Shaw and Linda Adams, five grandchildren, and ten great-grandchildren.

HARFORD – The year was 1907 and Harry Harford of Nemaha County decided to Homestead in Dundy County. Being the oldest of eight children and only 17 years old when his father died, he was no stranger to hard work. In December of 1908 he returned to Nemaha County where he married his sweetheart of several years, Margaret Hacker. On New Year's Day, Harry started out on the immigrant car with one horse, a cow, three pigs, a pump organ, chairs, a bed, a dresser and sack of apples which were frozen by the time he reached Haigler. A few days later his bride traveled by train where she was met by Harry with the horse and buggy for the 15 mile ride north to their home. Margaret said, "Oh, riding over these prairies to my new home took courage, I can tell you! I had one room with a lean-to kitchen to live in."

Harry and Margaret with Don (about 10-11) and Ray (about 7-8)

Don was born in 1917 and Ray in 1920. Life was not easy. Cow chips were collected for fuel in the cooking and heating stoves. Kerosene lamps were the evening light. Harry worked to build his cow herd and as neighbors got discouraged and moved on, he purchased more land to add to the 640 acres he originally homesteaded. Harry's health failed and in 1931, at the age of 52, he passed away in his home, leaving a wife with two young sons. Banks in Benkelman and Haigler went broke. Margaret needed to borrow money to continue with the operation. With her perseverance and determination, Margaret was loaned $500 from a bank in Wray, CO. Margaret Harford and sons continued to expand the ranch. Don married his high-school sweetheart, Virginia Clegg, in 1939. To this union two children, Karen and Gail, were born. Ray married Melba Boyd in 1945 and Linda,

Margaret Harford and sons in the 1940s

Glenda, Ted and Tom were born to this union. Margaret passed away in 1972 and for 10 more years Don and Ray worked in partnership, being good stewards of the land. Windbreaks were planted and they worked closely with the University of Nebraska and Extension and received honors in range management. All six of Margaret's grandchildren worked on the ranch stacking hay, branding calves, fixing fence and all that goes along with the business.

In the early 1980s, the ranch was divided so that the children who wanted to continue with ranching could work with their own fathers. Ray passed away in 1997 leaving his son Ted to continue with his share of the operation. Don and his son Gail continued with their own ranch with Don still living on the original homestead and in the home.

Needless to say, some changes had taken place over the years to the home. Don's daughter Karen moved back to the ranch in 2001 to help her aging parents. Gail passed away in 2009 and Don in 2011. Karen now lives on the homestead and leases out her acreage to her cousin Ted and to a neighbor. Life is good at Home on the Range! *Submitted by Karen Harford.*

HAWKINSON – Nels Hacansson, Swedish name Americanized to Hawkinson, filed the first papers on the farm Oct. 31, 1884. He had arrived at New York and came to Nebraska, settling seven miles southeast from Stratton, NE in Hitchcock County. Stratton was not incorporated until 1886. Nels was 29 years old when he arrived. The Patent was issued May 11, 1891 on the Northwest Quarter, Sec. 33-2-34.

He married Sophia Karlsdotter (also from Sweden) Sept. 19, 1892. Three children were born to this couple - son Andrew and two daughters - Ellen and Anna. Nels dug by hand water wells for neighbors in exchange for working the land. The barn he built is still standing and two rooms of the original house are still occupied.

Andrew later farmed the land alone and bought out the interest of his two sisters. Andrew and his wife Adda had two sons - Arthur and Loren. Later the two sons farmed the land and then Arthur obtained full ownership in 1994. The acreage at that time was 862. On July 2005 the land was deeded to Christopher and Kathy Reid and since then the acreage has grown to 1,473.

In 1984 the Hawkinson family was awarded the Nebraska Pioneer award for continuous ownership of the land. It is still in the family as Chris became a part of the Arthur Hawkinson family when he was in high school. Also in 1984 Arthur and Loren hosted a 100 year celebration which included games, rides in a Model T truck, rides in a wagon drawn by a team of horses, a barbecue, homemade ice cream, a dance, etc. Fourteen of the 16 grandchildren of Nels and all of Andrew's grandchildren, spouses and great grandchildren were in attendance. Guests came from 11 states and the local neighborhood. *Submitted by Arthur W. Hawkinson.*

Andrew Hawkinson on header harvesting wheat

Hawkinson farm

HAWLEY – Thomas Hawley Sr., born Oct. 17, 1803, Stansted County Kent, South of London, England. Married Rebecca Venner 1824 in England. They had six children (three sons, three daughters). They came to America in 1849. Thomas died in Nebraska at the age of 88. William Henry, born Feb. 22, 1830, married Eliza Walton in 1856. They came to Nebraska in 1863 where he homesteaded on 160 acres and built a home. They had seven children (five daughters, one son). William Henry was a prominent retired agriculturist raising corn and wheat. Richard married Anna Wyne Aug. 10, 1887. They had five children (three daughters and two sons). Charles R. Hawley, born Oct. 4, 1897, died Aug. 24, 1982; married Inez Lockwood, Nov. 24, 1920. They had four children: Leland Lockwood, Betty Jean, Glendon Dwight and Connie Rae. They lived in the home that was built in 1914. Charles and Inez received the 100 year Nebraska Pioneer Farm Award in August 1962. Charles grew corn, wheat, oats and hay. He raised chickens, hogs and cattle. Charles farmed where he was born until retirement in 1963. They then moved to Auburn, NE.

The home of Richard and Anna Wyne Hawley

Glendon Dwight Hawley, born Dec. 10, 1926, died June 9, 2015. Married Catherine Hoover March 5, 1946. They had five children: Carol, Sammy (deceased), Helen, John and Charles. Carol married Robert Schmucker (Brock), Helen married John Jeanneret (Brock), they had three adopted children - Jan, Melissa (deceased) and Thomas (deceased). John married Nancy Green (Brock) and they had four children: Angela, Sara, Eric and Aron. Charles married Janet Buethe (Brock) and they had two children, Charles Cody and Katy.

Dwight and Catherine bought the original homestead March 1, 1970. Their crops consisted of corn, milo, beans, wheat and hay. They raised chickens, hogs and cattle where Dwight had a feedlot south of the farm. Many meals were served from the livestock that was raised for family and friends. The farm has been in the Hawley name for 152 years. Catherine still lives in the original home today.

July 25, 1978 Dwight, Catherine and their boys, John and Charles, formed a family corporation. Their main crops are corn and beans. They also have a cow/calf operation. John and Charles have been keeping the Hawley homestead going and with the help of John's two boys, Eric and Aron who are also farmers, they hope to keep the Hawley farm going for many more years to come.

HAYES – Newton Hayes took title to the farm in September 1902 while living in the Chester/Hubbell vicinity. During the following years he successfully developed the farm and constructed a new house featuring one of the first rural applications of a self-contained generation/electrical distribution system. He continued to prosper and improve the land, but in the fall of 1937 at the age of 77 years, Mr. Hayes was fatally injured by a runaway team of horses while harvesting corn.

Newton Hayes Farm, ca. 1912

At that time the farm was passed to his son, H.L. (Lester) Hayes, and became part of the total farming operation owned by Lester and his wife, Louise. Mr. and Mrs. Hayes raised their family during the ensuing years and were privileged to be selected the 1947 Honor Farm Family of the State of Nebraska (sponsored by the *Lincoln Journal*).

Upon the death of Mr. Hayes in 1986, Mrs. Hayes continued the farming operation and raised cattle until she passed away in 1998. The Hayes children continue to own and manage the farm. *Submitted by C.L. Hayes.*

HAYES – My great uncle, Newton Hayes, bought the NE 4 of 25-47 on September 4, 1900. He gave $410 and had a mortgage for $100. Basically, all quarters are supposed to be 160 acres, but this quarter has 192 acres. He raised alfalfa, and had some milk cows, hogs, and chickens.

Upon his death, this quarter was transferred to my grandmother, Mina R. Griffith.

In February of 1948, Mina Griffith transferred this quarter to Valma Lou Griffith Odell and Opal Griffith Roth. In July 1948, Opal Roth sold her share of the quarter to Velma and Charlie Odell, my parents. In July of 1980, this quarter was deeded to Kathryn M. Hood and Norman L. Odell. Later that year, sole ownership was deeded to Kathryn M. Hood by Norman L. Odell. She drilled an irrigation well and had a center pivot erected, and raised wheat, beans, potatoes, and alfalfa.

In October of 2008, she sold the center pivot to Norman and Janene Odell, her nephew, and the NW corner of the pivot was sold to Earl Hood, her son. She has a small tract where her buildings are located, and raises winter wheat and she and John raise hogs. Trees have been planted across the south side of the quarter, and along the SW corner of the quarter, and are maintained.

Norman and Janene Odell farm the pivot on the NE 4 of 9-25-47; their operation is organic.

Kathryn had six children: Earl T. Hood, Peggy Sindelar, Patty Kulas, Cheri Phillips, Ted C. Hood, and John W. Hood. Ted passed away in December of 2001.

Kathryn was elected Box Butte County Clerk in 1982 and served in that capacity until January of 1998. She still lives on the farm and enjoys country living.

The northwest quarter of 10-25-47 is now owned by Denise Krajeska and the northeast quarter of 10-25-47 is owned by Crystal Powell, both daughters of Norman Adell Sr.

HEATHER – William Y. Heather was a Union Corporal in the Civil War who served in Company C of the 8th Regiment, Illinois Cavalry. During the war he was captured but was able to jump from a train bound for Andersonville Prison living to eventually settle in Otoe County, Nebraska. In 1869 William and Hester purchased the 160-acre farm from Heath and Susan Nuckolls for $760, located at NE ¼, Section 13, Township 7N, Range 10 East of the 6th principal meridian. In 1908 a large barn was erected on the farm, which is still in use today, in 2015, as is the home, which was built in the early 1900's. The farm is located five miles south and three miles west of Syracuse.

On August 16, 1909, the farm was deeded to William and Hester's son, Harry, who lived and worked the farm with his wife Molly and three children, Edna, Thelma and Delbert. On February 28, 1961, Harry deeded the farm to son, Delbert and wife Marie, who continued to live on the property and carry out the farming operations until 2003 when they moved to Syracuse. They raised corn, milo, soybeans, wheat, oats and alfalfa, along with cattle, horses, hogs and chickens. There was always a large garden and beautiful trees, shrubs and flowers. Delbert and Marie had six children; Lucille (Stuart Kitt), Jane, Jody (Walter Pohlman), Wayne (Patricia Heather), Keith and Sandy (RK Piper); 8 grandchildren; 15 great grandchildren. The farm was a great place for the grandchildren and great grandchildren to visit; and the grandchildren who lived locally were able to show livestock in 4-H because of grandma and grandpa's cattle herd. Health care costs led to the farm being sold in 2011 to William (Bill) Goering, a long time friend and neighbor who had originally leased the farm after Delbert and Marie retired. Delbert and Marie were able to continue their beloved cattle operation for many years after retirement largely due to Bill's help.

Delbert and Marie received the Pioneer Farm Award in 1969.

HEMMER – The family of Everett and Norma Hemmer received the "NE Pioneer Farm Award" at the Platte County Fair on July 12, 2012. Everett has since passed away, Jan. 2, 2015.

Francis, Norma and Marvin receive the Pioneer Farm Award

Our family consists of Francis, Sharon (Greg) Klug, Monica (Jeff) Kavan and Marvin.

Our farm is located in northern Platte County along Highway 91 in Grandville Township, 17-20-2W. It began in 1902 when Nickolas Hemmer purchased 160 acres of land for $50.00 an acre. He also was one of the first supervisors of Platte County. It was farmed by his son, Guido. Guido married Catherine Brandl in 1915. In 1921 they took ownership and began building a homestead on this land. A home was built, then a barn and corncrib followed by more buildings to accommodate their livestock operation. Guido and Catherine had six sons – Eugene, Gerald, Wilfred, Everett, Glenn and Carroll and a daughter, Sr. Eleanor. They retired to a home in Humphrey in 1953.

The home today

Everett completed a tour of duty in the US Army and returned to the farm with his wife, Norma Herink in 1953 and farmed the land. In 1967 they bought the land from his dad, Guido and in 1975 purchased an adjoining 120 acres.

An irrigation well was drilled in 1955 and in 19596 the task of irrigating with "hand-moved pipes" began. In 1970 when pivots were introduced in this area, the family bought their first pivot. The farm is diversified with crops of corn, soybeans and alfalfa and in earlier years, some milo and wheat was raised.

Aerial view of the Heather family farm.

Combining – Marvin and Francis Hemmer

Dairying has always been their major enterprise. In 1981 a milk parlor was built and Grade A milk was sold. They raised all their replacement heifers to build their herd as they continue. In the '70s they would farrow to finish several hundred hogs and fed their steer calves.

They all attended St. Francis Church and their children were educated at St. Francis School in Humphrey.

Everett and Norma moved to Columbus in 1995 and are members of St. Isidores Catholic Church. They have three grandsons, two granddaughters and two great grandsons.

Francis and Marvin have taken over ownership and continue to farm the land and work the dairy. They are the fourth generation to complete the 100 years and beyond. In 2014 they have added a new entryway and a large garage to the original house, as shown in the photo. *Submitted by Norma Hemmer.*

HENDRICKSON – Andrew J. Hendrickson (1861-1938) and Elsie C. Oberg (1859-1936) came to Phelps County, NE from Galesburg, IL in 1885. They were married Nov. 6, 1884. Each one was born in Sweden. Records show Andrew J. Hendrickson became a US citizen on Dec. 12, 1888.

My grandfather, Andrew J. Hendrickson, acquired 160 acres of land (northwest quarter, Section 25, Township 7, Range 20, Garfield Township) from the Union Pacific Railroad on July 6, 1885. The UP had gained land in Nebraska in 1862 when President Abraham Lincoln signed legislation creating the railroad. At one time, the Union Pacific owned 1/16 of all land in Nebraska, but as it expanded westward, it sold off parcels to pay for the expansion.

My grandfather was one of the first people in the community to install gas lamps in his home. Later, this was improved by a Delco light system and included lighting in the farm buildings.

Running water was piped to the home in 1912. Early income consisted of corn and wheat with a few acres of oats. Hogs, cattle and chickens, along with a few milk cows, supplied food for the family. Members of the family lived on the farm until 1937.

After my grandmother died in 1936, my grandfather moved to Holdrege. He gave one-eighth of the farm to each of his children. My father Carl's siblings wished to sell their interests. My dad bought out those interests, thus keeping the farm within the Hendrickson family.

In 1972, my father added another quarter section adjoining the first one (the SW quarter of Section 25). On Feb. 28, 1976, he incorporated the farm.

In 1986 the corporation purchased another 80 acres, which was across the road (N 1/2, NW 1/4 Section 36).

In 1990 we put a pivot on the southwest quarter. In 1995 we put a pivot on the northwest quarter. Before that, we used gravity irrigation. Our main crops have been a rotation of corn and soybeans. However, we have grown potatoes, popcorn, alfalfa, wheat, oats and sorghums.

In 1985, we were recognized by Aksarben for having a farm that had been in the family for 100 years. Today, a fifth generation owns stock in the corporation. *Submitted by Bruce Hendrickson.*

HENN – Our family farm was originally bought by my Great Uncle Anton Mescher on Dec. 13, 1898. This farm was 160 acres. The location of the farm is 3/4 of a mile west of Petersburg, NE. The official description is: NW 1/4, Section 26, Township 22, Range 7, west of the 6th Principle Meridian, Boone County, NE.

Original crops planted on the farm were corn, oats and hay. Sometimes barley was planted with the oats. Livestock raised were cattle, both stock cattle and milk cows, hogs and chickens. Horses were used to farm the land. My grandfather owned a pair of Belgians that he was very proud of.

Owners of the farm: Anton Mescher; his brother and family Fred and Agnes (Schnittker) Mescher; Fred and Agnes's daughter and her husband and family, Joseph and Agnes (Mescher) Henn; Joseph and Agnes's son, Jerome Henn; Jerome's daughter, Pamela (Henn) Sondag.

Even though they didn't own the farm, another son of Joseph and Agnes Henn and his family, Joseph "Jay" and Virginia Henn lived on the farm for 18 years. At this time, Joseph and Agnes owned the farm and Jerome farmed the land. Also at that time Jerome raised corn, oats and hay. Animals on the farm during these years were cattle, sheep, hogs, chickens, geese, ducks and guineas.

The Henn family farm

The original 160 acres are still retained today. It is a working farm. The crops raised are corn and soybeans. Randy Ketteler farms the land. The livestock raised today are stock cattle raised by Bob Henn. Both of these men are my cousins.

Many of the original buildings of the farm are gone now. Those still standing are a small grain shed with a smaller feed shed next to it located on the east edge of the farm buildings, a brooder house for chickens and the cement block bottom of the water reservoir near the house.

The original house was a two-story with a living room, dining room, one bedroom, kitchen and a pantry downstairs with bedrooms upstairs. There was an outhouse. This house partially burned in the 1950s and a smaller house was built in its place, which still stands. There was a large red barn across the yard east of the house. This barn had three large sections. The southeast corner housed the horses, the section north of that was where the cows were milked. Another large section was west of that for multiple uses. The entire upstairs of the barn housed a hay loft. Near the hay loft ladder was a small granary. My dad tore this barn down in the 1980s and built a three-sided cattle shed. Across the yard, on the south side of the farm buildings, was a machine shed, which was used for horse tack and wagons in the early days. Later, it was used for machinery until my dad tore it down and built a large Behlen Building closer to the house. Next to the machine shed was a garage. There was also a chicken coop, corn crib, storm cave and wind mill on the north side of the farm buildings. A smoke house was between the house and the chicken coop, and hog sheds were near the barn. All of these are gone. My dad bought a 1947 Farmall when he started farming the land in 1948. A grain bin built later is still in use today. Also in use today is a cattle tank.

This farm kept my grandma and grandpa and their family (including my dad) alive in the terrible 1930s when times were so bad people didn't have enough to eat and were losing their farms. My grandma raised a large garden and canned many vegetables as well as jams and jellies. My grandpa butchered cattle and hogs for the family's meat. He also traveled to Omaha to sign up for a government loan to save the farm.

In 2007 I applied for and received the Pioneer Farm Award honoring this farm, which has been in our family for over 100 years. After receiving the award at the Boone County Fair in Albion, NE, around 200 family and friends celebrated this achievement on the farm. *Submitted by Pamela (Henn) Sondag.*

HENNING – Freidrick (Fritz) and Dorothea Henning, German immigrants, purchased the West half of the Southwest quarter of Section 33, Township 6, Range 15, East of the Sixth principle meridian (120 acres) in Nemaha County, Nebraska from W. S. Workman and his wife on August 31, 1878.

Fritz and Dorothea Henning, with children Louise and Carl

Freidrick (Fritz) and Dorothea had two children, Louise Hedivig (Hattie L.) and Carl Johannes (John C.). Fritz was considered a

Generations of the Henning family, who gather every year on the 4th of July.

hard worker and raised corn, alfalfa, wheat, hogs and cattle and built a farmhouse, hay barn, stable and other farm buildings. He also raised grapes, which he made into wine that he stored in a deep wine cellar and served to visitors.

Upon Fritz's death in 1915, John C. took over the farm as requested by Dorothea and Hattie, who died in 1917. John C. married Leona Grace Groff in 1922 and inherited the farm when Dorothea died in 1923. John C. and Grace raised four sons on the farm: Charles E., L. Eugene, Johnielee and Jerry T. In addition to the crops and livestock the farm had an orchard of pear, apple and cherry trees. A fire destroyed most of the original farmhouse in the early 1920's and John C. built a brick house to replace the old farmhouse, sparing no expense. The house had running water, Delco electricity produced in the lighthouse, a full basement with a coal cellar, four bedrooms, a kitchen, a dining room and additional storage upstairs. He was one of the first to have a car and drove with the boys in the rumble seat. The Great Depression hit hard; John C. suffered financially but maintained significant assets, including the original 120 acres. John C. was a leader in his community with many talents. He and Leona Grace spent many evenings dancing where John C. would often be the square dance caller or the fiddle player.

As a young soldier, Charles E. Henning fought in the Battle of the Bulge and earned a Purple Heart. After World War II, Charles E. returned to farm with his father. He used the GI Bill to attend Peru State College for two years and then switched to on-the-job training (farm) for two years. He married Roberta Jane Applegate in 1946. They raised ten children: C. Gary (deceased), Bruce E., Alan L., Judy J. Peterson, Fritz, Janice Henrichs-Smith, Jodie J. Ferguson, Larry D., John E. and Eric G. Charles and Jane farmed with John C. for over twenty years until his death in 1970. They moved into the family farmhouse in 1974 when they bought the farm from Leona Grace who moved to Peru, Nebraska. In addition to farming, Charles had a realtor's license, which he utilized to sell farm land. Charles worked hard supporting the American AG Movement and participated in the Tractorcade to Washington, D.C. in 1979. He was the Nebraska DAV Commander for two years while Jane was the Commander of the Auxiliary. They have volunteered at the Veteran's Medical Center in Lincoln, Nebraska for over 25 years. Charles and Jane retired in 1995 and the farm is leased on a yearly basis where corn and soybeans are grown. They divide their time between the 1920's farmhouse and Omaha, Nebraska. They have 22 grandchildren and 29 great-grandchildren who gather for yearly family reunions on the 4th of July. In 2014 the family celebrated Charles' 90th birthday.

HENRY – Benjamin F. Henry purchased the Southwest 1/4 of Section 13, in Township 4 North, Range 1, East of the Sixth Principal Meridian, in Jefferson County, State of Nebraska in 1883 for $2,000.00.

Arlin and Marlys Koch with children Tim, Troy and Teri Ann

B.F. and his bride, Catherine (Burkey), from Bernville, PA were married Feb. 25, 1883, came to settle on this prairie land and make it their home. They raised five children: Pearl, Mae, Charles, William and Anna. In 1890, B.F. purchased the Southeast 1/4 of Section 13 for $3,000.00.

The prairie sod was broken up with a horse and one bottom plow. The couple first lived in a sod dugout, then a small timber home. A wood framed home was soon built to raise their family in.

Farming was done by hand labor and horses. Corn and oats were raised to feed the cattle and horses.

In 1913 the youngest son, William, married Lulu (Brinkman) and purchased 80 acres of the farm from his father. They too made their home here and raised four children: Walter, Wesley, Elinor and Wayne on the farm. In 1924 William purchased another 160 acres from his father. Farming was with horses and primitive equipment. Later, modernized equipment was purchased to make farming easier. Long hours

The Henning home and farm.

Aerial view of the Henry farm.

Hans and Myrtle Horst with her brother, Walt Stubbs, with the horses and mules used for all farm work, ca. early 1900s.

This photo shows the location of the barn after twin tornadoes destroyed the entire farm on June 15, 2004.

Aerial photo of the farm in 2007.

My grandparents farmed with mules and horses, raised a son, Pete and daughter, Marge. Pete married Millie Helms in 1939, and later moved to the Horst family farm when his folks moved into Wisner.

My folks raised three daughters, Judy, Cheryl and Lynn, but sadly lost their young son and daughter. We were busy with 4-H club calves, farm chores, house chores, and exploring along the creek as well as piano and dance lessons and country school. My sisters had horses and I had a donkey and lots of cats. Dad farmed and also was a cattle feeder. Mom was an awesome cook!

The folks moved to town, the house had renters...but sadly, the twin tornadoes on June 19, 2004 destroyed the entire farm—al the buildings, trees and crops, but not the memories. Fortunately, the renter and his little dog survived the terrible ordeal. Many came to help clean up the debris like out neighbors also were doing. But we have many great memories and know someday our daughter, Brittany, will be the 5th generation to carry on the legacy of the Horst Family Farm. *Submitted by Lynn Horst Sill.*

HORSTMAN-FISCHER – On December 3, 1841, William Charles Mecke was born in Trogen Hannover, Germany. Immigrating to the United States, he first located in West Point, Nebraska, where a German conclave had already settled.

William Mecke moved to the Creighton area in 1871. The following year he returned to West Point, where he and Katharina Faber were married on April 9, 1872. They returned to the Creighton area. Later that year they filed a claim to homestead 160 acres a mile west of Creighton.

Katharina was a native of Germany, born in Freisen, Regierungsbezik Trill, on May 28, 1848. While people from Wisconsin were visiting Germany, they invited her to emigrate from her native land. On October 26, 1869 at the age of 21, she accepted and traveled with them to Wisconsin. From Wisconsin she moved to West Point, where Mr. Otto Bauman employed her. Seven children were born to Katharina and William. Henry was the third child born in the Creighton area, but the first of the three to live his entire life in the Creighton community.

Mary, one of the Mecke children, was born on December 5, 1876 and grew up on the family farm. On February 11, 1898, she married Henry August Horstmann. The following year they purchased 160 acres of farmland southeast of Creighton, where they lived until Henry's death. To this union seven children were born. Of all the children, Harry remained on the Horstmann farm, helping his father farm so that his parents could remain in their home. Harry married Mae Taylor in 1944 in Winnetoon. They lived on the farm until Harry's death in 1964. Harry and Mae had six children.

Mary Ann Horstman and her husband, Lyle Fischer, purchased the farm in 1972 from her mother. Their farm consists of raising irrigated corn and soybeans. They also have a cow/calf operation. Lyle and MaryAnn are the parents of two children: Stacy and Robert. Although Robert does not live in Creighton, he helps his parents during the busy time and does farming of his own. Robert represents the fifth generation.

HRADEC – Frantisek and Antonia (Vavrfa) Hradec came to America from Bohemia in 1895. They had 11 children, eight were born in Czechoslovakia, three were born in Nebraska.

The Hradec boys, Norman, Dale and Larry, 1955

Frantisek Hradec purchased 160 acres from John A. Wisherd of Stanton County, NE for $10,400.00. The farm was the SE 1/4 of Section 16 in Township 21 North Range 2, Ramshorn Precinct, later known as the Sunny Slope farm. Possession was given March 1, 1908.

On April 7, 1909 Frantisek Hradec sold one acre to the Stanton County School District. District 56 was established. It was a historic one room schoolhouse and was in operation until the 1960s.

The Frank Hradec farm

The old Herbek farm, ca. 1900, still runs today.

Frantisek and Antonia sold the farm to their third child, Frank Hradec, on June 12, 1917.

Frank Hradec became a US citizen on April 29, 1905. After he bought the farm from his father, Frank began building the farmstead. He built a large barn, hog shed, chicken house, brooder house, corn crib, machine shed and garage. The house was built in 1920.

Frank Hradec began farming with a team of Belgium horses. He purchased his first tractor, a 1929 McCormick Deering regular. This was followed by a 1950 Ford, a 1954 Ford and a 1951 International Harvestor H Tractor.

Frank Hradec married Rose Marie Sykora from Omaha, NE in 1933. To this union three sons were born – Larry, Dale and Norman. The boys began helping their father farm at a very young age. Their crops included corn, oats and alfalfa. Frank bought three registered Brown Swiss heifers that grew up on the farm and were artificially inseminated until he had a registered Brown Swiss milking herd of 20 cows. Frank also raised feeder cattle and raised and farrowed pigs. Rose Hradec bought baby chicks form the Norfolk Hatchery and raised about 1,500 hens for a hatching flock. She sold the eggs back to the Norfolk Hatchery. Over the years the farm became quite a showplace.

It was very convenient having the school on their farm. The three boys, Larry, Dale and Norman, just had to walk across the pasture to go to school. When the weather was bad, the school teacher would stay with the family.

Frank Hradec passed away in 1962. Frank's widow, Rose Hradec, then owned the farm until 2006 when she passed away. One son, Dale Hradec, passed away in 1988.

Today the farm is owned by Larry Hradec and Norman Hradec. Larry's daughter, Jean Hradec Spale and her husband, Jon Spale live on the farm. Jean and Jon Spale are the fourth generation on the Hradeck Farm. *Submitted by Larry Hradec.*

HRBEK – In 1874 Jakob Hrbek (now Herbek) left the family farm near the small village of Sazomin in Moravia, modern day Czech Republic. He and his wife Maria (Svoboda) came through the port of Baltimore, MD and on to Nebraska. First they came to Crete then promptly to the Deweese area. They first settled three miles southwest of Deweese on a parcel of land that the railroad was allowing settlement on to encourage people to move west. They lived in a dugout sod-house and with their children - Crissy, August, Henry and Mary, they began a new life in Nebraska. It was in 1900 that Jakob's son Henry set out to purchase the 24 acres in Nuckolls County that the Herbek family still farms today. That same year Henry married Antonia Karmazin, whose father Frantisek was the previous owner of the land. Much happened in 1900 for the Herbeks as that was also the year Jakob, the family patriarch, died. In 1908 Henry W. Herbek finally paid off the 240 acres in Nuckolls County, Township 9, Section 4 on Range 7. This land was passed on to one of Henry and Antonia's sons, Godfrey A. Herbek, then on to one of Godfrey and Josephine's (Kotinek) sons, Donald J. Herbek and finally on to one of Donald and Cynthia's (Wagner) sons, Brian P. Herbek. Brian Herbek and his wife Blythe (Weir) still enjoy working the land today, not just for a living but for the lifestyle. With such a rich family heritage the land is more than just property and more than just a source of income. It's a home where the stories of a generation live. It's "grandpa's place." It's the old "home place" where the old house, barn and windmill used to be. It's "the Herbek farm." *Submitted by Donald J. Herbek.*

HUEFTLE – In 1884 Karl Thomas Hueftle acquired a total of 240 acres located in Frontier County, 14-8-24. Oats, corn and wheat were crops grown on the farm and also Hereford cattle.

The farm passed from Karl Thomas Hueftle to G.A. Hueftle, then to Edward and will pass to Kent Hueftle.

The acreage of this farm has been added to. Red Angus cattle are raised now; corn, beans, feed and wheat are the grain crops grown.

The barn built in 1908 is still in use. The boys slept in the huge barn loft during the summer and winter.

Edward (deceased) and Bonnie Hueftle had three children – Stacy, Kent and Ann (deceased.) *Submitted by Bonnie Hueftle.*

Wedding of Brian and Blythe Herbek and Herbek side of the family married on the heritage farm land.

Pictured l-r: Karl, Edward, Dora (mother), Lola, G.A., and Elmyra. In front are Doris and Dora

HUETSON-EDWARDS – Minnie Marie Sachtleben born July 10, 1886, married Chester Alonzo Huetson, a lifelong farmer, February 20, 1907. For their wedding present, Minnie's father Henry H. Sachtleben gave her 160 acres of land and a new house built in 1907 similar to his own. He purchased the land on April 20, 1906 and deeded it to her February 18, 1920. Eighty acres continuously remains in the family today. The legal description is Gage County, Nemaha Township, N 1/2 of the NW ¼ of Section 7, Township 6N, Range 7E of the 6th P.M. The house, which stands today, was one of the first homes in the area to have its own generator.

Front Row: Chester and Minnie Huetson. Back Row Left to Right: Orville, Helen, Raymond.

Minnie's father, Henry H. Sachtleben (born September 16, 1853) came to the U.S. from Hanover Germany in 1868 with his parents John (12/15/1822) and Minnie (Mollman) Sachtleben (8/15/1824) and siblings Wilhelmina (1850), Herman, Maggie, Jennie, Lizzie ("Marie" 1867) where they eventually settled in Menard County, Illinois. John passed away March 18, 1891 and is buried in Menard County, Sugar Grove Cemetery in Sweet Water, Illinois. His wife, Minnie Mollman Sachtleben followed her children to Gage County Nebraska. She passed away March 25, 1903.

Minnie's mother, Elizabeth ("Lizzie") Vesselman, born January 20, 1861, came by herself to the U.S. from Hanover, Germany in 1879. Lizzie's parents Henry and Mary (Middlekamp) Vesselman remained in Hanover, Germany.

Her parents Henry and Lizzie married in Illinois March 8, 1883 and also made their home in Menard County, where their first child John "Johnny" was born on December 7, 1883.

In the spring of 1885, Minnie's father Henry Sachtleben, now 32 years old, came to Nebraska and settled in Nemaha Township, Gage County. The nearest town would become Firth, Nebraska, incorporated on January 20, 1879. He bought one hundred and sixty acres of land, in Section 14. At the time of purchase only ninety acres of this land were under cultivation and set to work improving the property and buying more land. When hard times came, he purchased land bartering horses, chickens, produce or other livestock. By 1915 he owned nine hundred acres accumulated through his own efforts, having started life as a poor boy. He was able to give land/or farm to all five of his children; John "Johnny" (born 1883 in Illinois), Minnie (1886 on the farm Sec 14 Nebraska), Harry (1888), Emma (1891), Carl "Pete" (1894), all were fluent in German and English.

Henry passed away in 1925, Lizzie in 1939.

Minnie and Chester Huetson had four children, all born at home, Evelyn (July 5, 1909) Orville (1911), Helen (1917), Raymond (1921). Evelyn died fifteen days after birth and is buried in an unmarked grave in Sunny Side Cemetery, Nemaha Township Sec 17-T-6N R7E. In Minnie's youth she attended and was confirmed in the German/English Lutheran Church on March 23, 1902. In 1924 Chester and Minnie transferred their membership to the Congregational church in Cortland.

Front Row: Elizabeth "Lizzy" and Henry Sachtleben. Back Row Left to Right: Emma, Harry, Carl "Pete", Johnny, Minnie

Chester and Minnie accumulated more farm land and retired to the town of Cortland in June1950 their son Raymond and family then moved into the house and farmed the land. Their eldest son Orville and family lived and farmed in Highland Township.

Minnie Huetson was a marvelous cook and seamstress, her embroidered towels, elaborate crocheted doilies, runners and blankets are well known and treasured. Her two favorite horses were the ones she hitched to her buggy, Queen and Florey. They were happy to have all three children and their families living nearby. Minnie passed away May 10, 1953, Chester March 22, 1967. Both were active member of the community. Upon Chester's death the eighty went to their daughter Helen and her husband Neil.

Helen Vera Huetson was born January 16, 1917. She attended Good Hope Country School from kindergarten through eighth grade, riding her horse to school whenever possible and graduated from Cortland High School in 1935. She was baptized at St. John's Lutheran church east of Cortland and was confirmed on April 5, 1931 at the Pilgrim Congregational Church (Cortland United Church) where she was a member until her death.

From Left: Randy, Teresa, Neil and Helen Edwards.

Helen married Neil John Edwards on January 16, 1940 and they made their home on his family homestead two miles west from where she grew up, Section 10-6-6 in Highland Township. Neil's grandfather Joseph Krapp and family traveled to Nebraska in 1872 to homestead their land.

Helen and Neil loved farm life and seeing results of their labors. He was an avid golfer, participating in tournaments and a lifetime member of golf courses in Beatrice and Crete. Helen especially enjoyed watching sunrises and sunsets out her kitchen windows, dancing, traveling and card clubs, her favorite was bridge.

Their children, Teresa Joan (July 18, 1954) and Randall (Randy) Wayne (January 8, 1956) graduated from Norris High School. Teresa enjoyed outside of normal activities competing in archery target tournaments throughout the state, once going to nationals. Her career was in finance and accounting until her retirement. Randy knew from day one he wanted to be a farmer and especially enjoyed golfing, fishing, a game of pool on his antique snooker table, just any event with friends. Randy farmed with his dad and was killed in an automobile accident July 20, 1980.

Neil passed away December 5, 1980, Helen June 11, 2002. Neil, Helen and Randy are buried at Lincoln Memorial Park Cemetery in Lincoln Nebraska next to her parents.

Upon Helen's death the land passed to her daughter, Teresa Edwards McHargue. Teresa married Peter C. McHargue December 1, 1985 and they reside in Lincoln, Nebraska. Peter, a CPA, retired from Farmers Cooperative in 2011 as their Chief Financial Officer and Assistant General Manager. They have one daughter Rachel

Left to Right: Gary, Cody, Gerald, Brett Behrends

who is married to Ryan Butler and they have two children Nevin and Alison.

The land is farmed by brothers Gerald and Gary Behrends and Gary's sons Cody and Brett Behrends. Gerald and Gary are great-great-grandsons of Joseph Krapp.

Crops planted today are a rotation of corn and soybeans.

For more family history, refer to Joseph Krapp and Philip O'Brien sections in this book and *History of Gage County, Nebraska* published in 1918 by Hugh J. Dobbs — Henry H. Sachtleben pg 1043, William Kramer pg 881, Herman Penterman pg 774 and John Krapp pg 649.

IODENCE – In the late 1800s, William Iodence moved from Illinois to Seward, NE and a few years later to Hemingford. William Iodence would become the first county attorney for Box Butte County and purchase land that is now part of the Iodence Ranch. His son, Charles Gladstone, would begin a purebred Hereford herd in 1917 and pass that herd to his grandson, Charles, in the 1950s. In 1990, his great grandson, Brian, began managing the ranch and does so today, with his wife Carol.

William Joseph Iodence established Iodence Ranch in 1886.

In recent years, conservation practices have made major changes to the Iodence Ranch. Simple practices, like planting 15,000 trees, to complex projects like engineering a Federal 404 permit to dam the Niobrara River, have improved productivity and functionality of the original ranch. Today cattle numbers, on the same acres, have increased 150%. The quality of landscape and wildlife these conservation practices produced developed hunting, fishing and raising Chocolate Labradors. Thus, "Shared Outdoors" was designed to provide outdoor recreation to others. Guests can enjoy the ranch at the private lake on the river, with camping and RV hookups. This outdoor venue, with toilet shower and kitchen functions, works well for weddings and family reunions, etc. The "Lodge" provides additional comfort to guests that hunt or enjoy these outdoor activities. The Lodge on the Iodence Ranch, located less than 30 miles from both Chadron and Alliance, works well for guests arriving for city events wishing to enjoy rural accommodations.

Located just a few miles off payment, the Iodence Ranch (1886-present) has opened its doors to outdoor entertainment while continuing to function as a working cattle ranch. Anyone can find "Shared Outdoors" at www.sharedoutdoors.com or like us on Facebook.

ISHAM – Charley B. Isham arrived in Clay County, NE from Madison, Lake County, OH by train in July 1872. He would be 21 on Sept. 6, 1872. He began living on S 1/2 of SE 1/4 of 4-5-5. In 1875 he became the owner of this 80 acres. He married Mary Smock in 1886. Children Marshall (1889) and Frances (1896) were born and raised on this farm.

The Isham homestead, ca. 1950

In 1904 he purchased 80 acres on the north of the homestead, making 160 acres-SE 1/4 -4-5N-5W of Clay County, NE. The two story house built in 1907 is still in use.

Charley lived on this 160 acres until he and son Marsh moved to the town of Ong a few months before he died on Nov. 13, 1941.

Children Marshall and Frances Isham Storie were the owners from 1935 when a Federal Land Bank loan was obtained.

In 1963 granddaughter, Mary Alice Storie Foster and husband Wade became the owners. They have two sons, Morris and Clay Foster. Grandchildren are Emma, Henry, Rose, Ella and Mary Foster. *Submitted by Mary Alice Storie Foster.*

JASA – The Jasa family farm was first purchased by Peter and Antonia Jasa on May 2, 1910 from Chase and Clara Sliger. They purchased the land in Table Rock so they could be close to a school enabling their children to get an education. Peter and Toni lived on the farm until 1930. During that time they raised their four children; Ludvik, Millie, Louis and Loyd. While continuing to till the land, they rented the house out for a few years. Renters included daughter, Millie, and her husband, Roy. Also renting the house for a while was Doretta Burow, who later became Louis' wife. In 1950 Peter and Toni decided to return to this farmstead. They remodeled the house and moved back in time to celebrate their 50th Wedding Anniversary. Peter and Toni remained in the house until the mid 1960s, when they passed away within a year of each other. Son Loyd Jasa had lived with his parents and remained on the farm until 1980. Grandson Paul Jasa and wife, Lynda, moved their family onto the farm in August 1980. Paul and Lynda's family included three children: Caryn, Daniel and Joel. Paul and Lynda remain on the farm today. The farm house was again remodeled in 2015. The project was led by Jim Jasa Construction. Jim is a great-grandson to Peter and Toni.

Paul and Lynda hope the farm will remain in the family for many years to come.

JENSEN – Lewis C. Jensen homesteaded Section 28, Township 20 North, Range 39 West, Arthur County in 1913 and received the original patent April 29, 1918. Lewis Jensen died as the result of an accident in 1943. At the time of his death, Lewis had acquired an additional five sections. Lewis' widow Ruth F. Jensen and his children, John L. and Kaye F., continued to operate the ranch and Ruth added three more sections in 1946. Son John L. Jensen began operating the ranch in 1960 and acquired 3,960 acres in 1977, which brought the ranch to its present acreage of 9,720 acres.

Jensen sod house constructed prior to 1926; picture taken in 1928.

Lewis C. and Ruth F. Jensen

John and his wife, Shari, still live on the original homestead and part of their home is the sod house that was built by Lewis C. Jensen prior to 1926. *Submitted by John Jensen.*

JENSEN – The original Jensen family farm's first owner, Peter Jensen (Randy's great grandfather), acquired the 160 acre farm in Dixon County (NW 1/4 section, Section 23, Township T-27-N, Range SE) on June 28, 1910. In 1911 Peter's son, Henry Jensen, (Randy's grandfather) bought the farm. Randy's father, Lawrence Jensen, purchased the farm in 1970. In 1991 Randy and Lesa Jensen purchased the family farm. This 160 acre farm has remained in the Jensen family for 100 years in 2010. In August of 2011 we received the Nebraska Pioneer Farm Award.

The Peter Jensen home and outbuildings.

The original farmstead house burned down in 1933. Randy's parents lived in the original wooden corn crib on the farm while the current standing house was being built by Henry and Lawrence Jensen. This house was lived in by Lawrence/Lona Jensen for 46 years and Randy/Lesa Jensen have resided here since 1979. The 160 acre farm has been very well maintained throughout the past 105 years.

The 1920 big wooden barn still remains in great shape and is heavily utilized. The old wooden horse barn burned down in the late 1980s and hosted many neighborhood dances in its haymow during the 1930s thru the 1950s. Consequently, the haymow floor was very slick. This barn is depicted in the photo on the most right side. An underground cellar/cave with an arched brick ceiling is located outside the back door of the house. It is approximately 75 years old and still in use today. Also, a 1940s built hog house still remains intact on the farm.

Randy has restored his father's (Lawrence) first brand new John Deere 1937A tractor and it resides on the farmstead.

The family is proud and grateful of their legacy to be a Nebraska Historic Farm and hope for it to continue on for generations to come. *Submitted by Randall and Lesa Jensen.*

JESSEN – In 1900 Edwin and Barbara Ingwerson left their home in Avoca, NE and came to Lodgepole, NE. They stayed with friends for a few months, then in 1901 they purchased the current farm located 18 miles north of Lodgepole, NE, located N ½ of 14-16-47. The farm originally had a straw barn and soddie. The family of Edwin, Barbara and 8 children lived in the soddie while he first built the granary, then began building the current stone house in 1906. The older boys slept in the loft in the granary until the house was built. Edwin was a stone mason and limestone was plentiful just 3 miles north. He hauled the stone by team and wagon and cut the stone with hammer and chisel. The house is 1½ stories with 7 bedrooms and a large dining room. It was known as a mansion at that time. It also was known for their dances and parties, as the largest room, a dining room, contained a player piano and an organ. Edwin played the accordion and a neighbor played the fiddle. This farm was also used as an overnight stop when ranchers herded their cattle from Lodgepole to Lisco, where cattle were shipped on the railroad.

Jessen farm house

Edwin and Barbara left the farm to their 8 children, and Paul eventually bought the land from the other siblings. When he married his wife, Hazel Wood, he built a house on the farm, just 2 miles away. He left the stone house and half of his farm to his daughter, Joyce (Ingwerson) Jessen. She and her husband, B. Henry Jessen, lived there since 1955. They moved to Chappell for a number of years, then restored the house in 1984 and moved back to live there until 2005. In 2005, Henry and Joyce moved to Sidney, while their daughter, Nancy (Jessen) Rowlett, and her husband, Eldon Rowlett and their family moved to live in the stone house. Currently the farm is leased out growing primarily wheat, along with millet and milo. Eldon and Nancy, 4th generation ancestors, love living in the stone house, which, with just a few changes, has been kept primarily as it was when it was originally built. Their children and grandchildren (5th and 6th generations) love visiting "The Farm" to hunt, ride horses, collect eggs from the chickens, and just enjoy the peace and quiet of farm life.

JOHNSON – Otto Johnson immigrated from Sweden to the US in 1883. Alma Anderson immigrated in 1893. Otto's brother had married Alma's sister, and they lived in Malvern, IA. It was there that Otto and Alma met. They married in 1895 and Otto worked for the railroad. Two daughters and a son were born in Malvern.

Otto Johnson family, 1915. Back row: Frank. Middle row: Edna, Olga, Hilda, Alma Sophia holding Selma, and Otto. Front row: Margaret and Alma Wilhelmina

After living in Malvern and Oakland, NE they moved to Holt County, near Page where Otto farmed his older sister's land. Another baby girl was born there in 1902. He learned from two friends about the possibility of homesteading in southern Brown County. He and his friends traveled to the area to look it over. The friends did not see the possibilities, Otto did.

New frame house of Otto and Alma Johnson, ca. 1915.

When he arrived back to Page he corresponded with the US Land Office in Broken Bow inquiring about buying out the relinquishment to this land: the S 1/2 of sec. 15-26-21 and the N 1/2 of sec. 22-26-21, a total of 640 acres, about 20 miles south and four miles west of Long Pine.

The family traveled approximately 100 miles to their new home site in a hay wagon covered with canvas and pulled by a team of horses. When they reached the crest of the last hill on May 16, 1906 the only thing they saw was blackened prairie, there had been a prairie fire recently and the only thing left was the sod house that the previous homesteaders had built.

They soon dug a well and put up other buildings while still living in the sod house. A new four room frame house was built in 1912. Two more daughters had been born in the sod house, the seventh one was born in the new house.

The Sandhills ground is fragile, with only a thin layer of sod covering the sand. It is not suitable for farming, but Otto needed some cultivated crops. So as soon as they had arrived at their new home, he began to break up ground, using a plow pulled by a team of horses and guided by him. Each year he broke up a few more acres. Fifteen acres in 1906 produced 60 bu. of corn. The last year recorded was 1912 when he broke 15 acres and planted 70 acres to corn. He cut about 25 tons of hay each year and pastured about 20 head of horses and cattle.

In order to finalize the proof of residence Otto needed to report to the Land Office the amount of broken land, type of crops and livestock. Also required was a list of improvements and values: Frame house 16x24 (story and half) four rooms, cellar and cement foundation plastered, six doors and seven windows, brick chimney, shingle roof = value $600; frame barn 18x28 grooved roof = $75; frame granary 6x8 = $20; sod chicken house 12x14= $20; well and pump = $60; 1600 forest trees and 100 fruit trees and shrubbery=$500; 80 acres of breaking $240; 1-3/4 miles 2 and 3 wire fence - $75. Total = $1,590. This information was reported, begin signed by Otto and two witnesses. This was done in Ainsworth before the County Judge because of the long distance to Broken Bow or Valentine.

The "Final Proof" papers were finalized and Otto and Alma lived here until October of 1944 when Otto died and Alma moved into Long Pine, next door to their oldest daughter. *Submitted by Lorene Babcock Mercier.*

JOHNSON – In 1888 Stanislaus Krajeski homesteaded 160 acres located in Cherry County, Township 34N, 32 R, Section 24. Hay, clover, oats and millet were the crops grown and cattle, hogs, chickens and milk cows were the livestock.

In 1937 the farm totaling 680 acres passed to son Paul and his wife Ruth. In 1975 when the farm was incorporated as Krajeski & Johnson, Inc. it contained 2,550 acres. Shareholders were Paul and Ruth Krajeski, Susan (Krajeski) and John Johnson and Jim Krajeski. Mark Johnson (son of Susan and John) has purchased 1,100 acres of adjoining land. Current shareholders are Susan, John and Mark Johnson.

Now the farm raises cattle. Crops are corn, millet, alfalfa, oats, peas, beans and hay.

Stanislaus began sampling soil from Valentine to Nenzel. When he found harder ground, he chose it for the homestead, 3-1/2 miles southeast of Nenzel.

The 1903 Prairie square home was added on to in 1994. Complete renovation (gutted the original home) was completed in 2015 and an addition was built in 2015. The original granary, barn and cow shed are still in use.

Susan and John have four children: J.P. (Kimberly), Mark (Janelle), Chris (Dave) Christie and Katie (Dan) Blowers. Grandchildren are: Dalton, Danner, Addison and Aubree Johnson; Hannah, Jenna, Luke and Sara Christie; Cole, Carly and Cache Gracey; Bree and Brody Blowers. Susan and John both retired from education. *Submitted by Susan (Krajeski) Johnson.*

JOHNSON – William "Bill," the only child of Albert and Violet Johnson, was born in Holdrege, NE in 1938. Bill started farming with his father when he graduated from Holdrege High School in 1957 on the land his Swedish great-grandfather, John W. Johnson, bought from the Union Pacific Railroad in 1879 for $400.00. The 160 acres is located in Phelps County, NE at NE 1/4, 23-6-19 and owned secondly by John's son, Edwin, then his grandson, Albert and currently by his great-grandson, Bill. Bill's two sons and their sons currently farm with the family business thus making six generations of the Johnson family farming this same land.

Ruth and Bill (far left) with Brian, Julie, Justine, Ethan, Jeremiah, Blake, Grady, Quinn, Jett, and Holly, 2010

In 1960, Bill married Sharon McNiel, a high school classmate. In 1964 Bill and Sharon purchased W 1/2 SW 1/4 25-6-19. This property was the original site of a one-room school, Bunker Hill School District #77. Bill's dad, Al, attended this grade school. One original post from the school building remains on the property. The sign identifying the school was rescued from the burn pile by Al and hangs over the fireplace in the home built on the property by Bill and Sharon in 1965. Bill and Sharon have two sons. Brian, born in 1964, is married to Julie Olson, also a Phelps County native. They have three children - Ethan, Justine and Jeremiah. Blake, born in 1969, is married to Holly Peterson from Bertrand, NE. They have three sons - Grady, Quinn and Jett. The grandchildren currently range in age from 8-20.

At the present time, Bunker Hill Farms produces primarily corn and soy beans utilizing a no-till crop rotation plan. Irrigation is primarily with pivot systems. Brian operates a feed lot on the home place.

In the fall of 2008, Sharon passed away unexpectedly. Bill and Ruth Grove, also a 1957 Holdrege High School graduate, were married in the spring of 2010. Ruth's paternal grandmother, Alma Pauline Anderson Grove, attended Bunker Hill School in the early 1900s. Ruth's mother, Violet Abrahamson Grove, taught in Magill School District #26 located NW of Funk, NE in the mid-30s. Ruth attended District #26 for all eight elementary grades.

JOHNSON – John N. Johnson and Anna Peterson were married on Dec. 26, 1875 in Iowa. Several years later they came to Nebraska and settled in what is now northern Franklin County. On May 6, 1903 they bought 160 acres of land, the northeast quarter of Section 2, Township 4 North, Range 14.

The John and Anna Johnson farm.

Their first home was a dugout but when Anna became concerned to have coyotes come up to the entrance of the dugout and howl at night, she insisted they build a house, which they did.

They raised livestock, crops and a large family. Their 12 children were – Jens, John G., Falk, Olie, Fred, Hannah Christensen, Hulda Nielsen, Olga Nielsen, Tena (Frandsen) Madsen, Clara Hansen, Esther Ostergaard and Nora Jameson.

John N. died in July 1924 and when Anna passed away in 1936, their son Fred and wife Esther bought the farm, owning it until Esther died in 1992. Their son John N. and wife Euella purchased the land in 1993 and still own it.

This farm has been owned and farmed by a member of the family since 1903, currently being farmed by a great great grandson.

The family is proud of their legacy and hope it will continue for generations.

KAAN – On Nov. 4, 1911 Joseph and Martha Parson bought approximately 80 acres located in Sioux County, Township 31, Sec. 10, Range 56. Livestock was raised on the farm.

Frank and Vivienne Parsons acquired the property from Joseph. Frank acquired other land in Sioux County to build the existing ranch to approximately 16,000 acres. Beef cattle are now on the farm.

Frank died in 1987 and Vivenne in 1996. The farm then was inherited by their daughter, Cynthia (Parsons) Kaan.

Several homestead buildings still exist. The existing barn was built in 1950.

The Nebraska Pioneer Farm Award sponsored by Knights of Aksarben and Farm Bureau was presented to this farm.

Cynthia has two daughters, Amanda Kaan and Jennifer Van Dyk. Grandchildren are Dutch Van Dyk and Taylor Van Dyk. *Submitted by Cynthia Parsons Kaan.*

KAPS – Joseph Kaps (1867-1927) and Hattie Jilg (1874-1938) were both born in Sollnus, Austria and came to the US with their families as children. They were married in Kearney, NE on Dec. 7, 1895.

Bill Kaps

Joseph Kaps purchased 166.88 acres NW 1/4 S 1 T 12 R 18 in 1901 and 165.92 acres NE 1/4 S 1 T 12 R 18 in 1902 (Harrison Township, Buffalo County, NE). The house on the property was built by the Kaps family prior to 1907. Joseph and Hattie had 13 children: Hattie, Adolph, Mary, Amelia, Richard, John, Christine, Tracy, Joseph (died as an infant), Jessie, Clifford, Joseph and William.

William "Bill" Kaps (1917-1981) purchased the NW 1/4 on Sept. 6, 1938 from the Hattie Kaps estate, the S 1/2 NE 1/4 from his brother, Clifford Kaps, and the N 1/2 NE 1/4 from his brother, Joseph Kaps Jr. and wife in 1948.

After returning from serving in the Army Air Corps during WW II, Bill Kaps lived on the property raising cattle and sheep until shortly before his death in June of 1981.

In the summer of 1993 Alma (McCullough) Kaps, Bill's wife, and their son, David Kaps, remodeled the original Kaps house. David married Jill Hansen and has a son, Xavier William Kaps.

The Kaps family received the Nebraska Pioneer Farm Award from the Knights of Ak-Sar-Ben for 100 years of continuos family ownership in 2008 at the Buffalo County Fair.

KATHMAN – Joseph Kathman was born in Ludlow, KY in 1849. For 19 years he helped his father run the family dairy. He later worked in a foundry in Cincinnati, crossing the Ohio River daily on a ferry. Joseph later moved with his parents to Ferdinand, IN where they farmed. He married Elizabeth Heilers in 1870. They had five children: Ann (Sister Mary Afra), Mary (Mrs. Henry Schrandt) and Bernard. In 1876 Elizabeth died giving birth to twin boys who also died.

Joe married Katherina Goedtken of Festina, IA in 1878. To this union 11 children were born: William, Lena, Carolyn, Elizabeth, Dorothy, Herman, Edward, Emma, Joseph, Frank and Henry "Doug." With only 80 acres of land, they moved to Nebraska where land was more plentiful and cheaper. They had relatives by St. Stephens, so the family moved in 1893 purchasing land southeast of St. Stephens. The granary was first erected and served as their home. Second to be built was the barn, followed by a two-story farmhouse.

Their son, Henry, purchased the farm in 1936 and later sold it to his nephew, Irvin Kathman and wife Bertha in 1970. In 1988 the two-story house was taken down and great-grandson and son of Irvin, Don and his wife Usa moved a home built in the 1930s from the former Ralph Kometscher farmstead located one mile east, to the Kathman farm. They began farming the land in 1991. In 1993, Don and his three brothers built a Grade A dairy on their father's family homestead. Irvin passed away in 1995.

Early day picture of the Joseph Kathman farm. The building to the left served as the family home until the barn and house were later built.

Don and Usa purchased the 160-acre tract from Don's mother in 2005. They dairy, raise sheep, pigs and chickens along with dryland farming. Don and Usa are the parents of three boys - Brian (deceased), Seth, who is attending college in Branson and Lane, who attends Lawrence/Nelson High School. They are members of St. Stephen's Catholic Church and enjoy living close to their parish, family and hometown of Lawrence. *Submitted by Donald Kathman.*

KAVANAUGH – Owen and Margaret Riley Byrne from Roscommon/Leitrim, Ireland received a land grant in 1863 to homestead 680 acres in today's Plum Creek, Pawnee County, NE. Owen left the land to their four sons (Tom, James, John, Mike) and daughter Mary Byrne Donahue. He also donated the land for Burchard, NE's Sacred Heart Catholic cemetery. Tom, James, and Mike farmed side by side and their descendants still own the land. (Kavanaugh, Hart, and Sisco families respectively). John moved to O'Connor in Greeley, NE.

Mary and John Carson, Pauline Kavanaugh, and John Kavanaugh.

The Bryne – Carson – Kavanaugh home.

Tom and Susan Duffin Byrne farmed their acres with children Edward, Owen, William and Mary Pauline. When their sons moved on, Mary and husband John Carson continued to farm the land. Tom died in 1918, but Susan remained with Mary and John until her death in 1937. John and Mary successfully raised cows, pigs, chickens, vegetables and corn for the next 42 years. They lived their final years with their only child, Pauline Carson Kavanaugh, in Columbus, NE. Her son, John Kavanaugh, moved to the farm in the early '80s. He was well known throughout the county for his expertise in raising fruits and vegetables. John owned the farm from 2001 until his death in 2011. He willed it to his siblings Mary Kavanaugh Schaecher Brugman, Dan Kavanaugh and Maureen Kavanaugh Konwinski, who named it the Byrne, Carson, Kavanaugh (BCK) Farm in honor of its successive family owners. *Submitted by Maureen Konwinski.*

KEEFER – Paul Emmanuel "Peter" Keefer, with sons, Paul E. and Fred, and daughters Polly and Minnie, came from Germany in 1884 to Sarpy County near Papillion and purchased an 80 acre farm. His wife died shortly after arrival. Paul E. married Bertha (Schmidt) and they purchased a second 80 acre farm, one mile from the original. Fred, who never married, continued to farm the original 80. Paul and Bertha had seven children and had an orchard with many fruit trees. They hauled the produce to the market in Omaha with horse and wagon. Their son, Herbert, took over the farming at age 20 because of the declining health of Paul E.

Herbert Keefer with daughter Elinor, 1942

The original 1884 farm in 1994, owned by Harold and Susan Keefer

Herbert married Ruth (Sprague) in 1938. They lived upstairs in the farm house until 1945 when they bought the original farm from Fred, then added an additional 80 acres. They ran a grade A dairy there until they retired in 1979. They had one of the first milking parlors in the state and went to a bulk milk tank system after a disastrous barn fire in 1955. They had three children - Elinor (Groth), David and Harold.

Harold and his wife, Susan (Okeson), were in partnership with Herbert and Ruth from 1973 until Herb's retirement in 1979. They continued the dairy operation and increased the herd size to 65 cows until 2001 when they retired from the dairy business. They purchased the 160 acres (16-13-12) from Herbert and Ruth in 1984. They continue to farm 240 acres today. They have three children - Sheri (White), Amy (Kingston) and Debra (Mangano).

David and Harold bought the second 80 acres (17-13-12) when Ruth died in 2009, from Elinor and Ed's children - Mark, Steven, John and Jane (Tucker). Elinor died in 2006.

David and his wife, Sheryl Sue (Ehlers), live on a corner of that farm and run a few head of cattle. They have two children, Daniel and Kimberly (Fluckey).

The metropolitan area of Omaha is getting closer to the farms by the day, so we may be the last generation to farm here. *Submitted by Harold Keefer.*

KEHLBECK – Henry F. Kehlbeck was born March 7, 1857 in Bruchmuhle (Bruchhausen, Hanover) Germany. In 1885, he moved to America and worked for a farmer near Avoca (Cass), Nebraska. On November 17, 1892 he married Alvina A. Peters, daughter of Peter and W. Magdalena (Schomaker) Peters. She was born on the Peter Peters farmstead (Township 10, Range 12, Section 21, SE1/4 in Avoca Precinct, Cass County) on September 8, 1868. Peter Peters and his wife settled in territorial Cass County in 1859. Peter became a prominent stockman and farmer.

Henry Kehlbeck Descendants (mid-1980s). Front row: Kim Stohlmann Roecker, Craig Stohlmann, Gary Lenker, Becky Lenker. Middle row: Rosie Kehlbeck Rohlfs, William Kehlbeck, Minnie (Leefers) Kehlbeck, Clara Kehlbeck Steinhoff, Arthur Steinhoff. Back row: Lanny Lenker, Linda Rohlfs Lenker, Kirby R. Drake, Cindy Steinhoff Drake, Shirley Steinhoff Drake Stohlmann, Danny Stohlmann. Not Pictured: Fritz Rohlfs (deceased), Quentin A. Drake (b. 1990).

Henry and Alvina's only child William was born July 16, 1897 on a farm between Avoca and Nehawka. On January 3, 1898, the Kehlbecks purchased the SW1/4 of Section 22, Township 10 N, Range 11 E (Weeping Water Precinct, Cass County) from Jacob Schwab (today 149.96 acres). They moved to the farm as it was and by 1905 had completed building the present two story wood home. During these early years Henry transferred cedar seedlings from north of the house to along the driveway and around the house. The trees have survived and are a well known sight in the community.

The Kehlbecks were members of the First Lutheran Church of Avoca (North Branch) located across the Cass and Otoe County line in Otoe County on Highway 50. Their great-granddaughter and her family still attend this church.

On April 12, 1917, William Kehlbeck married Minnie Leefers (b. July 27, 1896, daughter of Henry and Margareta (Knickman) Leefers, Unadilla). On November 4, 1918, their only children (twin daughters) Rosie and Clara were born on the farm.

Alvina Kehlbeck died March 3, 1934 and the following day Henry deeded the farm to his son William. He lived with his son and family the rest of his life and died July 1, 1940. They are both buried in the North Branch Cemetery.

On January 24, 1940, Rosie Kehlbeck and Fritz Rohlfs, (b. March 17, 1916 d. February 5, 1980, son of Harry and Nellie (Stubbendick) Rohlfs, Otoe), along with her twin sister Clara and Arthur F. Steinhoff (b. March 7, 1920, son of William and Magdalena (Buckman), Steinhoff, Avoca) were married in a double wedding ceremony at First Lutheran Church.

In 1946, William and Minnie Kehlbeck retired and built a 1 ½ story brick home in Syracuse. William and Minnie made frequent trips back to the farm where William continued to trim the cedar trees just like he had done his entire life. They lived in Syracuse until age related ailments required that they move to a nursing home. William was moved to Nebraska City first and Minnie joined him years later. Minnie passed away September 24, 1994 and William September 17, 1995. They are both buried In North Branch Cemetery.

After William and Minnie moved to Syracuse their daughter Rosie and her husband Fritz, along with daughter Linda Kay (b. December 1, 1942, m. Lanny Lenker, December 20, 1963, b. January 17, 1942, d. December 27, 2008, son of Reed & Bernice (Hillman) Reed, Dunbar) moved to the farmstead. The Rohlfs retired from farming in 1973 and moved to a home in Syracuse. The Kehlbeck's rented out the farm until 1975 when they sold the farmstead to their daughter and husband Clara and Arthur Steinhoff.

The Steinhoff's had two daughters, Shirley (b. June 4, 1944, m. Danny Stohlmann November 15, 1963, b. January 14, 1943, d. March 13, 1994, son of Ellsworth & Norma (Luby) Stohlmann, Louisville-see Frederick Stohlmann Family) and Cindy (b. December 23, 1956, m. Kirby R. Drake October 7, 1978, b. January 10, 1954, son of Arthur & Gwineth Drake, Elmwood). Cindy and her husband Kirby moved into the Kehlbeck home after their marriage.

Arthur and Clara continued to live on the Buckman/Steinhoff farmstead (located in the opposite corner of the Kehlbeck farmstead) until the mid-1990s when they retired and lived in Lincoln, Syracuse, and finally a nursing home in Ashland. Arthur passed away June 21, 2003 and Clara passed away July 31, 2006. Her sister Rosie Rohlfs passed away November 15, 2006. They are all buried in North Branch Cemetery.

Cindy and her husband inherited the Kehlbeck farmstead and the farming operation was handled first by her sister Shirley and husband Danny, and today the land is being farmed by the Drake's nephew Andrew Stock from Murdock. The Drake's have one son, Quentin Arthur Drake (b. July 5, 1990). The surviving descendants of Henry and Alvina Kehlbeck are their three great-granddaughters, Cindy, Shirley, and Linda along with their families.

The farm has now been in the family for almost 118 years. The present home contains portions that are over 135 years old. The many buildings on the farmstead were built by/for Henry and William Kehlbeck and a new shed was added by the Drake's in 2011. The farmstead was placed on the National Register of Historic Places in 1985 and received the Ak-Sar-Ben Pioneer Award in 1998.

KEIM – Dennis Keim acquired 160 acres in 1888, located in Nuckolls County, Section 28-4-5. Crops were corn, wheat, oats and alfalfa. Livestock was cattle, hogs and chickens.

The farm passed from Dennis Keim to his son Will Keim, then to his son Darrell Keim and in September 2005 to Larry Keim. This farm now contains 608 acres and produces corn, soybeans, alfalfa and cattle.

Aerial view of the Keim farm

A tile silo and the house remain on the property. Darrell Keim received a Conservation Award from Aksarben. The family of Larry Keim received the Nebraska Pioneer Farm Award for over 100 years of continuous farming and ownership. The Keims are still actively farming.

Larry and Marilyn Keim have four children – Tammy (Monte) Trampe, Cindy Hutchinson, Amy (Kevin) Wenske and Ricky (Natasha) Keim. They have nine grandchildren – Tristen Trampe, Wendy Levering, Meisha (Nate) Jackson, Wade Wenske, Tanner Wenske, Ezra Keim, Finley Keim, Lennon Keim and Abel Keim. Great-grandchildren are Shaelyn Levering, Shamar Levering, Shontrell Levering, Shakeir Levering and Shyler Levering. *Submitted by Larry Keim.*

KERBEL – Prior to coming off the railroad in 1892, this property was an overnight stagecoach stop on the Gross-O'Neill route, a trip that took two days one way. The men slept in the haymow of the barn and the women slept in the house.

Barn on the Kerbel farm

Passengers and freight were brought up the Missouri River by steamboat and unloaded at Tower near the Iron Post landing, and freighted up the Nine Mile Gulch and Nine Mile Hill to Gross and then across the hills to O'Neill.

The outside wall of the original house, which is still standing and lived in, was concrete with a 10-inch dead space between the outer and inside walls. Two additions have been added to the house over the years.

Lydia Stearns was the first person to homestead this 165-acre farm in section 35, Township 34, Range 12 in Boyd County under section 2291 of the revised Statutes of the United States, on November 14, 1906. The land next passed through the hands of R.C. Peters, John V. Hellman, F.E. and M.V. Railroad Co., and William D. Bradstreet.

Our grandmother, Francis (Frances) Kerbel and her husband, Joseph, purchased the land from William and Bertha Bradstreet on June 21, 1909. The Kerbels were parents of eight children: Stacia, Agnes, Emma, Alice, Julia (stillborn), Mary, Rose, and son, Ray.

Besides farming, they operated a dairy—bottling and delivering milk in the nearby town of Spencer.

Ray was in the eighth grade when his dad had an accident while driving the water wagon down a hill and broke his back, which at time he quit school to operate the farm.

When the grandchildren graduated from country school and started high school in town, they lived with their grandparents. One grandson said, "Grandpa's main job was getting everybody up at 5 in the morning to do chores, milk cows, fill and deliver milk before going to school."

Ray married Mary Koerber on Dec. 2, 1941, five days before Pearl Harbor in O'Neill, Nebraska. Both Ray and Mary spent their entire married life on the farm. Their children are Marita, Joe, Judy, Jeanine, and Louis. They have eleven grandchildren.

Ray inherited the farm and continued operating the dairy until stricter regulations required too expensive upgrades. He kept busy raising cattle and hogs, and crops of corn, oats and alfalfa. He also learned and worked as an electrician when electricity arrived.

Mary kept a large flock of hens and sold eggs. Every spring she raised, butchered and sold about 500 fryers, ducks and geese.

Although Mary was looking forward to seeing the farm receive the 100 year Pioneer Farm Award, she passed away in 2007, two years too soon, leaving the farm to five children, which is now known as Five K Farm, LLC.

KILPATRICK – James Kilpatrick came to Antelope County from Phelps County. He purchased this 160 acre farm located in Antelope County, SW 1/4 Section 35, Township 27, Range 6 in March 1903.

He and his wife, Mary Jane, had three children – Fern, Cloyd and Clara, twins, age 6. They slept in the corncrib the first summer as there was not room in the house for the whole family.

The Kilpatrick homestead in 1902. Pictured are James (father), Cloyd, Fern, Clara, and Mary Jane (mother).

Crops were corn, oats and hay. Livestock was cattle, hogs and chickens. Cloyd inherited the farm after the death of his mother in 1953. He continued to live there.

Cloyd and his wife Hope had two children, Cloyd Jr. and Elaine. He farmed with his son. He purchased the farm April 1973. Cloyd Jr. was born on this farm.

Cloyd Jr. and his wife, JoAnn had two sons, James and Steve. James remained on the farm and farmed with his father. He purchased the farm in March 1996. He continues to farm this 160 acre farm where corn and soybeans are raised now.

Across the road is the Willowdale Cemetery. Many of the family are buried there.

Kilpatricks have lived here since 1903. *Submitted by James R. Kilpatrick.*

KIME – Clyde Kime left his ranch north of Mullen to search for new land. He rode 34 miles before he found the place to call home. The land was purchased Jan. 8, 1913 from part of Wyoming Cattle Company for $5,500 - $8.59 per acre. Clyde built a barn, which was the family home until a house was built. When Clyde returned to Mullen to gather his family, he was caught in the March 1913 blizzard. Clyde drove a freight wagon with four head of horses while his wife, Maude, pregnant with their fourth child, drove a buggy with another two horses and the milk cow hitched to the back.

When they first arrived, the entire valley was nothing but a prairie dog town. The Kimes immediately started building a house and by winter of 1913 the first part of the house was enclosed. Clyde had 640 acres and just one milk cow. During WW I he raised horses and sold them to the government at Fort Robinson.

Clyde Kime

On Oct. 27, 1916 Clyde was named postmaster of the Lund Post Office. The post office remained at the ranch until July 31, 1930. In 1917 they bought a one-wire hand crank telephone – the only telephone in the area. They also had an ice house, which was a hole in the ground about 12x12.

The ranch is now owned by cousins Steve and Don, both grandsons of Clyde and Maude. While they enjoy looking back on the last 100 years of their family's history, they are hopeful to keep it going for another 100 years. *Submitted by Donald E. Kime.*

KNIEP (WENSKE) – The original owner of the Kniep Family Farm was J. Henry Kniep, grandfather of the present owner, who purchased the NE 1/4 of Section 29-14 in Thayer County near Byron from John Friedrich Eitzmann and wife, Ida, on Nov. 21, 1902.

The Wenske family in October 2015. Pictured are three generations who grew up on this farm: Susan, David, Luetta, Kalen, Alan and Linda.

Some years passed and he began developing the building site in preparation for his son, August, to live there. Soon after August's return from serving in the US Army in WW I, he was married to Anna Ottemann in 1919. On May 19, 1920 J. Henry Kniep, a widower, passed the property on to them for $1 with love and affection. They both worked side by side farming and raising livestock, making the transition of farming with horses to tractors and other machinery. He also erected a wind charger to supply their own electricity. They raised four children: Wilber, LaVern, Loren and Luetta.

In March of 1953, they decided to retire and move to Byron. Their daughter, Luetta, husband Ralph Wenske and their daughter moved on to the place from the Edgar area. In 1959 August and Anna Kniep gifted the east half of the quarter to their daughter, Luetta and Ralph. In 1965 the west half was deeded to them. After some testing for irrigation and it deemed practical, the farm was converted from dryland to irrigated, thus making the land more productive and valuable. Later the pipe gravity method was replaced by a center pivot. Livestock was always a part of the operation. They raised four children: Linda, David, Susan and Alan. More land has been acquired over the years to expand the operation to 920 acres of cropland, pasture and hay. Presently, David is farming the land and Alan resides on the place.

KOEPKE – On July 19, 1889, Ludwig and Ulricke Koepke purchased 160 acres in Wayne County, seven years after immigrating to America from Germany with three children. Three more were born after arriving in America, with the youngest named Otto.

Otto grew up and married his wife, Ida Grimm, in 1915. They lived on the family farm with his parents. He drove a Model T Ford. They farmed with horses, using a walking plow and planting corn with a horse drawn planter. Crops planted were corn, oats, barley, and prairie hay. They raised cattle, hogs, chickens, ducks, and geese, while also feeding the horses. His first tractor, purchased in 1936, was a John Deere D to be used for plowing. They had three sons, with the youngest named Lester. Otto farmed with his father until the death of his mother in 1916. Ludwig then moved to Hoskins and sold the farm to Otto in 1932.

Lester married Irene Uttecht in 1942. They lived with Otto and Ida for the first four years, the first two of which Lester served in the Armed Forces during WWII. In 1946, Lester moved to Norfolk to pursue full time work. In December 1958, Otto and Ida moved to Norfolk, and they turned their farm over to Lester and Irene, who raised a son and a daughter. Irene raised 500 laying hens each year, from which she sold eggs to pay the bills. Lester fed out cattle while also growing crops. They farmed until they retired in 1982, and moved to Norfolk in 1987.

Aerial view of the Koepke farm.

The Koepke family in 1903.

Lester's eldest grandson, Darin and his wife, Kelli (Winder), were married in July of 1988 and rented the farm from Lester. They raised 200 head of ewes and had feeder hogs and cattle. They had two daughters. In September of 1998, they moved to Hadar to pursue full time careers.

In 2001, Lester's son, Larry and his wife, Diane (Koehler), moved to the family farm and primarily raised corn, hay and oats while running a successful Simmental cow/calf operation. Diane passed away in October of 2015. The farm currently has a 100 Simmental cow/calf operation and a 75 Ewe/Club lamb operation. Larry still farms with the help of his son, Darin.

Lester passed away in 2008, leaving Irene owner of the original 160 acres. She is 96 years young, and can tell stories of life on the farm. The original house still stands along with the original horse barn. We still do our own family butchering each year of cattle, hogs and chickens. We can't imagine life any other way.

KOHRS – My farm in Nemaha county, situated two miles northeast of Johnson was homesteaded by Benjamin & Mary Fredenburg in 1867, which they owned until 1900 when my grandparents, William and Emma Rogge Kienker purchased the farm for $8000 ($50 per acre). My grandparents owned the farm through good times and bad. They would not think of going through a bankruptcy during the time of the Great Depression. They were of the old school—if you made a deal you kept your word and worked it out. They both died in the 1940's and my mom and dad, Herman and Alice Kienker Kohrs, purchased the farm from their estate for $100 per acre [$16,000] for the 160 acre farm. Dad farmed the place until 1974 when he rented it to the Woerlen family. My folks died in 1993 and I inherited the farm from their estates. It is still rented to the Woerlen family. Farm taxes today are getting close to what my grandparents paid for the farm.

William and Emma Rogge Kienker

My dad was a pioneer in soil conservation practices being one of the first to use contours and terraces. The farm is completely terraced and tiled to control erosion and make it easier to farm. (In 1994 the families of Herman & Alice; Sterling & Naomi of Beatrice, David & Linda of Omaha, Keith & Lorraine of Pueblo Co established a memorial scholarship at the University of Nebraska. It is to help students studying agriculture science and natural resources).

Through the years the farm has produced wheat, corn, oats and soybeans. Now corn and soybeans are strictly rotated each year. There are no buildings on the farm. A 20 x 40 foot building was on the property when my grandparents purchased the farm. It was probably built to comply with homesteading laws. My mom was born there. It was moved to an adjacent farm in the 1930's. This farm is considered a raw quarter, has no fences and has had minimum livestock.

The farm has a name. We lived on another farm three miles away. When my dad would tell us to go to the other place to work, we knew where to go. Hence the name—The Other Place.

KORTE – As a 1986 recipient of the Nebraska Pioneer Farm Award, we share our story. Carl Korte Homestead located 12 miles northeast of Columbus, Platte County, NE was purchased by Carl L Korte on May 13, 1886 for $ 17.50 an acre. Korte, a native of Germany, came to the US when he was 17 years old. He and his wife, Elise Viergutz reared Otto, Ernst, Walter, Arthur, Ella, Frieda, Anna.

A son, Ernst, remained on the farm and married Louise Otte in 1917. The couple's children are Edgar, Elmer, Allen, Norman, Harry, Evelyn and Dorothy. In winter, a beef and three hogs were home butchered and hams, bacon and sausage cured in the smoke house, along with a big garden and fruit trees kept the big family fed. Milk cows were a major part of farm life. Cows were milked by hand until electricity came in the late 1930s and the milking machines.

Ernst's son, Norman, married Elaine Bremer in 1962 and acquired the farm and is living in the house that Carl built in 1905. Their children are Rita, Lisa and Jonathan. They have nine grandchildren, who are fifth generation - Amber, Dustin, Josh, Justin, Laura, Brant, Jordee, Katee and Rory. Cows were a major part of farm life from the beginning until the

Carl and Ernst Korte model some of the most reliable farm equipment in 1923, a few work horses and a spring buggy. Carl Korte founded the farmily's farm in 1886 and Ernst took it over after Carl retired.

The Korte family with the 3rd, 4th, and 5th generations taken on the site of the original farm homesteaded in 1886, on the occasion of Norman and Elaine's 50th wedding anniversary, 2012.

end of 2007, at age 80 when Norman semi-retired. The milk marketed thru Dairy Farmers of America by our small grade A dairy.

Jonathan married Shelly VonSegeren in 1998, inherited and continues to farm the original 80 plus 80 added in 1887 plus other owned and rented farm ground, which now totals 880 acres. After the milk cows left, Jonathan invested into a hog enterprise. The farm is still a diversified operation, using a soil conservation program of crop rotation and no till, growing corn, soybeans, alfalfa hay, some under pivot irrigation and 100 head cow/calf

Norman and Elaine continue to enjoy country living in their century old farm house. Their son Jonathan and wife Shelly and family built a home on the original 80 acres. The family is proud and grateful of their legacy and look forward to celebrating as a 150 year Heritage Farm in 2036. *Submitted by Norman and Elaine Korte.*

KRAPP – Brothers 17 year old Joseph and 12 year old John Bonifaz Krapp came to the United States in 1847 from Hessen, Germany. Joseph was born June 26, 1830 and John, February 8, 1835.

Joseph married Anna Marie Vollert in 1858 in Pittsburg, Pennsylvania. Anna "Mary" was born in Hessen, Germany April 8, 1830. Ten children were born to this union; Salome, Mary, John, Lena, Charles, Frances, Peter. Three died in infancy.

Joseph and Anna, along with Joseph's brother John were pioneer settlers in Minnesota, Wabasha County, West Albany Township, near Lake City in 1861. Joseph and Anna's children were born in Minnesota, except for their youngest, Peter, who was born in Nebraska March 24, 1873 on the homestead.

Joseph Krapp

In the spring of 1872, several years after Joseph first visited Nebraska and after Joseph and Anna's daughter Frances was born on September 13th, 1871, the family acquired a covered wagon and began their long overland journey to Nebraska, Gage County, Highland Township, Section 10, Township 6 North, Range 6 East to homestead.

John's homestead application for 80 acres (10-6-6 West half of Southeast Quarter) was signed 8/01/1872.

John B. Krapp

Joseph's application for 80 acres (10-6-6 East half of Southeast Quarter) was signed 6/28/1873.

By August 1872, John had built a 12 x14 foot one-story house, and a 12 x16 foot stable. Joseph built his house in the summer of 1873. They improved the land by building a granary, corn cribs, and a well; planted corn, wheat, oats, and barley; and set out about 100 fruit trees and one acre of forest trees.

In 1882, Joseph's home served as a chapel for the first Catholic Mass in the area. Father Wunibald Wolf came from Crete for this occasion. This was the beginning of St. James parish.

John never married and farmed with Joseph and family until his death from a heart attack on November 19, 1902. Joseph died at home January 20, 1916, Anna on February 18, 1903.

Joseph's son, John, was born April 18, 1864 and to whom Joseph and John's original homesteads were bequest. For clarity, he will be referred to as John Jr. He was seven years old when his family arrived in Gage County.

At the age of sixteen he made his first investment in land by purchasing eighty acres of railroad land in Section 11 Highland Township part of which was platted by the village of Cortland. By 1902 he had accumulated his holdings to 800 acres. Upon his death he owned 1,120 acres.

When John Jr. inherited the two homestead 80's the legal description changed slightly to become SEC 10-6-6 N ½ SE and SEC 10-6-6 S ½ SE. A hardworking, prosperous farmer and leader in the community, he was known as "Grandpa" to the people of Cortland and actual grandfather to fifteen grandchildren.

Krapp home place

On September 15, 1904, John Jr. married Clara B. Doyle, a former school teacher. To this union were four children, all born in the family home: Dorothy (Grady), Leland, Neil, and Alice (Weber). John Jr. passed away on February 21, 1942. Clara was born May 8, 1874 and died on June 24, 1944.

Neil John Edward Krapp, was born on January 19, 1910. A lifelong farmer, Neil enjoyed fishing and was an avid golfer. On January 16, 1940 he married Helen Vera Huetson at St. Mary's Cathedral in Lincoln, Nebraska. Helen, the daughter of Chester and Minnie (Sachtleben) Huetson was born January 16, 1917 on her family farm located two miles east in Nemaha Township, 7-6-7. To this union two children were born, Teresa Joan (July 18, 1954) and Randall (Randy) Wayne (January 8, 1956). The family changed their last name to Edwards in 1966.

The stately six bedroom old house with majestic roller doors separating the parlor from the dining room was torn down in 1952/1953 and in 1954 their one-story stone ranch was finished and furnished with a few treasured furniture pieces from the old house. The farm included two ponds, one creek (Indian Creek), three barns, feed lot, workshop, machine shed, and brooder house. Sweet clover, alfalfa, corn, soybeans, wheat, and sorghum have all been planted and until the 1970's. There were always a herd of Black Angus cattle.

Randy, like his grandfather John Jr., bought his first 80 acres at a young age. He was 20 years old when he purchased 80 acres of land one mile north of family farm on February 17, 1976. He also created the Edwards subdivision to Cortland, which was platted on March 9, 1977.

Neil passed away December 5th, 1980; Helen on June 11, 2002. Randy was killed in a car accident on July 20, 1980.

The oldest of John's children, Dorothy Krapp, was born Feb. 19, 1906. Married in 1926 to Richard Grady she had two sons, John (known as Jack) and Duane. Dorothy wore many hats supporting her family during her life including working at Bauer's Chocolates in Lincoln and as a vending area attendant at the Lincoln train depot. All that knew her considered her a gentle soul and kind person willing to do anything to help another person. She was especially fond of spoiling her grandchildren and miniature dachshund "Shorty." Dorothy's land passed to her son Jack and then to her grandchildren John, Dennis, and Patricia, the current owners.

Alice was the youngest of the Krapp children born on May 28, 1913. In 1934 Alice graduated from the University of Nebraska-Lincoln and worked for C.I.T. Corporation. Alice married Ray Weber and the couple spent their working years in Columbus, Ohio. Ray worked for Western Union and Alice as an underwriter for Midland Mutual Insurance. The couple returned to Lincoln for their retirement years. Ray passed away in 1986 at 74 years of age and Alice in 1993 at age 80. Alice's land was passed on to her nephew John (Jack) Grady and then to the current owners, her grandnephews and grandniece, John, Dennis, and Patricia Grady.

John Jr.'s land passed to his children. 720 acres remain in the hands of his descendants. 400 acres including brothers Joseph and John's homestead land (which has remained in the family over 140 years) is currently owned by Neil's daughter Teresa Edwards McHargue. It is farmed by descendants of Joseph's daughter Frances – her great grandsons, Gerald Behrends and Gary Behrends and Gary's sons Cody and Brett. Crops planted today are a rotation of corn and soybeans.

Dorothy's and Alice's land is cared for by Dorothy's grandchildren, John (Mick) Grady, Dennis Grady, and Patricia (Grady)

For more history on the Krapp family refer to O'Brien (Phillip and Frances) and Huetson-Edwards in this book.

KREJCI – George W. and Jane E. Berry purchased 640 acres in Homestead County, Sec. 35-22-50. This purchase occurred in 1906 but was not recorded until 1912. The farm was used for pasture and cattle and chickens were raised here.

Building the house on the homestead of George and Jane E. Berry. Pictured l-r: Mable Berry on bench, Walter Berry on Pet – the horse, Bro. Robinson the plasterer, Frank Berry with the hoe and shovel, and Bro. Smith on the side of the house. The chicken house in the background is 15' x 30'. The built the chicken house first and lived in it until the house was built.

The farm passed from George and Jane to sons Walter and Frank Berry. Frank bought Walter's share and then passed the farm to his son, Dale. In 1979 George and Belva (Berry) Krejci bought the property after Dale's death. The farm retains 630 acres and now produces wheat and is used for pasture.

A cement chicken house was built and lived in until a two story frame house was built.

Belva and George have three children – Dwight, Lavon and Sondra. Brenden, Tanner, Natalie and Caleb are their grandchildren. *Submitted by Belva (Berry) Krejci.*

KREJCI – In 1889 Joseph and Marie Krejci homesteaded 160 acres in Sheridan County, Section 6-27-46. Cattle and chickens were the livestock raised here and wheat and corn were the crops grown.

The farm passed to the Krejci's son Edward and then to George, Edward's son. George now owns a total of 940 acres where livestock is cattle and sheep. The land is used for pasture and wheat and alfalfa are the grain crops grown.

Krejci family, 1898-99. Pictured from left: Mary on horse, Father Joseph, Baby Frances, Bessie, Mother Marie, Rose, Anna Krejci (aunt), Joseph on horse, and Edward. In the back is Frank standing in the horse and buggy.

George obtained the farm on Feb. 23, 1971 but has lived here all of his 80 years. He is still farming and also restores antique tractors.

The dugout sod houses remain on the land plus the frame house that was moved from Antioch (potash boom) remodeled and built onto.

In 1989 the Aksarben 100 year centennial celebration occurred.

George and Belva have three children - Dwight, Lavon and Sondra. Their grandchildren are Brenden, Tanner, Natalie and Caleb. *Submitted by George Krejci*

KRIEBEL – Henry E. Kriebel and Elsie M. Daubert purchased the 160 acre farm in Burt County, described as SW 1/4 of Section 32, Township 21-N, Range 10E, from Henry E. Smith and Lydia H. Smith, H & W, in January of 1912. It was recorded in Book 19, page 78 of deed records on March 6, 1913. It was known locally as the "Glendale Farm."

An early threshing machine and crew. Seated at left is Henry with the dog, Buster. Standing in the middle is his brother-in-law, Frederick Daubert and other neighbors assisting with the harvest.

They had one son, Kenneth and one daughter, Roma. Henry was a breeder of purebred Poland China hogs and Plymouth Rock chickens. He attended the Fremont Normal School for three years. He served as Township Clerk, Treasurer and Assessor in Summit Township.

Their son, Kenneth, purchased the farm from his parents on March 1, 1943.

Kenneth F. Kriebel and wife Hildreth E. Kriebel conveyed the farm to their daughters, Elaine K. Chamberlain and Carolyn A. Nordberg Nov. 30, 1987. The daughters continue to own the farm today.

As was customary at the time, the owner of the machine would thresh the neighbors' grain in exchange for help with the harvest. The small grain would be oats or wheat. Henry and son Kenneth owned one of these threshing machines.

The Henry E. Kriebel farm received the Pioneer Farm Award from the Knights of Aksarben in 2012. *Submitted by Elaine K. Chamberlain.*

KROV – My grandparents, Anton J. and Anna M. (Malovec) Krov purchased the north half of the southeast quarter of section 8, township 16 north, range 4 in Butler County containing 80 acres on Dec. 27, 1909 from Julius and Frantiska Ustohal. Possession was granted March 1, 1910. They paid $52.00 an acre ($4,160.00); by 1933 the loan was paid. There were four children – Frank J. Krov, Rozvita B. Krov, Martha K. Krov, Joseph F. Krov. On May 24, 1919 construction of a two-story house began and was completed by Oct. 10, 1919. Anna's brother,

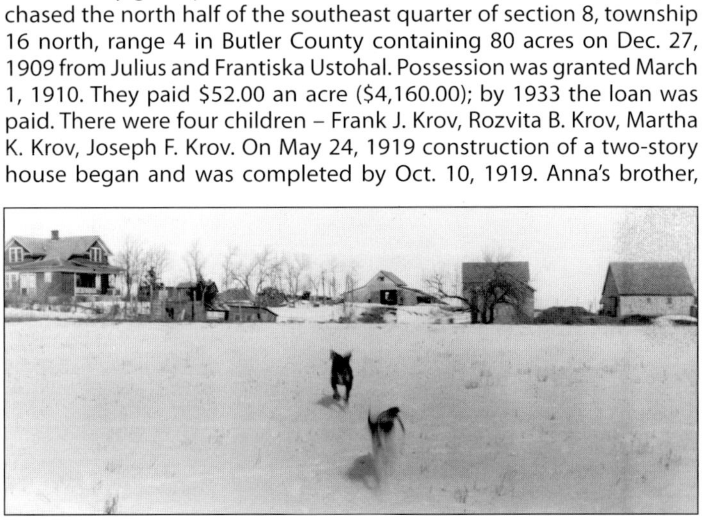

The Anton and Anna Krov farm

Louis Malovec, came to live with them to help with the farming. Crops raised were corn, wheat, oats, hay. Livestock included were cattle, horses, hogs, chickens, ducks. Eggs were sold for grocery money. There was a large orchard, vegetable and flower gardens. Although Grandma didn't like to share her "special flowers," she gave my mom the "Double" orange day lily by mistake! Oops! Today Grandma's lilacs, peonies, irises, day lilies, roses are still enjoyed.

My parents, Joseph F. and Adeline C. (Kutka) Krov were married April 10, 1944 in Fremont. We lived on the farm until the folks rented the Dr. J.A. Jelinek farm. On May 21, 1951 the 80 acres was deeded to my Dad by his folks. Woodrow and I inherited the farm in 1996 after Dad died. The farm is rented to a neighbor; corn and soybeans are raised.

We were awarded the Nebraska Pioneer Farm Award – Ak-Sar-Ben Foundation and Nebraska Farm Bureau on July 24, 2011 at the Butler County fair in David City. *Submitted by Woodrow and Kathryn (Krov) King.*

KRZYCKI – John Krzycki was born in Posen, Poland, Jan. 20, 1847. He was fourth in order of birth in a family of 10. As a young man John pursued his education, eventually becoming a teacher in the public schools of Chodzicz, Poland, but the reports which he heard concerning the opportunities and advantages of the new world led him to bid adieu to his native country and sail for the US in 1870. He first settled in Indiana, then coming to Columbus, NE in 1871.

On Aug. 1, 1871 he married Barbara Borowiak, also a native of Posen, Poland. To this union 13 children were born: Frank, Saloma, Alex, John, Kumecilla, Barbara, Laudislaus, Anton, Vincia, Anna, Elizabeth, Theodore and Phillip.

On Dec. 20, 1900 John and Barbara purchased a 160 acre farm near Farwell, NE from Elizabeth McCormick for $4,000. This was the beginning of the Krzycki Family Farm that would be passed down through the generations.

John and Barbara sold the farm to their son Alex and wife Maggie on April 4, 1902, for $2,000. They had four sons - Vincent, Marion, Alex and Roman. In June of 1914 Alex died, leaving Maggie to raise their four sons. Maggie later married Joseph Lukasiewicz

When Vincent married Lottie Curtis, they lived on the farm raising their two sons, Lavern and Adrian. On June 21, 1930 Vincent died on the front porch. His wake was held at the home.

After Lottie and sons left the farm, Vincent's brothers, Alex and Roman, batched there until Roman married Hazel Sonderup April 11, 1940. Roman and Hazel raised two sons, Robert and Edward, on the farm till 1945 when Roman's health required him to quit farming.

On Nov. 10, 1952 the son's mother, now Maggie Lukasiewicz, deeded the land to her three living sons - Marion, Alex and Roman and to Vincent's two sons, Lavern and Adrian.

The farm was rented out to several renters for the next few years, until May 14, 1966 when Roman and Hazel's son, Robert, married Patricia Kusek and they started their married life there.

Robert and Patricia had one son, Tracy, and twin daughters Penny and Peggy, whom they raised on the Krzycki Farm. Robert and Patricia still reside on the farm, still making memories and enjoying their retirement years there. Their eight grandchildren - Andrew, Nathan, Benjamin, Amy, Carson, Erica, Alex and Brogan - have all enjoyed spending time on the farm, playing in the dirt, many long walks exploring and just enjoying the open space in the country.

The farmland and pasture is now rented out. The crops grown now are corn and soybeans. The farm still maintains 30 acres of native grass.

Roman Krzycki with his first tractor with the farmstead in the background, 1941.

In the early years, they planted wheat, corn, alfalfa and milo and raised cows, hogs and chickens. Now we have a dog and three cats and of course, come summer the cows are back in the pasture for five months.

In August of 2000 the Krzycki Family Farm received the Nebraska Pioneer Farm Award. Life is good on the farm. *Submitted by Robert and Patricia Krzycki.*

KUESTER – Henry Kuester was born in 1860 and came from Wisconsin in 1866. He purchased 160 acres in Haymow Township, Stanton County, NE. He planted lots of trees and crops raised were corn and alfalfa. Livestock was cattle, horses and chickens. Henry and his wife had eight children.

Kuester farm, ca. 1996

The farm passed from Henry to his son Ewald and to Harlan Kuester. By 1968 when Harlan obtained the farm it had grown to 240 acres. Now corn and soybeans are the crops grown. The original house is still on the property.

Harlan and Doris have three children – Jeff, Peggy Munger and Teresa Robinson. Grandchildren are Melisa and Alison Kuester; Jennifer, Elizabeth and William Munger and Samantha and Andrew Robinson.

Harlan retired in 2014 but is still taking care of the homestead, enjoys fishing, grandchildren and playing cards. *Submitted by Harlan Kuester.*

KUSEK – In 1882 Peter J. Kusek came from Poland to America with his family, ending up in Columbus, NE. Peter's family then moved to Duncan, NE to farm some land there. Peter, when grown, went to work in Omaha for the Union Pacific Railroad, for 9 cents an hour. When the railroad life was not a success, Peter moved to Sherman County to start a family.

He purchased 320 acres in Sherman County from John Brynes for $8,000.00 on July 26, 1905. On Dec. 18, 1907, he sold 160 acres of the Kusek Family Farm to his brother, Frank.

At the time Peter bought the land, a small square house, several small buildings and a windmill stood there. Peter was a farmer, but also a carpenter and so he built on three rooms to the little square house and a screened-in porch on the west and south side of the house. It was said that Peter and his brother Frank worked together and they bought lumber by the railroad car load for $600. They would bring the lumber from Litchfield nine miles one way with horses and wagons. By the time they loaded the lumber, the round trip took a long day.

On Jan. 28, 1908 Peter married Katherine Iwan at St. Stanislaus Catholic Church in Duncan, NE. They raised their seven children on the farm: two daughters, Theresa Smedra and Rose Kuta and five sons - Adolph, Vincent, Paul (who died on the farm in 1928), Louis and Raymond.

On Feb. 18, 1944, Peter sold the farm to his son Vincent. Vincent and his wife Christine too raised their family of five children on the farm; two daughters, Geraldine Wardyn and Patricia Krzycki and three sons - Denis, Allen and Thomas.

Kusek farm, 1905

Being part of this family, I Patty have many fond memories. We worked hard all week, but Sunday was Sunday. Church was first, then we children would sometimes take a much needed nap or go for long walks with Dad. Mom never went; I think she just liked the quiet time! Mom loved her big garden and canned everything she could. Our cellar was full. Our swimming pool was a nice big dam out behind the house. We had a much envied, by our cousins, playhouse. It even had carpet, which the mice loved to raise their family under - until spring when we came back and cleaned house, them included. We went to District 11 School, which was about a mile from us. We have many memories of ball games, sledding and games played during our recesses.

Due to failing health, Vincent's son Thomas took over the farming operation in the fall of 1973. Thomas and his former wife Jane raised their two sons, Kyle and Kendel, on the farm.

Thomas sold the farm to his son Kyle and wife Kayla in December of 2010. They still reside on the Kusek Family Farm along with their two children, son Koy and daughter Kesslyn.

They raise mostly corn and some alfalfa and have recently expanded their cattle operation to include a feed lot that can house 1,000 head of cows.

They still reside in the original home built by Grandpa Peter, along with many updates.

The farm received the Nebraska Pioneer Farm Award in 2009. *Submitted by Patty (Kusek) Krzycki.*

KYES – John M. Kyes was born in 1838 in Oswego County in New York, the oldest of three children born to Russell and Nancy (Colgrove) Kyes. He grew up there and in 1855 he moved to Iowa. In 1860 he joined the westward movement to Denver, CO where he drove stagecoach for two years. After arriving in Merrick County in August 1862, he drove stagecoach for two years for the Western Stage Co. It was at a stage stop southeast of Central City operated by Jason Parker that he must have met his future wife, Viola Parker, daughter of Jason Parker.

The Kyes family, ca. 1895. Front row, l-r: Jessie, John M. (father), Viola (Parker) (mother), and Ella. Back row: Allie, Edward, Mary (Mayme), John (Jack), Annie, and Russell J.

Aerial view of the homestead ca. early 1950s.

John married Viola Parker of Merrick County on Dec. 25, 1864. They had 10 children, two of whom died as infants. They were the first couple married in Merrick County. On June 26, 1873 they purchased the south 80 acres of the southwest quarter of 10-14-7 in Merrick County for $400. In 1878 they acquired the north 80 acres of the same quarter by government patent.

John Kyes built the first frame house to be erected in Midland Township in Merrick County. He and his wife Viola helped organize one of the first rural schools in the county, the Gardner School.

On July 2, 1913 they sold the quarter to their youngest son, Russell J. Kyes. Russell was born March 11, 1878 and married Belle Copeland on Jan. 29, 1902. They had five children, one died as an infant. In the early days this farm had the only bridge across Prairie Creek that people from the north and west could cross during high water when they wanted to get to Central City.

Russell Kyes died in February 1943. In his will he left the farm to sons, Dean C. Kyes and Marion J. Kyes. In November 1945 Dean C. Kyes and wife Erma (Gleason) filed a Quitclaim Deed leaving Marion J. Kyes as the sole owner of the farm.

In 1988 Marion J. and his wife Maxine (Henderson) moved into Central City and sold the farm to Marion's nephew, Marvin G. Kyes and his wife, Mary Ann (Wells) who are the current owners of the farm. Marvin retired from farming in 2003 and Marvin's nephew, Steve E. Kyes, farms the ground and Steve's brother, John Z. Kyes, lives on the property. *Submitted by Edward A. Kyes (father of Steve and John).*

LAMPRECHT-SCHROEDER – Johann Wilhelm Lamprecht, born in 1857 in Branken, Germany, immigrated to Omaha, NE in 1882. He took out a homestead claim in 1894 east and south of Bloomfield, NE and lived there before purchasing the northeast quarter and northwest quarter of Section 2, Township 30 North, Range 3 West in Morton Township, Knox County, 1/2 mile north of Bloomfield, NE from James and Lizzy Baker March 9, 1909 for $30,456.00. Johann and his wife, Martha Bals, born 1875 in Germany and their children – Alvina, Wilhelm, Otto, Theodore and Harry – moved to this farm. The home originally was a one-story home and when Johann and Martha moved there, the roof was raised to make it a two-story home for their growing family (daughter Marie was born in 1915).

St. Marks Lutheran Church, organized in 1906, held mission festivals in the early days at this farm. When the Lamprecht family moved there in 1909 they raised the roof to make it a two-story farm house. The small Norway spruce tree seen in the foreground was still standing in 2015.

Theodore Lamprecht with his 1925 Ford Coupe that he bought for about $700. His friend Marvin McKay is also pictured. Two of these same farm buildings are still standing today.

Lou Ann Landholm Farm

Theodore, my father, married Mabel Harms June 20, 1928 and the newlywed couple moved into Bloomfield. Through the years the land was transferred to Theodore and Mabel and more land was purchased.

Marian, the eldest of three daughters of Theodore and Mabel, and Marian's husband, Marvin Schroeder and their children, Dennis, Roger, Mary and Lowell, moved to the farmstead February 1960 and Theodore and Mabel moved into Bloomfield.

Through the years the land was transferred to Marvin and Marian. Marvin and Marian moved into Bloomfield in 1991.

Marvin and Marian deeded the east half of the N.W. Quarter section to their sons, Roger and Philip, January 2001.

Roger and Philip are farming the ground, which was originally purchased in 1909 by their great-grandfather, Johann Wilhelm Lamprecht. Their mother, Marian Schroeder, owns the N.W. Quarter section where the farm buildings are located. Roger and his wife, Liz, live in the farm home. Philip lives in Bloomfield.

The Lamprecht-Schroeder farm borders the Knox County Fairgrounds. The Nebraska Pioneer Farm Award was presented to us at the 2009 Knox County Fair. Marvin died in 2010. *Written by Marian (Lamprecht) Schroeder. Johann Wilhelm Lamprecht was my grandfather. He died in 1939.*

LANDHOLM – Charles Landholm came to Boyd County in March 1893. He first homesteaded on land two miles north and one mile east of Bristow. He also purchased 160 acres from Andrew Lantz on August 20, 1908. This tract, the NW ¼ of Section 35, Township 34, Range 11, was first owned by Erik Hedlund who homesteaded it on November 2, 1895. Ownership was transferred to George Roth later in 1895, then to Erik Hedlund, and then to Andrew Lantz in 1906. Herman Landholm purchased this land from his father, Charles Landholm, on February 21, 1934. Ownership was transferred to Edit Landholm, wife, in June 1966 upon Herman's death and to Lou Ann Landholm, daughter, upon Edith's death in April 1976.

Charles Landholm came from Sweden to America in 1869. He first settled in eastern Iowa and then came to Dodge and Cuming Counties in Nebraska, where he homesteaded. The first home was a dugout in a side hill with dirt floors and a hay roof. He met and married Ellen Peterson. She also emigrated from Sweden to the United States in 1868. She and her family first came to Illinois and later to Nebraska. Charles and Ellen Landholm were parents of thirteen children.

The Charles Landholm family came from Cuming County to Boyd County by train in two railroad cars, and then by team and wagon to the Rosedale community north of Bristow. Eleven trips were made to O'Neil that spring moving lumber, livestock, machinery, and household furniture. Each trip involved one day going and one day coming back with team and wagon and fording the Niobrara River, as there were no barges. In early days, the Rosedale community had a post office, a store, a school, and several churches. Charles Landholm became an active citizen in Boyd County. He was one of three first Boyd County Commissioners.

Lou Ann Landholm continues to own the farm of about 318 acres, as additional land abutting the original tract was purchased by Herman Landholm. Corn, beans and wheat are the main crops grown. The Wayne and Diana Wendt family have been faithful stewards of the farm for fifty years. *Submitted by Lou Ann Landholm.*

LANGE – Charles and Auguste Lange purchased 160 acres of land from Tom Allison September 1897. This land is located in the shadow of Chimney Butte south of Whitney.

Their son, Richard, and his bride, Mildred, moved to the farm in 1920 and am assuming this is when the house and barn were originally built. They added other buildings - chicken house, milk shed, hog houses, etc. and lived there until 1941 when they moved to town. Another son, Carl, farmed the land for his parents. Charles died in 1949 and Richard inherited this farm and had the house and barn moved to Alliance where he was Deputy Sheriff. The other buildings were either moved or torn down. After his death in 1955, his daughter, Evelyn Lange Vorovka, inherited the farm and sold it to her Uncle Carl and Lillian Lange about a year later so it continued in the family. At this

From the 1920s when Richard and Mildred were living on the farm. The iris and spirea that Mildred planted beside the house are still alive, as is the yellow rosebushes she planted. Richard also planted many elm trees (most have died) as well as a grove of plum trees that are still producing.

time, approximately three acres of the land was sold to a neighbor who needed a road to their landlocked home. Upon Carl's death in 1981, the farm was inherited by his daughter, Carol Lange Doyle, who has it rented to a neighbor. The only improvement on this farm now is a working windmill in the pasture. *Submitted by Carol J. Doyle.*

LANGE-McGILL – Frederick Ernest Lange Sr. was born in 1827, Geringswalde, Mittweda, Saxony, Germany. To escape military conscription in 1848 he left for America. From 1849 to 1850, worked his way west, arriving in Burt County, NE in May 1855. He was the first permanent white settler in Burt County. He located the Golden Spring, walked in and claimed the spring and 160 acres. A death plot over ownership was foiled. Frederick built a cabin and started farming. Additional land was acquired in 1860, 1862, 1867, 1880 and 1882, resulting in a total of 1,020 acres.

In 1859 Frederick and Mary Horsky (1845-1876) married. They had six children: Augustus (1862-1895), Min (1864-1963), Clara (1866-1934), F.E. Jr. (1870-1930), Oscar (1876-1969) and Catherine (1876-1876).

Frederick died in 1905 from a horse accident. Son F.E. Jr. took over the farming operation. In 1906 Frederick's will was contested by daughters Mina and Clara, resulting in a family rift. By court order, sons F.E. and Oscar retained 498 acres. In 1930 F.E. Jr. died in an accident and the farm passed to Oscar. When Oscar retired, son-in-law Carson Anderson with hired help operated the farm.

From 1986 to 1992, land was acquired, now totaling 960 acres – only 114 acres less than F.E. Sr. acreage at his death.

In 1955 at a Centennial Celebration, a monument was erected with family names. In 1957 Aksarben awarded 100 continuous years family ownership. In 1969 Oscar Lange died and the farm passed to his daughters, Pearl Lange-Anderson and Mable Lange-McGill. In 1972 Mable Lance-McGill purchased her sister's share. Since 1977 this is a family farm corporation of four family members. Larry Mussack continues to farm to this date. *Submitted by Kenneth H. Lange-McGill Jr.*

LECHTENBERG – Frank Joseph Lechtenberg, a widower from Carroll, IA, bought 160 acres on Jan. 25, 1910 from William and Ida Hislop for $11,200. Frank's son, Frank Ambrose Lechtenberg, and his bride, Anna Dextro, arrived by train to this farm in the spring of 1913 from Iowa. On June 30, 1919 Frank and Anna bought this land plus an adjoining 160 acres from his father for $11,200.

Pictured back row: Lawrence, Agnes, Joe, and Eugene. Front row: Leo and Fritz.

Frank raised hogs and cattle. His family milked some cows which provided them milk and cream. The extra cream and eggs were sold in Butte. With this money groceries were obtained. Wednesday and Saturday were town nights. All supplies needed on the farm could be purchased in these rural towns.

Anna had a large flock of chickens. She also had a large garden and canned hundreds of jars of vegetables. There were no refrigerators or freezers until later in the 1940s when REA came to the area.

Lechtenberg family, 1940s. Eugene was in the U.S. Marines overseas and is not in the photo.

The Lechtenberg's did have lights earlier in the house provided by a 60 foot tower (a wind charger) with batteries in the basement. They also had water in the house, which was a luxury. An artesian well and earthen dams supplied water for the livestock.

Corn was picked by hand in the '20s and '30s with a horse and wagon. The day began early in the morning. At noon there was a load of corn, which was unloaded with a scoop shovel by hand. Oats were put in bundles and shocked. In August neighbors would get together to thresh the oats with a threshing machine, the oats would be put in a bin and the straw used for bedding.

Frank and Anna raised six children – Lawrence, Joseph, Agnes, Eugene, Leo and Fritz. Their religion was a very important part of their lives.

Fritz and his wife, Darlene, bought this farm from his folks in 1960. Frank and Anna retired and moved to Butte when Fritz returned home from the Army in 1954 to farm. They also raised six children – Patricia, Susan, Louis, Frank, Thomas and William. Over 100 acres of this farm is irrigated supplied by a spring fed dam.

Fritz passed away in 1996 and his two youngest sons operate the farm plus additional acres which were purchased through the years.

This farm is three miles east of Butte. The Nebraska Pioneer Farm Award was awarded for 100 years of ownership by the same family. Ownership will be passed on to the Lechtenberg's son, Frank, and his wife Lisa. *Submitted by Darlene D. Lechtenberg.*

LEWIS – In 1879 Joshua Jones acquired 160 acres located in Polk County, SW 1/4, 32-15-2. On this farm pigs, chickens, horses and milk cows were raised. Crops were milo, wheat and oats.

Timm family farm, 1957

Joshua Jones was the great uncle of the present owner, Kay Timm Lewis. In 1892 Joachim Timm, Kay's great grandfather, gained possession of the farm. In 1898 the farm was passed to Elias Timm, Kay's grandfather. In 1957 Philip Timm, Kay's father, gained possession. In 1963 Kay's mother, Agnes Jones Timm, owned the farm and in 1980 possession passed to Kay.

The farm still retains 160 acres and now produces corn and soybeans.

Elias Timm, Philip Timm and Kay Timm Lewis lived in the farm houses.

Kay and John Lewis have three children – Jeffrey, Alan and Julie (Lewis) Ellis. Grandchildren are Todd Lewis, Brett Lewis, Megan Lewis, Katelynn Lewis, Jack Ellis and Claire Ellis. *Submitted by Kay Timm Lewis.*

LIEDTKE – Frederick W. Liedtke was born on Sept. 7, 1836, in Ragnif, Prussia and in 1860 migrated to this country as a stowaway on a ship. In 1863 he enlisted in the Army in Pennsylvania and eventually became a Captain and served in the Civil War.

Robert and Melba Liedtke

While Frederick was in the Army, Otto B. Liedtke was born in Charleston, SC in 1867. After being discharged from the Army, Frederick returned to Pennsylvania. There he left his family to go west and start a new life.

In March of 1872 he came to York County, NE and timber claimed the northeast quarter of section 2, township 12, range 2, five miles west and one mile north of Gresham. He stopped here because there was a hill to make a dugout for his home and a pond at the bottom to water the oxen.

In the spring of 1873, his wife and two children, Otto and Clara, joined him. By this time other settlers had homesteaded in the same section. Eventually they all built sod houses in the center of the section and dug an open well in the exact center of the section for each of them to use.

Frederick eventually was elected York County Clerk. In 1878 he was elected State Auditor and this is when Otto B. Liedtke took over the homestead.

For a timber claim, 55 acres of trees had to be planted. Some of the original cottonwood trees are still standing today in the pasture. When the early settlers came here, there was nothing but prairie and prairie chickens. They had to go to the river bottoms to find trees.

Frederick W. Liedtke

The early settlers endured many hardships. The first one was the Easter Blizzard of April 1873. Another hardship was when the grasshoppers were so thick they blocked out the sun.

Otto told of the times they made the 15-mile trip to York for supplies with oxen. They would leave at 4:00 in the morning and by the time they returned home, it would be close to 4:00 the next morning.

Otto farmed the homestead until about 1917. His children were Frank, Fred and Vesta. Frank farmed it until 1949. He raised four children there: Howard, Lucille, Maude and Robert. Robert has lived on the property his entire life. Robert farmed until 2000 and raised two children there, Pamela and Rex. Robert and wife, Melba, built a new brick ranch home in 1980. Rex farms the land today and is the fifth Liedtke generation to farm this quarter of land. He raises commercial corn and soybeans. Robert's son-in-law, Gerald George, uses the 40-acre pasture for feeder calves.

LJUNGDAHL-HENDRICKSON – The original 160 acre farm was acquired by the great-grandparents of the current owner, Robert Hendrickson, in 1911. Troed Ljungdahl and Hilma Johnson both immigrated from Sweden in the 1880s, Hilma in 1882 at age14. As a teenager, Hilma came to work in Shickley NE for a well-known Swedish immigrant and founder of Stockholm Lutheran Church, Squire Anderson. While there, she met and married Troed and they had two children, Lillie (1891) and Carl Edwin (1895). The family traveled on an extended journey to Sweden in 1910. Family lore has it that Troed and Lillie enjoyed Sweden and would have settled there, but Hilma and Edwin did not like living there. They returned to Nebraska in 1911 and set about constructing a modern home and farmstead on the southeast quarter of section 32-5-4 in Fillmore County. They built a seven room farmhouse, a barn with wooden pegs, and erected a windmill. They established a typical diversified farm with milk cows, a big garden, other livestock, and crops such as corn, wheat and pasture. Two years later, Lillie married Albert Hendrickson and moved 1.5 miles south, where Robert's father, Maurice, was born in 1923. He was one of seven children in the family. Edwin and Lillie each inherited 80 acres of the farm upon the death of Hilma Ljungdahl in 1957. Edwin farmed the family farm until his tragic death in 1964, when an auto jack failed during a car repair. His wife Esther held the south eighty until her death in 1995. Robert Hendrickson purchased 40 acres, including the homestead, from her estate in 1996. Robert and his siblings acquired the north 80 acres from the estate of Norma Hendrickson (one of Albert and Lillie's seven children) in 2006. The farm now includes center pivot irrigation and produces primarily soybeans and corn.

Ljundahl family

LOUDON – David Boggs Loudon was born in Huntington County, PA in 1865 and moved with his parents to Page County, IA when a young boy. With his wife, the former Harriet Apple, and five small children they later moved to a farm near Brainard, NE. While residing there Hattie and an unborn child died during childbirth. A couple years later he sold the farm and moved to Wymore, NE where he operated a hotel.

David Roy Louden returns home from WWI, 1919. Pictured include David Boggs Louden (seated) with children David Roy (back left), Vernon, Bessie Dawson, Winnie Beckhoff, and grandchildren Mildred Dawson, Harriet Dawson and David Beckhoff.

In the late fall of 1909 Mr. Loudon, his second wife Laura and the smaller children moved to a homestead in Calf Creek valley northeast of Thedford, NE where he raised cattle and a huge garden that he sold the produce to neighbors. He would dig a large hole in a hill side, line it with straw or hay and fill with potatoes, turnips, carrots, etc., then cover with more hay and dirt and uncover as needed during the winter months.

David Boggs Loudon was granted a patent on 640 acres on Nov. 24, 1913 in Broken Bow, NE. In 1916 Mr. Loudon and his son (David) Roy also bought an adjoining 360 acres. On Mr. Loudon's death in 1921 his land was divided among his heirs--sons (David) Roy and Vernon receiving 160 acres each and his wife and daughters--Laura, Bessie Dawson, Winnie Beckhoff and Lucille Barlow receiving 80 acres.

(David) Roy and wife (Elma) Fern Allen Loudon lived on the original building site their entire married lives and raised their four children here. Roy and Fern, as they were known, raised Hereford cattle and turkey, sometimes as many as 1200 birds. In the earlier years they New York dressed the turkeys and shipped them to Chicago by rail, but later sold them on foot to a packing plant in Gibbon, NE.

Roy passed away in 1985 and Fern followed in 1987. On their death the land, including the original 160 acres, passed to their three surviving children - Doris Hamilton, David Loudon and Joellen Phillips with Joellen receiving the land that contained the original homestead.

Joellen and her husband, Lynn, reside on the ranch about nine months of the year and during the summer pasture cattle for their son, Lyle and Twila Phillips of Mullen. Lyle's cattle carry the original D lazy L brand that belonged to his great grandfather, David Boggs Loudon. *Submitted by Joellen Loudon Phillips.*

LUETH (LUTH) – Andreas Wilhelm and Carolina Maria Elisabeth (Horn) Luth paid thirty eight hundred and eighty dollars on March 11th, 1892, for the 160 acre homestead. In 1919, Andrew and Carolina retired in Emerson. Henry Chris and Martha Henrietta (Bokemper) Lueth, took over the farm for the next 30 years. (During WWI, Luth was changed to Lueth). In 1949, Henry and Martha retired in Emerson. Robert Charles and Ardeth Mary (Ottoman) Lueth lived on the homestead from 1949 to the fall of 1951.

First part of 1952, Elmer Henry and Laurene Sophie (Sebade) Lueth began their 37 years of living on and farming the homestead. November 1989 Elmer and Laurene retired in Emerson. January, 1990, Lewis David and Janie Lou (Lueth) Gutzmann moved onto the homestead. Lewis, with his brothers, had already been farming the homestead for approximately five years. Lewis (& brothers) and Janie completed the purchase of the homestead on March 9th, 1990. Lewis and Janie are in their 25th year on the Lueth homestead and the homestead has been in the family for 123 years as of 2015. Lewis (also known as Butch) and Janie Lueth Gutzmann have plans to move into Emerson in December of 2015.

Their son, Austin Lewis Gutzmann, moved back to the homestead earlier this year, and plans on making the homestead his residence and eventually enter a farming career.

Land Description: Southwest Quarter (SW π) of Section Twenty (20), Township Twenty-seven (27) North, Range Six (6), East of the 6th P.M. in Dixon County, Nebraska.

LUNBERY – In April of 1886, Sanford Lunbery purchased an 80 acre farm, one and one-half miles east of what was to become Stamford. Born in Pennsylvania in 1865, Sanford, his parents and three brothers, moved across Ohio, Illinois and Iowa, prior to settling in Harlan County.

Sanford Lunbery and grandson, Roland Lans

For many years Sanford's family lived in a dugout on the hillside above the Sappa Creek bed that ran through the farmstead. A daughter, Viola, was born there in May of 1897. That fall the family moved into a frame house on the other side of the creek bed. That house is still the basic farm home. The old dugout site is still visible on the hillside north of the house. A large bank barn was built just west of the dugout site and north of the house and creek. It also still stands on the site. The Sappa Creek was rerouted by the railroad in the early 1930's. Family members have lived on the property continuously since it was purchased in 1886

In 1940, Viola Lunbery Lans, the daughter born in the dugout and moved to the new farm house, became the owner. She owned the acreage until 1974 when it was acquired by her son, Roland Lans. Roland and his family resided on the property from 1946 to 1981 when his son, Bruce and family took over occupancy. Bruce acquired the property in 2012. Bruce's sons, Ben and Alex, are involved in the farm operation giving optimism for a heritage Family Award in 2036.

Lunbery-Lans Farm, 1959

Lunbery-Lans Farm, 2009

The original farm was made up of 80 acres, including pasture, cropland, creek bed and timbered areas. The timberland on the south side of the property was once used as a park area called Lunbery Grove in the 1920s and '30s. Over time timbered areas have been converted to cropland. Today, adjoining land has been acquired to make the farm unit 320 acres of irrigated cropland, pasture and the buildings site.

The Lunbery-Lans farm received its Pioneer Farm Award in 1986.

LUNDAHL – Andrew Lundahl was born in Skane, Sweden in 1853. He immigrated to America in 1879 and lived in Gibson, Illinois, and Stanton, Iowa, for a period of time. In 1895, he married Ida Bodelson in Stanton, Iowa. After a few years they felt this area was getting crowded, and a number of Swedes began moving to a Swedish community at Wakefield, Nebraska. Andrew made the trip to Wakefield a couple of times, and then was employed by the railroad there. He bought a farm north of Wakefield, known later as the Herbert Lundahl Farm. They put all of their earthly possessions in a railroad car and moved with their two sons, Herbert and Albert, to the farm. Their daughter, Lillie Lundahl Bates, was born at this farm. In April 1913, Andrew Lundahl and his wife purchased the adjoining farm, later known as the Albert Lundahl Farm. In 1931, Albert married Bertha Nagel from Walnut, Iowa. They lived in the old house on the farm until Christmas of 1931, when they moved into a new house featuring running water and electric lights, powered by a Delco plant with batteries in the basement.

The Lundahl farm

Alberta Lundahl Panning and Marilyn Lundahl Dutton were born in this house and they attended District 75, Sunnyside School, which is on the southwest corner of the farm.

Ida Lundahl died in 1950. After Ida's death, Louis and Lillie Bales purchased the homeplace, Albert and Bertha Lundahl bought the Albert Lundahl Farm, and Herbert and Ruth moved to Wakefield and purchased the old Wakefield Hospital.

Bertha Lundahl died August 1990 at the age of 96, and Albert died April 1997 at the age of 98. Alberta Lundahl Panning and Marilyn Lundahl Dutton inherited the farm after Albert's death, and they continued to rent the farm land to Lowell Newton and later his son, Wayne Newton.

In 2004, the family decided to sell this property and it was put up for auction. The building site was purchased by Michael Rischmueller, and Gene and Alberta Panning purchased the farm ground. Gene Panning passed away May 6, 2009 and Alberta continues to rent the farm ground to Wayne and Sandy Newton, who have farmed it for a good number of years.

MARSHALL – The Dawson County farm currently known as Buzzard's Roost Ranch at Eddyville, NE is now owned by the fourth generation of the Marshall family. The ranch surrounds a portion of the Buzzard's Roost Canyon. It is said the area gets its name because of the buzzards flying overhead.

Folklore suggests the rugged land once provided a hiding spot for outlaw Jesse James. There is a cave where James reportedly took cover. While it is questionable if the James legend is true, many stories have been told over the years.

This land was purchased in 1913 by W. H. (William Henry) and M. Ellen Marshall from Burton Noyes. This original section of 190 acres was the beginning of the land owned by the Marshall family. W.H. and his son Marion raised registered Hereford cattle on this land.

W.H. Marshall sold the land to his son, Marion Marshall in February of 1942. Marion and his wife Nellie Lahm Marshall raised ten children, Maurice, Doris, Arlene, Helen, Betty, Bonnie, Allen Lee, Lois, Melvin, and Bill on this land northeast of Eddyville. The Marshalls lost a son Vernon when he was child. Ten of their eleven children were born in the house that still houses family members today.

William Henry and M. Ellen Marshall

Marion and Nellie Marshall

Melvin and Jean Marshall

Marion and Nellie Marshall and family with children, l-r: Lois, Bill, Betty, Allen Lee, Melvin, Arlene, Helen, and Doris

Melvin and Jean Marshall and family,, back row (l-r): Kent Frost, Mike Wendorff, Suzanne Wendorff, Tom Kennedy, Christine Kennedy, and Tim Marshall. Seated: Shari Frost, Jean Marshall holding Ben, Jacob Frost, Melvin Marshall holding Matthew Kennedy, Cindy Marshall. In front: Leah Wendorff Beckenhauer and Mary Wendorff Malcom.

Pictured l-r: Cindy Marshall, Tim Marshall, Suzanne Wendorff, Trevor Beckenhauer, Alayna Beckenhauer, Zoe Malcom, Ashton Malcom, Tim Beckenhauer, Leah Beckenhauer, Mary Malcom, and Darin Malcom.

Marion and Nellie sold the land to his son Melvin and his wife Jean Mohler Marshall in 1972. They raised four children in Melvin's childhood home, Suzanne, Tim, Shari and Christy. They lived on the land until Melvin's death in 1988. Jean continued to manage the land until her death in 1999. Since that time, the land is owned and managed by their living children, Suzanne Marshall Wendorff, Tim Marshall and Shari Marshall Frost McClure.

The family dream of receiving the Pioneer Family Award came true in 2013.

MARTIN – In 1869 Joseph Martin walked from Wisconsin to Madison County, NE. There he obtained 156 acres through the Land Grant Program. We have the original paperwork signed by Ulysses S. Grant. Joe was elected Sheriff of Madison County and he had a 1/2 mile road cut from his home straight into town. Joe and his wife, Sarah Best Martin, had five children - Grace, Meda, Joseph Edgar "Ed," Robert and Chrissie who died in infancy. Joe built his house high on a hill so that with binoculars he could watch the hired men work.

Joseph Martin, 1890s

While Grace and Meda married and moved away, my grandfather, Ed Martin stayed to farm. He married Jennie Nyland of Norfolk in 1908. He purchased 400 acres and built a home for his wife, daughter Jean and himself. The 400 acres are in Stanton County, but adjoin the Madison County land. Joe continued to purchase land so that at his death each of the children inherited 80 acres. The" home place" was sold after Joe and Sarah passed away.

Ed's health did not allow him to continue farming, so in 1920 the family moved to Madison and he opened an insurance agency. However, he continued to actively participate in management of the farm. Crops raised were corn, milo and oats. Ed passed away in 1948 leaving Jennie, a 50 year old widow. She had only a fourth grade education, but she did a wonderful job of managing the farm. Daughter Jean Martin married Clinton Williams on Sept. 14, 1940 and they had one daughter, Carolyn Jean. Jean inherited 80 acres of Madison County land from her father and 40 acres from Meda Martin Bell.

Jean Martin Williams passed away in 1971 and two years later, my grandmother, Jennie passed away. I, Carolyn Jean Williams, married Robert Reade on Aug. 9, 1969 and inherited the farmland upon their deaths. There are a total of 520 acres - 400 that my grandfather purchased and 120, which my mother had inherited. We have had the same excellent tenants, the Stevens, since the late 1980s and the grandson, Cody, presently lives in the century-old house, which my grandfather built. The last 30 years we have grown corn and soybeans. There are three Reade sons - Chris, Cory and Chad and each of them with their wives has three children. Cory is maintaining a strong interest in and participation in helping to manage the farm. *Submitted by Carolyn Williams Reade.*

MASCHMANN – This land was granted to the State of Nebraska by the US in 1868 and then granted to the North Central Railroad Co. in 1871. The railroad sold the land to George Schaiffenacker in 1874. It is believed that Schakiffenackers built the Big White House, which is still standing and the big barn which was recently taken down. Seven years later the property was sold to John and Louise Warnecke. It was 22 years later, on Feb. 3, 1903, when the farm was sold to Carl and Anna Hansjurgen.

Carl and Anna Hansjurgen and family, 1904. The big white house has been restored, but the barn was taken down in 2015.

Carl owned the land for the next 20 years. He and Anna had seven children: Louise, Henry, Carlena, Ida, Anna, Ella, and Emma (Dale Maschmann's mother).

In 1923 Carl sold the land to his son-in-law, George A. Poppe, husband of his oldest daughter, Louise. Louise and George were married in 1909. They were the parents of three daughters: Emma, Lydia and Merna.

George Poppe died in 1952. Louise continued ownership until her death in 1973. At that time the property transferred to the living daughters, Lydia Cole and Merna Penz. Following the death of Merna Penz in 1985, the land transferred to Lydia Cole and husband Carroll Cole. It became the property of Dale Maschmann upon the death of Lydia Cole in 1994. Dale's mother, Emma, was Carl and Anna Hansjurgen's youngest daughter.

Dale Maschmann married Beth Mendenhall in 1955. They have three children: Kim Mari, James Reid and Andrew Charles.

It was in 1959 that Dale began farming this quarter section and he continued farming it for the next 37 years, until his retirement in 1996. The crops raised were corn, milo and soybeans.

Lydia and Merna had begun the restoration of their childhood home in 1981. It was most certainly a labor of love. Dale and Beth continued the restoration in 1995 and moved into the Century Old Restored Farm Home in 1999. Upon Dale's death in 2010 his Trust transferred the land to his three children - Kim Sheppard. James Maschmann and Andrew Maschmann. Beth has continued living in their restored retirement home. *Submitted by Beth Maschmann.*

MASCHMANN – Daniel G. Mendenhall was born near Ridgeville, IN on July 7, 1846. At age 18 he enlisted in the 147th Indiana Infantry and was sent to join the army of the Potomac in Virginia, but didn't arrive until Lee had surrendered. After duty in the Shenandoah Valley he was mustered out at Harper Ferry.

Eldon and Lucy Mendenhall with daughters Marilyn (15) and Beth (20), about 1954.

He farmed the home place for three years. Then deciding to come further west, he came to Stove Creek Precinct in Cass County, NE in 1871, where he bought 80 acres of NE Quarter section 25. He broke the land, farmed it for two years then sold it and bought the NW Quarter section 25. He broke this broke this ground, set out a large grove of trees and six acres of orchard and vineyard.

He had married Jane Smithson in 1868. They had seven children, one of whom, John Wesley, our grandfather, took over the farm when his parents, Daniel and Jane, moved to California around 1899.

John Wesley Mendenhall married Nettie Anna Hendricks in 1900. They had four children: Bernice, Eldon, Inez and Fern. Eldon began farming the land in 1929 after his marriage to Lucy Ripley on June 4, 1929. He purchased the land in 1943 and added the NE 80 acres which Daniel first bought, plus the NW 80 acres

Eldon received awards for his farming practices and in 1968 he received an award for outstanding accomplishments in soil conservation. Crops raised were corn, wheat, soybeans, milo, alfalfa and oats. In the early years there were a few dairy cattle, sheep and chickens. Later there was a cow-calf herd and hogs, raising them from farrow to finish.

Eldon and Lucy had two daughters, Beth Janet and Marilyn Jean. Beth married Dale Maschmann in 1955. They have three children - Kim Mari, Janes Reid, Andrew Charles and four grandchildren. Marilyn married Dean Crewdson in 1960. They have four children: John Oliver, Cynthia Jean, Scott Dean and Timothy Allen (deceased) and five grandchildren. Marilyn later married Dwight Knapp in 1985 and after his death married Leo Walko in 2012.

Eldon retired from farming in 1972, but remained on the farm until 1982 when he built a home in Elmwood, NE and moved into it just six weeks before his death, Aug. 15, 1982. He sold a 6-1/2 acre tract where the buildings were located but retained all other land.

Beth Maschmann and Marilyn Knapp-Walko are the present owners of the Century Farm. *Submitted by Beth Maschmann.*

MAUST – In 1892 Albin and Mary Maust purchased land north of Carleton in both Thayer and Fillmore Counties. In 1897 72 acres, the north half of the northeast quarter of section 5-4-3 Carleton Precinct was added. In 1908 the east half of the southeast quarter 33-5-3W Hamilton Township, Fillmore County and also the northeast quarter 33-5-3W was purchased. Albin's son, J.W. and wife Pearl acquired all of the land.

Maust farm, 1954

JoAnn (Maust) Row and husband Doug purchased the land in 1972 from J.W. and Pearl where they raised five children: Mark, Michael, Jeffrey, Valerie and Michelle. They continue to farm with sons Mike and Jeff. Jeff and Sally's sons, Jeremy and Ben, are now fifth generation farmers. The farm was renamed Bear Creek Farms Inc. in 1984.

J.W and Pearl were married in the house in the picture and lived there for over 70 years. JoAnn and twin sisters, Doretta and Doris, were all born in this house. After farming through the '30s, J.W. was determined to have irrigation, putting down his first well in 1944. He had registered Holsteins and registered Durocs along with a hatchery. J.W. named his farm Plain View Farm and it is on the barn along with his name. In 1990 a tornado entered just east of the driveway, passing just east of the house shredding the cedar trees and destroying trees around the house. The house escaped with minor damage. A wildfire entered the northwest part of the section in January 2013. Driven by a 60-mile-per-hour wind, the fire was headed directly for the farmstead. With the aid of neighbors and 11 rural fire departments, the fire was put out 100 yards north of the trees.

Mike and Diane continue to live on the farmstead. Jeff and Sally live on the other 100-plus-year farm visible in the background. Doug and JoAnn also live on the same section of land in a home built in 1963.

McCONNELL – The McConnell family farm is located directly south of the now, nearly forgotten village of Sarben in eastern Keith County. William D. McConnell and his wife, Mary Jane "Dixon" McConnell, moved to the area from a previous dryland farm located southeast of Paxton just across the southern Keith County border in Perkins County. Drawn to the valley by the prospect of developing irrigation potential they, and their family at the time, took up residence in the Sarben area in 1899 and lived a short time in a sod house on an adjoining property to what later was to become the family farm.

Brothers Corbin G. (left) and Lester A. (right) with Lester's two oldest sons Cecil (left) and Vinis "Vince" (right) at Lester's home on the East 80 in 1924. Cecil was a B-17 pilot in World War II.

From the "soddy", they were able to move to another farmstead located close by and finally on January 15, 1903, they were to purchase the NW ¼ Section 26-14N-35W from William J. McDonald, with that original deed begin recorded on February 14, 1903.

William and Mary Jane split the east 80 of the farm out and sold 40 acres each to sons Lester A. in 1918 and Corbin G. in 1919 to help get their own families established. During the height of the Great Depression in 1933, William D. McConnell died suddenly, leaving the sons Lester and Corbin scrambling to preserve the entire 1/3 section family farm. They were able to shift ownership of the west 80 to Corbin and the entire east 80 to Lester so William's estate could be settled, and

William D. (standing left) and Mary Jane (seated back seat, dark hair) with other family members and sisters, admire son Lester's (standing right) 1917 Model T. Lester and new wife Fern drove this car to Cheyenne, Wyoming on their honeymoon in September of 1918.

McConnell family farm original main homesite on the West 80, taken in 1982 by son Mark S. from the Piper Tri-pacer kept on the family farm and flown by both Merle L. and Mark at that time.

the farm was left intact. These two brothers owned and operated the farm up until the early 1960s, when Lester's youngest son Merle L. and wife, Delores, took over the farming duties and were able to purchase the east 80 after Lester's sudden death in 1963, and later the west 80 from Corbin in 1982.

Merle and Delores and their son, Mark, continued to operate the farm until the ownership was transferred once again in 1998 to current owner Mark S. and wife Alice Susan McConnell, as the fourth generation of continuous family ownership. We are proud of the heritage that has been established on this farm and grateful that our ancestors had the courage and tenacity to meet the challenges through the troubled years and share in the joys of the rewarding years and that it continues for future generations.

McCORMICK – On April 12, 1905, George B. and Maggie Orr purchased the 80-acre property located at 25-16-42 in Lewellen, NE. It was later sold to John G. and Lula (Orr) McCormick. After their passing, the property went to Harold and Lila (Spillman) McCormick. After their passing in 1985, the property went to Ron and Darlene (Wolford) McCormick, who currently own the property. The produces cattle, corn and alfalfa. Ron and Darlene have three children—Michael, Steven and Mark (dec.)—and two grandchildren—Blake and Abby.

The original farm house

The McCormick family

McLAUGHLIN – On Feb. 23, 1900, Peter Lautenschlager and wife Amelia (Schimpf) Lautenschlager purchased the Northeast Quarter of Section 22, Township 4 North; Range 4 West of the 6th Principal Meridian in Thayer County, NE from Andrew B. and Margaret Emerick.

They moved from east of Carleton, NE with their two children, John and Margaret. Another son, Rolla, was born in December 1900 and a

Pictured l-r: Victor Sorge, Robert Sorge, Christine Rezek, Carol Sorge, Larry McLaughlin, and Leon McLaughlin.

son George in October of 1903. They built a new house in 1907, where their last child, a daughter Elsie Amelia, was born in October of 1907.

Peter passed away in 1943 and Amelia passed away in 1954.

Elsie (Lautenschlager) and husband Robert McLaughlin, having rented the farm since 1939, purchased the land at the estate auction and became owners on June 28, 1955. Their children - Leon R., Larry D., Carol A. and Leslie A. - were all born while living there. Leslie was killed in May 1951 by a train.

After Bob's death on Jan. 25, 1966 in a train accident, Elsie continued to live on the "Home" place until about a year before her death in April of 2003.

The farm then passed to her children - Leon of Lincoln, Larry of Davenport and Carol (McLaughlin) and Victor Sorge of Deshler.

In 2010 great granddaughter Christine (Sorge) and Jeffery Rezek purchased Leon's share. And so the legacy continues. *Submitted by Carol Sorge.*

MEISTER – The land, located in Monterey Township, Sec. 24, was acquired by Herman Meister Sr. in 1879.

Herman was a blacksmith who immigrated from Eckenhausen, Prussia, in 1871. He made contact with his cousins in the Monterey area and then moved on, earning money as a blacksmith in mining and lumbering camps in the West.

Herman returned to Cuming County in 1879. He purchased the land for $5 an acre from EL and TL Co. He married Gertrude Grovijohn at St. Anthony's Church, St. Charles, and built a house and farmstead here. Their children: Justina Heimann, Mary Heimann, Herman Jr., Bernadine Heimann, Helen Knievel, Bertha Kreikemeier, Henry, Josephine Burger and orphan train-adopted Viola Pospishil.

In the 1890's Herman Sr. maintained a full farmstead of milk cows, cattle, horses, hogs, chickens, ducks, corn, oats, garden, orchard and grape vines until his retirement.

In 1920, youngest son Henry married Kunigunda Schmuecker at St. Anthony's Church, St. Charles, Henry took over the farming. With soil conservation in mind, Henry installed five dams. Their children were: Monsignor Andrew Meister, Hermines Ulrich, Doreen Meister, Veronica Emanuel, Edward Meister, Melvin Meister and Henrietta Pane.

Youngest son Melvin terraced the farm in 1957. He married Delores Prinz in 1958. Melvin was activated in the Army Reserves in 1961-1962 during the Berlin Crisis. On his returned to farming he adopted minimum tillage in 1964. He received the area JayCees Outstanding Young Farmer Award and numerous conservation awards including 1973 Lower Elkhorn Natural Resources District Award.

Their children are Arthur, Donald, Patrick, James and David Meister, Jean Meister Westermann, Barbara Meister Sylvester, Janet Meister Gronert and Julie Meister Shay.

Soil and water conservation practices and minimum to no-tillage was done. For feedlot pollution control a vegetative treatment system was installed in 2009.

Melvin and Delores retired on the farm in 1998 renting the feedlot, hog barns and farmland to a young area farmer. Conservation of soil and water continues to be a high priority.

MERRIHEW – Dr. William Earl Merrihew, being a dentist from Bethany, MO, decided to try for a homestead in the sand hills of Grant County, NE. He wasn't the first one to apply. In fact, he got the last one that was available. This happened in 1914. W.E.'s father, Curtiss, had already acquired a homestead in Sheraton County, north of Ellsworth, NE near the old Spade Ranch. They moved here and set up a sod house. There were four boys and a girl – Bert, Victor, Curt, Joe and Margaret.

With W.E. doing dentist work in Hyannis and a lot of hard work from the boys, he built the ranch into a place of over 12,000 acres. In later years W.E. sold the ranch to Victor and Curt and moved back to Nevada, MO where he acquired a farm. Joe and his family also moved to this farm. Bert had started an electrician business putting up wind chargers for people in the hills as there wasn't any electricity in the country then.

Victor and Curt kept operating the ranch as a partnership. Victor and Dorothy had two children, Walter and Harry. Curt and Mary had three children – Bob, Knight and Margeret. As their children grew up, we needed more to do than just work on the ranch, so a haying contract was worked out with the neighbors. Most of these children also went to a boarding high school at Curtis, NE as did a lot of kids in the hills. Most of us ended up marrying our spouses from people that we met at school.

Partnerships have kept the ranch together over the years. At the present time, Harry and Jerry, second cousin, are operating the ranch. Harry's boy, Rocky, is also here on the partnership.

Harry has served on the local school boards for almost 20 years, a member of Nebraska Cattleman most of his life along with National

Merrihew farm in 2015

Cattlemen, served on the brand committee and state brand board for eight years. Also was on the Panhandle REA board for 25 years. *Submitted by Harry B. Merrihew.*

MEYER – March 18, 1884, Frederick Meyer purchased a parcel of land known as, "Common School Land" from the State of Nebraska. This land was described as the "South East Quarter of Section 36, Township 5, Range 11, West of the 6th P.M containing 160 acres, more or less, according to the U.S. Survey".

Frederick Meyer had come from Posen, Germany to America in 1868 at the age of 18. He became a U.S. Citizen in 1876. He moved from Illinois to Nebraska in 1880 and in 1884 purchased the land that has been in the Meyer family for over 100 years. The original purchase price of the 160 acres was $1,120.00.

Frederick and his wife, Emilie raised eight children. Five of their children died of tuberculosis. In 1907, Emilie also died of TB. This left three children, Fred, Hugo and Annette. Their father, Frederick Meyer died in June of 1920.

Sale Day, March 13, 1945.

Frederick Meyer gave this parcel of land to his son, Hugo Meyer. When the Great Depression began in the early 30s, the family could not grow crops. The lack of rain left the soil too dry. In order to survive the "dirty 30s", Hugo Meyer took out several mortgages against the farm.

With the Depression ending, the Meyer family was once again able to grow crops. Hugo had completely paid off his mortgages by the year 1945. Hugo and his wife, Rosetta, had raised 4 children on this farm. Agnes, Evelyn, LaVerne and Marlin. Hugo passed away in 1974.

In 1979, Hugo's son, Marlin Meyer and wife Louan, purchased this SE quarter from his mother, Rosetta Meyer. They raised four children on this farm. Wesley, Patricia, Susan and Ronald.

After retiring from farming, Marlin and Louan's youngest son, Ron, farmed the ground for more than 20 years, improving the land by planting hundreds of trees, applying proper agricultural practices, and winning awards for his efforts.

In 2005, the South East Quarter of Section 36, Township 5, Range 11 was sold to Kenneth and Susan Parr. Susan is the daughter of Marlin Meyer. Ken and Sue currently own the SE quarter – which Marlin and Louan Meyer named, Lone Oak.

MEYER – I live on the original Wells Homestead, 12-7-5, where the log cabin still resides. Some logs have fallen out and we had logs laying beside it to replace the fallen ones, then it flooded three times this spring and washed our logs down the creek, but the cabin is still standing.

My name is Linda Meyer, maiden name Wells. I married Ron Meyer on April 30, 1988 and we purchased my family's farm in September 1988 from my Grandmother Beulah Wells. My father, Raymond Lee Wells, farmed the land and operated a 100 head dairy operation from 1959 to 1988, but he never owned the property. I have two sisters and four brothers. I am the second oldest child and I grew up helping with the farming and the dairy operation. My husband Ron and I are raising our two children on the family farm. We milked up to 100 dairy cows until 2003 when we sold all but a dozen Holstein to another family dairy. Since then we continue to dairy on a small scale. We also raise chickens, rabbits and have horses that keep us busy and our children enjoy exhibiting them in 4-H and FFA.

The original Meyer log cabin

We also produce CORN, soybeans, alfalfa, prairie hay and sometimes oats on our farm ground. Ron works full time in town and farms on the side. I stay home and hold the farm down, I enjoy taking care of the animals, milking the cows and helping with the farming. The children help out when they can.

We plan to continue farming the land and raising our animals and keep it in the family. Many people ask me why I ended up with the farm, because I have four younger brothers. Well I was the one who milked the cows before and after school from age 7 throughout my school years. When I took a job off the farm, I still came and helped milk and complete the other farm chores, I also helped with all the farming. I was my dad's right hand man…that is what he used to say. Several months after my husband and I were married, we were asked to purchase the farm. Grandma was diagnosed with cancer and other family issues arose, so we jumped in and purchased the property to keep it from being sold at auction.

We officially named our farm "Legacy View Farm," we used this as our pre-fix to register our Holstein dairy cattle. We have received numerous honors such as: 1995 Outstanding Young Dairy Cooperators, 1999 Young Dairying Award, 2000 National Dairy Shrine, 2002 Farm Journal Spokesperson of the Year, Ak-Sar-Ben Nebraska Pioneer Farm Award and many more. *Submitted by Linda Wells Meyer.*

MIRATSKY – In August 1874 Wenzel Wittera obtained 160 acres located in Butler Creek, Skull Creek, 15-4. Crops grown on the farm were corn, wheat, oats and alfalfa. Livestock was cattle, hogs and chickens. The farm went from Wenzel to Agnes Wittera, to Vaclav and Mary Wittera, to James and Estella Wittera-Miratsky, to the Miratsky family.

Vaclav and Mary Wittera with daughter Estella Wittera-Miratsky.

This farm still retains the 160 acres. Now corn, soybeans and alfalfa are the crops grown and cattle are raised.

This farm was awarded the Knights of Aksarben Pioneer Farm Award.

James has four children – Jim, Angela, Matthew and Elizabeth. His grandchildren are Katie, Emily, Marissa, Leah and Damian. James retired Dec. 31, 2011 and now does odd jobs in addition to farming. *Submitted by James Miratsky.*

MOLCYK – On March 1, 1915 Andrew and Veronica Molcyk purchased 160 acres in Platte County located in the NE quarter Sec. 2, T-19N R-3-N. At Joliet, IN in 1991 the land was passed from Veronica to daughter Rose and her son Phil. In 2004 Rose passed on leaving the farm to Phil, present owner. After purchasing the land, the existing farmstead was moved from the bottom on the NW corner of the land 1/4 mile east where a well was drilled, making the farmstead about 1/8 mile from a road. They lived in a granary. The first year the house, a large shelter belt, was started which exists today. The cedar trees on the south side of this house are over 90 years old. The barn was brought up from the bottom along with the old house (split into a granary and chicken coop). They never lived in the old house. An orchard was planted south of the house, which dried out during the drought. The hog shed was built from cedar poles brought from out west from a forest fire where the government allowed people to go in and cut burnt trees. The pasture still shows the old fence line buried by the dust of the '30s.

The Molcyk farm

This farm wouldn't exist today if not for my Grandmother who raised six children through droughts, depression, storms, wars, the loss of her husband and threat of foreclosure. She brought it back and kept it in the family. To my mother, Rose Molcyk, who came back to the farm and joined forces with Grandma to do the chores, housekeeping and care for Grandma in her senior years and for raising me.

To all the farm WOMEN, you made this state what it is today. Thank you. God bless you all. *Submitted by Philip A. Molcyk.*

MULCAHY – Patrick (born 1859) and Ellen (born 1858) Nolan Mulcahy, born in County Clare and County Mayo, Ireland, traveled to Chicago, IL in 1875 and 1874. They met and married in Chicago on Feb. 10, 1891. They immediately took the train to Clay Center, NE. Patrick's sister, Ellen Mulcahy Joyce, and her family recommended coming to Nebraska to farm. My father, Roger (born 1893) and his sister, Mary, were born in Clay Center. After two fires in the house, Pat and Ellen moved to Greeley County in 1899. Other folks from Clay Center had moved to Greeley, NE so they had friends in farming. They lived in a small wood house on the side of the hill facing southwest. A son, Joseph, was born in 1900. They purchased the ground in 1901. The children attended the county school one mile west. They moved from the farm house in 1918 to a home in the village of Greeley (still owned by Mary Ellen Mulcahy). Roger and Alice Kavanaugh of O'Connor married Feb. 12, 1934. Roger died in 1965 and Alice in 1986. Both interred in Greeley Sacred Heart Cemetery. *Submitted by Mary Ellen Mulcahy.*

MURPHY – On March 6, 1903 Watson and Sarah Rogers purchased 160 acres of land in Loup County, NE, SW 1/4, sec. 27, twp. 22, range 20 from Thomas and Eunice Miller for $300. Watson and Sarah built and lived in a sod house until a frame house was built in 1907. It was here where Watson and Sarah raised their eight children. Sarah passed away in 1925 and Bessie, a daughter, remained in the home to care for her father until his passing in 1930. His will stated that daughter Bessie Rogers receive the property described, she having cared for and made a home for me during my old age.

Watson and Sarah Rogers

Bessie Rogers married Daniel Murphy soon after and she and Dan made this their home. Daniel Roger was born Dec. 11, 1932 and Donald Leon was born March 17, 1934. Dan passed away in 1970 and Bessie in 1972.

Dan and Bessie Murphy with sons Daniel Roger and Donald Leon

Sons, Daniel and wife Ellen and Donald and wife Neva were co-owners until Jan. 20, 1995 when Donald and wife Neva purchased Dan and Ellen's share. On March 16, 1995 in consideration of love and affection, son Bryan K. Murphy was granted as joint tenant.

Through the years many family members have called this place home. Dan and Ellen had a trailer house on the corner of the yard for nearly two years when his parents were still there. Dan worked for a construction company to build the nearby highway. Dan followed the construction for a number of years that eventually led them to Kansas. Goddard has been their home in their retirement. Dan and Ellen are the parents of three children - Danny, Donna and Patty. They also have five grandchildren.

Don and Neva have two children, Bryan and wife Juli have four children: Ashley, Hayley, Marley and Brandt and a grandson Teague. They live in Ainsworth. Brenda and husband Mark Knight live near Ord and have two children, Jordyn Jean and Tanner.

In 1995 Don and Neva moved to the home place and remodeled the house, and the old outhouse was put out to rest. Water was trenched from the windmill to the house, the term "running water" took a new meaning... running through the pipes and not being carried in a pail. REA was brought to the home in the late '40s but was taken out in the '70s. Neva wanted a phone, so in the mid '90s a phone line was installed and is still in use today. Today electricity is provided through wind and solar panels, a project that is maintained by Don. Propane and wood provide heat for the home.

Through the years the land provided income for the families with cattle in the pasture and hay meadow providing the winter feed. Bessie raised chickens and a large garden. The Murphy family was truly a pioneer family, living close to the land. Dan and Bessie never owned a car so depended on other family members for their monthly trip to town for coal, chicken feed, sugar and other grocery staples.

Dan and Ellen Murphy with children Danny, Donna and Patty

Cattle still graze the pastures and large

Don and Neva Murphy with their children and grandchildren

round bales are seen on the meadowland. Just as the land provided for generations in the past it will provide for generations to come. *Submitted by Donald and Neva Murphy.*

NELSON – In 1878 Anders Nelson, his wife Kajsa, and two sons, Peter and Charles, settled in Kearney County, NE. On Feb. 2, 1880 Anders went to the land office in Bloomington, NE and paid $18.00 for his homestead application. In order to take full title he had to remain on the land for five years, clear at least 50 acres and maintain his cabin. In addition to all this, he had to become a citizen of the US, which he did on Dec. 14, 1882 after he renounced forever all allegiance to the King of Sweden.

Steve, Sam and Scott Nelson

On Aug. 24, 1885 Anders Nelson paid the sum of $8.00 in the land office in Bloomington, which was the balance required by law for the entry of his homestead. By this time the homestead was worth over $2,000.00. Two years later on July 26, 1887 the President of the US, Grover Cleveland, assigned forever, through a Patent Record, the 160.68 acre homestead located in the southwest corner of section 18, Sherman Township, Kearney County to Anders Nelson and his heirs.

In June of 1887 Peter and his wife Ida began their life together on the Nelson homestead. In 1894 Peter and Ida purchased a farm across the road and they and their three small children moved, where they went on to have five more children. Anders and Kajsa lived alone at the homestead until 1900, when ill health caused them to move in with Peter's family.

In 1916 Ander's grandson Walter and his wife Edna built a new house at the homestead, continued to farm the land and raised their three children. In 1954 Walter's son, Dennis and his wife, Shirley Ann, made their home at the homestead. They improved the farm with irrigation and row crops and raised their two children. At the present time, Shirley Ann continues to live at the homestead.

Steve Nelson, son of Dennis, and Steve's son, Scott, currently operate the homestead. Steve and his wife, Elma, live on a farm three miles northwest of the homestead, where they raised their three children. Scott, great-great-great grandson of Anders Nelson, is the sixth generation in the Nelson family to farm the Nelson homestead. Scott, his wife Amy, and son Sam live on the farm across the road from the old homestead. *Submitted by Steve and Elma Nelson.*

NELSON – Bengt Nelson (Sept. 3, 1842, Horby, Sweden – Sept. 24, 1924, Lincoln, NE) immigrated to the USA in 1869 and applied for a homestead on E 1/2, NW 1/4, 24-15-8 in Marietta Township, Saunders County, NE. He received the homestead deed on June 24, 1878.

Bengt worked on the Burlington R.R. to support himself and Marlin Larson looked after Bengt's place in his absence.

Bengt Nelson

Anna Jonson Nelson

Bengt Nelson farm, ca. 1926

In 1872 Anna Jonson immigrated to the USA from Horby, Sweden. She was born Aug. 25, 1844 in Horby, Sweden and died Jan. 30, 1893 in Wahoo, NE. She kept house for her brother, Lewis Johnson. Anna and Bengt courted for a period of time.

On Oct. 15, 1874 they were married. On that day Bengt hitched his two best horses to the lumber wagon and went to pick up Anna. They headed off to Fremont, NE and found the courthouse and got a marriage license. Then they found a Probate Judge, Wm. C. Ghout, to marry them. After that they then went to the general merchandise store to buy household furniture and then back home – What A Day!

On July 12, 1880 Bengt and Anna bought 160 acres SE 1/4 13-15-8, Marietta Township, Saunders County, NE. They paid $1,800.00 for the land. He later sold it to his son, Fredrick, in the 1920s for $8,000.

They had six children – four of them died in an epidemic. Two boys survived, Fredrick Julius Nelson (July 27, 1879 – Dec. 16, 1951) and George Harold Nelson (Nov. 12, 1884 – Feb. 22, 1952).

In 1891 Bengt and Anna built a nice large home. In 1892 Anna sadly died – she only got to enjoy the new house one year. *Submitted by Glenn Nelson.*

NELSON – George Harold Nelson (Nov. 12, 1884 – Feb. 12, 1952). His father, Bengt and his mother, Anna, were both immigrants from Sweden.

George married Esther Helena Jeppson on Feb. 12, 1912. They had two children, Harold Ernest Nelson, born Dec. 3, 1912 and died of a blood disorder Dec. 12, 1912. Roland Gordon Nelson was born on Nov. 15, 1913 and died on Feb. 1, 1980.

George purchased his first farm, W 1/2, NW 1/4, 12-15-8, Marietta Township, Saunders County, NE. He was 18 years old at the time. He purchased 140 more acres in 1905, this made a total of 220 acres. Bengt, his father, helped with the purchase and George repaid in the fall in 1915.

Esther and George were members of Alma Lutheran Church in Mead until they moved to Fremont, NE in 1942 and joined Sinai Lutheran Church. George served on the council of both churches. George was also on the Bank of Mead board.

During George's lifetime he operated a diversified farm and built a farmstead that stood the test of time. He and Adolph Penner built

The original George Nelson family home, 1902.

6% interest rate. Total property taxes in 1916 were $82.11. In contrast, total property taxes in 2014 were $4,921.12.

Cattle on the Gary and Lynette Nelson farm

My grandfather, Casper Lange, owned the land for 35 years, and then Casper's daughter, Loretta Piening and her husband Adolph owned it the next 34 years. Casper's granddaughter, Lynette Nelson and her husband Gary have been the current owners for the last 37 years. Lynette and Gary Nelson grow corn and soybeans on their farm in the Davey, NE area. They have two children - son Stacey Nelson of Davey, NE and daughter Crystal Perrozzi of Brooklyn Park, MN. They also have four grandchildren and two great grandchildren.

This farm is a gently rolling native prairie with multiple ponds, streams, woodland areas and some CRP land overlooking the Pawnee Lake area. A lot of the hedge post fencing is still original from 100 years ago. Consistently, this farmland has been primarily used for grazing beef cattle and hay - my grandfather ran cattle, my parents ran Hereford cows and we've continued the tradition with a registered Angus herd over the past 37 years. A decision that my grandfather made and all the hard work of my grandparents and parents has left a legacy for the future generations of our family. *Submitted by Lynette Nelson.*

NEWELL – The Newell family came to America from Ireland in 1765. After the American Revolution they began to move west from Pennsylvania to Ohio and into Iowa. William W. Newell was born in Louisa County, IA in 1858. W.W. and Sallie Parkin were married in 1883 and traveled to Wyoming, settling near Laramie Peak along with other family members. In 1897 the need to educate their children brought the family down the North Platte River. W.W. and son Fred drove the livestock and Sallie drove the covered wagon, crossing the Laramie River just below the guard house at old Ft. Laramie. They purchased land in Mitchell Valley below the Mitchell ditch. He raised alfalfa to feed his shorthorn cattle. The children herded them along the grassy meadows of the North Platte River. W.W. leased land in Sioux County. Delighted with the way his cattle gained on Sioux County grass he took up a homestead in Sioux County southwest of Agate, NE. Five of his children: Fred, Louie, Vinnie, Bill and Carrie filed on homesteads. In 1914, Carrie filed for a 640 acre homestead on (Section 14, Township 27, Range 57W). Carrie always ran cattle on her land and leased it to her father. He kept his steers until they were four years old. He then trailed them to the railroad at Andrews or Mitchell, NE. Carrie later sold her homestead to her sister, Nellie Newell Hatch. The land then came to Nellie's daughter, Helen Hatch Spencer and is now owned by Helen's daughters, Rebecca Eisenbarth of Torrington, WY and Bonnie Cawiezel of Morrill, NE. Bonnie's children, Pete Cawiezel and Casey Debus, are the sixth generation to run cattle on this excellent Sioux County grass. *Submitted by Bonnie Cawiezel.*

O'BRIEN – Brothers James (October 31, 1850) and Philip (August 15, 1852) were born to James and Rose (Naohan) O'Brien in Middle Town, Co. Monachan Ireland. They came with their family to New York in 1864 and lived in Fort Wayne, Indiana before coming to Nebraska. James arrived in 1872; Philip and the remainder of the family in 1874. The brothers worked for the Union Pacific Railroad. James worked building the transcontinental railroad and was present at Promontory, Utah for the driving of the Golden Spike. Philip worked for three years as a fireman out of Rawlings, Wyoming, returning to Nebraska in 1877.

In 1879 the brothers purchased 160 acres (N ½ NE ¼ and E ½ NW ¼ Section 19, South Pass Precinct, Lancaster County) from the Burlington & Missouri Railroad for $1,344.17. They later purchased the SW ¼ of Section 18. This land passed to Philip and Frances' son Joseph and then to Joseph's daughter Kathleen who sold it to the current owners Gary and Gerald Behrends (Philip and Frances' great grandsons).

In 1892, the O'Brien brothers purchased 160 acres (S ½ NE ¼ and N ½ SE ¼ Section 24, Buda Precinct, Lancaster County) for $5,600 and another 80 acres in 1893 (N ½ NE ¼ Sec 24, Buda precinct Lancaster County) for $2,000 from Erastus Rhykert. During this time period James and Philip worked at the Roca Rock Quarry, walking to work before dawn and returning home at dusk or after dark (eight miles each way). The land was later passed on to Philip and Frances' daughter Marie and her husband Kenneth Clough. And then to the Clough's daughter, Marian L. Clough Krueger, the current owner. Marian's son, J. Kenneth Krueger, currently lives on the homestead. It is being farmed by Philip and France's great grandsons Gary and Gerald Behrends along with Gary's sons Brett and Cody.

Philip and Frances Krapp, daughter of pioneers Joseph and Anna (Vollert) Krapp, were united in marriage October 1, 1895 at St. James Catholic church in Cortland. This was the first wedding performed in the new and first Catholic Church in Cortland. (The first Mass was held in the farm home of the bride's parents, which was the beginning of St. James Catholic Church.)

Philip and Frances made their home on the farm (Section 24 Buda Precinct, now owned by Marian L. Krueger) where their five children – Rose (Clare), Marie (Clough), Joseph, Lelia (Likens) and Veronica (Vern) Fuhrman – were born and raised. James resided with his brother Philip and family.

In 1914 the O'Briens purchased SE ¼ and NE ½ NE ¼ Section 25 Buda Precinct Lancaster County. This passed to daughter Rose and husband Leo Clare and then to their daughter Dorothy Behrends. Dorothy's sons Gary and Gerald are the current owners.

In 1917, the O'Briens purchased S ½ SW ¼ Section 33 Lancaster Co Buda Precinct and NW ¼ Section 4 Highland Township Gage County. This passed to their daughter Lelia Likens and then to her sons Jerry and Larry who are the current owners. Jerry and his wife Donna live on this property.

As an infant, Frances traveled with her family to Nebraska in a covered wagon in the spring of 1872. They lived on the farm (SE ½ SE ¼ Section 10 Highland Township. Gage County) in the covered wagon while building their sod house which was completed that fall. This land is owned today by great granddaughter of Joseph & Anna (Vollert) Krapp, Teresa Krapp Edwards McHargue.

Pictured l-r: Children of William and Sallie Newell on horseback.

James and Philip O'Brien

Philip and Frances O'Brien

In 1903, the O'Brien brothers expanded their holdings when they acquired 655 acres of pasture land (Section 3, Highland Precinct, Lancaster County) 2 ½ miles South of Denton, NE for $12,000. After Frances death, son Joseph purchased this land from the estate and his daughter Kathleen sold it to Audubon Society. It is now known as *Spring Creek Prairie*.

Gary, Cody, Gerald, and Brett Behrends

In 1921 Philip, Frances, and James moved into their newly built home (Blk 2-Lots 1, 2, 3 Cortland Original) where they spent their remaining years. They all died at home - James, June 29, 1931; Philip, December 10, 1939; and Frances January 1, 1949. They were lifelong Catholics and are buried at St. James Catholic Cemetery in Cortland.

Descendants of Philip and Frances are still living on land acquired by the O'Briens. This includes grandson Jerry Likens and his wife Donna and great grandson Ken Krueger. Owning and living on O'Brien land are great grandsons Gary Behrends and his wife Sue, Gerald Behrends and his wife Judy, and great great grandson Cody Behrends and wife Jill own and live on the O'Brien land. Descendants owning O'Brien land but living elsewhere in the community are grandchildren Marian L. (Clough) Krueger and Larry Likens.

The O'Briens acquired 1,920 acres during their lifetime. Throughout the years crops included wheat, oats, corn, milo, alfalfa, and soybeans. They also raised cattle and hogs. Today the crops are generally a rotation of soybeans and corn.

Their legacy is being carefully maintained by great grandsons Gary and Gerald Behrends and Gary's sons Brett and Cody. Together, they farm 1,360 acres of land that flows through the O'Brien and Krapp families.

For more family information see the Joseph Krapp in this book.

O'MARA – The O'Mara Homestead was acquired by William O'Mara on October 23, 1879. This consisted of one hundred sixty (160) acres, he purchased the rights from John Duffy.

William O'Mara, at age four immigrated to the United States from Ireland with his parents to Holland, Brown County, Wisconsin. Later he married Rose Kennedy, an emigrant, and they farmed near Holland.

After the death of his wife and daughter, he came to Plum Creek, NE to visit his sister, Mrs. Ann McNamara, who lived near where the present town of Eddyville now stands.

He bought the present farm, which had a four room frame house on it. He returned to Wisconsin for his four remaining children, Elizabeth, Rose and Julie and his only son, John W. who was an infant. He loaded an emigrant car and came to Plum Creek and they moved into their home on New Year's Day, 1880. He had sold his two improved farms in Wisconsin before coming to NE. This farm consisted of one hundred and sixty acres of good level land. William O'Mara died in 1880 at the age of thirty-five. After his death, his sister, Ann McNamara with her five children moved into the house to help care for the children.

When John W. became of age, he purchased his three sisters' shares of the estate and farmed there many years. John W. had five children William, Esther, Blanch, Emma (1904), and J. Earl.

When his youngest son J. Earl married in 1929, he and his wife LouElla moved to the farm and resided until the death of LouElla in 1994 and Earl in 1999. After the death of John W. in 1949, J. Earl purchased his sisters' and brother's share of the estate, this made 647 acres two irrigation wells were added making the land more productive.

J. Earl and LouElla had eight children: Earl J., Angela, James D., Roger, Theresa, Richard (1942), Kenneth, and Paul.

After the death of J. Earl, Kenneth and his wife Sarah (Kinish) purchased his brothers' and sisters' share of the estate. Kenneth and Sarah (2005) had three children: Richard, James, and Julie. Kenneth has one granddaughter Jean and one great granddaughter Afton.

Kenneth married Marjo (Blair Kort) who had two children Randall and Rita and four grand children, Logan, Becca, Andrea, and Allyson.

Kenneth has added cross fences reducing the big pasture into 80 acre plots with underground water lines for two 10 foot water tanks in each 80 plus water for winter range. Most of the fences have been replaced. One additional irrigation well with a wiper has been added. Several thousand cedar trees plus hundreds of broadleaf trees have been cut by hand and removed to improve grazing for the cattle and heating the homes for the winters.

The farm became a century farm in 1979.

OSBORN/MORGAN – In the spring of 1911 armed with only two mules, a plow and a tent, young Ernest Osborn left his family's beautiful Maple Grove Farm in Iowa to board an emigrant train for Sidney, NE. When he arrived he met Al Jorgenson, co-owner of the Al Jorgenson-Leo Osborne Real Estate Company and his future brother-in-law, to start a new life and home on the ground he had purchased in 1910 (SW 1/4-1-16-51) in Cheyenne County, NE. Possessing much common sense and knowledge of crops, Ernie searched the countryside looking for land where the prairie grass grew the tallest, knowing that soil producing healthy grass would in turn produce good crops.

Left to right: Wesley Morgan, Dana Morgan, Brody Morgan, Chris Morgan, Stuart Morgan, Linda Morgan, Elizabeth Higgins, Lianne Morgan Higgins, Rob Higgins, Danielle Higgins and Katie Higgins. Not shown, grandson Tyson Morgan.

Together Ernie and Lillie (Jorgenson) Osborn worked their farm, gradually increasing the acreage and raised five children - Loretta, Ruth, Marion, Clifton and Oakley on the farm. In the 1960s his son-in-law, Kenneth, Morgan began farming the Osborn farm. In 1967 Ernie passed away and the original quarter passed to his son, Oakley Osborn. Oakley conveyed that quarter to his sister, Ruth (Osborn) Morgan in 1977 and the land was purchased by Ruth's son, Stuart, and wife Linda Morgan in 1979. They began farming in 1978. On May 23, 1983 the land was conveyed to Morgan Farms Inc.

Stuart and Linda are 100% owners of Morgan Farms and currently live on the land and in the house Ernie Osborn built. Stuart and Linda raised two children, Chris and Lianne, on the same land as Ernie raised his children. Their son, Chris and family have moved back to the family farm. Chris is the fourth generation farmer on the Osborn/Morgan farm. Daughter Lianne and family live a few miles away on land purchased by Stuart and Linda.

OSLER – Joseph Osler filed on his homestead claim in April 1885. This was on the NE 1/ 4 of Section 35-9-36 in what was then Keith County, NE. He married Catherine Lambert in St. Paul, NE in April 1886 and they immediately came by train and team and wagon to the new "soddy" that spring.

Joe had been working with a railway survey crew and the word was that the new RR line was to be built about one mile north of his homestead. This all changed and the line was built nine miles north and was called the "HI-Line" of the Burlington RR.

Joe brought a small herd of cattle with him and raised corn for cow feed as well as wheat, rye and millet. He also cut native grass that grew

Back row: Josie, Amanda, Scott, Debra, and Victoria. Joan and John are seated.

in the draws for feed. Joe also filed for a 160 acre Timber Claim at this time. It was the NW 1/4 of 35-9-36.

Joe's son, Frank, moved onto the homestead along with his wife, Bertha, in 1921. They raised their family of six children here. The children's names were Katherine, Frank Jr., Barbara, Robert, Dorothy and John.

In 1953 Frank's youngest son, John, and wife, Joan, moved onto the home section and raised five children there. Their names were Susan, Scott, Douglas, William and Amy. John and Joan bought the home section in 1957 and an irrigation well was put down. They started raising Nebraska Certified Seed for which the family was honored as Premier Seed Grower of Nebraska in 1973. Certified crops grown were wheat, grain sorghum, chaffy warm -season grasses, cool -season grasses and alfalfa seed.

In 2004 John and Joan's son, Scott, and wife, Debra, bought the home section (35-9-36) and proceeded to raise their three girls - Victoria, Amanda and Josie at that place. All of them have been very active in the affairs of the farm. *Submitted by John Osler.*

OSTERGARD – In 1883 John Ostergard came from Denmark and homesteaded the northeast quarter of section 14\25. He was followed to America by his family in 1984. His wife and six children met him in Omaha. There John took sick and his wife nursed him back to health. She took sick and died and was buried in Omaha. The children were farmed out to other Danish immigrants excepting James, the eldest, and Crist, one of the youngest. The children kept in contact throughout their lives. As soon as possible the oldest boy took a Homestead. James Estergard took the Southeast quarter of Section 22 14\25. This was added to by a tree claim in NE 14\25. These two quarters are still in the family. The second quarter was proved up in 1891.

The tree claim contained honey locust and mulberry trees as well as other species of tree. The mulberry trees spread into some of the canyons around the area. Within this tree claim, James put in two rows of hog wire with a roof and filled it with straw. This created a protected enclosure to raise hogs after they were weaned. Additional tracts of land were added and when John died some of his land was auctioned to settle the estate. James and Crist returned to their land and purchased his father's land. Samuel retained a share of the land holdings. Land was then acquired when available and some traded hands with the settlement of estates over the years. All but one quarter remain in the Ostergard Family.

In the 1930s land was obtained 35 miles north in the sandhills. This resulted in a horseback cattle drive several times a year between summer and winter pastures.

On arrival to the US in 1883, John Ostergard lived in a cave. He then lived in a sod house. In 1911 James built a frame house. He and his wife Mary had three children who survived to adulthood. Harry alone stayed on the ranch with his wife Ilene, both full-blood Danes. They had six boys, four who have remaining interest in the ranch. Harry's wife Ilene came to the Etna Community by way of Omaha and LaGrange, IL. She became a crucial cog in the wheel of the community. Because of her varied experience and knowledge, Ilene would be invited to meetings to explain various government programs. She came from a life of all the modern conveniences, transitioning to a situation of no electricity or water in the home. No telephone. No car. Her husband worked from dawn to dark and the nearest neighbor was 1.5 miles away. Soon the couple had six boys.

Succeeding generations made some additions to the ranch and it was still intact at the turn of the new century with several sons retaining ownership. In 2013 the family held an ice cream social to celebrate 130 years of the ranch. Over 400 people came from Valentine to McCook, from Alliance to Omaha. Jack's kids, Ann Weekley and Ross Ostergard, plan to celebrate 150 years of the ranch in 2013. Jack's wife, Mary, is 1/8 Dane, so the Majority Danish blood line continues. *Submitted by Jack Ostergard.*

OTTO – My name is Lucile Carlson Otto and my husband is Glen Otto. My Great Grandfather homesteaded 80 acres of our farm on June 14, 1870. He had emigrated from Sweden in 1869; his wife, Nilla, with three children came to America in 1870. The homestead document states the W l/2 of the SW 1/4 of Section 10, Township 13 of Range #7E was the homesteaded land where he had built a house 12x16 with one door and two windows. John had also built one stable 12x20 ft., one granary 16x16 ft. and one well 29 feet deep and planted two acres of forest trees and several fruit trees. He also purchased another 160 acres from the Railroad. This land was located in the SE 1/4 of Section 9, township 13, Range 7. Eventually they raised corn, wheat, oats, barley, hay and hogs and cattle. John and Nilla had seven children of which my grandmother, Elna, was one of them. Elna and her husband, Hugo Carlson, took over the farm approximately in 1898 and lived in the larger house my Great Grandfather had built. They also had which children of which one was my father, Walter Carlson. When he returned from WW I after serving in France and Germany, he and my mother, Ellen, took over the farm. They farmed until 1954 when my husband, Glen Otto, and I took over the farm. Our children were Kristi, Mitchell, Kim and Beth. Mitchell started partnering with us in 1986 until two years ago when we retired. Mitchell and Valerie now live on the home place and are farming the land. Each family had lived in the house built in 1898, updating and modernizing it through the years. In 2014, Mitch and Valerie built a new house on the same location. The crops raised now are soybeans and corn rotation with some hay and no livestock. We have all enjoyed this way of life and the legacy that was left to us by my Great Grandfather.

We received the Pioneer Farm Award from Aksarben May 15, 1979. *Submitted by Glen and Lucile Otto.*

OWENS – Rose Ann Owens is a granddaughter of Hubert and Mary Lorens. In 1904 Hubert bought the southeast Quarter Section, 14 township, one North of Range 35 west of 6th P.M. situated in Union Precinct, Hitchcock County, State of Nebraska. Hubert bought this land from a Mr. Hugh Gillick, an unmarried man. Consideration was $1000.00. The deed was recorded on the 14th day of November 1904.

This land had a dug-out house, (three sides of the house were built into a bank of a hill.) It had a roof with board rafters and the ceiling was covered with tar paper, than sod. To the front was a wooden door and window. Mr Gillick gave Hubert his red polled milk cow. Cows were selling at the time for $11.00 to $15.00 each.

Lorens farm in the 1920s

Hubert built a new home from lumber in 1910. Hubert and Mary had five boys: Alois, (Rose Ann's father), Harry, John, Fred and George. George was born in the new house in 1910. This is the home that the family lived in and still stands today on the farm. *Submitted by Jerry and Rose Ann Owens.*

PAESL – Kenneth R. and Margaret A. Paesl Farm was purchased from the State of Nebraska (University Land), located in Knox County Nebraska to Mary E. Armstrong to Casper J. Paesl and Barbara Paesl on July 2, 1903.

The Casper and Barbarba Paesl farmhouse and barns.

Northeast Quarter of Section 26 Township 29 Range 5 West of the 6th principal Meridian, containing 160 acres more or less from Mary E. Armstrong & her heirs.

On December 12, 1941 Casper and Barbara transferred the farm to their son and wife, R.E. (Raymond) and Margaret T. Paesl.

In March of 1972, R.E. (Raymond) and Margaret T. Paesl sold the farm to their son, Kenneth R. Paesl and wife Margaret A. Paesl.

On October 22, 1955, Kenneth R. Paesl married Margaret A. Armbruster. Kenneth and Margaret moved into the original house on the farm, which was built in 1910 by Kenneth's grandparents, Casper and Barbara. Kenneth and Margaret, lived on the farm for 51 years. They raised their four sons, Gregory (deceased), Gerald (deceased), Daniel of Yankton, SD and Stephen and wife Linda of Omaha, NE.

Kenneth and Margaret have three grandsons, Jeffrey and wife Sara, Lance and Gregory, all of Omaha, NE.

Kenneth and Margaret raised cattle, hogs and farmed crops - corn, soybeans, and hay.

Kenneth R. and Margaret A. Paesl received the Nebraska Pioneer Award from the Knights of Ak-Sar-Ben Foundation for 100 years of continuous family ownership.

Members of the Paesl family

PARDE – On March 1, 1900 John W. and wife, Jennie, Parde purchased this 160 acre tract of land in Hanover Township, located in the southwest quarter of Section 22, Township 5, Range 7 for $4750.00

He farmed it till his death in 1911. Ownership of this property remained with Jennie and provided for her and their six children.

John W. and Jennie were the parents of Teete, Willie, Emma, Trientie (who died as an infant), Henry, Heye and John.

John J. Parde was born Jan. 17, 1909 and on June 17, 1931 married Harmina Johanna Miller. They were blessed with two daughters, Joan and Marian.

In 1951 the opportunity was provided for them to purchase the family farm and they continued farming it until retirement.

During his farming career corn, wheat, oats, alfalfa, sweet clover and later years also milo were produced.

John and Jennie Parde on their wedding day, March 24, 1893

John J. and Harmina Parde

Ray and Marian Mulder

Upon the death of John J. and later Harmina J., the property was inherited by their two daughters, Joan and Marian. At that time the two maintained joint ownership with Marian's husband, Ray, being the tenant.

In 2000 the family received the Pioneer Farm Family Award.

Currently, the farm is owned by Diane, the daughter of Ray and Marian Mulder.

PATTERSON – Wilbur Fisk Patterson moved to northwest Nebraska from Jacksonville, IL in 1885, establishing a homestead in Box Butte County. By 1889 he built a sod house on the property north of Alliance, and that year married Mary Jane Small who taught school two miles north of the homestead. Wilbur grew corn, oats, potatoes and flax. They had three children: Wilber Jr., Erwin S., and Mary.

Wilbur and Mary Jane lived on the homestead until their deaths, Wilbur in 1934 at the age of 79, and Mary Jane in 1966 at the age of 101. Their daughter Mary married John David Watson, an employee of the farm, and the couple took over the Patterson farm until 1982. J.D. and Mary Watson died three days apart in April 1986. Their son, John Howard Watson and his wife, Shirley, worked the farm until 1987 and their sons, Mark, Bruce and Paul, now operate the farm, growing garbanzo beans, corn, peas and wheat.

Standing l-r: Bruce, Paul and Mark Watson. Seated: Shirley and John Howard Watson.

Pictured l-r: John Daniel Watson, Jean Watson, John Howard Watson, and Mary Watson.

Wilber Fisk Patterson homestead

PEPPERL – In 1876 Mr. Pepperl bought 160 acres located in the SE 1/4 of Sec. 12, Range 9E, Township 2N, Pawnee County, NE. Corn, wheat and oats were the grain crops grown and cattle were raised on this farm.

This farm passed down from parent to child and from 1984 is now owned by Edward Pepperl as a life estate.

The farm currently retains 150 acres. The west half of Burchard, NE (10 acres) was carved out of the east corner of the property when Burchard was incorporated.

At present, corn, soybeans and prairie hay are grown on the farm and cattle are raised. *Submitted by Donald J. Pepperl.*

Second home, a sod house, 1881

Third home, 1890

Fourth and fifth home, 1968

Present day farm, 2015

PETERSON – Nels Peterson arrived in Illinois in 1864 from Sweden, and hearing about free land moved to Nebraska in 1870. His first home was a dugout on School Creek, later homesteading on a 150 acre farm located in section 25 of Emerson Township in Harlan County, known as the Nels Peterson farm. Nels married Carrie Peterson in 1881 and to this union six children were born: Theodore, Lavinna, Joel, Edward, Lillian and Myrtle.

Joel, Nels' son, farmed with his father. Joel married Daisy Olson in 1912. Two children were born, Marvin and Velma, to this union.

Marvin married Hazel Pearson in 1936. To this union four children were born: LaDonna, Gary, Marlys and Rodney. Marvin and Hazel purchased the estate in 1937. Gary chose to join his father in the farming operation.

Gary was united in marriage to Susanne McInturf in 1967. To this union three sons were born: Kyle, Kevin and Kenton. In 1975 the operation became incorporated and was then known as Peterson Livestock Inc., which stands today. The operation consisted at that time of raising cattle, hogs, corn, soybeans and alfalfa.

Gary's son, Kevin, who has a love for the land, chose to farm with his Dad. Kevin married Sarah Esslinger in 2009 and three children were born to this union: Mavrik, Karson and Korie. Gary and Kevin continue to raise cattle, corn, soybeans and alfalfa using no till farming practice and the aid of modem day technology, consisting of GPS guidance, tractors equipped with auto steer, variable rate pivot irrigation, the use of soil moisture sensors and pivot control from cell phones.

Gary's son, Kevin, and his family live on the family farm making the sixth generation to reside there. A quote from Marvin, now deceased "We present-day farmers, knowing the problems our ancestors faced daily, should be encouraged by the courage and determination they showed in times worse than we are experiencing now." The family is proud and grateful for the courage and determination of their legacy and hope for it to continue on for generation to come. In 1985 our family received the Nebraska Pioneer Farm Award sponsored by Ak-Sar-Ben, Marvin and Hazel were honored as the Harlan County Great Nebraska Family in 1992 and Gary and Susanne were honored as the Harlan County Great Nebraska Family in 2005. *Submitted by Gary Peterson.*

Joel with his first crop of irrigated corn, 1954

PIENING – Louie and Lizzie Piening bought their farm of 120 acres in Lancaster County, NE (29-10-5) in December of 1910 and the deed was filed in March of 1911 from W.C. and Loula R. Gunn of Bourbon County, KS and John and Marie Connor of Fremont, IA for $10,240 or $80 an acre. He owned it for 55 years. The land was then passed on to his son, Adolph and Loretta Piening, in the year of 1966. He owned

Lizzie and Louie Piening with Adolph

the land for 12 years. In 1978 the land then passed on to his son, Burdette and Virginia Piening. They are still current owners of this piece of property. They have planted soybeans, corn, grain sorghum and wheat. Burdette introduced gravity irrigation to the land and is now doing some drip irrigation. They ran a 100 cow/calf operation until 2003. They still live in the original two-story home built in the 1880s of which is held up by hedge posts. They have two sons, Troy and Terry and four grandchildren - Brandon, Dustin, Logan and Caitlin. Troy and grandson, Dustin, are farming the ground and have a welding/fabrication shop on the farm. Terry and family reside in Holdrege, NE. *Submitted by Burdette Piening.*

PIEPER – Wilhelm Pieper immigrated to America from Westphalia, Germany in the 1860s. In the northeast corner of Colfax County, just 1/4 mile weest of Dodge County line, an 80 acre tract of land was purchased by Wilhelm from Benjamin Kelly on March 27, 1888. Cattle, hogs and chicken were the livestock raised on the farm and corn, wheat, oats and alfalfa were the crops grown.

Wilhelm and Maria Pieper, ca. 1860

The original 80 acres remain in this farm which was passed from Wilhelm to his son, William in 1899. In 1947 the farm passed to William's son, Paul Pieper Sr. In 1988 the farm passed to James Pieper, son of Paul.

Corn and soybeans are grown now. Buildings remaining on the farm include the horse barn built in 1912 and restored in 2009, a cow barn built in 1922, a cement double silo corn crib built in 1949, brick home built in 1949, hog barn built in 1957 and a shop machine shed built in 1978.

Horse barn on Pieper farm, built 1912

In August 1988 the family was honored as a Nebraska Pioneer Farm Family at the Colfax County Fair.

Jim and Nancy (Thiele) Pieper have two children, Lori Pieper and Kelly (Pieper) Jensen. *Submitted by Jim Pieper.*

Pieper farm, 1980s

PIITZ – Leon and Nancy (Schmid) Piitz of Brainerd, Nebraska are the proud owners of Leon's family farm and were named the recipients of the 2015 Ak-Sar-Ben Pioneer Farm Family Award.

Leon's grandpa, Ed Dostal, purchased the ground in 1902. The farmstead's house, which was built by Leon's grandpa in 1905, served as the site of his wedding to Leon's grandma, Rosy Kovar. Leon's mom, Rose (Dostal) Piitz, grew up on the farm and she was born in the house. Rose and her three brothers Ed, Anton and Louis continued

Front row, r-l: Lady from Pioneer awards, Lindsey (Piitz) Fehlhafen, Shelby Fehlhafen, Nancy (Schmid) Piitz, Leon Piitz, and Rose Piitz. Back row: Josh Fehlhafen, Jeff Piitz, Alex Tietjen, and Chelsea (Piitz) Tietjen.

to farm the ground. It was passed on to Leon and Nancy in 2002, and they continue to farm with a rotation of corn and beans to this day.

In the past, however, other crops have been grown on the farm, such as wheat, oats, alfalfa, milo, and prairie grass. Cattle, hogs, horses, ducks, and chickens were also raised.

Leon has many fond memories of the farm, when he was growing until his uncles passed away. The entire extended family would come and celebrate Christmas in the house. They would also sled down the hill next to the house in the wintertime. Everyone would also get together at Eastertime and hide eggs around the house.

Leon and Nancy handed a three acre area around the house to their son Jeff Piitz, who is currently renovating the old house, and he plans to live in it someday. He lives on the yard to this day in a camper.

The farm ground will be some day handed down to their three kids, Jeffery Piitz, Lindsey (Piitz) Fehlhafer, and Chelsea (Piitz) Tietjen. Hopefully it will be handed down to further generations of the family.

The farmstead is located in Butler County, township 16-N, R-4-E Sec. 31.

PINE GROVE STOCK FARM – The Lancaster County land that is now Pine Grove Stock Farm (SE ¼, Section 19, South Pass Precinct, SE 1/4 19-7-7) was purchased by my Great Grandfather, Frederich Jacob Boesiger (10-18-1845 – 5-4-1928) from Wilhelmina Rausch Cramer on February 5, 1906 for $12,000.00. It was a wedding present to my Grandparents, Frederick Jacob Boesiger (10-14-1886 – 5-17-19964) and Maude Sargent Boesiger (10-8-1887 – 9-29-1978) on March 13, 1913. It passed to my parents, Donald Fredrick Boesiger (12-31-1915 – 11-27-1999) and Alice-Leone Ambler Boesiger (4-23-1915 – 11-10-2014) on Grandmother's death. After Mother's death, it passed to her children, Pamela (5-2-1943), Donna Mankin (5-24-1945), Fredrick (6-28-1947) and Susan Landau (6-25-1951) who established the LLC in 2015. My Great Grandfather also purchased the NW ¼ of the adjacent section to the east, which contained the family orchard lost in the 1930s drought. This section eventually went to my Madalyne Boesiger Latshaw (10-29-1921 – 11-24-2008) and then to her three children; her son Earl Latshaw (10-3-1961) currently farms both quarters.

My grandparents raised corn, wheat, oats, hay (red clover, alfalfa, prairie, brome), cattle, hogs, chickens, guinea hens with horses and mules. We assume they continued what Great Grandfather had raised. The orchard provided apples, mulberries, pears and cherries. Milo replaced wheat and oats in the 1960's and was replaced by soybeans; alfalfa is the sole hay crop. There is currently no livestock.

Mason A.T. Cooper and carpenter Chris Dohe created an unusual tile-block and wood barn in 1918. Pilasters reinforce and support a completely freestanding roof. Doubled rafters make up the double-gabled profile and are supported by wood buttresses anchored over the tile walls and floor joists embedded in the tile walls. Neighbors were skeptical and thought it would blow over in the first windstorm.

PIPER – James Buchanan was president of the US in 1860 when Frederick Piper bought the nearly quarter section of land from a soldier, John W. Pate, who had fought in the Indian Wars. The land had been part of his pay from the government for serving his country.

The land is located between Brock and Johnson in Nemaha County.

Frederick and wife Matilda farmed the land until retirement, and then sold it to his daughter, Mollie (Piper) Horrum. After her husband Steven's death, the farm was passed on to her brother, Otto Piper.

When Oklahoma was opened up for homesteading, Otto rented the land to George Meyer from Johnson and Otto headed to Oklahoma to stake his claim. He settled in Faye, OK and married Leona (Crawford) Piper. They raised four children: Mary, Fredrick, Gladys and Nellie.

George Meyer rented the land for 27 years. In 1936 Otto and Leona's son, Fredrick, purchased the farm. In March 1936 Fredrick and his wife Velma (Ruhl) Piper, children Marion and Doris left Oklahoma for Nebraska. Fredrick drove a 1929 Chevy pulling a trailer with all of their belongings. He hired a trucker to haul their horses and cattle.

After moving to Nebraska, Fredrick and Velma had three more children – Vera, Jack and Kenneth.

Throughout the years Kenneth farmed with his dad until he and his family moved to Otoe County. Then Fredrick and Velma's son, Marion and grandson, David rented the land.

After Marion's death in 1992 grandson, David, continued to farm the land until 2000.

In 2000 Craig Rieken, son of Fredrick and Velma's daughter, Vera (Piper) Rieken, took over the farming operation. Today he owns and continues to farm the land that has been in the Piper family for 155 years.

Fredrick and Velma farmed for 56 years until he was 92 years of age. He was recognized as the oldest active farmer in Nemaha County. *Submitted by Vera Rieken.*

PLACEK – Antonin J. Placek (A.J.), a master merchant tailor and musician, was born in Sadska, O Kresu Podebrdy Bohemia on November 5, 1856. His wife, Mary, was born in Sadska on August 12, 1860 to Josef and Mary Horak Noll. They were married November 10, 1880. The couple had three children: Marie, Anton and Josef.

Five years later, the family set sail for America, where they took a per-emption on 40 acres of bottomland at the tiny village of Sparta near Niobrara, Nebraska. A.J. joined Professor Frank Lenger's military band, the leading band in North Nebraska, as a drummer. Little Josef died and was buried on the farm. Three more children were born in Knox County: Anna, Karolina (Carrie) and Joe.

When the Ponca Indian Reservation was opened for settlement, they moved to Boyd County and settled on a 160-acre homestead in the SE ¼ of Section 5, Township 33, Range 10, northwest of Lynch, where they resided for the rest of their lives. The barn was the first building erected and they lived in it with the livestock until the house was built.

Here four more children were welcomed into the family: Amalie (Mollie), Anastazia (Daisy), Ludwig, and a stillborn child. Daisy died and was buried on her 26th birthday.

The Placeks were charter members of the Lynch ZCBJ Lodge and Mr. Placek was also a member of the school board of District 35 for many years. He also played in Dr. Ira's band.

They celebrated their golden wedding anniversary on November 13, 1930, at the home of their daughter, Mollie, and husband, Frank H. Lenger, in Niobrara. Frank's brother, Professor John F. Lenger, Gregory, South Dakota, 'tied the knot' — the shape of a quarter-inch rope tied about their wrists — and conducted the same wedding ceremony for the couple when they were married 50 years earlier. One of the most unique gifts they received was an old-fashioned pipe imported from the old country.

A.J.'s death on February 11, 1948, broke the banns of marriage for over 67 years. Mary died April 12, 1953 and the land passed into the hands of their son, Anton Jr. A.J. and Mary's grandson, Willard, purchased the land in 1964. He and his wife, Marita, have two sons, Eugene and Keith, two grandchildren, and a great-grandson.

The land became a Pioneer Family Farm in 1992.

PLANANSKY – Josef and Veronica Planansky and family arrived in the US from Bohemia in 1872; settled in Pawnee County, NE, later moved to Box Butte County in northwest Nebraska in 1887. They filed on land in Box Butte County. Veronica and son, James, age 18, filed on N 1/2 Section 18-27-51. James then bought 80 acres, E 1/2 of SE 1/4, 18-27-51 from the US Government for $1.25 per acre; Benjamin Harrison was President at that time. Over the years he bought SE 1/4 Section 7-27-51 and from a different owner the SW 1/4 Section 7-27-51 and NE 14 Section 13-27-52 and SE 1/4 13-27-52. Total acres of the farm is 1040, all in Box Butte County. They had a cow herd and raised wheat and cattle feed.

Present owners Vern and Helen Planansky standing in front of the house built by James Planansky in 1922, taken in 1986.

James married Sophia Caha in 1910. They raised three sons and one daughter, all born on the farm. Vern, the youngest son, born Nov. 19, 1917, remained on the farm. He married Helen Jezek in 1947. They bought the farm from his father in 1950 at a reasonable price, with the stipulation that his parents live the rest of their lives on the farm under the care of Vern and Helen. James died in 1954 as a result of a car accident. Sophia was bedfast with arthritis, died in 1961.

Vern and Helen raised three sons - Gary, Dennis and Rodney and daughter, Christine. None stayed on the farm after graduating from college.

Land owners of this farm were Veronika Planansky, then her son James, and then James' son, Vern, who is the present owner.

Vern was awarded the "Family Recognition Certificate for Pioneer Families of Nebraska" on March 6, 1989.

The original house and outbuildings of the A.J. Placek farm

James and Sophia in buggy. Sophia is holding their first child, Victor, born March 20, 1912.

Early picture of the farm site. The center rider is James Planansky; Veronica Planansky is by their home in the background.

Vern retired from farming in 2001 at age 83. Vern and Helen are still living on the farm in the house that James had built in 1922, and hope to live there the rest of their lives.

PORR – Leopold Porr Sr. was born in Duchroth, Rheinpfalz, in 1812. He married Elizabeth Hoffman, and they had four children: Leopold Jr. (d. 1925), Katherine (d. 1910), Charles (d. 1915), and Elizabeth (d. 1907). On March 7, 1857 at the age of 45, he and his family left Duchroth, Germany for America, leaving from France. They arrived in New Orleans, LA, on May 11, 1857, after a 67-day trip on the sailing vessel named, The Adams. The family journeyed up the Mississippi and its connecting rivers by steamboat, landing at St. Joseph, MO. They stayed for three months, and next, they crossed the Missouri River by ferry to Nebraska City, NE. In the spring of 1857, the Land office initially opened in Brownville, NE, and in 1859, Leopold Sr. was granted a land patent situated eight miles south of present-day Humboldt.

Seated from left: Margaret, Charles Porr Sr., Minnie (with hat), and Leopold Porr Jr., ca. 1900.

Next, the family travelled about thirty miles south on foot and by oxen cart to the new land, where the family built a small log cabin. The men farmed with the team of oxen, having no horses early on. Whether they hauled wheat, bacon, butter, etc. to St. Joseph, MO, or Brownville, NE, they usually received very low prices for their grain and meat. Often the family exchanged their goods for wearing apparel.

Great-great-grandfather Leopold soon became an American citizen in 1862. Upon his death in an accident at age 67, his land holdings were then split between his two sons. Leopold Jr. stayed at the home place, and Charles Sr. settled on the south half of the land patent. As the sons' families grew, their descendents bought more land. It is here in this story that we leave Leopold Jr., his sisters and their families.

As time passed, my great-grandfather Charles Porr Sr. and his wife, Margaret (Schmitt) Porr, had nine children. Later in life, they deeded their new land to their son, Charles Porr Jr. (1852-1953). By now, Nebraska was a state, and younger Charles had a new "home place" located near the old post office/store then known as Athens. The land is described as South Half of the SE ¼ of Sec. 17, Township 1, North, Range 13, in Speiser Precinct, Richardson County, State of Nebraska. It is a mile and a half west of the original patent. Charles married a Swiss neighbor girl, Christina Fankhauser. In 1906 they had one son, Karl Porr, who was my father. He walked early nearly a mile to the Athens country school and attended the German Bible School in the summers. The Porr family raised chickens, pigs, and had milk cows, and also grew corn and alfalfa, but now farmed with work horses on their subsistence acres. In his turn, Charles Porr Jr., in 1944, in a Quit Claim Deed ceded his land to his son, Karl Porr. Karl changed to mechanized farming and began raising Hereford cattle in the 1950s.

This log cabin was the original Porr family home.

Karl left the farm to his wife, Margaret Porr, who left the land to her daughter, Polly Porr Duryea, on 13 December 2003. I am the current living land owner, now in 2016. This property was approved for the Nebraska Pioneer Farm Award on May 13, 2016, as the direct descendants have owned this property for over a hundred years. The Porrs were founding members of the German Evangelical Church, later known as the Four-Mile Church. All the Porr family mentioned above are buried at Four-Mile Church Cemetery, south of Humbolt, NE, on the intersection of NE Highway 8 and 105. *Submitted by Polly Porr Duryea.*

POTMESIL – Alois Potmesil arrived in the USA in 1895 and worked as a farm laborer until 1900 when he purchased land in Shell Creek area, S 1/2 of NW 1/4 Sec. 25 in Township 21, Range 4, for $1750.00 from P.E. McKillip. In 1910 he built the house that still stands. Alois married Marie Blecha in 1911 and had four children: Mary, John, Bessie and Rosie, all born in the house. In 1953 Alois and Marie moved into Newman Grove and son, John and wife Thelma moved to the homestead house. John and Thelma raised seven children: John, Don, Karen, Delbert, Theresa, Helen, Kenneth and Margie (who died in infancy). Alois walked to the farm everyday to help his son on the farm. In 1968 when Alois died, John and Thelma became owners of the land. In 2004 the homestead became a Pioneer 100 year farm. In 2004 John died, the land was put up for sale and John's son, Don and family bought the Homestead and continue to own and keep the farm for the family. Making it a four generation family farm. Everyone in the family worked hard to maintain the farm through the depression, cholera, grasshoppers, tornados and droughts and through their efforts the farm is still in the family. The families raised animals: chickens, hogs, cattle, and horses, to name a few. They raised a garden so they could be self relent and then crops. *Submitted by Donie and Earleen Potmesil.*

The Potmesil farm

PRINZ – A young German immigrant, Anton Prinz, came to Nebraska in 1881, married Mary Ann Schinstock in 1894, and bought two 80 acre sections in Cuming County, Cuming Township, Sec 17, NE 1/4.

Anton and Mary built the farmstead with an orchard as they raised corn, oats and alfalfa and livestock. They had nine children, four of whom lived to adulthood,---Frances Prinz Brown, John, Arthur and Victor Prinz.

In 1933 Anton and Mary retired and moved to town as Arthur married Elizabeth Meiergerd and took over the farm. The farm transitioned from horsepower to tractor power. They had two daughters,

Delores Meister and Norma Jean who is Sister Mary Beth Prinz. Upon the death of Arthur in 1959, Elizabeth moved to West Point. The farmstead and land was rented.

Elizabeth retained ownership until she passed it on to her grandchildren, Meister Family Farms LLC, who includes Arthur, Donald, Patrick, Jean, Barbara, James, David, Julie and Janet Meister.

Soil conservation, minimum and no-till practices and maximum care of the soil has always been practiced. Pivot Irrigation was added. Later the building site was removed. A hundred-year-old lone cottonwood tree stands on the corner of the property. *Courtesy of Delores Meister.*

RADER – Jesse Edward Rader married Josephine (Hobbs) in 1907 at Lincoln, Nebraska. To this union 15 children were born: seven sons—Vernon, Kenneth, Edward, Harold, Jesse Jr., Glen, and LeRoy—and eight daughters—Lucille, Clara, Evelyn, Goldie, Grace, Elaine, Uralee, and Faye. In 1907, Jesse Edward Rader and wife Josephine Rader with Brother George William Rader moved to Custer Township in Antelope County, Nebraska. On October 2, 1913, they purchased 240 acres of land. The NW ¼ and NW ½ of the NE ¼ of Section 11-27-6 in Ellsworth township, Antelope County, Nebraska. The Jesse Rader family raised crops of corn, oats and hay. They raised livestock of horses, cattle, hogs, and chickens. They butchered their meat. They grew and canned a big garden. These things were done by horse drawn equipment and manpower. Later in life, Jesse Rader moved part of the family to the Los Angeles, California area.

On September 3, 1946, Edward Louis Rader married Lois Mae Evans in O'Neill, Nebraska. To this union two children were born: Galen Edward Rader and Sandra Kay (Fuchtman). In September of 1950, Edward Rader purchased the SE ¼ of Sec. 9-27-6, Ellsworth township, from Josephine Rader. In November of 1952, Edward Rader purchased the NW ¼ of Section 17-27-6 from Lyle and Maxine Proctor. On September 14, 1957 son Edward Rader and Lois Rader purchased the NW ¼ and the SW ¼ of Section 11-27-6 from Jesse and Josephine Rader. In March of 1962, Edward Rader purchased the NE ¼ of Section 15-27-6, Ellsworth township, from Herman Dicks. The Edward Rader family raised cattle, hogs, chickens, and did cropland farming. Eddie and Galen Trucked did custom harvesting and sold Funks Seed. They managed a Funks Seed warehouse in Nebraska.

On November 9, 1974, son Galen Edward Rader married Marilyn Jean (Hemenway) in Neligh, Nebraska. To this union three children were born: a son, William Joe Rader, and two daughters, Kristine Catherine and Catherine Mae. On March 1, 1982, Galen Rader and Marilyn Rader purchased the NW ¼ of Section17-27-6, Ellsworth township, Antelope County, from Grace (Rader) Miller. In March of 2002 Galen purchased 480 acres from father Edward Rader, the NE ¼ of Sec. 15-27-6, SE ¼ of Sec. 9-27-6, and the NW ¼ of 17-27-6, all in Ellsworth township, Antelope County, Nebraska. The Galen Rader children: William Rader, married Anne Slahn of Norfolk, Nebraska on October 3, 2000; Kristine Rader, married Steven Wacker of Plainview, Nebraska on May 24, 1998; and Catherine Rader, married Luke Johnston of Lincoln, Nebraska on February 17, 2009.

The Rader family homestead in 1946. The barn and outbuildings have been preserved for future generations.

In 2016, Galen and Marilyn Rader still live on the property and actively farm. We are blessed with six granddaughters—Brittani Wacker, Carissa Wacker, Ashley Wacker, Sidney Rader, Sierra Rader, and Allison Johnston—and one grandson—Abram Johnston. The family is proud and grateful of their legacy, and hope it continues for generations to come.

RAMEY – 160 acres in Nance County, NE was granted by President Garfield in 1884 to Mr. Ramey.

He lived in a cave while building a one room house in 1890. Evan Francis bought the farm with borrowed money in 1891. He immigrated from Wales in 1881 to Livingston, IL. In 1885, age 25, he married Elizabeth Travers in Manhatten, Will County, IL. They moved to Clay County, IL. Then to Beaver Township, Nance County, NE in 1891. They had six children: Charles, Thomas, Albert, Ivor, Lillie and Elizabeth. Mrs. Francis died when Elizabeth was born and Mr. Francis continued farming with horses and raising the children alone.

Ivor and Hannah Ramey with their seven children.

Ivor married Hannah Larkins in 1930. They had seven children: Richard, Marilyn, Wendell, Shirley, Beverley, Blanche and Ivor Jr. "Bud."

Evan died in 1933. Ivor and Charles took over the farm. Farming of hay, corn, wheat and oats was done with horses and later with a steam tractor. Some livestock was also raised.

In 1948 electricity came to the area. Ivor and Hannah remodeled the home in 1949. Two additions, indoor plumbing and basement with a tunnel to the storm cellar, were added to the house.

In 1956 a tornado demolished the barn, garage (replaced with large Quonset) and chicken house.

In 1960 son Wendell married Arletta Bantz and took over farming of the land. They had four children: Larry, Lanny, Loren and Lori.

Just before his death in 2006 Wendell sold the farm to his son, Loren. Loren lived in the old farmhouse until 2014 when he built a new one. Loren's main crops are soybeans and corn. *Courtesy of Arletta Francis.*

RATH – Michael H. Rath is the great great grandson of Johann Rath. Johann homesteaded the original 80 acres in October of 1883. It is located in Fillmore County, Grafton Township, located at section 20-Range 8-4. Johann arrived in the Sutton area in 1875 from Glueckstai, Russia. Before he and the others could immigrate, they had to give up their Russian citizenship. His first wife, Elizabeth Braun, accompanied him over from Russia, bringing two children with them, Elizabetha and John Jr. (William's father). Two more children were born once here in the sod house, Christian John and Christina. Johann remarried three more times, having a son Heinrich "Henry" with his second wife, Katherine Reuther Ederhardt, he was also born in the sod house. Henry was a well known and liked person in the area.

The sod house was inhabited till the 1940s. The Bill Bauer family rented it from Christian Rath. This family had three of their nine children there. They recalled the dirt floor, walls and window wells 1-1/2 to 2 feet deep. Also how they would make huge batches of popcorn using a screen cage on the top of the wood stove.

Another 20 acres was added in 1967, which had a house and other outbuildings, bought by Lawrence and Janet Rath. In 1982 Michael tore down the old house and built a new house; both of his brothers, Rick and Todd Rath, helped in the process. In 1995 Mike started raising cattle so he leased the pasture from his parents, which included the small fields, which we farm. In 2006 Shelley Seager moved to the

Pictured l-r: David Miller, Sherry (Miller) Buck, Rick Rath, Janet Rath (in front), Shelley Seager, Mike Rath, unknown, and Todd Rath

property, Mike's fiancee; the cows have never been tamer. This year we bought 24 laying hens and ended up with about 16 guineas. We repurposed an old corn crib and moved an old metal building from a old farmstead to make a palace for our girls. They will be supplying the town of Grafton with eggs. Mike purchased the 80 acres from his Mom in 2012.

In 2014 we received the Nebraska Pioneer and Heritage Farm Award, at the Fillmore County Fair. We grow crops on that piece of land. We still find pieces from the old homestead after rain, pieces of rusty equipment, pottery shards, old glass and even marbles can still be found up there. We grow corn and beans using about 16 acres. The rest is still pasture, were we raise cow-calf pairs. The pasture has a gravel and sand pit, it was used to build a lot of Highway 6 around Sutton. *Submitted by Michael Harold Rath.*

RATHBUN – Job Beebe Rathbun born 1829, Steuben County, NY, married Olive H. Buck in 1853. They farmed in Rochelle, IL, had three daughters and three sons: Fanny, Clarence, Isabel, Ellsworth, Edward and Esther. In 1871 sold their land and traveled by covered wagon to Crab Orchard, NE where they purchased land.

Aug. 22, 1876 they invested in land in Gage County, moved one-fourth mile east of DeWitt, NE near the Big Blue River, Section 18, Twp. 5, Range 5, Grant Township, 240 acres.

Son Clarence married Mary Nail of Irving, IL, Sept. 20, 1885, later moving to this farm. Three children - Henry L., Mae and William. Building a second home on this farm, with his brother, Ellsworth.

Henry Rathbun married Ada Meier of DeWitt, NE, Nov. 23, 1910; moved to the original home. Three daughters - Vera and twins Meryl and Beryl. A huge cave was built 1909, place for potatoes, home canned fruits and vegetables.

1914 plum, cherry, apple trees and a black walnut grove were planted; crops were corn, wheat, oats, barley and alfalfa; farming with horses. Raising hogs, milk cows and chickens. His first tractor a 1920s Case. Sawing wood and hauling ice from the ice house on the Blue River.

1937 application to irrigate from the Big Blue River was obtained. Henry built an elevated earthen dike to pump water into for irrigation. Using four horses and a "fresno," two horses and a "slip" plus shoveling by hand to build the grade.

Rathburn farm, 1930s. Pictured are Vera, Ada, Meryl and Beryl

Aug. 24, 1947 - Meryl Rathbun married Russell Zimmerman, living at the east place. Children Dennis and Linda moved to the west place in 1964.

Irrigation by flood using gated aluminum pipe. Crops later were corn, milo, changing to soy beans and raising Black Angus.

Dennis married Mary Kelley of Omaha; two sons, Ryan and Scott. Linda married Charles Burns of Plymouth; three children - Matthew married Stephanie Fjordbak, two children Eli and Wyatt; Joshua and Rachel married John Kleppe.

Russell and Meryl have lived on this farm 68 years, retired in 1990, having received the Aksarben Nebraska Pioneer Farm Award. The farming is now done by a renter. *Submitted by Meryl Rathbun Zimmerman.*

RATHJE – Bernhard "Ben" Christian Martin Rathje was born in Germany on Oct. 18, 1870. He came to the US in 1897. Ben married Anna Fiedler on Feb. 14, 1900. They raised four children – three sons and one daughter: Arthur – Hubert – Freda and Walter.

On Jan. 17, 1913 they purchased 160 acres in Nance County, NE. The southeast quarter, Section 23, Township 17, Range 7, west of the Sixth Principle Meridian. Livestock and many crops were raised on the farm.

Ben and Anna Rathje's home and their family, 1914. Pictured l-r: Walter, Ben, Anna, Freda, Art, and Hubert.

Ben passed away in September 1926. His wife, Anna, lived and worked on the farm until their youngest son, Walter, returned home from military service during WWII.

Walter took over this 160 acres farm and added additional acres to his farming operation in the following years.

Walter married Laverne Keith on Jan. 31, 1946. They had two children – Karen and Larry. Walter's mother, Anna, passed away in March 1956. Walter moved his family to Fullerton, NE in 1959 and slowly started to sell off the farm buildings – house – and equipment. As the buildings sold, this added extra farming acres to the productivity of the farm.

In 1967 Walter and his wife moved to Fremont, NE. Walter passed away in May 1996. His wife, Laverne, continued to operate the farm till she passed away at 97 years old in May 2014. Their daughter, Karen Rathje, inherited the farm April 10, 2015.

Over the last 10 years a number of conservation improvement projects were completed. This has opened more acres for farming.

We are thankful and very proud to continue the legacy our grandparents and parents started. We hope the future generations continue the path for us.

RATHJE – The Rathje farm is located at NE 12-13-1, Hackberry Township, Polk County, west of Surprise, Nebraska. The east half of the quarter was homesteaded by John A. Griffin in 1875, and the west half by William Henshew in 1874. In 1876, Giffin bought the west half from Henshew. The quarter was then sold to Henry J. Crapenhoft in 1885.

In August 1892, Henry and Talleta Rathje, immigrants from Germany, purchased the quarter. Henry had brothers living in the area.

Home of Henry and Talleta Rathje, 1908

In German tradition, Henry built a large barn. A large house was built later. The inside of the barn had room for ten horse stalls, milk cow area, farrowing and feeding hogs, grain storage, and a hay loft above everything. A driveway went through the barn for loading and unloading hay or grain. A six-sided wooden cupola rested on the center of the roof. A corn crib and chicken house were built. Crops for the time were corn, wheat, oats, alfalfa, and timothy grown to feed livestock and cash crop.

Henry died in 1919 and the land was rented. Taletta lived on the farm without electricity or running water until she died in 1944. They had no children, so renting now went to an estate. Henry and Taletta are buried in Faith Lutheran Cemetery, Butler County.

In the fall of 1951, Harold and Hannah (Rathje) Miller purchased the farm. Hannah was a niece of Henry. Farming the land began in the spring of 1952. Summer of 1956, Harold, Hannah and children Maynard, Arlan and Alta moved from a Butler County farm to Polk County. The house was upgraded to modern standards and conveniences.

Crops raised were corn, milo, wheat and alfalfa. Cattle, hogs and chickens were raised. They was always a large garden. Irrigation began in 1964 with gated pipe and sprinkle pipes. Today it is gated pipe and pivot. Corn and soybeans are cash crops.

Hannah passed in 1967, Harold in 1984. They are buried in Faith Lutheran Cemetery, not far from Henry and Taletta. The farm was willed to Arlan Miller and Alta (Miller) Beltz.

The farm has been in the Rathje family since 1892 and continues today in 2016. Maynard and Kevin Miller operate the farm under Homestead Farms.

REEKER – In 1910, Ferdinand and Minnie Wisch purchased 480 acres in Madison County, Highland Township, 17-23-3. The land was used for grassland, cattle and horses. The farm passed to their son, Frank Wisch, and then to Gene (son-in-law) and Ruby Reeker, to maintain 200 acres today raising corn and soybeans. Still standing is a barn built in 1902.

Following the death of Ruby, Frank married Annie Reeker in 2014. They have five daughters — Suzanne, Rebecca, Genie, Mari, and Shervy — fifteen grandchildren, and fourteen grandchildren. Frank retired from farming in 2014.

RETZLAFF – Albert L. Retzlaff, my great, great uncle, purchased the SW 1/4 35-T10-R8 farm from Mr. and Mrs. Edward C. Marx on Sept. 18, 1905. This farm was previously purchased from the Burlington and MO River R.R. Co. Albert was a breeder of Shorthorn cattle and Poland China hogs as well as growing various row crops.

Riding the farm machinery at Uncle Albert's farm

My father, Dale E. Retzlaff, farmed for Albert and upon his death, my father and mother acquired the farm commonly referred to as "Albert's." As a child, my sisters and I visited "Uncle Albert's" frequently.

I have called this farm my home for many years, raising my children, Mike and Haze, here. My husband, Doug, and I continue to live on the farm and raise Shorthorn cattle, sheep, goats, chickens, alfalfa, corn and soybeans. We are blessed to have our grandchildren, Michael III, Izek and Henry, visit often to help with the care of the animals and love riding on the farm machinery.

Upon the death of my parents, Dale and Rhoda Retzlaff, the farm was passed down to me and my children.

Albert was the son of Charles F. Retzlaff who settled in Lancaster County in 1858 and established the Stevens Creek Stock Farm.

In 2014, the Nebraska Pioneer Farm Award sponsored by the Knights of Ak-Sar-Ben Foundation and the Nebraska Farm Bureau was presented to the family. *Submitted by Cheryl R. Johnson.*

REWERTS – On July 22, 1904 155 acres was purchased in Adams County, located in NE 1/4 3-8-10 by Frank Rewerts and William A. Rewerts, Steve's grandfather. The farm passed to Grandmother Anna B. Rewerts on May 28, 1945. On March 3, 1965 the farm passed to her son, Lloyd W. Rewert (Steve's father), then to his wife, Dorothy L. Rewerts (Steve's mother), on March 7, 1980, On Aug. 15, 2007 Steve (grandson of William) and Jeanne Rewerts purchased the farm from Brad A. Rewerts. The house yard is owned by Bradley A. Rewerts, grandson of William.

Anna B. Rewerts and son Lloyd at the farm house in 1925.

Aerial view of the farm in October 1978.

The brick 1-1/2 story house built in 1925 remains on the farm and is the hub of family gatherings.

Steve and Jeanne have two children, Connie C. Grabil and Brian E. Rewerts. *Submitted by Steve and Jeanne Rewerts.*

RHODIG – I acquired the farm in June 1981 when my mother passed away.

George Washington Seybold, my great grandfather, and his son, George William Seybold, my grandfather, purchased the 160 acres farm from Daniel McKay and his wife Feb. 26, 1902.

In 1907 George Washington Seybold sold the property to his son, George William, my grandfather, and he moved to Trenton, NE. Farming the land with his four sons. The land was transferred to my grandmother, Alice Carskadon Seybold upon George William's death.

In March 1955 my father, Leslie R. Seybold and his half-sister, Addie Masters, executors of the last will and testament of Olive Carskadon Seybold, deceased, sold the 160 acres farm to Althea Dickerson, my mother Seybold and Phyllis Seybold Golter Rhodig in equal shares. (My first husband, Arlyn R. Golter, was killed in Korea in 1953.) In 1981 my mother, Althea Dickerson Seybold, passed away. As of 1982 Phyllis and present husband, Eldon H. Rhodig, are owners of 149 acres of the SE 1/4 of Franklin County, Township 31, Section 4, Range 16, raising mostly corn and soybeans with an irrigation system on the Seybold-Rhodig farm, located seven miles south of Wilcox, NE. The other 10 acres belongs to my brother's family.

I am married to Eldon H. Rhodig since August of 1954. We have two sons – Allen L. Rhodig and Brian S. Rhodig; five grandsons – Joshua Jacob, Ethan, Benjamin and Nicholaus Rhodig.

I retired in 1993 after working for a used car dealer for eight years, 25 years for a dentist and eight years at a small animal serum plant.

Now I've taken up bowling two times a week, doing genealogy for six families, gardening in spring and summer. We did our traveling when health was better. *Submitted by Phyllis Seybold Rhodig.*

RIGGS – On July 31, 1915, Frank and August Riggs purchased 320 acres located in SE 1/4 S21 T32 R 51 and NE 1/4 S28 T32 R51, Dawes County, NE.

Frank and Augusta Riggs lived on the farm from 1915 to the late 1950s. They raised six children, four boys and two girls, with no electricity at all and never had indoor plumbing until the last few years. While they were trying to get established, they would trail the animals north to the river to water them at least twice a day. Water was also hauled six miles from town to provide water for the family and chickens.

Sandy and Sam Riggs, August 2, 2015

They soon hand dug a well and a windmill was installed for watering the animals and household. A wooden storage tank that sat on stilts followed. This tank gravity fed water into the house providing indoor plumbing. A dam was dug near the well, which later took the life of the youngest son. They were dredging out the dam with a team of horses pulling the big scraper bucket when the youngest son was pulled under getting caught up in the reins, weeds, and mud causing him to drown in the water and mud. There are no original buildings left on the farm.

Ralph Riggs (l), son of the second owners, with Frank and Augusta Riggs (first owners).

Frank Riggs was very involved with 4-H. He was exceptional at raising Grand Champion hogs. Frank was an exceptional orator of stories and always had an audience everywhere he went.

Ralph was Frank's next to the oldest in the family. Ralph received an award posthumously for "The Most Persistent Farmer."

Sam is Ralph and Leona's fifth child out of seven, one girl and six boys. *Submitted by Sam and Sandy Riggs.*

RISH – One of the best known and most prosperous retired farmers to be found in Merrick County, is John M. Kyes of Central City. He is also to be numbered among the oldest settlers, having come to Merrick County in August 1862.

On Christmas Day two years later, he married Miss Viola Parker of New York state and they enjoy the distinction of being the first couple to be married in that county.

Mr. Kyes, born in Oswego County, NY in February 1839, was the oldest of three children born to Russell and Nancy Colgrove Kyes. Mr. Kyes grew up to manhood in New York state and remained there until about 1855, when he went to Iowa. He remained there until the spring of 1860, when again he joined in the never ceasing westward movement, going this time to Denver, CO. He remained there two years, prospecting for a time and afterwards driving on a stage line.

When he first came to Merrick County, Mr. Kyes drove for the Western Stage Company for about two years. A short time after his marriage, he and his wife moved to a farm south of Chapman. About 1861 he took up a homestead of 80 acres on Prairie Creek, northeast of Archer. He later purchased 320 acres adjoining. For many years this place was known as the "old home place."

Mr. Kyes was instrumental in organizing his school district number 23 and for four years served as treasurer of the same.

Mr. and Mrs. Kyes had 10 children; seven of whom are living. Those still living are as follows: Mary L.; Ella Wickham of Furnas County; Anna, now Mrs. Willard Halsey of Nance County; Allie, now Mrs. Benjamin Lantz of Chapman; Russell J.; John H. and Jessie. The last two named are twins. *Submitted by Carolyn Kyes Rish.*

RITER – The farm was originally bought by Patrick and Catherine Hamilton 1890. The 360 acres is located in Dixon County, Daily Township. Crops were corn and small grains. Livestock was cattle, hogs and horses.

The farm passed from Patrick to son, Joseph, and his wife, Victoria. They were on the farm from 1919 to 1950. They had three sons and one daughter. All three sons enlisted in WW II. The oldest son, Patrick, was killed in action in Luzon, Philippines so Joseph retired in 1950 and the farm has been leased since then. In 1950 the farm passed to Lloyd, Mike and Margaret.

Joseph and Victoria had 13 grandchildren, 18 great grandchildren and six great great grandchildren. It is leased at this time.

The garage built in 1924 remains on the property. The remaining 240 acre farm now produces soybeans and corn. *Submitted by Margaret Riter.*

ROBERTS – Three Roberts brothers – Robert, Thomas and Joseph – came to American from Wales in the 1700s. One brother settled in New York, one went south and Robert bought land in West Jersey.

Thomas, Robert's fifth child of eight children, 1739 – 1784. Joseph, Thomas' first child of seven, 1763-1831. Robert, Joseph's fifth child of eight, 1806 – 1884, came to Ohio with his parents in 1814.

Theodore, Robert's 11th child of 11, 1854 – 1933, married Alice Scott. On Oct. 8, 1909 Theodore and Alice purchased 200 acres in Hamilton County, Monroe Township, 31-11-7. Corn, wheat, hay and oats were grain crops. Dairy cattle were raised on the farm.

Thomas C., Theodore's second child of six, 1887 – 1962, was born in Morton, IL. He came with his parents to Hamilton County, NE when he was very young. He married first Myrtle Bailey and upon her death, married second Eleanor Caulkins. He added 64 acres to the property.

Leonard A., Thomas C.'s sixth child of six, was born 1937 in Phillips, NE. Leonard is now the owner of the 264 acres that his grandfather and father bought. Corn and soybeans are the crops now grown.

Leonard and Ramona M. Budler were married in 1968. They had four children – Richard (deceased), William, Rhonda and Jeffrey. Grandchildren are Jennifer, Myriah, Landon, Gavin and Jacob. *Submitted by Leonard A. Roberts.*

ROBERTS – Richard Roberts Sr. entered into a purchase agreement with John J. Johnson on Nov. 22, 1879 for E 1/2, SE 1/4, 35-16-8, Saunders County, getting the deed on Nov. 30, 1881 and possession on Feb. 14, 1882. He also added 40 acres to the north before 1905. Richard Sr. and Anna's children were Hugh, John, Harry, Richard Jr., William, Mary, Margaret and June.

Hugh purchased 80 acres to the south on Oct. 2, 1891 from Arend and Hilta Schmidt.

Richard Jr. purchased the 120 acres from his mother June 21, 1905. He had married Josephine Hammel July 11, 1895. Their children – Freddrick died in infancy and Gladys on June 4, 1899. At that time Hugh and mother, Anna, were all living together. Gladys was very small and was kept in a shoe box on the door of the warming oven of the cook stove. She was very fussy but managed to survive. Anna left the home then. They moved to Wahoo in 1906.

Gladys married Harry Gorman Jan. 31, 1923. They had a son who died in infancy and daughter Catherine. They lived mostly in Wahoo and Lincoln.

Catherine married Leo Coady Feb. 14, 1946. They had an infant son died in infancy, Micheal, Richard and Patricia.

In the 1950s a wildcat oil company drilled for oil on the land, leaving at night with a capped pipe in the ground, after telling Leo things looked good.

Richard farmed the land from around 1973 until his death in 2010 when Micheal and Patricia started farming it.

Richard Roberts Jr. and his family, Josephine and Gladys.

Micheal married Melvena Ruhe June 1998. Their children are Patrick and Catherine. Hopefully they'll be the next generation to love the land. *Submitted by Patricia J. Coady.*

ROBOTHAM – Robert Rowbotham purchased 80 acres on Oct. 12, 1872 under acts approved by Congress May 20, 1862 and March 21, 1864 to secure homesteads to actual settlers on the public domain.

The land went to Robert's son, Pliny Rowbotham. Pliny removed the "w" from the last name. Pliny and his wife, Mary, had two sons, Frank and Harry.

Henry Robotham with his dog, mules Joe and Jen, in front of his new barn, ca. 1910

Harry inherited the land. Harry Robotham married Etta Brown in 1920. Initially they sold butter that they churned from cream, eventually acquiring a route in Lincoln selling butter, bottled cream and eggs. Harry built a silo in 1931. In 1966 Harry was honored as a pioneer dairyman by the Nebraska Iowa Milk Association. Harry and Etta had five children: Frank Plin, Lenard, Don, Earl and Harriett.

Leonard inherited the original homestead. He married Phyllis Masker on July 3, 1947. They have five children: Evaline, Raymond, Mary, gene and Bruce. Leonard is currently retired and his son, Raymond, does the farming.

ROCKHOLM-LUDEMANN-ANDERSON – In February 1908, Fred and Anna (Rogers) Rockholm and their six children moved from Avoca, IA, to the 148 acre farm located in Section 20, Township 35 North, Range 15 East, two miles north of Naper in Boyd County. They purchased the farm from Alice M. Gardner, who bought it from Bert Zimmerman, the man who patented the homestead.

Anna's cousin, Thora Nelson, lived a mile east of this farm and encouraged them to buy it because of the live stream that started just west of the buildings. The live stream allowed the family to raise a huge garden and provided fresh water for their horses, cattle and

The Fred Rockholm farm, 1939.

hogs. In the next eight years, Fred and Anna added six children to their family. In 1914, a large barn and in 1919 a modern hog house were built, both of which are in good repair and still in use today. The Rockholms raised corn, barley, oats, soybeans, and hay.

In 1946, Fred and Anna's daughter, Marguerite and her husband, Reo Ludemann, bought the farm and moved there in March 1947. They built a new house in 1949 to accommodate their family, which grew to include four children. They added many machine and grain storage buildings. Reo received several conservation awards as well as the Ak-Sar-Ben Good Neighbor Award in 1968.

In 1979, their daughter, Ann and her husband, Larry Anderson, moved to the farm and took over the operation in 1982. They raised hogs and cattle, corn, soybeans, and alfalfa, and added several other buildings. Reo and Marguerite retrained a life estate interest in the farm. Reo died in 1999.

Marguerite accepted the Ak-Sar-Ben Nebraska Pioneer Farm Award in 2008. After Marguerite's death in 2011, title to the farm was transferred to Ann and Larry. They have five children and nine grandchildren. *Submitted by Ann Anderson.*

ROHWER – Claus Rohwer (1824-1911) and Catherine Simm Rohwer (1828-1888) moved to the United States from Holstein, Germany, with their four children, Henry, Carl, Marie and Gretchen, in October of 1868.

The Rohwers purchased a farm on Long Creek, four miles west of Fort Calhoun, Nebraska. Claus built and operated a horse-powered mill, which he later changed to water power. During a period of drought, in the 1890s he converted to mill use to a hay barn. He formed an insurance company and served as president of this company for twenty-five years.

Their son, Henry, grew up to be president of the Fort Calhoun State Bank. Among his many endeavors, he served as a member of the Washington County Board for six years and was a member of the Nebraska legislature for two years from 1901-03.

Another son, Christian (1869-1945) was born in Washington County and graduated from the University of Nebraska with a degree in civil engineering. He was Washington County Surveyor, worked for the Nebraska State Highway Department, and, for several years, for the Wabash Railroad in Arkansas.

Henry's and Christian's brother, Carl (1866-1942), carried on the farming tradition in Washington County. He married Mary Henrichsen in March of 1890. They farmed northwest of Blair before purchasing the present day Rohwer Farm from Mary's father, John Hen-

Photo of a painting of Rohwer Farm done by Judith Rohwer before the buildings were torn down in 2000. Title: Road Past Grandpa's Farm.

ricksen, in 1914. John had purchased the 200 acre farm in 1884 from George Rathman.

Carl and Mary raised seven children: Minnie, Claus, Carl, Henry, Ann, Kate, and Harry. The elder Carl died in a farm accident in 1942. He fell from a hayrack when a board broke as the rack was going over the creek behind the barn. With the help of her son, Henry (1900-1994), Mary continued to operate the farm until her death in 1946.

When Mary died, Kate Rohwer inherited eighty acres of the county road. She sold her land to Marvin Rohwer (Harry's son) in 1966, but continued to live on the farmstead with her brother, Henry, until her death in 1972. Marvin built a house and out buildings on the eighty acres, and raised corn, soybeans and cattle.

Carl Rohwer (1903-1997) inherited forty acres. He lived in Omaha and rented his land to Henry. Marvin purchased his forty acres in 1988. Henry inherited eight acres, which included the farmstead. He farmed the land and lived in the original house until he moved to Good Shepherd Home in Blair, Nebraska, in 1986. Marvin purchased his farm in 1990.

In 2000, the farm house and outbuildings were considered unsafe and torn down. The barn is still in use. Just before Henry passed away in 1994, he reminisced that the best time of his life was when, in the mid-1920s, the neighbors got together with the family to build the barn, then had a dance in the loft.

The original 200 acre farm is all one piece again, and, as of 2016, Marv and his wife, Judy, continue to live on the farm, but rent out the land. They have seven children in their combined families, and eight grandchildren, none of whom live in the area. *Submitted by Marv and Judy Rohwer.*

ROLL – On March 1, 1871 Nickolaus and Elizabeth (Sand) Roll purchased 160 acres in Otoe County (22-8-14) for $5,900.00. They moved from Watson, MO with their sons, Fridolin and John. They planted an orchard, raised livestock, corn, oats and hay.

Suzann (Lechner) Adams, Carl Roll, and Bobbi (Roll) Lechner, 2009

John and his wife, Theresa (Goehl) inherited the farm after Nickolaus died in 1897, Elizabeth in 1907. They raised 11 children on the farm – Fridolin, Marie, Anna, John, Helen, Joseph, Gertrude, Edward, Emma, Hilda and Carl. John was also growing vegetables and sweet corn for the local canning factory, along with a saw mill.

Theresa passed in 1932 leaving John to survive the depression with sons Joe, Ed and Carl. They scraped and struggled to save the farm when corn was 10 cents a bushel.

John passed in 1939; the farm now belonged to Joe, Ed, Carl, Helen, Hilda and Emma, who were all still living on the farm. The boys learned to fly and started their own airstrip around 1948. Many remember "Rolls Brothers Airport," Joe was an instructor, crop duster and all three were involved with the Civil Air Patrol. Joe was killed while crop dusting in 1965 leaving the farm to Ed and Carl.

Nickolaus (standing) and Elizabeth Roll homeplace, late 1800s.

Carl married Rhobie (Hauptman) in 1955, raising daughters Rhobie (aka Bobbie) and Carlene. The farm was divided in 1966, Ed the north half, Carl the south half.

Today Bobbi and husband Dan Lechner, along with their children Robert and Suzann (Lechner) Adams, own 126 acres of the original 160 acres. Bobbi and Dan live on the home place, which is producing corn and soybeans.

Carlene and husband Dave Hoyle own 18 acres.

In August 2003 the farm received "Nebraska Pioneer Farm Award." 2015, the farm has been in the family for 144 years. Eventually Dan and Bobbi's grandchildren will own the farm. The sixth generation! *Submitted by Rhobie Lechner.*

ROSENBURG & SELLERS FARMS – The ranch was homesteaded in 1887 by Swedish immigrants Andrew and Johanna Rosenburg on 7,000 acres where the railroad was going to be built. While homesteading with Andrew's two brothers, the family worked as blacksmiths and provided food for the railroad workers. Andrew and Johanna had a daughter named Hilda who would later marry Ben Norman and have eight children.

Andrew Rosenberg was born July 20, 1849 in Linderod, Sweden, and came with his brothers to settle in Burt County, Nebraska in 1881. Andrew married Johanna Nilson in Tekamah, Nebraska on November 21, 1885. Johanna was born May 6, 1846 in Christianstodalean, Sweden and came to work in Chicago in 1884 only to move to Tekamah less than a year later. The two had their daughter Hilda on December 5, 1886.

In 1888, following Andrew's brothers, Andrew, Johanna, and Hilda came to Crawford, where Andrew and his brothers had adjoining land claims. After building a house of their own, a depot and water tank were built on the new railroad by the Rosenberg's. A store and small residence followed making the village of Adelia. The brothers went to work helping lay the railroad tracks running past their homesteads from Crawford, Nebraska to Ardmore, South Dakota.

Rosenburg Family - As more people moved to the area to homestead, social gatherings were held in the schoolhouse built east of the railroad tracks at Adilla. Hilda was one of the first students. A Lutheran minister came from Hemingford for church services hosted by surrounding families, including the Rings, Petersons, Rabens, Mengs, Hansens, Rosenbergs, and Friels.

Rosenburg family

A few years later Alelia would burn down from a fire started in the school's pot belly stove. By 1907, the railroad moved the depot three more miles north. The new settlement was called Joder, and it became a train-loading place for livestock. After Joder was created, the railroad built another northern town known as Orella. When the railroad began carrying mail, Orella got a post office.

In 1907, Ben Norman came to work for Andrew, and on February 15, 1909, Ben and Hilda were married. Hilda has lived on the ranch all but the first 5 years there. Ben and Hilda lived on the ranch all their married life, and they lived in her home built just yards from the original log home. They had 8 children born to this union. Seven of them lived to become adults.

Bernard Ben and Hilda's middle son, married Evaline Johnson on July 14, 1954. They had 3 children, Gary, Ben, and Jean. Ben married Diane Kunz on May 23, 1981 and they have 4 sons Shane, Clint, Luke and Brett. Ben died in 1999. Diane and the boys live on an addition

SELLERS – Charlie Sellers acquired this 80 acre farm in Dixon County, northeast of Wayne, Nebraska, from his good friend Oliver Graves. He had not owned it very long when he decided to try some work in Wyoming. He packed up his family- wife, Lydia, and children, Georgia and Joy, and headed west in a covered wagon. His life was cut short early, thus finding Lydia and children back in the Wayne area. She remarried and lived on another farm, where she raised many chickens and gardened. She gave the Sellers farm to her children at passing and they held it until their passing, leaving it to their children, Charles Herron, Steve Girardot, and Jean Girardot. Although Georgia and Joy never spent much time on this farm, and left Nebraska in their later lives, they continued to keep this small farm their mother worked so hard to keep for them, as it was their Papa's and they had loved their Papa dearly.

Georgia, Joy, Lydia, and Charlie Sellers, 1912

SELTING – The farm, 80 acres, was originally school land purchased from the state of Nebraska in 1907. This land is located in Antelope County, Elgin Township, 16-24-7.

Corn and soybeans were the crops grown and the land was also used as pasture. In the 1940s an additional 80 acres were acquired. The farm passed from William and Rosalie Paul to their children, Russell and Nina Paul. This 160 acre farm now produces corn and soybeans.

The current owner, Nancy Selting, is a granddaughter of the original owners, William and Rosalie Paul. Gene Selting, a great grandson and his wife, Angela, are presently farming the ground, making it four generations of family stewardship. *Submitted by Nancy Selting.*

SHAFFER – George C. Shaffer purchased a 120 acre farm southeast of Beatrice on March 7, 1910. He and Cora A. Shaffer raised six children: Carrol, Roy, Dwight, June, Cora E. and Berwin here.

When George died in 1934, the farm went to his widow and children. Son Berwin began to farm the land.

Sue Shaffer (age 5) held by mother Kathryn with father Berwin Shaffer Sr. By the fence are Berwin's brothers Dwight and Carroll Shaffer. Photo taken ca. 1946.

In 1944 when George's widow Cara A. died, the children transferred their shares to sister Cora E. Morten. Berwin, now married to Kathryn, lived in the house raising children Berwin Jr. and Sue. Two years later in 1946, Berwin Sr. and Kathryn became owners.

Berwin Sr. was severely injured in a car/train accident in 1957. Son Berwin Jr. was called home from the Army in Germany to plant the spring crops.

In 1973 Berwin Sr. and Kathryn deeded the farm to their children, Berwin Jr. and Sue Shaffer Frahm.

The following year Berwin Jr., wife Roselyn and their four children moved to the farm and into the big house. Berwin III, Peggy, Joy and Brian became the third generation to grow up in the house.

By 1976 Berwin Jr. and Roselyn became owners. Although Berwin Jr. died in October 2014, Roselyn continues to live on the farm.

While cooking for the thrashers in the fall of 1913, a chimney fire caused total destruction of the original house. The family had to live in the garage until a new, large, two-story Prairie Square house could be build.

At the time of construction, the house was the first in the rural area to have electricity and indoor plumbing.

As a result of the fire, and not a way to fight the flames, George built a very deep, partially buried steel tank near the barn. Used for livestock, it also provided fire protection. *Submitted by Roselyn Shaffer.*

SHEARER – August Buske Sr., and wife Wilhelmine "Minnie" and two young sons immigrated to the US from Germany in 1882. The two boys contracted measles on the trip over and died shortly after landing in the US. They first settled in Michigan. In 1887, the family, which now consisted of August, Minnie, and three children, moved to Louisville, NE. The family arrived in Garden County in 1892 and took a homestead eight miles northeast of Lisco. That site was later traded for other land. On April 25, 1901 August Buske Sr. registered his deed to 152.8 acres for $200. Utilizing both the Homestead and Kinkaid Acts, along with other purchases, there are at present approximately 4,200 acres owned and operated by Buske lineage. August Sr. passed the land on to his son, August, Jr. who then passed it on to son Preston Buske. Upon Preston's death the land was passed to his daughters, one of whom, Marlis (Buske) Shearer and husband Ron, have passed it to their daughters, Tiffany (Shearer) Zorn and Allison (Shearer) Roberts. Tiffany and husband Scott are currently operating the business and in 2014 their son Michael Zorn joined the operation, making him the sixth generation of Buske ag producers in Garden County.

In 1909 a home was built of magnesia rock cut from the hills at Rocky Point west of Lisco. This house remains the operation headquarters. The fifth and sixth generations of Buske descendants eat in the same kitchen, relax in the same living room and enter through the same doors as August Sr. and his family. The original barn remains in use in 2015.

Thomas and Selma (Stadig) Belieu, parents of Myrtle Buske (August Jr.'s wife) were also Garden County pioneers and their original homestead in Garden County remains in the family's operation as well. *Submitted by Marlis Shearer.*

SHERMAN – After leaving Chicago, where they operated a milk delivery business, Henry and Rosella Sherman with their family of four, moved to Dawson County, in Nebraska. In March 1908, they moved to a farm four miles northeast of Lexington, on Section 26, Township 10, Range 21.

Irving, Dale's father, helped his father, Henry, farm by working the cow herd and planting and picking corn. The Sherman's were well-known for raising white corn. The best ears were kept for seed until about 1940, when yellow hybrid corn was planted.

Aerial view of the Sherman farm

They always had about 50 cows, which were driven by horses to the pasture in the spring. In the fall, they would be driven home to feed on the corn stalks left in the fields. When the baby calves arrived they would be ear-notched rather than branded. In the late summer the calves would be driven to town and loaded on railroad cars to be sold in Omaha. Sometimes Irving or his father would ride along on the train.

About 1930 Irving and his bride, Elsie, moved onto the Sherman Family Farm. They raised four children: Thelma, LeRoy, Dale and Eileen. Irving continued to farm with the help of his two sons.

In March of 1961, Dale and Jo with their first son, Ken, moved into the home where Dale has spent 78 out of his 79 years. Dale and Jo purchased the farm in 1978.

The main crops raised on the farm have been corn and alfalfa.

Dale retired from active farming in 2002, but he and Jo still live on the farm and rent out the land.

Their daughter, Cindy, and son, Tim, have provided seven grandchildren who enjoy visiting the family farm where there is plentiful room to play and explore. *Submitted by Dale L. and Elinore Jo Sherman.*

SIEBERT – In June 1879, the Johan Siebert family left the agricultural life of South Russia (now Ukraine) to seek less government demands of military might in solving problems. Landing in Philadelphia, they boarded the train for Nebraska. Since the most desirable land in the close knit Mennonite Community had already been established, the Sieberts settled on a hilly location.

According to the Warranty Deed, on Aug. 22, 1879, Johan Siebert purchased and in April 1882 received the homestead agreement acquiring 160 acres in York County, Township 12; section 9; range 4 near Lushton, NE. Details of what crops were farmed immediately after this time is unknown.

On Sept. 16, 1897 Johan Siebert transferred or sold the farm and land to his son, Peter Siebert. Peter and his wife Katharina established a new farmstead on the acreage in the early 1900s. According to records, mostly corn, oats, wheat and hay were grown. Trying to live a mostly self sufficient lifestyle, Peter and Katharina also had cattle and chickens. Raising a family of six had its challenges since Katharina was blind, however, it did not prevent the family from getting things done. After their son, Peter P. Siebert, got married in December 1928, Peter and Katharina moved to Henderson and Peter P. and his new wife, Anna, stayed on the farm.

According to a contract for Real Estate, Peter P. purchased the land in May 1945. Peter P. and Anna raised corn, milo, wheat, soybeans and alfalfa at various times and they had a herd of cows that were milked every day until 1963. After that, Peter P. kept the herd but changed from milking to feeding out steers for market. He very often got the top dollar when his steers were taken to the livestock auction. They also raised pigs and chickens, again trying to live a self sufficient lifestyle. When they moved to Henderson in November 1969, they continued to rent out the land and buildings. An opportunity came to sell the farm buildings to Calvin Hiebner in 1987. Peter P. passed away in 1996 and when Anna died in 2004, all six of their living children inherited the land. Since none of the children were farmers in the area, or didn't farm at all, they continued to rent out the land to Calvin Hiebner until November 2014 when they sold the land to Calvin.

Although Calvin is not a son or grandson to Peter P., he is a great-great-grandson of Johan Siebert, the original owner of the land. Calvin and his wife, Melissa, are raising their four children on this farm.

In 1981 the Johan Siebert family farm plaque was given from the Knights of Ak-sar-ben honoring the fact that the farm had been in the same family for 100 years.

Submitted by Aldena Siebert - one of Peter P. and Anna Siebert's children and the only one living in this area and great-granddaughter to Johan Siebert, the original owner.

SIELER – August F. Schindler came from Germany, settled in area, and applied for a homestead about 1888. He married Martha Clark and they raised a family of five. In 1902 August obtained 167.9 acres located in Bush Township, Boyd County, NE. Corn, cane and hay were the crops grown while August retained the farm.

August and Martha Sieler and family

Son Albert took over the farm from 1931 to 1972. He and his wife Regenia raised their family and a grandson. After Albert's death, wife "Genie" was the owner. Her daughter, Alberta "Bertie" and husband Dayton Sieler then took over the farm and raised their daughters, Belinda and Vickie there. After Regenia's death, the farm went to "Bertie" and Dayton Sieler in 1994.

The farm retains the original acreage and now produces hay, pasture, corn and beans. Turkeys, geese, chickens and pigs are raised on the farm.

The two story home with carved wood work, open stairs and four open porches was built in 1902 and remains today.

August, as well as Albert, farmed with horses until 1939 when he bought a John Deere tractor. Life became easier then. The farm is in a valley along Ponca Creek, so water was always in the home. Electricity was turned on in August 1949 and electric lights replaced kerosene lamps. TV became a way of life.

One porch was converted to a bathroom in 1953. Electric appliances were added.

Farming equipment made life easier for all. *Submitted by Alberta L. Sieler.*

SIEMS – Frank Siems purchased the southwest quarter, Section 26, Township 3, Range 6 East 6 Principal Meridian, Gage County, NE in 1916. Frank farmed this ground and in 1936 his son, Raymond Siems, was given ownership of the east 80 acres of this quarter to farm.

Raymond and his wife, Gertrude (Baumfalk), were married in 1929 and made their home on this farm. Two daughters, Marilyn and Elaine, were raised here.

Arlin and Marlys Koch with children Tim, Troy and Teri Ann

Raymond died in a car accident in 1949. Frank Siems moved to the farm and lived with his son's family to help manage it. The cropland was then rented out to various neighbors over the years.

After Frank's death in 1955, the east 80 acres of the farm was inherited by Raymond's wife and two daughters. Marilyn (Henry) and Elaine (Kilday) and Frank's daughter, Anita, had interest in the remaining land to the west and Gertrude had life estate in the dwelling. Marilyn and Elaine sold their interest to their aunt, Anita, after their mother, Gertrude, moved to town.

In 1981, Gertrude gifted her daughters, her share of the east 80 acres. Elaine (Kilday) sold her land interest to her sister in 1986, leaving Marilyn as the sole owner of this property. Arlin Koch, Marilyn's son-in-law, rented the ground and started farming it.

In 2005, Marilyn gifted the farm to her daughter, Marlys (Henry) Koch. Currently, the farm is owned by Arlin and Marlys Koch Land & Livestock LLC with their children - Tim, Troy and Teri Ann.

SOLLER – William "Bill" Soller came from Wisconsin in the 1880s. He was looking for land to settle on and make his home. Bill traveled as far as Idaho, but said the best land he had seen was in Nebraska.

This barn was in the 1930 tornado, which took the top off. Bill had just got it built when the storm hit.

He bought 160 acres from Samuel and Hannah Koot in 1889 and built a soddy. Bill brought with him to Nebraska his new wife, Doris. They went on to have nine children and sadly lost two little boys.

Bill and Doris worked hard and eventually got enough land to leave their seven children 80 acres each of farmland. Bill also accumulated pastureland.

The family endured many hardships including a fire in the original soddy, a tornado in 1930 and the devastating dust bowl.

The Soller farm, 1911

The Soller family in front of their house built in place of their original soddy home, 1903.

The picture shows the house they built in place of the soddy in 1903. The Soller family has two farms in the 100-Year Pioneer Heritage alumni. The farm that my father received was 100 years old in 2011. Bill purchased that farm in 1911 from Irvin Brooks. Albert added pastureland and raised 500 head of cattle, and also buffalo.

I, Don Soller, inherited this farm in 1985. My wife Beverly and I live here still and our sons, Pat and Mike, also use the farm now. *Submitted by Donald Soller.*

SPILGER – The Howard County George and Margaret (Jahn) Spilger Farm, owned by Sandra Spilger Baird, was named a Nebraska Pioneer Farm in 2011. Over 30 Spilger descendants attended the award ceremony at the County fair. The 160 acre farm located on the border with Merrick County, near Palmer, has been continuously owned and operated by the Spilger family since Feb. 15, 1911. Their son George Jr. and wife Helen (Obermiller) Spilger purchased the farm in 1953 from Margaret's estate and their daughter, Sandra, inherited it from them in 1995. They had two other daughters, Gladys Patton and Donna Elliott (deceased). Douglas Patton, a grandson of George Jr. has farmed the land since 1980.

George and Margaret Spilger

The Spilgers purchased the farm for $12,000. The land had been homesteaded and title was transferred by the USA to Charles Simpson on March 25, 1885. George and Margaret Spilger were married in Edgington, IL on March 25, 1889 and moved to Nebraska in 1890. They owned farms in Rogers, Dodge County and Atkinson, Holt County. George and their other children - twins Charles and John, Eva, Margaret, Fred and Paul - were born in Rogers. A son Henry was born in Atkinson and died at the age of 7 months. When they moved to Palmer in 1911, they had planned to go to Oklahoma, but this was their last home.

The family arrived at the farm with their worldly goods on three wagons, eight horses, two mule teams, a pony and a dog and a gift of a bag of seed corn from a farmer in Wolbach where they spent their last night before arriving at the farm, which had a three-room house with a lean to porch and a few trees.

Family lore claims that the Spilger family planted 1000 trees, including Catalpa trees and Osage orange trees, because George believed they made good posts, strong and unbreakable. Many of the fence lines on the farmstead remain with some of the original tree posts. A windstorm that spring destroyed everything but the house and a small building. A two story house was moved in and the family built a barn, granary, garage and other buildings. They also laid cement sidewalks connecting the buildings, the livestock tank and garden. The sidewalks, the garage, the barn, granary and a smoke house remain on the farm. The farmstead area is comprised of 20 treed acres. Elk Creek runs through the southeast corner of the property. The Spilgers planted alfalfa, wheat and oats on the land that was well suited for alfalfa, heavy soil, some of it mixed with clay. George (Sr.) was creative, and he held a patent for a "Gate Mounting and Operating Mechanism" which was granted on Nov. 23, 1926. The original gate remains on the farm.

George and Helen Spilger

It is important to note that all of George and Margaret Spilger's children owned and operated farms, including the daughters Eva and Margaret, as unmarried women. The farms are in close proximity to the Spilger farm and four are presently owned by descendants: Gladys Patton, her son Doug Patton and wife Karrie and Lennis Spilger. Paul and his wife Ann lived on the Spilger farm in a second house for a time and their children Gary, Gilbert and Connie were born there.

When George Jr. purchased the farm, he dismantled the house and left the other buildings intact. Cornices, the beveled glass from

Spilger descendants receive the Nebraska Pioneer Farm award at the Howard County Fair, July 2011.

the front door and columns were preserved. He raised corn, wheat, alfalfa and soybeans. The farm was irrigated by two wells, located on George's adjacent farm, by a siphon gravity system, and the alfalfa land by sprinkler system from a small holding pond. Sandra Baird installed a pivot irrigation system in 1998, and corn, soybeans and alfalfa are raised. Helen Spilger worked the farm with George. George planted walnut trees on the farm, continuing the land's stewardship.

George took pride in his herd of Aberdeen Angus cattle. He and Sandra were members of the American Aberdeen Angus Association. Sandra and his grandchildren Tracy, Doug, Janice and Jerry Patton showed cattle from his herd in the 4-H beef and beef breeding divisions of the Howard County and Nance County Fairs, winning many prizes.

Sandra lives in New Haven, CT with her husband Kirk Baird. Her children are Kate Sosnoff and Ben Sosnoff, and she has one granddaughter, Sofia Rojas. She is retired from her position of Chief Family Support Magistrate in the State of Connecticut Judicial Branch. The family is proud of their heritage and the future it holds. *Submitted by Sandra Baird.*

Picture of the family at the 100-year farm celebration, front row, l-r: Conor, Quinn, Barton, Bailey, and Bradon Spilker. Back row: Preston and Stephanie Spilker, Barton, Robert, Vicki Spilker, and Chad and Alicia Spilker.

SPILKER – On March 1, 1911, Ernst and Anna Spilker purchased 240 acres in Jefferson County, located in Washington Precinct (Section 19 and 20).

Picture of the farm ca. 1920-30.

The barn on the property that was erected by previous owner still stands. Dances were held in the barn back in the 1930s and 1940s. Ernest Spilker started farming the tract in the late 1930s after his father, Ernst, passed away and did so until 1970. He raised wheat, alfalfa, corn, cattle, hogs and chickens. The current owners, Robert and Vicki Spilker, continue to live in the home and raise corn and soybeans on the farm along with other land they have acquired over the years. Their son, Chad Spilker, will be the fourth generation of Spilkers to continuously farm this ground for over 100 years.

The Spilker family is proud of their legacy and name. They hope their farm will be in their family for many generations to come. *Submitted by Robert Spilker.*

SPRACHER – Adam H. Spracher received a homestead grant on 161 acres on Feb. 27, 1879 at the land office in Bloomington, NE for $4. The land is in Webster County, section 30, range 3-9, Stillwater precinct. The land passed to his daughters, Macy and Susan. Macy's land was willed to Roger S. and Velma (Stutesman) Albin, her nephew and

Farm equipment on the Spracher farm, 1998

Aerial view of the Stauffer farm

wife and later passed to Robert Albin and his wife Marlene (Dalton) in 1972. Robert would be the great grandson of Adam Spracher, the original homesteader. Robert and Marlene reside on the farm and are still involved in farming. Some of the farm ground is now farmed by their son-in-law, Bob Conway. Crops grown are corn, wheat, sorghum and soybeans. Cattle are also grazed on the grassland. The original farm has expanded to more than 2000 acres. The farm received the Nebraska Pioneer Farm Award in 1979.

Robert and Marlene have two daughters; Jody, who is married to Terry Niles, they have two daughters, Kim and Kate. Kelli is married to Bob Conway; they have four children: Kailey, Corey and twins Lane and Casey. *Submitted by Robert Albin.*

STATLER – In February 1874 at Norfolk, NE the US Government sold 160 acres in the northeast quarter of section 7 in Township 23-N-3, Kingsburg Precinct to Henry K. Watson. In 1899 another 80 acres was acquired, making a total of 240 acres in this farm.

Originally the farm produced prairie hay, corn and was used for pasture. Horses and cattle were the livestock raised. The farm now produces corn, soybeans, prairie hay and is used for pasture.

This farm passed from Henry K. to his son, Herbert L. Watson. Herbert's daughter, Tonie I. (Watson) Statler was the next owner and in 1989 Henry T. Statler inherited the farm from his mother, Tonie.

The Nebraska Pioneer Farm 100 years award from Ak-Sar-Ben was presented to this farm. *Submitted by Henry T. Statler.*

STAUFFER – Norman and Jacqueline Stauffer are the current, fifth generation landowners of the farm located in Section 32, J Township in Seward County. In 1884 Norman's great great grandfather, Joseph Stauffer, purchased the original 80 acres for $2000. His son, John Stauffer, took over the farm in 1908.

In front of the house Daniel Stauffer built, he and his four brothers are pictured hauling wheat into town. He farmed the original 80 acres and subsequently acquired an additional 160 acres from 1910-1912.

In 1946 Clarence A. Stauffer began farming the land. In addition to producing oats, wheat, corn and alfalfa, he raised hogs, chickens and dairy cattle. As the farm was being transferred to the fifth generation, Norman and Duane Stauffer, soybeans were introduced in 1979. In 1980 Norman and Jacqueline Stauffer implemented ridge till farming and were early proponents of this agronomic method.

Photo of the original Stauffer family farm

They raised three daughters: Charlene, Gwendolyn, Brenda and continued to farm in the tradition of the family. Today they live in the original farmhouse built by his grandfather and enjoy frequent visits from their grandchildren Riccardo, Jameson, Andrea, Vanessa and Gary III. *Submitted by Norman C. Stauffer.*

STEINHOFF – August Steinhoff was born September 25, 1854 in Wetteborn, Hanover, Germany (according to Naturalization Record and German Church Record-tombstone says 1857). He was the son of Henry and Johanne (Jurries) Steinhoff. (He had one sister Caroline who died in Germany). In 1868 he left Germany and settled in Platteville, Wisconsin where several other Steinhoff families resided. On March 6, 1879 he was married to Bertha Genthe, daughter of Joachim E. and Rebecca (Buckerman?) Genthe. Their first son William H. (d. February 29, 1944, m. Magdalena (Lena) K. Buckman April 28, 1909, b. September 26, 1879, Platteville, WI, daughter of Fred D. and Wilhelmina (Bunsen), sons Ervin & Arthur, see Fred D. Buckman Family) was born June 26, 1880. They resided on a farm two miles south of Platteville until 1880 when they moved to a farm 2¼ miles southeast of Lorton (Otoe), Nebraska. Their son Fred E. (d. January 12, 1968, m. Sophia R. Grafe January 30, 1908, children Clara Duebler, Agnes Zeigler, Mabel, & Edwin) was born on October 15, 1881.

On October 2, 1882 the Steinhoff's purchased a farm in T9N, R11E, SW ¼ of Section 21 (North Branch Precinct) in Otoe County. The rest of their eight children were born on this farm. They were Henry (b. January 16, 1883, d. March 6, 1967, m. Sophia A. Albers March 11, 1908, children Harvey, Alvin, Irene Strong, Lorenz, Bertha, Ralph, and Lois Hoke), Mary (b. July 20, 1885, d. May 17, 1971), Louis Ernest (b. July 21, 1887, d. October 19, 1972, m. Sophie Ehlers February 12, 1913, children, Alvina Thormahlen, Gladys Davis, John, Lorene Rosenthal, Esther Wessel, and Warren), Eddie (b. July 26, 1891, d. March 3, 1950, m. Alma Windels February 24, 1915, children Ruby Copenhaver, Delores Bruns and Aldean Antes), Meta (b. November 28, 1895, d. June 19, 1939) and Minnie (b. July 20, 1897, d. February 16, 1985, m. Alphonsa Welch May 19, 1946). Delores Bruns is the only surviving grandchild of August and Bertha Steinhoff.

August and Bertha resided on the farm until May 6, 1908 when they moved to Syracuse. Their son Fred resided on the farm and inherited it in 1918. On March 6, 1929, August and Bertha celebrated their 50th wedding anniversary.

August and Bertha were members of the First Lutheran Church of Syracuse. On January 3, 1934 Bertha died of pneumonia. In later years August owned property in South Dakota. It was on the return of one of his trips to that state that he was involved in a hit and run accident in Omaha and died from his injuries on November 14, 1937. August and Bertha were buried in the Park Hill Cemetery in Syracuse.

Fred Steinhoff left the family farm to his son Edwin (b. March 27, 1916 d. August 1984, m. Hazel Sugden April 17, 1939, b. June 5, 1915, daughter of Rudolph D. and Elizabeth (Jacobs) Sugden, d. April 15, 2011). Ed and Hazel had three children, Phyllis J. (b. January 20, 1939, m. Edgar Mueller January 21, 1957), Edwin, Jr. ("Sonny" b. April 1, 1942, m. Ronda (Carrigan) April 30, 1963), and Kathy S. (b. January 16, 1955, m. Roger Royal). Edwin and Hazel originally moved to the Steinhoff

August Steinhoff Family, front row: Bertha (Genthe) Steinhoff, August Steinhoff, Mary Steinhoff. Back row: Fred Steinhoff, Minnie Steinhoff Welch, Henry Steinhoff, Meta Steinhoff, William, Louis, & Eddie Steinhoff

farmstead in 1941 where they lived until they retired in 1976 and moved to Syracuse. Their son Sonny and his family than moved to the farm. Sonny passed away on April 26, 2013 and his wife Ronda continues to live on the Steinhoff farmstead. The Edwin Steinhoff family received the Nebraska Pioneer Farm Award from Ak-Sar-Ben in August of 1982.

STETSON – Eugene Stetson homesteaded in 1882 and received the certificate in 1895 from President Grover Cleveland. The Stetson Homestead is located in Dawes County, T 31, R 51, 160 acres. Wife Annie obtained 160 acres. They bought more land and the farm grew to 1240 acres by 1914; all in section 18, E 1/2 and SE 1/4 section 19, NE 1/4 of section 7.

Some Stetson Homestead owners. Front: Anna Soester Stetson, Orville Stetson. Back: Frances Stetson Mosher, LaVerne Stetson, Jewell Stetson Shanley.

The farm produced wheat, oats, hay and pasture. Eugene and his brother, Isaiah, operated a custom threshing business.

Succeeding owners: Annie Stetson et al estate, 1914-1937. 1937 – January 1944, brothers Clarence, Orville, Lester Stetson and sister Annie Pearl Stetson Seegrist were owners. From 1944-1984, Orville Stetson was owner. The Anna Stetson et al estate was the owner, 1984-1989. 1989-1996 Frances Stetson Mosher, Jewell Stetson Shanley and LaVerne Stetson. 1996-2011 Jewell Shanley and LaVerne Stetson. Dec 5, 2011 LaVerne and Shirley Stetson. They sold all of the land in section 19 except for 15 acres separated by a railroad. Today it is 693 acres, all of section 18 and 15 acres in section 19.

Crops today are hay and pasture with a history of wheat and oats acreage.

The present house was built in 1910 by Eugene and his family. It is a two-story concrete block. Orville, who was 12 years old, hauled the blocks from Crawford (five miles) with a team and wagon.

Wife: Shirley Wasserburger Stetson.

Children: Patricia Stetson-Warning, E.D. "Corky: Stetson, Ronald Stetson and Helen Stetson Fankhauser.

Grandchildren: Skip Dean, Brian Dean, Emily Dean, Kirsten Stetson, J.D. Stetson, Patrick Stetson, Bradley Stetson, Andy Stetson, Sonny Fankhauser, Chelsea Fankhauser Klein and Casey Fankhauser.

Present Stetson Homestead owners: E.D. "Corky" Stetson, Patricia Stetson Warning, LaVerne Stetson, Shirley Wasserburger Stetson, Ronald Stetson, Helen Stetson Fankhauser.

Great Grandchildren: Skylar Dean, Royale Dean, Hanna Dean, Haylea Dean, Ruby Dean, Demi Jo Stetson, Ezekial Stetson, Gemma Stetson, Paxton Klein and Paisley Fankhauser. *Submitted by LaVerne E. Stetson.*

STEVENS – A.D. and Sarah (Heiderstadt) Stevens brought their family by covered wagon from Filmore County to the Maywood vicinity in the early 1890s, following his second term in the State Legislature. In 1895 they took a 160 acre homestead on the Hayes-Frontier County line: NE 1/4 Section 25, Township 8, Range 31 West. Two of their grown children also got homesteads and they later bought two adjoining quarters. The family has the Patents for those five quarters, dating back to 1891.

The Stevens' youngest daughter, Esse, was born in a dugout on the homestead in 1899. The present frame house was built when she was about 8 years old. This was her lifetime home except for the first three years following her marriage to Clyde M. Brown. In 1921 they moved onto the Stevens homestead and later bought the ranch. Here they raised three children, Melvin, Marion (Hoschouer) and Mackay and celebrated their 60th anniversary before Clyde's death in 1978. Esse lived till 1997.

Clyde and Mackay farmed with horses, picked corn by hand and depended on windmills for water. The first tractor was purchased in 1944 and a one-row cornpicker in 1950. REA came soon after. Now most of the stock wells are on electricity and tractors and machinery are much larger.

Mackay has spent his life on the ranch except for two years in military service. In 1949 he married Betty Evans, another Hayes County native. Their home is also on the Stevens homestead and their sons, Jim and Ralph, were raised here. Through the years more land in Hayes, Frontier and Lincoln Counties has been added and Mackay has built a reputation herd of Red Angus cattle. Ralph, wife Linda (Bose) and their son Dillon (the fifth generation) are part of the family operation, Canyon Cattle Co. Jim and Nena (Lippold) Brown, Mound City, MO also share in this heritage.

In 1995 Esse, Mackay and Betty received the Nebraska Pioneer Farm Award from Ak-Sar-Ben. This 120 year old ranch has been in total control of family members except for two years it was rented out before Clyde and Esse moved onto it in 1921. Mackay and Betty still live on the original homestead. *Submitted by Mackay and Betty Brown.*

STOEHR – Phillip and Mary (Lohnes) Stoehr were married on January 30, 1879 in Tazewell County, Illinois. On September 28, 1889, they purchased the southwest quarter in section 8-12-12 in Cass County, Nebraska from Robert and Mary Anderson. When moving from Illinois to Nebraska in 1889, Phillip and his young son, George, made the journey first with a team of horses and a top buggy to live on this farm. Mary and their daughter, Margaret, came a short while later with their possessions. Louise was born after settling in Nebraska.

Neighbors on a nearby farm were the Adam and Wilhelmina (Nolting) Kaffenberger family. Their daughter Friederica (Rica) married

Friederica & George Stoehr at their Cedar Creek farm in the 1950s

Phillip, Mary, & George Stoehr, mid-1880s

Stohlmann family

George Stoehr on March 9, 1904. After their marriage, Phillip and Mary moved to an adjoining farm so that George and Rica could live on the home place, which they first rented from Phillip and Mary. Then, in 1922 George purchased it from them and bought his siblings' shares.

Two sons were born to George and Rica, Elmer (Tuffy) Stoehr and Victor Stoehr. Both sons were life-long farmers in Cass County, just as their father and grandfather before them.

In 1959 this farm passed from George and Friederica Stoehr to their son, Elmer. Upon Elmer's death in 1961, his wife Rose obtained life estate use of this land with the remainder interest going to his brother, Victor's descendants.

Victor married Matilda Stohlmann in 1929. They had one daughter, Dorothy, who married Loren Duever in 1961. Dorothy and Loren had one daughter, Denise, who married Alan Horn in 1991. Denise and Alan have two children, Michaela Horn and Eric Horn.

The current owner is Phillip and Mary's great-great-granddaughter, Denise (Duever) Horn. This land was farmed for 72 years by the Stoehr family. It has been rented to other farmers since Elmer's death in the fall of 1961.

STOHLMANN AND KUNIGUNDE (SCHOENMANN) – Fredrick P. Stohlmann in 1858 walked to Nebraska and saw a light and decided this was where he was to settle. He stayed with Patrick Blessington, who lived in a cave at that time (prior to building the 'stone house'). Patrick and Fredrick walked to Nebraska City and filed their homestead papers for $1. He walked back Waukesha County, Wisconsin (just west of Milwaukee) that same year, bringing six ears of sod corn as proof of Nebraska's soil fertility in order to convince Andreas Schoenmann to allow his daughter, Kunigunde, to marry him. The following May 19, 1859 he and Kunigunde were married in Watertown, Wisconsin.

Both Fredrick and Kunigunde had been born in Germany, he on March 14, 1832 in Blomberg, Lippe-Detmold (about 30 miles southwest of Hannover). He was the son of Friedrich Bernhard Arnold Stohlmann, an Indigo dyer, and Marie Louise Recker. He was christened August Philip Friedrich and was the fourth of nine children. Kunigunde was born on June 17, 1838 in Rabelsdorf in northern Bavaria, the daughter of Andreas Schoenmann and Eva Barbara Andres. She was the second of seven children.

They and their families immigrated to America, he in 1849 and she in 1854, coming to Wisconsin and settling on farms in Waukesha County. Shortly after their marriage (her father had given her as a dowry, two oxen, two cows, two hogs, several chickens and a dog), the young couple left Wisconsin for Nebraska in a covered wagon drawn by the two oxen and leading the two cows behind the wagon. Although she was not happy at leaving her home in Wisconsin, trying several times to induce her husband to turn back, and despite many hardships and trials on their long journey (a need to replace the original yoke of oxen, the wagon slipping off a crude bridge into a stream with all of their possessions getting soaked, wagon canvas torn to shreds in a storm, requiring the bride to sew the strips back together), the pioneer spirit prevailed and they finally arrived at Plattsmouth, Nebraska, where Kunigunde refused to cross the Missouri River on the dilapidated ferry but agreed to being ferried across in a boat. After remaining here for five days in the home of John Holchu (Holcher or Holschuh), they moved on to the area of their farmstead in Cass County, a few miles south of present day Louisville.

The legal document, dated December 5, 1861, describes the property as follows: "the south east quarter of section thirty five, in township twelve of Range eleven, in the District of lands subject to sale at Nebraska City, Nebraska, containing one hundred and sixty acres". The property warrant was assigned to Fredrick Stohlmann from Elizabeth Brickey, widow of Peter Brickey, Private, Revolutionary War. The document carried the signature of President Abraham Lincoln. Their first winter was spent in the home of Patrick Blessington and the year (1860) they planted their first crop of corn, beans and watermelon.

Their marriage was blessed with eleven children: George J. (6/5/1860-12/22/1912) married 4/23/1889 to Louise Kupke (5/31/1869-12/27/1956); John (10/5/1861-2/1/1865; William J. (11/28/1864-3/4/1929) married 6/6/1889 to Theresa O. Thimgahn (8/10/1870-

Fredrick and Kunigunde Stohlmann

A&L Stohlmann Wedding

Stohlmann family, 1954

12/23/1923); Fredrick A. (4/28/1866-2/6/1959) married 9/12/1894 to Bertha Kupke (6/10/1873-12/17/1963); Louise (5/8/1869-7/5/1914 married to Herman Kupke (9/23/1866-4/7/1919); Andrew (5/13/1872-4/15/1938) married to Martha Winter (8/27/1879-11/3/1965; Infant son (4/11/1875-4/11/1875); Amelia (4/11/1876-12/16/1945) married 6/26/1910 to Charles Voigtman (7/16/1887-6/13/1978); Kunigunde (1/27/1879-2/4/1879); August (1-16-1880-10/19/1964) married Mary Louise Jochim (2/20/1883-9/16/1978); Henry (1/29/1884-9/14/1905).

In April 1872 School District 79, Platte Valley School (later changed to College Hill School) was formed. On April 5, 1872 Fredrick Stohlmann was appointed to a committee of three to locate a suitable spot to build the schoolhouse. After a site on the Henry Blume farm had been selected (across the road from his farm), he was appointed, on April 26, 1872, to a five man building committee. He served as moderator (secretary) of the school from 1878 to 1898. He was also active in the formation of the Immanuel Evangelical Lutheran Church, serving as trustee on March 28, 1876 when land about two miles west of his farm was purchased for the location of the church.

On February 7, 1880, Edward Murphy was the first adult Catholic to die in central Cass County. Since there was not a church or cemetery in the locality, the local Catholic community gathered to discuss their problem the Sunday after the death. Fredrick Stohlmann came to the meeting and offered to donate one acre of his land to be used for a Catholic Church and Cemetery with an option to purchase a second acre at a fair price. They promptly voted to accept the generous offer. The cemetery is still in use today and the church has been relocated to the nearby town of Manley.

On March 26, 1904 the farm was deeded to their son, August. Fredrick died on January 23, 1903 and Kunigunde died on April 19, 1906 and they are buried in the Immanuel Lutheran Church Cemetery. On April 27, 1904 August was married to Mary Louise Jochim. Their marriage was blessed with eight children, all of whom were born on the farm: Louise (6/13/1905-3-9-2003) married 6/9/1929 to Harold A. Dumke (11/19/1905-5/8/1984); Matilda (1/13/1907-3-9-2005) married 6/9/1929 to Victor J. Stoehr (3/23/1906-2/26/1972); August A. (11/25/1908-2/11/1971) Married 6/25/1933 to Esther Albert (8/19/1907-2/8/2001); Arnold H. (4/14/1911-9/13/2003) married 8/31/1939 to Charlotte L. Coatman (1/12/1910-10/30/1994); Clarence H. (5/1/1913-4/10/2003) married 9/4/1938 to Laura L. Thomas (7/20/1916-9/13/2009); Dorothy M. (3/13/1915-12/13/1926); Ellsworth F. (12/31/1917-9/30/2011) married 4/11/1942 to Norma G. Luby (4/30/1923-5/22/1990) and married 11/20/1955 to Doris M. Noerrlinger (7/9/1921-); Orland H. (11/25/1920-10/18/1979) married 8/15/1942 to Ruth M. Luby (10/28/1920-11/14/1984).

On October 23, 1967 the farm was deeded to Danny E. Stohlmann (1/14/1943-3/13/1994), (son of Ellsworth), married 11/15/1963 to Shirley A Steinhoff. She lives on this original Stohlmann farm. They have two children; Kim E. married 2/25/1989 (m.ds. 1992) to Brian Roecker (4/2/1964-11/20/2010), and Craig A. married 6/12/1999 to Stephanie Fjell.

Kim lives west of the Stohlmann farm near the site of the early Stohlmann Reunions on part of the old Kraft farm. She and Brian have a son, Logan, (great-grandson of Ellsworth), who currently lives in Louisville having purchased Ellsworth and Doris Stohlmann's home in 2013. Craig and Stephanie live on the Stohlmann farm east of the main farm house with their three daughters, Mackenzie, Rylee, and Emery.

The descendants of Fredrick and Kunigunde began having annual family reunions in 1930, continuing to the present day. Fredrick A. Stohlmann was instrumental in organizing the first reunion, all except one has been held in close proximity of the original farm and Immanuel Lutheran Church, with which Stohlmann family members have been affiliated since its founding around 1866.

In 1936 a compilation of the family membership statistics was begun by George C. Stohlmann, grandson of Fredrick and Kunigunde; starting in 1940 this continued, as it does to the present, through the efforts of the following grandchildren and greatgrandchildren of the pioneer couple: Louise Dumke, Julia McCafferty, Martha Stohlmann Gaebel, Anne Kroeger, Martin and Opalle Stohlmann, Kenneth G. Stohlmann, Eleanor Stohlmann Box and John and Judi Stohlmann. Kenneth Stohlmann initiated a semi-annual family newsletter in 1983, which is presently being mailed to over 520 family member addresses by Eleanor Stohlmann Box and John and Judi Stohlmann. John Stohlmann has researched the family ancestry back to Johann Stohlmann (ca.1500-ca.1600) in Herford, which was an Imperial City of the Holy Roman Empire and part of the Hanseatic League. It is also about 30 miles southwest of Hannover. The Stohlmann ancestors were for the most part merchants, civic leaders, and church leaders. Clarence Stohlmann had been instrumental in researching the civic and church activities of the pioneer Stohlmann to arrive in Nebraska, Fredrick P. Stohlmann.

By the end of 2010 there were, including spouses, approximately 1500 descendants of Fredrick and Kunigunde, with 1200 then living, and they are located in 38 states and several foreign countries.

STOLLE – There have been four generations of Stolles living on the NW ¼ 29-24-3 farm by Meadow Grove, Nebraska. The father, Bernhard Stolle, was born in Germany on January 16, 1845, arriving in New York City on March 20, 1869. He died November 30, 1938.

Pictured are Bernhard Stolle (seated), John Henry Stolle (left), Johann B. Stolle (right), and Orlin J. Stolle (child). Calvin and Devon Stolle not pictured.

John Henry Stolle was born December 12, 1873. He died May 7, 1939.

Johann Bernhardt Heinrick Stolle was born November 3, 1901. He died May 21, 1965.

Orlin Stolle was born December 2, 1930. He died May 10, 2013.

Bernhard applied and received a homestead for 160 acres south of Battle Creek on March 18, 1871, later purchasing the S ½ - SE ¼ - 2-23.

He married Anna Schulenberg on November 17, 1872. Four children were born to them: John Henry, Mary, Ernest, and Augusta. John Henry married Anna Hobus on February 8, 1899. In 1900 they moved to a house on the now Stolle farm, which had been homesteaded by Patrick O'Neil on May 7, 1872 and signed by President Grant. The title had changed three times. Then it was deeded to Charles and Evelda Walter in 1889. John Henry purchased it in 1902 by paying $2,700 and taxes that was the mortgage on it. They had six children: Wilma, John, Ted, George, William, and Walter.

Louie Schulenberg had built a new house by 1919 and the John Henry family moved there and lived until 1929, when they moved back to the 80 acres he inherited from his father, when his mother passed away.

Johann Bernhardt and Martha (Zohner) married February 23, 1930 and lived in this house until 1932. Their son, Orlin, was born in the house December 2, 1930.

The John Henry family had decided to move back to the Stolle farm. So, Johann Bernhardt, Martha and Orlin moved to the small house for 10 years or more.

This was the start of the dry '30s, and the John Henry family had mortgaged the farm and couldn't pay the loan, interest or taxes. He died in 1939 and as a widow, Anna decided to sell it to Johann Bernhardt and Martha in 1942. They had three more children: Joan, Mary Ann and Shirley. Johann Bernhardt passed away in 1965. Martha continued living on the farm until 1968, when she sold it to Orlin J. and LeOra Stolle. They had four children: Nickii Sue, Calvin J., Devon M., and Patsii Ann. They farmed and owned a mobile grinding business. Orlin passed away in 2013.

The highway has taken land away twice, so there are only 148.21 acres left.

All the Stolles were hard working Christian farmers who raised oats, wheat, corn, rye, hay, cattle, hogs, chickens, big gardens, apple and cherry orchards, and later soybeans.

As the farm had a creek leading to the Elkhorn River, there were many big floods and very cold, snowy winters.

Bernhard was a founder of St. John's Church in Battle Creek; John Henry of the St. Matthew Church in Meadow Grove. John Henry and Johann Bernhardt died in this house. Orlin died at Battle Creek.

The Stolle farm was honored by the Knights of Ak-sar-Ben Foundation at the Madison County Fair in 2003.

At the printing of this in 2016, LeOra still lives on the farm.

STOLTZMAN – Christian Stoltzman was discharged in 1865 from the Civil War and came to Nebraska in search of land to procure under the Homestead Act. He homesteaded 160 acres seven miles northeast of West Point in Cuming County in 1870. He married great grandmother, Mary Feyerherm, that same year and these two pioneers planted a grove of trees, built a house, farm buildings and orchard. They had five sons and one daughter: Fred, Charles, Otto, William, Henry and Mary. They all helped raise the corn, oats and alfalfa, cattle, hogs and chickens and through the years my great grandparents were able to purchase enough land so each son had a farm of their own when they married. After Christian's death my grandfather, Henry, married Matie Prawitz in 1908 and took over the Homestead. They had two daughters, Delilah Abendroth, my mother, and Dorothea Peterson. They farmed the land for many years until they retired in West Point. They rented the farm to neighbors for several years until Delilah married my father, Fred Abendroth. My sister, Jo Ann Fullner, and I, Caroline Zuhlke, grew up on this Homestead.

Christian Stoltzman family homestead, 1948, pastel painting by owner Caroline Abendroth Zuhlke

They farmed the Homestead from 1939 until their retirement in West Point in 1969. They rented the farm to neighbors until 1983 when their grandson, Curt Zuhlke, my son, farmed the Homestead until his death in an auto accident along with his father, Willis Zuhlke, my husband, in 1999. My mother, Delilah, then rented the farm to her nephew and great grandson of Christian, Neal Peterson. He continues to farm the land for me today. After the death of my mother, Delilah Abendroth, in 2013 the farm was inherited by my sister and me. Neal continues to farm my share and JoAnn's son, Russell Fullner, farms her acres so it continues in our family.

STOUT – Clarence Cleveland "C.C." Stout came to Nebraska from Fayette County, OH with his mother, Elvina Griest Stout and two siblings, Florence A. and Mary J., after the death of his father, James, in 1873. Elvina purchased the NE 1/4 of 11-4-12 in Harmony Township in 1881. An elder brother, John, is on record as having purchased 160 acres of land in Section 10 in 1880. A two story farm house was erected on Elvina's land, which was to house four succeeding generations of the family before it was demolished.

The two-story farm house on the Stout farm.

Clarence married Lydia Duncan in Dunbar, NE in 1888. They had three children - Vern James, Ivan Irl and Leila. In 1912, Clarence purchased the SE 1/4 of 2-4-12 from Grace E. Hicks Cousins and Ward H. Cousins, her husband. On this land he raised corn and wheat.

C.C. was a faithful member of the Congregational Church where he served as Sabbath School Superintendent. When this church disbanded he affiliated with the Bladen Methodist Church. He was elected President of the Webster County Sunday School Association and served for five years. He was greatly interested in the affairs of young people and the problems confronting them.

Upon C.C.'s passing in 1928, his property was transferred to his widow, Lydia Stout and his son, Vern Stout. Lydia transferred the land back to Vern Stout in December of 1937. Vern and his wife Edith Mattock Stout shared this property as joint tenants. Vern passed away in 1956 and the property remained in the hands of Edith until her death in 1985.

Vern and Edith had two children, Marlys Darlene and Harland Vern, who preceded their mother in death. Upon her passing the property was passed to Harland's daughters, Rebecca Stout Pitman, Tamela Stout and Pamela Stout Kelley McAmis, who hold it dear as a family legacy of over 100 years. Farmed for many years by the Paul Krueger family, it is producing crops of corn and soybeans.

STREIT – On April 23, 1884 VPRR obtained 160 acres located in NE 1/4 3-9-19. On July 23, 1886 William Cooper purchased 160 acres located in SE 1/4 Section 34 Township 10. From 1903 to 1915, Severin Streit owned and farmed this 320 acres. Joe Streit owned the property from 1915 to 1970 and in May 1970 Larry Streit came the owner.

Originally the farm produced corn, cattle, hogs and chickens. Now this farm produces corn, soybeans, forage, alfalfa and cattle, using a two pivot system. The old main part of the house built in 1880 remains on the farm.

In 2003 Larry S. Streit received the 100 Year Farm Award.

Larry and Janice M. Streit have eight children: Mike, Mark, Lori, Joe, Lisa, Linda, Melanie and Amy. They have 15 grandchildren. *Submitted by Larry S. Streit.*

STUBER – On March 7, 1898 William G. Stuber purchased 160 acres in Buffalo County, located in the southeast quarter of Section 30, Township 12 North, Range 13 West of 6th Principle Meridian. On Aug. 3, 1951 the farm was deeded to William G. Stuber and his wife, Louise Stuber. After W.G. Stuber's death on Oct. 22, 1953, it remained in his estate until it was deeded to Harvey and Minnie Stuber on April 20, 1955. Harvey Stuber died on Jan. 15, 1999 and the farm was then deeded to his surviving wife, Minnie Stuber. After Minnie Stuber's death in January 2003, the farm was deeded to Charles and Marilyn Stuber, the present owners.

Harvey and Minnie Stuber lived on the farm from 1925 until 1984, when they moved to the town of Ravenna, NE. During that time, the primary crops that they grew were corn, wheat, barley and alfalfa. They also produced some beef cattle and hogs. Their two children, Charles and Ramona, were raised on the farm and attended a one-room elementary school and then high school in Ravenna, NE.

In 1984 Randy and Elizabeth (Buzz) Basnett moved to the farm and have continued to live there more than 30 years. Center pivot irrigation was added to the farm in 1975. The Basnett's produce corn, soybeans and beef cattle. Their three sons, Chris, Rick and Curtis were raised on the farm. *Submitted by Charles W. Stuber.*

SUGHROUE – Edward Sughroue, farmer, born Stephenson County, IL Feb. 14, 1871; son of T.J. Sughroue and Alice Doran. Educated at Illinois Rockford Business College, Rockford, IL. Married Barbara Tines May 1, 1895, Indianola (deceased 1931). Sons: Paul, Clarence, Frank, Timothy, Martin, John, Leo; daughters Mary (Mrs. A.R. Rothmyer), Alice (Mrs. E.S. Thomas), Nellie.

In 1892 Edward obtained 160 acres located in Red Willow, Alliance, SW 1/4 Section 21. The farm produced corn, wheat, milo and livestock was cattle and hogs.

Edward Sughroue was elected in 1919 to the Nebraska Constitution Convention of 1919-20 as Red Willow County representative.

Edward was the no. 1 loan borrower from the Federal Land Bank. He submitted his loan application on April 20, 1917 and Federal Land Bank closed the loan on June 25. He was elected director and president of Indianola, NE Federal Land Bank on Feb. 1, 1917 and served continuously for 37 years. He died at his home in Indianola, NE on Oct. 16, 1963.

In 1979 William Eugene Sughroue, present owner, inherited the farm from his father, Martin, who acquired the farm in 1964 on the death of his father Edward.

This farm has grown to 480 acres and now produces corn, soybeans and wheat. *Submitted by William Eugene Sughroue.*

SWANSON – Louis Swanson bought 160 acres located in Dixon County, Logan Township (6-27-4) in 1904. Corn, alfalfa and oats were the crops grown here and hogs, chickens and cows were the livestock raised. Sometime later, six acres were taken from the farm for a highway.

The farm passed from Louis to his son, Emil and then to his son, Ernest. The farm still retains 154 acres and now the farm is in CRP.

All of the original buildings still exist except the machine shed that was built new. The Swanson Bed and Breakfast existed here for 15 years until 2003.

The only time there hasn't been a Swanson farming the farm was a short time when Ernie was in service and it was rented out.

Ernie started farming in 1953. After retirement in 1996 they worked in the church, visited people in hospitals and care center.

Ernie and Lyla were married June 15, 1952. Their children are Regg Swanson, Ann Scholl and Lon Swanson. Grandchildren are Casey Gruber, Brit Swanson and Eve Adcock – Lisa Scholl and Scott Scholl – Blakey, Hunter, Parker and Will Swanson. Great grandchildren are Miles and Charlotte Gruber – Tyler and Holt Adcock. *Submitted by Ernest and Lyla Swanson.*

SWEEM – Joseph Sweem and Robert Bowen Sweem, father and son, both homesteaded eleven miles east and north of Bayard, NE in the spring and summer of 1910. Joseph filed on 191.55 acres on May 27th, and Robert B. filed on 153.26 acres on July 30th. The legals are: NESW, NWSE & SESW, S1/2 SE Section 13, W1/2W1/2NE, West of Highway, W1/2, PT NWSE, Section 18, all in Township 21N, Range 51 West of the 6th PM in Morrill County, Nebraska.

Joseph and his wife, Sarah, and their son, Robert B. came to Morrill County from Bussey, Iowa. They made their living raising Shorthorn cattle and dry land farming. After Joseph and Sarah passed away, their land was left to their two children, Robert B. and Esta (Billy) Marsh. Robert B. later bought Esta's share.

Robert married Elva (Fiscus) Grigsby. In 1942, they were blessed with one son, Robert Eugene "Bob"! After the death of his dad in 1959, and his mother in 1996, Bob inherited the property. He faithfully took care of his mother until her death, lived on the same place farming and ranching like his parents and grand parents did. He never married. He was a great musician and played the "fiddle" at various community events and nursing homes. It was quite unusual that Bob was only the second generation to own his dad's share of their homestead over 100 years old. He passed away March 6, 2013, and the property was sold at auction on June 14, 2013 to Darrell and Leanne Neu.

In 2010, the Nebraska Pioneer Farm Award from the Knights of Ak-Sar-Ben Foundation and the Nebraska Farm Bureau presented a plaque and an outdoor sign to Robert E. "Bob" Sweem for 100 years of continuous ownership of the farm.

Robert Bowen and son, Robert Eugene Sweem, homestead house and shorthorn cattle. Courtesy of Joyce Russell

SWEENEY – Henry and Bridget Sweeney were the original owners of 320 acres located in Custer County, 16-18-22. In 1915 the farm passed to Edward J. and Delia Sweeney and in 1974 to their son, Edwin J. Sweeney.

The crops were corn, milo, wheat and cattle were raised here. Today the farm retains 320 acres and now produces soybeans in addition to the previously mentioned crops. The same house is still on the property. Current renters are Don and Deb Cantrell, Merna, NE. *Submitted by Edwin Sweeney.*

Edwin and Kathryn Sweeney

Aerial view of the Sweeney farm

SWENSON – On March 16, 1903 Olaf Swenson purchased this quarter, the SE quarter of section 34-4, for his son Arthur to farm. When Olaf died in 1927 the land was left in his will to Arthur. Later other land was added to the original quarter. Arthur then willed the land to his grandchildren with his son Orville having life estate. Orville died in 1992 and now the quarter belongs to his daughters, Lucile and Kathleen, and son Richard. Richard and wife Barbara live on the farm. The land has been rented to the same farmer since 1984. The cropland is rotated between soybeans and corn.

In 1909 Arthur married Lois Rousch and three children were born to them in the farmhouse. In 1923 Arthur and Lois built a new house, and the old one was moved to Shickley, NE. The oldest building on the farm is the big barn built in 1910.

Swenson farm around 1951-52

Arthur used to divide his cropland as follows: 2/5 corn, 2/5 wheat, 1/5 oats. He had a steam engine and threshing machine and hired out to thresh for the neighbors. Orville added a dairy of Ayrshire cows in the 1940s. In February 1950 Orville had a farm sale and moved to Colorado. The first irrigation well was drilled in 1954. The farm was rented out from 1950 until 1980 when Richard and his family returned to Nebraska and moved onto the farm. Richard and Barbara have three living children and three grandchildren. Richard is a retired teacher and Barbara, a retired social worker.

In 2003 the Swenson farm received the Pioneer Farm Family award by Ak-sar-ben. A celebration was held on the farm with guests and family members coming from as far away as the West Coast, Tennessee, Illinois, Texas and Sweden. *Submitted by Richard Swenson.*

SWINNEY – William Wallace Swinney was born March 2, 1857 in Cleveland, OH. At the age of 19, following a severe case of pneumonia, he came to Fremont County, IA. While there he experienced the great Missouri river flood in 1881 and the severe blizzard in 1888.

Back row, l-r: Margaret Swinney Sybrant, Christina Swinney (daughter-in-law), and Gilber Swinney. Front row: William Wallace Swinney and Howard Swinney. Photo taken in 1942 before Howard went into the service in 1943.

In 1885 he married Miss Francis Young at Nebraska City, NE. To this union was born five children: Perry, Gilbert, Guy, Gretchen and a stillborn. Only Gilbert and Gretchen lived.

Mr. and Mrs. William Swinney moved their family from Nebraska City, NE to south of Atkinson, NE area in 1904. In the spring of 1905 he moved his family to southern Rock County on a 640 acre tree claim on the Kinkaid homestead in Pewaukee Precinct. Along with his wife, son and daughter they made their home. He continued to live there until the time of his death on Sept. 13, 1943. Mrs. Francis Swinney passed away March 10, 1920 after a short illness. She was laid to rest with her other children at Nebraska City, NE.

Gilbert was born Dec. 25, 1889. At the age of 17 in 1906 Gilbert Swinney homesteaded a quarter of land just east of his father's land which is call Antelope Hill. He married Christina Hall Grimson on Sept. 9, 1920. As a young bride, Christina came to the Swinney home. William, Gilbert, Christina and Gretchen lived in the house together.

Howard and Leona (back center) and members of their family in 1994.

Gretchen Frances Swinney was a school teacher. She taught in public schools in her home community. She also bought a small store for her father that set just east of the present house so that her father would have something to do. The old Highway 183 went by their place going south.

Howard milked cows for 45 years. He sold off his last cows in April 1987.

Gilbert and Christina were blessed with three children: Howard (Oct. 25, 1921), Margaret (July 24, 1925) and Loren (June 25, 1927), who died at the age of 8 years old. Gilbert took over the place in 1935. They raised cattle and milked a few cows. Christina was appointed the postmaster of the Rose Post Office on July 13, 1937, which was housed in their home and served until her retirement on Aug. 1, 1966. Their home was a home to everyone that stopped by. They were very active in the Rose Community and the Rose Church. Christina passed away on March 30, 1967 at her home. Gilbert lived by himself until 1972. Because of his failing health, Howard and his family moved in with him and took over the responsibility of the place. Gilbert passed away Sept. 11, 1974.

Howard married Leona Kaufman on July 19, 1953. To this union four children were born: Gretchen (Aug. 24, 1954), Maretta (Nov. 3, 1955), Harold (Aug. 28, 1959) and Ward (Nov. 26, 1962. On April 26, 1968 the Lord took the boys home to be with him. They were killed in a cave-in accident. Howard and Leona bought a quarter of land southeast of the house that added to the other land to make 960 acres. He put in an irrigation system and planted corn and alfalfa to feed the milk cows and stock cattle. Howard passed away on Oct. 13, 1998. Leona Swinney is still running the ranch today.

Christina Swinney mowing hay ca. 1940-51. She was just learning to use the tractor at the time.

Howard's sister, Margaret, married Lester Sybrant on June 11, 1949. To this union three children were born: Judith (Jan. 17, 1950), Kristie (July 24, 1954) and Daniel Blaine (Oct. 5, 1958). Margaret and Lester live in Bassett.

SWOBODA – In 1907, Frank Sr. and Mary (Vavra) Swoboda purchased 111.78 acres in Saunders County, southwest of Wahoo, from

Jacob and Katerina Hoffbauer, who homesteaded the land (SE qtr SW qtr Sec 19 and the NW ½ NW qtr 30-14-7 E of the 6th P.M.) More land was purchased in 1913, which completed the 140 acres of the farm. Frank Sr. and Mary had four children: Mary, Sadie, William and Frank. A new house was built in 1917 and a new barn in 1927. Mary married and moved to a farm by Colon. Sadie moved to Wahoo. William moved to a farm north of Ceresco.

Frank Sr. and Frank continued working the farm. Corn, wheat, oats, alfalfa, cattle, hogs and chickens were raised.

In 1938, Frank married Hattie Jirovsky. Frank Sr. and Mary moved into Wahoo. The farm was transferred to Frank and Hattie in 1941. Two daughters, Kathleen and Sharon, were born. Kathleen married Dennis Stuchlik in 1963 and have one daughter, Denise. Sharon married Michael Robb in 1972 and resides in Scribner. They have three children: Tami, Jay, and Darci.

Frank and Hattie lived on the farm until 1976, then moved into Wahoo. This same year, Dennis and Kathleen moved to the farm, on which they still reside. Hattie passed away in 1977. After Frank passed in 1990, each daughter received 70 acres. Sharon sold her half to Dennis and Kathleen. They currently raise corn, soybeans, alfalfa, and meat goats. Dennis also does custom hay baling and owns a welding/machine shop in Wahoo.

Ak-Sar-Ben farm family award at Wahoo Rodeo, July 2008. Dennis and Kathleen Stuchlik, Ron, Denise Kali, and John Hauschild.

Sisters Sharon Robb and Kathleen Stuchlik with Ak-Sar-Ben farm family award, July 2008

Swoboda-Stuchlik farm in 1947

Swoboda – Stuchlik farm in 1991

Denise married Ronald Hauschild in 1991. They have two children: Kali and John. They reside in Wahoo.

Tami resides in Freemont, Jay in Scribner, and Darci married Sean Person in 2009. They have two children: Macy and Conner. They reside in Omaha. 4-H was an important part of life, as Ron, Denise, Kali, John, Tami, Jay, and Darcy were all actively involved.

TATRO – In January 1900 F.C. Tatro purchased 40 acres located in Fillmore County, 18-7-2. The farm has remained in the Tatro family since 1900, passing from F.C. Tatro to his son, Donald H. Tatro, in 1944. Irene Tatro, spouse of Donald, became the owner in 1990. In 1992 Jason H. Tatro, grandson of Donald and great grandson of F.C., became the owner.

The farm has grown from the original 40 acres to a size now of 480 acres. Corn and soybeans are the crops grown here now.

The oldest structure on the farm is the house, which was built in 1948.

Jason and Michelle Tatro have three children: Austin, Brock and Carson Tatro. *Submitted by Jason and Michelle Tatro.*

TAYLOR – On Jan. 23, 1914, James and Margaret Taylor purchased the SW 1/4 and S 1/2 of the NW 1/4 9-7-2 from Frank and Eleanor Owens. They purchased 240 acres at $112.50/acres for a total of $27,000.00. James and Margaret's son and daughter-in-law, Herbert and Nettie Taylor, and their three sons lived on the acreage until Herb's death on Feb. 19, 1920. Nettie and the boys moved to Fairmont, NE and James rented the farmstead out to other families.

Pictured l-r: Sandra (Taylor) Girmus, Shirley (Taylor) Sluka, Sharon (Taylor) Elznic, and Lindsey (Taylor) Stone

On July 14, 1939, James Newton Taylor passed away and left the farm to his son, Charles Taylor. In 1942 the state of Nebraska wanted 44 acres of the original 240 acres to build the Fairmont Air Base Hospital. They offered Charles $46.00 per acre for the land and the farmstead was part of that 44 acres. Charles deeded the remaining 196 acres to his brother, Ira Taylor, on March 3, 1948. Ira Taylor passed away on March 3, 1963 and the ground was passed to his sister, Daisy Taylor, on May 25, 1964. Daisy owned and rented the ground to her nephew, John L. Taylor Sr., until her death on Oct. 28, 1975. Daisy's wishes were to leave the ground to her nephews, James and Herbert Taylor Jr., the two surviving sons of Herbert and Nettie Taylor. James and Herbert Jr. were co-owners of the property until their deaths and then their seven children decided to sell the farm ground in September of 1996. John L. Taylor Sr., grandson of James and Margaret Taylor, purchased the farm ground for $1,650.00 per acre. John deeded the property to his children Sharon Elznic, Shirley Sluka, Sandra Petersen and John Taylor Jr. on Jan. 2, 1997. The Taylor Family Farms Partnership was established in 1999. Sandra Petersen married Norm Girmus in May of 2000 and John L. Taylor Jr. passed away on Aug. 18, 2003 and left his 1/4 interest to his daughter, Lindsey Taylor. The Taylor Family Farms Partnership continues to date between Sharon Elznic, Shirley Sluka, Sandra Girmus and Lindsey Taylor Stone.

Improvements to the Farm: 1978 - Well and pivot. 2008 - Well changed to electricity. 2010 - New electric pivot to replace the 1978 pivot.

The Fairmont Air Base Hospital was torn down in 1945 and the land was used to pasture cattle.

The state of Nebraska decided to sell the 44 acres in 1980. The Taylor family decided not to purchase the property back because of the cleanup costs. The 44 acres were purchased by Roger Lundburg of Fairmont and has been converted to farm ground.

TAYLOR – James Newton and Margaret Taylor purchased the NE 1/4 (160 acres) of 17-7-2 in the fall of 1891 from John Holmes for $3,300.00 ($20.625 per acre). James Newton Taylor passed away on July 14, 1939 and his wife Margaret Taylor on Sept. 25, 1941 and the 160 aces were placed into the Taylor Estate. In 1942, the state of Nebraska decided to build the Fairmont Air Base south of Fairmont, NE. The W 1/2 of the NE 1/4 was purchased from the Taylor Estate in the amount of $3,715.00 (approximately $23.22 per acre). The farm ground remained part of the Fairmont Air Base for the next 42 years. The state of Nebraska decided to start selling some of the farm ground in December of 1984. John L. Taylor Sr., grandson of James and Margaret Taylor, purchased 93.5 acres at the sale for the amount $70,500.00 (approximately $754.00 per acre). John L. Sr. and Wilma Taylor deeded the land to their son, John L. Taylor Jr. on Dec. 5, 1984. John L. Taylor Jr. passed away on Aug. 18, 2003 and by his final request the 93.5 acres was transferred to the John L. Taylor Jr. Trust until his daughter, Lindsey Dawn Taylor, reached the age of 28. The trust was dissolved and the farm ground was transferred to Lindsey Taylor Stone on Jan. 16, 2013. Lindsey remains the current owner of the property

Lindsey (Taylor) Stone and son Alex Stone with Dale Stengal of the fair board.

Improvements of 93.5 acres - W 1/2 of NE 1/4 17-7-2: new well and pivot – 1988. New pivot – 2002. Changed well to electric – 2007.

James N. and Margaret Taylor purchased the NW 1/4 (160 acres) of 17-7-2 on March 18, 1903 from Mr. and Mrs. Charles Burnett in the amount of $7,000.00 ($43.75 per acre). After the passing of James Taylor on July 14, 1939 and Margaret Taylor on Sept. 25, 1941 the 160 acres was deeded to their son, Charles Taylor, on Dec. 14, 1941. The state of Nebraska purchased the entire 160 acres from Charles on Nov. 16,1942 for the amount of $7,704.00 ($48.15 per acre) to become part of the new Fairmont Air Base. The 160 acres remained part of the Air Base for 42 years until John L. Taylor Jr., great grandson of James N. and Margaret Taylor, purchased 93.4 acres for the sum of $34,000.00 (approximately $364.00 per acre) on Dec. 3, 1987. John L. Taylor Jr. passed away on Aug. 18, 2003 and by his final request the 93.4 acres was transferred to the John L. Taylor Jr. Trust until his daughter, Lindsey Dawn Taylor, reached the age of 28. The trust was dissolved and the farm ground was transferred to Lindsey Taylor Stone on Jan. 16, 2013. Lindsey remains the current owner of the property. The remaining 66.6 acres are still owned by the state of Nebraska as part of the Fairmont Air Base today.

Improvements on the 93.4 acres - NW 1/4 of 17-7-2: new Well and Pivot – 2005.

TAYLOR – James Newton Taylor was born Sept. 30, 1850 in Wayne Township, Adams County, OH to James and Mary (Wright) Taylor. James Newton's father died when he was only 9 years old.

James Newton Taylor was married to Margaret Helmly, daughter of Christian and Christina (Dieder) Helmly, who came from Wurtenburg, Germany in 1846 to settle in Adams County, OH. The first eight years of marriage were spent as a hotel keeper and later as a farmer. In 1880 the family moved west to Lancaster County near Ashland and Greenwood, NE. In the fall of 1891, James boarded a train for Minden, NE to see about some land he had heard was for sale. The train was delayed in Fairmont, NE. While waiting, he started visiting with a gentleman that told him about some farm land that was for sale southeast of Fairmont. James rented a buggy and with a realtor he went to see the land. Before the day was over he had purchased the SE 1/4 of 17-7-2 for $4,500.00 from Jacob and Sarah Cable and the NE 1/4 of 17-7-2 from John R. Holmes for $3,300.00. James returned to Fairmont and took the train back to Lancaster County to make plans to move his wife Marqaret, sons Charles, Ira, Herbert, George and daughter Daisy to their new home in the spring of 1892. The SE 1/4 was chosen as the new farmstead. In 1903 a two story house was built consisting of three rooms on the main level and three bedrooms on the upper level. They moved two rooms from the old house and attached them to the west end of the new house. The old house also had two open porches. In 1906 a large barn was added and during the next 15 years they added grain storage, a hen house, smoke house, shop, corn crib and garage. In 1921 the hog house was built.

Pictured l-r: Debra Hourty of the fair board, Sharon (Taylor) Elznic, John L. Taylor Sr., and Wilmah Taylor.

In the spring of 1919 James, Margaret and daughter Daisy moved to Geneva leaving the farming to their son George who was married in February to Gladys May Dyer. They raised three sons: John, Kenneth and Donald on the family farm during the depression years. George's oldest brother Charles also remained on the homestead and raised horses and farmed. Charles lived in the old house and part time in the new house with George and his family.

James Newton Taylor passed away July 14, 1939 and Margaret (Helmly) Taylor on Sept. 25, 1941, leaving the SE 1/4 to their son, George and the NE 1/4 was placed in the Taylor Estate. In 1942 the state of Nebraska decided to build the Fairmont Airbase south of the town of Fairmont. Many acres of farm ground had to be purchased. The W 1/2 of the NE 1/4 of 17-7-2 was designated to be part of the new airbase. The Taylor Estate was given $46.00 an acre for their land. George was given $500.00 so a drainage ditch could be placed on the SE 1/4 to drain water from the above ground due to all the runways that had to be built.

During the construction of the airfield, Gladys rented out rooms to the men that worked on the buildings at the airbase. She used the money to close in the north porch of the house.

When the base completed it was discovered that the farmstead might be too close the planes to safely take off and land. It was feared the farmstead might need to be torn down. Through the efforts of surveying, it was decided to have the planes fly further west of the farmstead when taking off.

George and Gladys retired in 1946 and moved to Geneva when their son, John, returned home from the service and took over the farming operation. John and Wilma raised three daughters: Sharon, Shirley and Sandra and one son, John Jr. on the Taylor Family farmstead.

On Feb. 24, 1948 George and Gladys purchased the E 1/2 of the NE 1/4 for $4.800.00 from the Taylor Estate. With this purchase, George now owned the original 240 acres that his father, James Newton, had purchased in 1891.

On May 1, 1979 George passed away and left the SE 1/4 to his son, John. On Sept. 25, 1979 John purchased his brothers shares of the E 1/2 of the NE 1/4 for $1,000.00 an acre. This gave John the 240 acres that his grandfather, James Newton, had originally purchased.

In 1983 John Jr. purchased the 80 acres back from the state of Nebraska that was once part of the Fairmont Airbase. John Jr. paid $750.00 per acre for the land. A substantial amount more than the Taylor Estate had received 41 years earlier.

John and Wilma semi-retired in June of 1983 and moved to Geneva. John Jr. moved to the farm and made the fourth generation of Taylor's to live on the original 240 acres.

In 1988 pivots were added to the SE 1/4-17-7-2 and N 1/2 NE 1/4-17-7-2.

Wilma Taylor died Oct. 4, 1999. John L. Taylor Jr. died Aug. 18, 2003. John L. Taylor Sr. died April 14, 2012. The 240 acre farm is now owned by S.S.S. & L. Farms Partnership, owners: Sharon (Taylor) Elznic, Shirley (Taylor) Sluka, Sandra (Taylor) Girmus and Lindsey (Taylor) Stone.

TAYLOR – A. D McCorkhill was the original owner of this 80 acre farm, acquired on Jan. 15, 1887. The 80 acres are located in the north half of the north west quarter of Section No. 15, Township No. 6, Range 6, west of the Sixth Principal Meridian.

Although the original crops are unknown, a row of catalpa trees were planted along the north side.

The farm passed from A.D. McCorkhill to Catherine McCorkhill Stephenson, then to Allyn Stephenson Taylor, to Crittenden J. Taylor, Jane Taylor - his wife and William J. Taylor and John C. Taylor - his sons.

The N l/2 of the N l/2 (40 acres) remains in the family ownership. The S l/2 of the N l/2, (40 acres) were sold to pay for college expenses of Ruth Stephenson

The only award given to the family was the AKSARBEN HERITAGE FARM AWARD. Crit Taylor spent 30 months in the Civil Conservation Corp. in 1939-41.

General Information

Widow Jane and Crittenden J. Taylor had two sons, William J. Taylor and John C. Taylor, one

Grandson, Monte A. Taylor and one great grandson, Ryan Taylor. *Submitted by Jane M. Taylor.*

TAYLOR – The original owner of this 80 acres is Crit Stephenson who purchased it on April 4, 1884 and Jan. 20, 1903. The original crops are unknown. But corn is raised on the 80 acre farm now.

Crit Stephenson passed the farm to his daughter and son-in-law, Claudia and Clarence Buerer, who in turn passed it to their son, Kieth Buerer, Buerer Estate, then to John C. Taylor cousin of Buerers and great grandson of Crit Stephenson.

There are no buildings left on the farm. However a Graduate of Nebraska University Award was received. *Submitted by John Taylor.*

TEPLEY – Around 1895 Joe and Kathren Tepley purchased 160 acres located in Colfax County, 20-10-2E, NE. Corn, beans, oats and alfalfa were the crops grown on the farm.

The farm still retains 80 acres. Grain crops are corn, beans and alfalfa. Cattle, horses and chickens are raised now.

The farm still retains the house, a barn, garage, grain bin and hog shed. *Submitted by Adolph and Lois Vitek.*

TETEN – Harm Janssen Teten purchased the farm northeast of Talmage from Leonidas J. Holland and his wife, Sidney E. Holland, in July 1867.

An interesting note is that this purchase was soon after Nebraska became a state. On the deed "Territory of Nebraska" is scratched out and "State of Nebraska" is written in.

Harm and his wife, Talke Margaretha, deeded the farm to their son, J.H.J. Teten in 1880. John Herman Johnson Teten and Taalke Marie (Henrichs) were married in 1880.

Talke and Harm Janseen Teten

A two story, four room home brick house was built in 1889. A four level barn was built in 1905. J.H.J. hauled the limestone rock for the foundation and basement of the barn from a quarry near Coryell Park (south of Brock). He contracted pneumonia during this process and never completely recovered.

During a diphtheria epidemic in 1887, three children - Mamie, Harry and Maria - all died within a six week period. They later had two more children, Alide Marie and Reinhold H. Taalke was the owner until her death in 1941 when her son Reinhold became the owner.

Reinhold had married Minnie Emma (Kolsch) on March 23, 1927. He passed away in 1958. Minnie continued to live on the farm until 1970 when she moved into Talmage.

In 1971 Reinhold Jr. and his wife, Mary Elizabeth (Adams), and their children - Timothy, Thomas and Jelayne moved to this farm.

Minnie transferred the farm to son Reinhold in 1974. Reinhold and his wife, Mary, continue to live on the farm. *Submitted by Mary E. Teten.*

THEILEN – Frederick and Frederica Theilen came to Nebraska in the 1880s from Olderburg, Germany. They settled on a 40 acre plot of land in Grand Prairie Township of Platte County. They lived here in a dugout until they later acquired another 80 acres of land in the same section, where they established the farmstead where they raised their five children: Frieda, George, Emma, Ida and Fred Jr.

After Frederick's death in the 1920s, sons George and Fred Jr. continued to farm the land. In the 1930s Fred married Helen Ludetke and they moved into the farmstead and continued to operate the family farm. They used gift money from their wedding to purchase corn seed as life was difficult in the 1930s due to the depression. Fred and Helen continued to live in the frame house; many other building were added to the farmstead through the years with the help of family and friends.

This picture of Fred Theilen Jr. and his son David with one of his John Deere tractors was taken after the 1949 blizzard, which left huge snow drifts. The windmill in the background supplied water for the farm.

In time they were able to purchase another 80 acres of land in the adjoining section to the east. They raised all kinds of livestock including workhorses, Brown Swiss milk cows, beef cows, various breeds of hogs, chickens and lambs. Crops included corn, oats, wheat, alfalfa, sorghum, milo and soybeans. With two large produce gardens, meat animals and fruit trees they supplied most all of their food supplies with plenty to share. Hired hands were well fed with a hardy meat and potatoes meal with homemade pies to follow. Produce was canned or frozen to preserve for winter months.

Fred and Helen had two children, David and Marilyn, who grew up on the farm and helped with chores. In 1974 as Fred's arthritis worsened they moved to Columbus, although they quite regularly worked on the farm place. They rented the farmland to brother George's grandchildren, Glenn and Rodney Theilen. Three grandchildren were added to Fred and Helen's family - Nicole, Ryan and Sarah Vrana.

The farm ownership passed on to Helen with Fred's death in 1982. In 1991 with the passing of Helen, their children - the current owners, David Theilen and Marilyn Theilen Vrana - inherited the family farm. They rent out the land but Frederick's great grandson, Glenn Theilen, continues to actively farm part of the land.

The Fred Theilen Farm was recognized by Aksarben for over 100 years of farm ownership in the 1990s. *Submitted by Marilyn Vrana.*

THOMPSON – Joe Thompson, born in Illinois in 1879 came to Richardson County by covered wagon in 1881 with his parents, William and Amanda Jane Martin Thompson. They bought a farm northwest of Falls City and later purchased a farm and moved southwest of Falls City. In June 1905, William and Amanda purchased an additional 80-acre farm two miles west in Section 26 of Salem Precinct.

Joe bought it from his parents in 1906. In February 1908, he married Mabel Macomber, a pretty school "marm" who was boarding with his parents and they continued living there until their deaths in 1963 and 1964. In 1919, they purchased an adjoining 40 acres. The drought and depression of the '30s was a struggle to keep from losing everything, but they provided their son and four daughters with physical, educational and spiritual needs. Joe used horse equipment to farm,

Back row, l-r: Jason Baldwin, Ryan, Delman and Brady Trampe, and Dalton Loeffelholz. Front row: Cooper Trampe, Hunter Baldwin, Breanna and Payton Trampe, Jolene Baldwin, Rachel, Marlene (holding Waylor) and Megan Trampe, Holden and Hannah Baldwin, and Mason Trampe. Not pictured: Oakley Trampe.

It was then that Delman Trampe and his wife, Marlene (Geisler) Trampe, purchased the property from Carl's estate, finalizing the sale, Feb. 21, 1989. They added onto and remodeled the original farm home and lived there until they moved to Kearney, NE in 2004. They have three children - Ryan, Jolene and Brady. Both boys are involved in the farming and ranching operation. The family was honored to receive the AK-SAR-BEN Pioneer Family Farm award in 1994.

Since 2004, Ryan Trampe, his wife Rachel (Budt) Trampe; their children, Breanna and Cooper Trampe, and Rachel's son, Dalton Loeffelholz, have resided on this property. Brady Trampe, his wife Megan (Knoble) Trampe, and their children; Mason, Payton, Waylon and Oakley Trampe, live north and east of the homesteaded property. Jolene (Trampe) Baldwin, her husband Jason Baldwin, and their three children, Hannah, Hunter and Holden Baldwin make their home in Normal, IL.

TYRELL – In 1888, Miles and Lucy (Blair) Tyrrell headed to Nebraska from Northwest Illinois and settled in the Wisner area. In 1902, they moved to the present day 160 acre farm, southeast of Pender, in Cuming County, Nebraska. Needing a larger house, which this farm had, made it an easy choice for them since they had nine children. It even had a cupola on top, for the former owner, Warren Fales, to check on cattle in his distant pastures. Miles grandchildren had many a fun day in this cupola playing, hiding and sliding on the roof.

The Tyrell farmhouse, with a cupola on top

School teachers from Rural District #53 used to stay at this house occasionally and Miles even taught some of the classes for a few years. His children attended Rural District #53 along with Owen and Sara Tyrrell's family (grandchildren), and Margaret (Tyrrell) and Calvin Dunn's family (great-grandchildren). In all, the Tyrrells had four generations attending Rural District #53, which is now closed.

This farm has raised corn, alfalfa, red clover, soybeans, grass pasture, cattle-both beef and dairy, hogs, chickens, turkeys, ducks, geese, pea cocks, guinea fowl, horses, mules, burro's, goats and buffalo over the years.

Today (2015) this farm is owned by Randall and Diann Dunn, Great Grandson of Miles, and is being farmed by Travis and Jamie Ann Dunn, Great-Great-Grandson of Miles, raising corn and soybeans. Miles died on this farm in 1934, in the old house, from a farming accident, but his spirit continues to help guide the family into the future.

UECKER – On March 12, 1867 Carl Uecker bought 160 acres located in Madison County, 24, 1 West. Corn was the crop grown and livestock was beef cattle, hogs and chickens. Carl Uecker was one of the first of 42 families that settled in Norfolk, NE.

The original horse barn

The farm passed through five generations: from Carl Uecker to his son, Louis. To Louis' son, Herbert, then to Herbert's son, David. Mike Uecker acquired the farm in 2004 from his father, David.

The farm now contains the original 160 acres plus other surrounding farmland. Soybeans, corn, rye, sorghum and beef cattle are produced by this farm now.

Buildings still on the farm are the original house built in 1913, a horse barn and a cow barn.

In June 1967, the family was awarded the Pioneer Family Award.

Michael and Gina Uecker have two children, Conner and Ashley. *Submitted by Michael Uecker.*

ULDRICH – The name of the original settler of the farm was John Q. Harris, obtaining land from the state of Nebraska on Dec. 9, 1897. Owner John P. and Bertha Davis obtained the land on July 1, 1901. On Feb. 26, 1906 John D.J. and Lillie Davis acquired ownership. The west and east 1/2, section 36, NW 1/4 Glengary Township, Fillmore County, NE passed to Adolph Uldrich on March 10, 1907, 2:15 P.M., W.C. Peterson, County Clerk.

There was no relationship between the Davis' and Adolph Uldrich. Adolph and Antonie (Buzen) had three children – Frank, Antonio, Frances. Then to Anton W. Uldrich, 1946. Anton married Mamie (Chudly) Uldrich. Anton had two children, Ronald Dean and Marlene Mae Uldrich Agular.

Ronald D. married Mary (Engle) Uldrich June 2, 1957. Ronald and Mary had four children: Ronald Perry, David Rae, Janine Marie and Alex Dean. The farm went into Ronald D.'s name in 1990 with his wife, Mary L.

After Anton's death on Jan. 24, 1992 and Mamie Uldrich's death March 27, 2001, the farm went to Ronald D. Uldrich and wife Mary L. At present time the farm is in Mary L. Uldrich's name after the death of her husband, Ronald Dean on June 11, 2015.

Ronald D. and wife own 120 acres and his sister, Marlene Agular and husband Teo-Felo Agular own 40 acres.

The house, windmill by the house, a round bin and half of a new crib, outhouse, cattle shed in the feed lot and part of a barn remain.

The Uldrich farm owned by Adolph, Anton, and Ronald Uldrich

There are two newer hog confinements and a garage. The house was lived in after Anton/Mamie moved to town in 1989. Water was put in the house in 1986 as their grandson, Ronald P. and wife Diane lived there three years.

Ronald D. Uldrich farmed with his father when he was still in grade school and after graduation except for two years when he served his country in 1953-1955, most of the time in Korea. Ronald returned to farming with his father.

Alex has been farming since his sophomore year with his dad after having a heart attack and later Parkinson's. Alex is married, wife is Debra (Wilkins), he has three children and two stepsons. They live on their farm a mile west of Milligan, NE.

Our other children:

Ronald P. and wife Chris (Denton) Uldrich have three children – Michael, Rrech, Kaylee.

Janine M. Uldrich, husband Rick Buchanan, has three - Clarsea, Michele, Ethan.

David Rae has two children, Corey Rae and Alyassa.

Ronald D. and Mary L. have 34 grandchildren and great grandchildren.

Mary and Ronald D. lived on their farm 3/4 mile west of Milligan, NE for 58 years. Mary still lives there after Ronald D.'s passing June 11, 2015.

No one has lived on the home farm since 1989.

In 1997 Ronald D. and Wife Mary L. Uldrich received the award from AR-SAR-BEN for the farm belonging to Adolph, Anton W., Ronald D. and Mary L. Uldrich. We are grateful for their legacy and hope to keep it for generations in years to come. *Submitted by Mary L. Uldrich.*

VALENTINE – The Valentine family received the Pioneer award in 2006. The farm was called Blue Valley Stock Farm by Daniel Nettleton. In 1872 Nettleton, who was an injured civil war veteran, came to Clay County to homestead the NE 1/4 of section 18, Township 5, range 8. He added 320 acres, which he purchased, the SE quarter in 1880 and the SW 1/4 in 1881. He built an elegant, two-story square, five bedroom house in about 1897. The farm yard contained many buildings for his livestock operation. The farm is located on the Little Blue River and the Missouri Pacific Railroad went through the property. It was close to the Pony Express and Oregon Trail. The farm was sold on Feb. 19, 1906 to Remmer Johnson, who had purchased 70 acres in the NW 1/4 of the same section in 1905. Remmer owned the farm until 1941 when the depression had taken its toll and he was about to lose it. In 1941 he sold it to his daughter and husband, Lucy and Herman M. Valentine, for "One dollar and other good and valuable consideration." By raising Shorthorn cattle, pigs, sheep and chickens as well as grain and hay, they kept the farm.

View of the house in the 1970s on the south side. To the left is Lloyd walking down the sidewalk toward the summer kitchen, which was part of the old home and damaged by a storm in 2012.

Lucy Johnson Valentine is on the porch of the old home, originally built in 1912 and remodeled in 1948.

Herman was killed in a tractor rollover accident in 1949, but Lucy kept the farm going with the help of her children - Lavon, Herman J, Wendell, Lloyd, Lucille. Lucy died in 1988; Lloyd and Donna bought the farm from the estate. Their children are Cindy Hinrichs (Ron), Candy York and Darwin Valentine (Denise). Darwin and Denise live in the original square house although the other original buildings are gone.

The farm still produces grain and hay. The cattle are still red but Gelveih breed and Lloyd now relies on his son and son-in-law. Lloyd and Donna live 1/4 mile north of the farmyard. *Submitted by Donna Valentine.*

VANDERSLICE – Augustus VanderSlice moved from Illinois in 1882 to semi-retire from his medical practice. He was born Oct. 23, 1833 in Pennsylvania, graduated from University of Pennsylvania in 1860. Married Rebecca Rauch in 1860. They had 10 children, the fourth and seventh children became MDs and the others were in farming in the area. Augustus died Sept. 25, 1887 while on a house call for a sick neighbor. Rebecca stayed on the farm with the help of two of the older sons. She remained on the homestead until her death in May 1926. After the death of the two sons who farmed there, my father, Howard Roy, with his wife, Susannah Ring VanderSlice, moved to the homestead where they remained until their deaths in 1961. At that time I, Rebecca VanderSlice Rippe became Executress of the Estate and rented it to nearby farmers until the City of Lincoln took it for a city park by eminent domain. It is currently being farmed with corn, milo and wheat crops by the same I had rented it to. *Submitted by Rebecca VanderSlice Rippe.*

Children of H.R. Vanderslice, 1942. Picturd are Rebecca, Raymond, Edwin, and Lloyd.

H.R. Vanderslice and grandson Mark, 1956.

VAWSER – On March 3, 1902 Wesley Vaughn purchased 160 acres west of Bloomfield in Knox County, 17-30-4 E2, E 2 SE. The homestead had a one-room house. Later he built four rooms onto the house. The lumber cost $1,500.

The Vawser homestead, October 11, 2011

Wesley Vaughn passed away in 1948 and willed the farm to his daughter, Ella (Mary Ellen) Blankenfeld. Henry and Ella moved the granary and chicken houses about 1/4 mile south, where they had put in a new well. They moved there in the spring of 1950 when the built a new house from the lumber of the original house. They slept in a chicken house and cooked in the granary while building their new house. Later they also built a large barn.

In June of 1970 by quit deed claim, Ella transferred the farm to Helen Mansfield, her daughter, and husband Dale Mansfield.

In December of 1977 Leland and Rosellen Vawser purchased the farm from Helen and Dale Mansfield. Ella Blankenfeld was Leland's aunt and Helen Mansfield was his cousin.

Crops raised over the years were oats, corn and hay. After the Vawsers bought the farm, they had cattle and paint horses. We no longer had grain but had hay for the animals.

Leland and Rosellen moved to their farm in July of 1994. Before that time they lived there summers and many week-ends. In 1996 Leland built a large living room and upstairs bedroom onto the house.

On Dec. 7, 2009 we sold our farm to our sons, Steven Vawser and Wesley Vawser. My husband, Leland, passed away July 21, 2015. I continue to live on the farm caring for three horses, chickens, guineas and four cats.

Steven and wife, Anne, live in Gretna, NE. Wesley and wife, Elizabeth, live in Omaha, NE. They have two children, Nicole and Joseph. *Submitted by Rosellen Vawser.*

VORDERSTRASSE – In 1885 William Vorderstrasse, at the age of 16, came to America from Germany. He was the second member of his family to come to this country. The remainder of the family followed one at a time until enough money was available for the remainder of the family to come over. As a young man, William worked as a farmhand in the Plymouth, NE area and then came to the Deshler area as a farmhand in 1894. One summer day while threshing on the Hofts' farm near Kiowa, he went to the water well to get a drink. By chance Sena Hofts was there getting water for cooking the noon meal. This meeting led them to later being married in 1899. William and Sena saved their money and purchased a farm northeast of Ruskin. In the winter of 1910 they had a chance to sell that farm and purchase two quarter sections a mile west of Stoddard. There they farmed and built a two story house along with numerous outbuildings and raised 11 children. William passed away in 1946 and Sena passed away in 1966. Upon settling the estate in 1967 their son, Lawrence, purchased the home quarter and son William (known as Uncle Bill) purchased the other quarter across the road. Bill rented his quarter to his brothers, Lawrence and Clarence, and later to his nephew, James. In 2000 Uncle Bill sold his land to his nephew, James. *Submitted by James Vorderstrasse.*

William and Sena Vorderstrasse on their wedding day, 1899

The Vorderstrasse family with the 100 year award in 2011

VOTIPKA – The S of SE of 35-7-2 was granted to the B.M.&R. Railroad Co. on June 15, 1872. On April 5, 1884, the 80 acre farm sold to Timothy Hourigan. In Feb. 17, 1884. Timothy and wife Bridget Hourigan sold the farm for the sum of $6500.00 to Thomas and Catharine Bland. In Sept. 29, 1904, the land transferred to Tandy Bland. In that time span the farm had been mortgaged 13 times.

The Votipka family receive the Ak-Sar-Ben Pioneer & Heritage Farm Award.

On April 1, 1915, Vaclav Ben Votipka purchased the farm for $8000.00 from Tandy and wife Dora Bland.

With the passing of Ben Votipka on July 1, 1945, the farm passed to his wife, Anna Votipka and their ten children.

Out of the ten children this farm passed to his two daughters, Helen Bernasek and Alice Bealer. On May, 12, 1949, Helen Bernasek and husband Cyril M. Bernasek sold an undivided to Frank B. and Pauline F. Votipka. With the passing of Frank B. Votipka on Dec. 29, 1949, the interest passed to his wife Pauline Votipka. On June 22, 1956, Pauline purchased the other undivided of the farm from her sister-in-law Alice Bealer and husband Donald Bealer. With the passing of Pauline F. Votipka on April 24, 1980, since her son Rueben F. Votipka proceeded her in death on Feb. 17, 1955, the farm passed to her grandson James R. Votipka, who is the current owner. *Submitted by James R. Votipka*

WAGER – George and Lorissa (Cooper) Wager homesteaded the Wager farm, which is 2-1/2 miles west and 1/2 mile south of Brock on Feb. 25, 1876. The family first lived in a very primitive house (no glass in the windows).

George and Lorissa had two children, Frederick and Mary Ann. In the late 1880s a new house was built. The lumber was hauled from Nebraska City.

New house on Wager Homestead, built about 1880. Pictured l-r: George Wager, Lorissa Wager, Fred Wager, and Mary Ann Wager.

After the marriage of Mary Ann to William Oscar Adams and the death of Lorissa, George and his son, Fred, continued to live on the farm.

In 1936 (after George's death in 1935) the farm was transferred to Mary Ann (Wager) Adams.

Fred's son, William Neal Wager, told that George loaned money to many people - with nothing in writing. When George passed away Fred said to his sister, Mary Ann, that she could have the 160 acre farm and he would take the "loans." William Neal also said that for many years people would drive in their yard and tell his dad, Fred, that George had loaned them money and they were there to repay it. Fred felt that he had made a satisfactory deal with his sister.

In 1937 William Roy Adams, a grandson and his wife, Luella (Kohrs) moved to this farm. In 1958 they acquired the deed inheriting part of the land and purchasing the remainder from family members after the death of his mother, Mary Ann.

After Luella's death on Feb. 16, 1985 William continued to live on the farm. He passed away on Nov. 8, 1992. After his death, the land was deeded to their daughters – Mary Teten, Donna Reimers and Corinne Robinson. Mary's son, Thomas, now lives on the farm. *Submitted by Mary E. Teten.*

WAGNER – In 1886 Matthias Wagner purchased 160 acres located in Knox County, Miller Township. The farm produced corn, oats, hogs, cattle and horses.

The farm still retains 160 acres and now produces corn, beans, hay and cattle. All the buildings were re-built between 1910 and 1918.

The children of James and Bernice Wagner are Meha Val, Clem, George, Ralph, Ted, Carol, Angie and Randy.

The family has received several awards including a Conservation Award from *Sioux City Journal* and Sioux City Chamber of Commerce and two awards from Ar-Sar-Ben Foundation for 100 year farms. *Submitted by James Wagner.*

WALZ – Johann Walz and his wife, Franzieka (Gauggel) were born in Germany in the 1860's. Later they sailed to America where the couple settled in Lincoln County near Stapleton, NE in 1894 on land under the Homestead Act. It was here that they raised 9 children. Two of these children, John and Nicholas Walz, moved to Arthur County and staked claims under the Kincaid Act on 2 homesteads November 18, 1913. These two sections are located seven miles north of Arthur in what the Indians named "The Big Valley". Homestead final proof is dated May 11, 1918 and documentation on the Bureau of Land Management map is dated Jan. 30, 1919. Nicholas went into the military to serve in World War I and sold his section to John in 1919 and these 2 sections have remained in the Walz family for 100 years.

The Walz family in August 2013. Pictured l-r: Adrienne, Weston, Wheeler, Kelton, Troy, Marv, and Theo.

John and Mae (Stearns) Walz raised 5 children: Elizabeth, Alfred, Harlen, Gilbert, and Clarence on these sections. They lived in a sod house until 1938 when they had a house moved in from the Kingsley Dam evacuation area. Marv resides in this house located 20 feet from the sod house site.

John's grandson, Marv, son of Alfred, spent hours as a young boy helping his grandparents and uncles on this ranch. After serving his Country with the U.S. Army and in the Vietnam Conflict, Marv returned to Arthur County where he purchased the Walz Homestead Ranch in 1969 from the John Walz heirs. Marv and Theo (Wilson) Walz raised 2 children, Del Rae and Troy, on this ranch and now share these deep roots with their son, Troy and his family, Adrienne (Benz), Wheeler, Kelton, and Weston. It is with great pride that the Walz Family continue to live on and work this land, raising Angus cattle and Quarter horses while carrying on a 100 year legacy.

WATCHORN FARM – Richard Watchorn immigrated to the United States from County Carlow, Ireland in the 1880s. He returned to Ireland and brought back his sweetheart, Jane Hawkins, from County Kilkenny, Ireland. They married in Schuyler, NE in 1887. They raised four sons in the Leigh, NE area. It was Richard's desire to secure a farm for each son, which he did. The home place eventually became our father, Arthur's, and we were raised there. The farm is located 1-1/2 miles north of Leigh in Colfax County, NE quarter of Section 6, Township 20, Range 2 east of the 6th P.M. It was purchased in 1905.

Art Watchorn Farm

Arthur married Bertha Claussen in 1922 and they had two sons. Adrian died in infancy. Arlo was 4 when his mother died in 1932. Two years later Arthur married Eda Beitz. Two daughters were born to this union, Arlene and Ardith.

Due to declining health, when Arlo was drafted in 1951 the farm ground was rented out. Art and Eda continued to live on the farm until Art's death in 1960. Eda moved into Leigh and the land continued to be rented out. Eda died in 1993.

In 2005 the Watchorn Family received the Pioneer Farm Family Award. Recipients of the award were Arlo (Carlene) Watchorn of Yankton, SD; Arlene (Ed) Gnirk of Hoskins, NE; and Ardith (Gene) Schueth of Humphrey, NE. They received a plaque and gatepost marker.

The farm was sold to Roger Fuhr and Mike Fuhr in 2008. *Courtesy of Ardith Schueth.*

WENTWORTH – J.O. Wentworth came to Franklin County in 1872. He was a blacksmith by trade. In 1880 he bought the farm from the Shoemaker family. He went back to Michigan and his son, Louis N. Wentworth, took over. Louis married Elva Ronson and four children were born to them: Lady, Clare, Ray and Lucille. Louis died at age 51. Elva stayed on with the help of two sons. Then later J. Ray took over the family farm. He married Emma Eggleton and five sons were born to them: Robert, Paul, Dean, Neal and David. Neal was the one who stayed home to farm. Neal married Gaylene Kennedy and five sons were born to them: Merlin (deceased), Norman, Douglas, Marvin, Leland` Neal Jr.

The Wentworths are known for watermelons. Douglas is the fourth generation to raise them. We were awarded for 100 years several years ago. *Submitted by Leland Neal Wentworth.*

WESTERHOLD – Frank Westerhold purchased 136.81 acres in Wayne County, Leslie Township 34.25.5 on July 6, 1891. On the farm he raised corn, oats, milo, cows, pigs, horses, and chickens. After Frank the farm passed down to his son Herman, then to Herman's son Frank, to Frank's son Marvin, and to Marvin's wife Pauline.

Today the farm includes 150 acres and produces beans and corn. An old horse barn still stands that has blue-green steel siding. The Westerhold's have a patent on the land and a GLO seal signed by President US Grant.

Marvin (deceased) and Pauline have two children, David and Daniel, three grandchildren, Nicole, Jeremy and Brian, and two great-grandchildren, Samuel and Max. Their son David and his wife live in the farm and continue to work the land. *Submitted by Pauline Westerhold.*

WHEELER – John H. Wheeler purchased 640 acres located in Sioux County, Section 29 in Township 28. The year was 1911. This farm produced corn, potatoes, cattle, horses, hogs and chickens.

Seated are John H. and Dora Wheeler with 9-year-old Calvin M. Wheeler. In back (l-r) are Addie Wheeler Pedersen, Roy Pedersen, Mertie Wheeler, and Gladys Wheeler

This farm, containing the original 640 acres plus another 800 acres, was placed in the Calvin M. Wheeler Family Trust.

This farm had the first irrigation well in Sioux County. The original barn, corrals and cattle shed are still in use today. This farm now produces cattle, alfalfa and corn.

Calvin M. and Alice (Malone) Wheeler had six children – Keith, Karl, Glenn, Gaylord, Merilyn and Mary Alice. They had 17 grandchildren. Calvin did not retire; he passed away in 2004. *Submitted by Karl Wheeler.*

WITT – Michael F. Witt and two other families arrived on the banks of Swan Creek in 1861 to claim their homesteads. They lived in their covered wagons until a log cabin was built. The three families resided in that cabin until each had built a home. On August 8 of 1863, Michael was given a receipt for $12.00 from the Brownville Land Office for the entry of the NW ¼ of Section 13, T 5 N, R 2 E in Saline County under the Homestead Act of 1863. The other settlers signed an affidavit swearing that Michael had resided there prior to the 8th day of August 1863. In 1868, Michael built a 1 ½ story house in the German style, called a Fachwekbau, of hand-hewn timber and clay and straw brick. This structure still stands and is on the national register of historic places. In 1869, Michael paid the second half of his ($2.00) fee for the homestead and received his patent on Feb 15, 1870.

In January of 1882, Michael and Christina Witt sold the farm to their daughter Augustina Paulina Witt and her husband Charles F. Zabel. From there the farm would pass to his son John Henry Zabel, to his son Roger Zabel. Roger put the Witt House on the National Registry and did much to preserve the history of the farm by showing Steam engines and by demonstrating the art of spinning wool for many years. Roger's son Floyd Zabel farmed the land after Roger and in 2010 his cousin Kay Kottas purchased the farm. Kay and her husband now farm the land. Kay Kottas's mother Jessie (Zabel) Harrington, Roger's sister, also grew up on the farm as the daughter of John Henry Zabel. Kay (Harrington) Kottas is the great great granddaughter of Michael and Christina Witt.

The farm began as a traditional farm and continued to be so until 2010, when Kay purchased the land and began to convert the small farm to native plant and seed production. She is also currently attempting to restore the Fachwerkbau to its original condition.

WOEBBECKE – A 160 acre farm two miles east and two miles south of Seward has been in the Woebbecke family since June 5, 1913.

William Woebbecke bought the farm on June 5, 1913. William's son August inherited the farm. August's daughter Doris and husband, Bill Hartmann bought the farm on May 20, 1977 and are the current owners.

The Hartmann's provided the following information about the farm.

William Woebbecke paid $20,250.00 for the 160 acres. He died on Nov. 6, 1914 and his sons, Carl and August, farmed the land. They lived in a small house near the road. August planted an evergreen tree that still stands proudly today.

August inherited the land in 1945 when his mother, Catherine, died. During 1927 a new house was built and it is still being used, although it was extensively remodeled during 1984. Note, however, that the original slate shingles are still in place and doing their job. August married Martha Eggerling on Dec. 25, 1927. They had two children, Lloyd and Doris. Horses were the source of heavy power on the farm until August purchased a new Allis Chalmers WD tractor in 1950. He had some tractor power before 1950, however, in a Allis Chalmers Model C tractor. He also bought a new Allis Chalmers model 60 combine in the 1950s. All of these machines are still on the farm and are quite ready to operate today if asked to do so, Doris wrote.

Bill and Doris Hartmann bought the ground on May 20, 1977. In 1985 they moved to the farm and brought their construction company from Garland.

The Hartmann's have two sons, Andy and Jon, Both are active in the Hartmann Construction Company. *Submitted by William F. Hartmann.*

WOHLERS – Heinrich Gerhard (Dutch Henry) Wohlers, born in Germany in 1839, married Helene Marie Stulken in 1867 and owned a farm near Neuenwege that had been in the Wohlers family for 120 years. They had nine children. In 1882 after his wife's and one son's

Pictured l-r: Heinrich (Harry), Friedrich, Alma, Mary holding Anton, Anna, Heinrich (Dutch Henry) holding John, 1906

Wohlers homestead

death, Heinrich came to America to see what opportunities there were here, going to Iowa where he had relatives. Then he returned to Germany, sold his farm and returned to the US with four sons and two daughters. One son died and was buried at sea. One son did not come to the US.

They stayed in Iowa with relatives for a while and Heinrich applied for citizenship. Heinrich and a friend rode to Valentine by train then walked out to the tableland southeast of Crawford, which is now Dawes County. Heinrich selected a 160 acre homestead with timber and farmland thinking timber could be used for buildings, which it was for the house and barn. Land Description is NE 1/4 Sec 8 T30 R50 Heinrich and one son returned in the spring of 1885 making and living in a dugout for a couple of years. The rest of the children came a year later.

In 1906 Heinrich sold the homestead to his son, Friedrich and wife Mary along with surrounding 640 acres of land that he had purchased. They had 13 children. The first wood frame house was built around 1887. The brick part was built around 1904. The wooden portion of the present house was built in 1910 replacing the first wooden house.

In 1914 work was started on the barn, using all of the rough lumber from trees on the place, using no blueprint. The barn is 40 X 80 feet with a gable roof and an 18 foot lean-to. We still have all the receipts and time sheets for the barn. 30 kegs of nails were used with a total cost of the barn $3,000. The barn and house still stand in structurally excellent condition. The barn sits on the divide of the Pine Ridge, water from east side of roof runs to the Niobrara River and west side runs to White River going 400 plus miles through Nebraska before merging again.

In 1947 after Friedrich's death, John and Marguerite Wohlers purchased the place. John died in 1999 and Marguerite in 2006. Merton and Eldon Wohlers and Wanda Wohlers Leeling then inherited the property with Eldon's son, Bruce Wohlers operating the property till 2014 when Bruce and Marie's son, Taylor Wohlers purchased the property. Taylor is a great-great-great-grandson of Heinrich (Dutch Henry) Wohlers. He is now operating the place as a cattle ranch and grass hay and hunting lodge. It is the second oldest homestead in Dawes County continually owned by the same family. Over the years, cattle, hogs, sheep, horses, chickens, potatoes, wheat, barley, oats, corn, and alfalfa were all raised on the place. The timberland was 100% logged and thinned in 2003 and 2004.

Awards: Nebraska Pioneer Farm Award Wohlers Homestead, June 15, 1985. Dawes County Agriculture Hall of Fame Friedrich and Mary Wohlers. Upper Niobrara White Natural Resources Dist. Conservation Award for the Wohlers Family Excellence in Forest and Range Management, 2004. Nebraska Forest Stewardship Award, Eldon Wohlers, May 25, 2004. National Register of Historic Places, Oct. 15, 2004. *Submitted by Eldon Wohlers.*

WOLFENDEN – A farm or ranch is more than a tract of land; it is a way of life that creates a heritage of hard work and blessings to be passed down throughout the generations. John and Annie Kline began this heritage for what is now Wolfenden Ranch. John and Annie, along with their 4-year-old daughter, Lizzie, traveled by covered wagon bringing all of their belongings and a vision for a better life. Arriving 60 miles southwest of Valentine, NE in the center of Cherry County, T. 29 R. 33 S. 13, the wagon box was set off and a hole dug into the side of the hill. This is where they spent their first winter. A sod house was built the following spring.

Cyrus Robert Wolfenden, age 16, trailed cattle to the Dismal River south of Seneca, NE where he stayed and herded cattle on the open range while living in a covered wagon. Cyrus would later meet and marry Lizzie Kime, the daughter of John and Annie Kime. Due to Annie's failing health, John and Annie sold their holdings to Cyrus and Lizzie in 1907. Cyrus continued to buy out kinkaiders and homesteaders until nearly 20,000 continuous acres had been put together.

Wolfenden Ranch raised many horses for both ranch work and for selling to the Remount Station at Fort Robinson, Crawford, NE. A 100 head of the horses were used during hay season for putting up hay with a 21 man crew. Today a three person hay crew is used.

Cattle were trailed to Cody, NE then later to Mullen, NE to be shipped by rail to the Omaha Stock Yards. Both locations were a three-day trip. When trailing to Cody, the Gordon Creek was forged and the Snake River crossed along with a bridge on the Niobrara River. The trail to Mullen contained the North and Middle Loup Rivers to be crossed. Nights found everyone camping out, sleeping on the ground.

The Wolfenden family is very proud of the ranch and the heritage it represents. Wolfenden Ranch has been home to six generations, from John and Annie Kime to the youngest Wolfenden. LeRoy and Kay Wolfenden have three married sons, nine grandchildren and one great-grandchild, who either live on or enjoy coming to visit the ranch. God has blessed the family to live and abide in the beautiful Sandhills of Nebraska. *Submitted by LeRoy and Kay Wolfendfen.*

WOLFF – On March 2, 1907 Frank and Antonie Travnicek bought 160 acres located in Stanton County, Ramshorn Precinct, NE 33-21-2. Corn, oats and alfalfa were grown on the farm.

The farm passed from Frank to son Anton and wife Julia Travnicek on May 13, 1914. On Jan.l 5, 1955 it passed to Joseph and Vlasta Travnicek on Feb. 25, 1963 to Joe W. and Dorothy M. Travnicek and on Jan. 17, 1992 to Judith (Travnicek) and Eugene Wolff.

Travnicek farm in 1984

The wash house was the original house built on the homestead after the sod house. The silo is still standing. The granary was originally the District 31 Glantt School and is no longer standing.

The farm retains 160 acres and now produces corn, soybeans, alfalfa, cattle and hogs.

This farm is one of four farms Frank Travnicek bought when he came to America and is the only one that is still in the Travnicek family. He bought a farm for each of his sons.

Judith and Eugene Wolff have two children, Michelle (Wolff) Foster and Michael Wolff. Grandchildren are Jennifer Foster, Jocelyn Wolff, Christina Wolff, Jack Wolff and Dylan Wolff. *Submitted by Judith Travnicek Wolff.*

WOODMAN – John Lippincott, born 1821 in New Jersey, and Mary Dillon, born 1825 in Pennsylvania, were married April 14, 1844 in Pennsylvania. Thirteen children were born to this union, eight of whom died in infancy. The four boys who grew up pioneered in Nebraska.

Woodman farm

Wrich farmstead

John Francis "Frank" Lippincott and his brother Jeramiah Fullerton "Jerry" Lippincott came to Fillmore County, NE in 1870. Then all four brothers came to Elm Island Martin Township 32-9-12 and 33-9-12 Hall County, NE. William Bigler "Will" Lippincott in 1882, James Forman "Jim" Lippincott in 1883, Jeramiah Fullerton "Jerry" Lippincott in 1885, John Francis "Frank" Lippincott in 1892.

Raye Woodman, great grandson of Jeramiah Lippincott, bought the Lippincott 400 acres on April 4, 1971 from the Lippincott heirs. Raye's dad, Clifford Woodman, had rented this ground and farmed it since 1936. You might say Raye has farmed this land his whole life. The first crops were corn, wheat, alfalfa and native hay. Raye has leveled all of the land, put in irrigation wells and now it has center pivots and raise corn and soybeans except for 13 acres of that same native hay.

The Woodman farm has been in the family since 1912, Emerson and Ella (Lippincott) Woodman, 1959 to Clifford Woodman, 1995 to Raye Woodman. Clifford born 1913 and Raye born 1934 both on this farm. Raye Woodman married Carolyn Kroll March 11 1956. This farm has been our only home for our family of four sons: Denis, Douglas and Cheryl, Kenneth, Kevin and Sharon. We have four granddaughters: Carrie and Richard, Danielle and Adam, Ashley and Amy. Great grandchildren are Wyatt, Riley, Kylie and one due in December 2015 and one on our 60th Anniversary March 11, 2016.

Our four sons now farm the land, and we still enjoy living in the 130 year old farm house. The family is proud and grateful of their legacy and hope for it to continue on for generations to come. *Submitted by Carolyn Woodman.*

WRICH – Henry Wrich purchased 80 acres in Washington County, located in the east half of the Southwest quarter of section 17, Range 11, Township 17, East Sixth principal meridian, on September 1, 1906, from James Berry for $7,500.00.

Henry and his wife, Catherine, raised corn and hay and three children, Charles, Hilda and Harry. In 1916, their youngest son, 16-year-old Harry Carsten Wrich, plowed the land with a walking plow, so that the current house and barn could be built. After Henry died in January 1939, Catherine sold the land to Harry that March.

Harry, along with his wife, Helen (Hansen), raised corn, alfalfa, hogs, cattle and chickens. Helen had a vegetable garden and many beautiful flowerbeds. All three of their children, Jeanette (Schneider), Virgil and William were born in the farmhouse. Harry and Helen retired in 1968 and moved to Blair. The farm was leased to their youngest son, William (Bill), in January 1969.

Bill and his wife, Joyce (Larsen), raised corn, soybeans, alfalfa, hogs and cattle. They purchased this land in August 1978 and continue to farm it today, living in the original farmhouse. Joyce has always had a large vegetable garden and hosted many celebrations on the front lawn. They had six children: Deborah (Kerber), Jacqueline (Salerno), W. Dwight, Gayle (Russell), Randy Larsen and Richard Larsen. They have 12 grandchildren: Jamie (Kauffman) Johnston, Jeffrey Kauffman, Morgan Wrich, Carsten Wrich, Stephanie Kerber, Tony Salerno, Anna Salerno, Katherine Salerno, Maria Salerno, Laura Salerno, Emily Larsen and Delaney Larsen. They also have four great-grandchildren.

Bill and Joyce will celebrate their 60th wedding anniversary this June with yet another gathering on the front lawn. This farm and house hold many memories, rich and poor, happy and sad. It has been a privilege and honor for the Wrich family to care for this small parcel on God's big earth.

WULF – Henry Wulf bought the quarter section W 1/2 SE 1/4 - E 1/2 SW 1/4 13-5-2 land, Fillmore County, NE in 1901. When the farm was bought by Henry Wulf, the house that was there still remains. The house was remodeled in 1926. The upstairs and a kitchen was added at that time. The old school house in Ohiowa was torn down. This lumber was used to remodel the house.

In 1903 Henry married Carrie Thole. Four children were born to them: Mabel, Edward, Arthur and Edna.

The barn was built in 1905. The barn was put together with wood pegs.

Henry Wulf passed away in 1962. Carrie died in 1936.

In 1962 Arthur and Bessie (Nedza) Wulf bought the farm. Arthur farmed with his father. Arthur and Bessie raised their five children (Kathleen, Dorothy, Robert, Ronald and Patricia) on this land.

The Wulf homestead in 1917. Pictured are Caraline Wulf (Henry's wife), Arthur Wulf on tricycle, Mabel Wulf, and Edna Wulf.

Robert farmed with his father until Arthur retired. Arthur passed away in 1981. In 1992 Robert and Louise (Deke) Wulf bought the farm. They reside on the farm.

In 2001 Bessie Wulf and Robert and Louise Wulf were honored with receiving the Nebraska Pioneer Farm Award through the Knights of Ak-Sar-Ben Foundation.

The barn has had metal siding put on. The house has been remodeled some but hasn't had extensive changes.

The corn crib was built in the early 1900s. The hot shed, large chicken house and brooder house were all built with lumber form the old school house in Ohiowa. *Submitted by Henry and Louise Wulf.*

ZOUBEK – This farm was purchased on June 27, 1904, by Frank J. and Marie (Svajgr) Zoubek, both Czech immigrants, with a dream and a borrowed dollar to make it legal. Five years later, Frank was killed in a farm accident, leaving his wife with five children between the ages of 3 and 9. With the help of a brother and later a stepbrother, Marie learned to farm and managed to hold on to the land. She continued to farm it until 1914.

Frank J. Marie (Svajgr) Zoubek and family

Her son, Frank P. Zoubek, married Bessie Homolka and Marie rented the farm to them. Life looked good until the depression hit and Bessie became ill. They struggled like everyone else with not enough money and too many financial needs. When he tried to borrow $60 for 90 days from a local banker he was told he wasn't a good risk. Instead of giving up, he found another banker willing to loan him more money for a longer term and went on to make his dreams come true. Frank continued to farm on this land for 37 years until his sudden death in 1967.

At that time his son, Frank J., and wife Maud (Hayes) Zoubek bought the farm. Frank had farmed with his dad until he was drafted for the Korean Conflict. He served 18 months in Japan, then returned to farming. They raised four children and he farmed until his death at 79 years of age.

His son, Myron Zoubek. and wife Sandy farmed with Frank J. and continued to farm it after his death. They also raised their three sons and their son Brandon currently works with his father to farm this land. He is the fifth Zoubek generation to farm this land. *Submitted by Maud M. Zoubek.*

ZUHLKE – August Zuhlke came in 1872 from Kremmin, Prussia to America after serving time in the German Army. He married Louise Rades in 1872 and purchased and settled 80 acres of this farm that is 3-1/2 miles south of Bancroft, NE in Cuming County in 1879. 160 more acres were bought in the next three years. They endured many hardships those early years, the worst was losing five children during the diphtheria epidemic in 1886. After their death they built a new farmstead east of their present home. Here is where I, Caroline Abendroth Zuhlke, am now living. They had four more children: John, Robert, Lottie and August. But in January 1897 tragedy struck again when August was fatally injured in a hunting accident. Louise was a courageous woman and with the help of her second husband, Ernest Raabe, who she married in 1898, maintained the farm. One son, Martin, was born to this union. After their death the farm was divided between August and John. John farmed the land I live on. John married Anna Vogt and they had two children, Willis, my husband, and Loraine Martin. They farmed the land raising corn, oats, alfalfa, cattle, hogs and chickens. In 1949 John suffered a fatal heart attack and my husband, Willis, began farming. In 1953-54 Willis served in the Korean War. After his return we married in 1955 and began our married life here. We raised corn, oats, alfalfa and soybeans plus chickens, hogs and a 30 cow dairy herd. We had three children - Curt, Thresa and Lenn.

We farmed until 1999 when my husband and son, Curt were killed in an auto accident. I now rent the farm. This farm will be passed to my grandchildren so the farm may live on as the Zuhlke Family Farm.

House on Zoubek farm

August Zuhlke family farm in 1980, an oil painting by owner Caroline Abendroth Zuhlke

A Roster of Heritage Farms by Aksarben in Celebration of 150 Years of Continuous Family Farming (First Awarded in 2014)

Roster is by County, formatted by the Farm Name, City, and Year Awarded

Burt County
Stork, Robert, Tekamah, 2016
Hansen, Susan, Tippery Farms Inc, Decatur, 2015

Cass County
Behrns Family Farm, LLC, Louisville, 2016
Knabe, David, Nehawka, 2016
Vallery, Steve, Plattsmouth, 2016

Cedar County
Haberman, Wilfred & Joyce, Fordyce, 2015

Cuming County
Fullner, Stanley/R.L./Daniel/Terry, Madison, 2016
Kanter, Barbara Kay, Minneapolis, MN, 2015
Harstick, Glen Leo, West Point, 2015
Wieneke, Adana, West Point, 2015

Dodge County
Segers/Ditter/Garman, Alice/Betty/Mary, Fremont, 2016

Hamilton County
Harrison, Dorothy & Robert Harrison, Lincoln, 2015

Lancaster County
Wittstrick, Janice, Roca, 2015

Nemaha County
Dunekacke, David F., Johnson, 2015
Hawley, Catherine, Brock, 2016
Jensen, Evelyn Stoddard, Liberty, MO, 2015

Otoe County
Fey, Bob and Barb, Lincoln, 2016
Rowen, David J. & Jeri L., Lorton, 2015

Richardson County
Ahern, Anthony and Janet, Shubert, 2016
Duerfeldt, Marvin & Norma, Falls City, 2015
Dunn, Elizabeth/Andrew/Nicole, Falls City, 2016
Lunginbill-Cummins, Hazel, Humboldt, 2016
Lunginbill-daSilva, Elizabeth, Lincoln, 2016
Lunginbill-Warner, Frances, Eden Prairie, MN, 2016

Saline County
Kottas, Kay, Lincoln, 68502, 2015

Sarpy County
Fricke Jr., Milton & Joann and Charles & Marian Fricke, Omaha, 2014

Washington County
Greunke, Ellen Family Farm Trust, Arlington, 2015
Stork Lane LLC, Arlington, 2015
Sully, Mary Brislin, Fort Calhoun, 2015

A Roster of Pioneer Farms by Aksarben in Celebration of 100 Years of Continuous Family Farming, 1956-2016

Roster is by County, formatted by the Farm or Family Name, City and Year Awarded. This list was compiled using available information. The publisher apologizes for any errors or omissions.

Adams County

Anderson, Edward L. & Donna L., Ayr, 2004
Anderson, Chester E. & Agnes, Hastings, 1982
Asmus, Robert R. & Jo Ann, Glenvil, 1974
Augustin, Wesley & Lenora, Hastings, 1988
Augustin, Thelma, Juniata, 2006
Aulrich, Harvey & Lola, Hastings, 1985
Bachman, Conrad J & Nancy L / Paul A. & Mary L, Papillion, 1987
Baehr, Karl & LaVerda, Hastings, 2007
Berg/ Lukow, Erwin/ Sharon, Kenesaw, 1989
Blomenkamp, Clarence, Hastings, 2000
Blomenkamp, Ervin, Hastings, 2000
Blumenthal, Albert H., Ayr, 1974
Bohlke, Lloyd, Kenesaw, 1987
Bonifas, Ray & Sherri, Roseland, 2011
Brader, Dale & Dorothy/Charles & Cindy/ Richard, Glenvil, 1990
Brown, Marlynce (Krueger Family Farm), Hastings, 2005
Burling, Carroll & Bonnie, Kenesaw, 1988
Classen, Mildred & William, Blue Hill, 1987
Classen, Wayne, Glenvil, 1981
Classen, Orville, Glenvil, 1985
Consbruck, Leo B. & Alice, Juniata, 1988
Conshruck, Sharon, Juniata, 1999
Crosby, Mr. & Mrs. Walter F., Glenvil, 1971
Dale, Orville, Holstein, 2002
Davis, Albert, Hastings, 1987
Deveny, Tom, Ayr, 1999
DeWitt, John & Rosalia, Roseland, 1984
Dittrick, Joe & Betty, Kenesaw, 1984
Dyer, Robert E. & Alice M., Ayr, 1973
Eckhardt, Dennis, Betty Eckhardt, Ayr, 2010
Eltz, Elnora Eltz & Henry, Juniata, 2004
E & L Werner Family Limited Partnership, Santa Teresa, 2010
Fox, James, San Antonio TX, 1999
Frahm, Lois, Hastings, 2005
Frank, Sondra, Hastings, 2001
Fricke, Wilma R., Ayr, 1991
Gangwish, Howard/Doris, Juniata, 1996
Gangwish, Howard & Doris, Juniata, 2001
Gaymon, Ray E. & Guinivere E., Hastings, 1981
Gentert, Tammy (Gerhard Hoefer Farm), Juniata, 2005
Gentert, Tammy & Dennis, Juniata, 2012
Gingrich & Siblings, Janice, Bladen, 201
Goble, Gaylord & Martha, Ayr, 1985
Grabill, Robert and JoAnn, Hastings, 2003
Grothen, Eldon, Glenvil, 1994
Ground, William L., Juniata, 1995
Haselbarth, Clifford H., Kenesaw, 1994
Hesman, Gary, Hastings, 2003
Hinkrikus, Myron & Daniel, Prosser, 1981
Hinrichs, Jerry & Jimmie, Glenvil, 2003
Hoffman, John & Leslie, Juniata, 2003
Hohlfeld, Robert H. & Frieda E., Hastings, 1982
Hupf, Phillip, Redmond WA, 2001
Jacobitz, Delano & Frances (Kroll), Prosser, 2010
Janssen, Mark & Ruth M., Hastings, 1994
Johnson/ Jones, Hazel/ Shannon, Bladen, 1972
Jones, Pierce C., Blue Hill, 1978
Junker, Howard & Belva, Juniata, 1993
Juranek, Donald E., Hastings, 1997
Kaizer Pittz, Elizabeth, Hastings, 1985
Katzberg, Evelyn, Juniata, 2000
Kaufman, Edward F., Bladen, 1988
Kennedy, Vera, Juniata, 1989
Kimle, David, Kenesaw, 2009
Kissell, Mr. & Mrs. Gaylord K., Hastings, 1993
Koos, Donald Joseph, Roseland, 2008
Kort, Neil & Marlys, Blue Hill, 1991
Kral, Gary D., Juniata, 1987
Kroll, Ruth I., Juniata, 2004
Lay, Albertus N., Glenvil, 1984
Lisius, Milton & Velma, Juniata, 2002
Loudon, Donald M., Hastings, 1978
Markley, Gladys L., Juniata, 1980
Martin, John, Kenesaw, 2000
McLeod, William & Klara, Juniata, 2004
McMahon, Stanley & Janice, Hastings, 1999
Meyer, Vernon A. & Phyllis E., Blue Hill, 1976
Meyer, Anna C. L. and Orville, Blue Hill, 1977
Mott, Genevieve M., Hastings, 1996
Mulder, Burdette & Lela, Doniphan, 2002
Neeman, Robert E. & Margaret M., Hastings, 1993
Neinhueser, Kevin A., Holstein, 1994
Nienhueser, Marvin E. & Jessie, Juniata, 1983
Nienhueser, Eugene, Juniata, 2003
North, Sam L/ Virginia, Hastings, 1992
Novak, Larry & Barbara, Larence, 2007
Olsen, Marie E., Heartwell, 1983
Parr, Ken & Sue, Kenesaw, 2012
Peterson, Ward Marion, Glenvil, 1981
Pfeiffer, E. Marguerite/Robert/W. Dean/Kenneth, Trumbull, 1989
Plambeck, Duane & Janice, Kenesaw, 1985
Plum, Lenora, Hastings, 2002
Rainforth, Lylle A., Wood River, 1981
Reiners, Johnny G & Gertha, Juniata, 1993
Reininger, Mrs. George, Leigh, 1984
Riese, Anna Marie, Holstein, 1998
Rewerts, Steve & Jeanne, Hastings, 2010
Ruhter, Milton E. & Evelyn E., Juniata, 1983
Ruhter Family, Delmont, Prosser, 2003
Schmid, Mr. & Mrs. Carl, Papillion, 1986
Schmidt, Roger & Jeannette, Wood River, 2007
Schukei, Verona, & the Arthur E. Schukei Family Trust, Gibbon, 2010
Schukei, Herbert G. & Hazel M., Juniata, 1993
Seeman, Kathryn, Blue Hill, 2011
Sheffield, Bernard L, Holstein, 1985
Shriver, James, Hastings, 2002
Shurigar, Wilson W., Kenesaw, 1977
Stein, Willard & Della, Hastings, 1989
Stichweh, Lawrence & Frieda, Kenesaw, 1986
Strasburg, Godfrey & Ethel, Ayr, 1994
Struss, Mrs. George, Juniata, 1994
Uden, August & Marie, Juniata, 1987
Uhrmacher, Eddie/Sharon, Hastings, 1996
Veburg, Ronald N., Darlene B. (Prante) Veburg, Hastings, 2006
Weaver, Harold A., Juniata, 1979
Weber, Mrs. Paul, Juniata, 1978
Weeks, Michael, Juniata, 2000
Weseman, Randall A. & Kennedy, Vicki L., Juniata, 2013
Whitcomb, Mr. & Mrs. Carl R., Hastings, 1983
Whitesel, Belle, Juniata, 1983
Whitesel, Kendall and Gail, Juniata, 2003
Wiens, Glen & Jeannette, & Leola Wiens, Ayr, 2010
Wiltrout, Mr. & Mrs. Chester Judson, 1971
Woods, L. Alan & Marcia, Blue Hill, 1992
Woodworth Monahan, Lucy D., Hastings, 1978

Zidko, Frank A. & Connie, Spencer, 2001

Brown County

Bejot, Stephen, Ainsworth, 2005
Hitchcock, Edna, Ainsworth, 2003
Hitchcock, Gordon W., Ainsworth, 2003
Magary, Gordon & Gayle, Johnstown, 1983
McCullough, Elva, Long Pine, 1983
Mercier, Lorene, Grant, 2006
Moore, Mr. & Mrs. Everett F., Johnstown, 1984
O'Hare, Kenneth, Ainsworth, 1984
Reinert, Bertha M., Johnstown, date not available
Schenk Family, Ainsworth, 2007
Sherman, Dave & Ronda, Johnson, 2013
Weyer, Dean & Glen, Ainsworth, 1983
Williams, William & Louise Ann, Ainsworth, 1984

Buffalo County

Altmaier, Sharron, Kearney, 1993
Applegate, Mr. & Mrs. E. M, Gibbon, 1972
Ashton Burton, Helen Maxine, St. Helena CA, 1996
Axmann, Diane, Ravenna, 2008
Azale/ Line, C./Irma, Miller, 1987
Behrendt, Dee, Pleasanton, 1983
Bernet, Oliver S. Estate, Ravenna, 1991
Betebenner & Sidwell Land & Cattle Co. Inc., Gibbon, 2013
Betke, Royal & Patsy, Ravenna, 2003
Bish, Jane Ann, Miller, 1997
Bock, Eldor & Marilyn, Pleasanton, 2015
Brooks, Beverly, Shelton, 1971
Cagler Trust, Everett, Gibbon, 2007
Calvin, Rose F., Ravenna, 1983
Campbell, Harry & Hazel, Kearney, 1994
Carter/ Walenta, Margaret/ Herman, Omaha, 1993
Chramosta, James and Shirley, Gibbon, 2016
Cook, Leland & Ralph, Shelton, 1983
Cool, Michael & Diane, Amherst, 1999
Daake, Robert I., Kearney, 1997
Davis-Krueger, Carol E., Pleasanton, 2010
Day, Dennis & Sandra, Gibbon, 2011
Day, Laura, Kearney, 1979
Day, Mabel, Miller, 1985
Dealey, Jeannette L., Donald D. Dealey, Holdrege, 2009
Dean, Alice O., Shelton, 1977
DeNeui, Sharon, Pleasanton, 2012
Dixon, Russell, Pleasanton, 1992
Dixon, Phillip and Corliss, Pleasanton, 2010
Duncan, Glen & Helen, Kearney, 1998
Eckhout, John & Erva, Amherst, 1999
Eurek, Marvelyn, Hastings, 1998
Fisher, Ann, Amherst, 1995
Florence Bissell Family Trust, Kearney, 1987
Geisler, Eugene, Lincoln, 1981
Gillming, Ralph, Kearney, 1991
Glen Day Farms Inc., Mason City, 1990
Hadwiger Dauel, Barbara, Pleasanton, 1996
Hanisch, LaDonna (Calvin), Ravenna, 2006
Harmoney, Arthur, Amherst, 1994
Harnagel, Jay, Miller, 2015
Harris Stamm Farms Inc., Eddyville, 1993
Hauschild, David, Grand Island, 2002
Hauschild, Ruth, Ravenna, 2002
Hawthorne, Adeline, Ravenna, 2006
Heermann, Charlotte, Milford, 2014
Henninger, Franklin S. & Laura, Shelton, 1971
Hervert, Ronald, Ravenna, 1983
Heusel, Herbert, Grand Island, 1983
Hill, Garra & Virginia, Elmwood, 1983
Holoubeck, Sr., Archie More, Kearney, 1981
Holtz, Dennis, Shelton, 1993
Homan, Larry & MaryAnn, Gibbon, 2005
Hurt, Jack W., Kearney, 2010
Ingram, Nancy L., Kearney, 1982
Irvine, Bonnie, Grand Island, 1999
Jacobson, Clarence & Helen, Kearney, 1981
Jacobson, Larry & Colleen, Riverdale, 1985
Jacobson, Irene H., Riverdale, 1986
Jacobson, Mrs. Evelyn, Riverdale, 1988
James L. Miller Trust, Elm Creek, 1990
Janitscheck, Glenn & Rose, Pleasanton, 1989
John Farm Joint Venture, Janelle Grabowski, Linda Boucher, Connie Gregory, Ravenna, 2011
Johnson, Charles & Opal, Pleasanton, 1989
Juhl Carmann, Dorothy, Riverdale, 1979
Kaps, Alma M., Miller, 2008
Keens, George H., San Pedro, CA, 2013
Keep, William & Barbara, Elm Creek, 1995
Keep, Charles & Meredith, Elm Creek, 1995
Kegley, Glenn & Donna, Kearney, 1979
Kegley, Gary, Jana Kegley, Kearney, 2006
Kenney, Perry & Lester, Amherst, 1982
Kirschner, Lois, Pleasanton, 1987
Kirschner, Kenneth, Ravenna, 1987
Loeffelholz, Mr. & Mrs. Jerome, Pleasanton, 2013
Lowe, John G. & Andrea S., Kearney, 2004
Luth, Stephen Kenneth, Ravenna, 2004
Mann, Robert, Omaha, 1982
McKean, Jerry & Marlene, Kearney, 2006
McMahon, Robert, Amherst, 1994
Minnie M. Cook Family Trust, Amherst, 1986
Muhlbach, Bruce, Ravenna, 1981
Muhlback, Donald, Ravenna, 1983
Musil, Rose, Ravenna, 1990
Nelson, Emily, Gibbon, 2015
Nickman, Cecil L. & LaVaghn, Kearney, 1988
Niemack, Marvin & Gail, Shelton, 1998
Oliver, Mr. & Mrs. Vernon, Shelton, 1965
Oliver, Donlad & Lucille, Shelton, 1980
Open Circle R Inc., Pleasanton, 1985
Parks, Larry D., Kearney, 2008
Pettett, Gordan Paige, Elm Creek, 2000
Pfeiffer, Elsie, Amherst, 1981
Phillips, Ercil, Pleasanton, 1987
Plautz, Michael & Susan, Ravenna, 2003
Plischke, Neta, Kearney, 2000
Pritchard, Robert, Pleasanton, 2000
Quaring, Darrell R, Shelton, 2008
R & M Acres, Inc., Gibbon, 2009
Randall Land & Cattle Co, Gibbon, 1982
Reese, Duane & Opal, Pleasanton, 1980
Richards, Roy A., Elm Creek, 1979
Richards, Wayne O., Elm Creek, 1986
Richter, Mrs. Neva, Kearney, 1980
Richter, Edward & Edna, Kearney, 1991
Richter, Kenneth & Elaine, Kearney, 1991
Riessland, Connie, Pleasanton, 2014
Riessland, LaVern & Marles, Riverdale, 1988
Rosenthal, Don, Gibbon, 1990
Royle, Agnes, Amherst, 1992
R.T. Rogers Farms, Inc., Gibbon, 2007
Rumbeck, Edward, Amherst, 2011
Sadler, Patricia D., Kearney, 1996
Schanou, Robert & Sharon, Shelton, 2015
Schiemann, Bessie L., Shelton, 1984
Schuller, Mildred, Ravenna, 1986
Schulte, Lawrence J., Pleasanton, 2008
Sealing, Mrs. Charles & Gaylord, Shelton, 1976
Shafto, Robert & Doris, Pleasanton, 1987
Sheen, Harold & Mardell, Kearney, 1977
Slater, Shirley, Shelton, 2001
Smith, Mrs. Horrace, Gibbon, 1983
Smyth, Greg & Angela, Shelton, 2012
Solomon, Melvin & Sandra, Shelton, 2011
Speidell, Richard L., Grand Island, 1995
Stafford, Dorothy, Riverdale, 1999
Stark, Donald, Ravenna, 1990
Stittle, Evelyn, Ravenna, 2004
Stuber, Minnie, Raleigh NC, 2001
Sullwold, Donald, Elm Creek, 2004
Thornton, S. Cary, Kearney, 1972
Tillson, Charles & Helen, Ravenna, 1992
Trampe, Delman & Marlene, Amherst, 1994
Unick, Lloyd, Litchfield, 1985
Urwiller, Henry J., Ravenna, 1978
Urwiller, Wayne & Velma, Ravenna, 1995
Vest, Walter & Lola, Pleasanton, 1999
Vohland, Evelyn, Gibbon, 2009
Voss, Harold, Ravenna, 1985
Webb, Lawrence M. , Kearney, 1979
Wensel, Violet E., Gibbon, 1983
Weston, Dorsey W. , Gibbon, 2004
Whitcomb, Richard, Barbara Whitcomb Trust, Arvada, 2007
Wilke, Loren & Lloyd, Hazard, 1995
Wilkie, L. Floyd, Gibbon, 1973
Wing, Mary, Kearney, 2002
Zimmer, Sylvia, Pleasanton, 2006
Zimmerman, Esther L., Kearney, 1991
Zwiner-Rodehorst, Betty & Children, Pleasanton, 1987

Burt County

Ackerman, Mrs. Julia, Blair, 1978
Anderson Jr., Ralph M., Tekamah, 1993
Anderson, Doris, Oakland, 2011
Anderson, Hollis Bruce & Ann M., Oakland, 2012

Anderson, Robert & Joan, Tekamah, 1999
Anderson, Verna T. (Mrs. Clifford E. Anderson), Tekamah, 2004
Bacon, Donald & Betty, Lyons, 1998
Bacon, Randall & Mary Fran, Lyons, 1998
Baker, Vera B., Tekamah, 1994
Beaver, A. R., Tekamah, 1965
Beckman, Carla, Fremont, 1976
Bengston, Frederick & Shirley, Oakland, 1992
Bigelow, Bradley W., Kearney, 1994 Bigelow, Mary, Fremont, 1994
Bowland, Lee, Tekamah, 2012
Bray Hanslip, Mary Jane, Decatur, 1982
Bryant, Jack & Suzanne, Tekamah, 1988
Bucy, C. Earl, Tekamah, 1990
Cain, Donald R., Arlington, 2014
Chamberlain, Elaine K., Tekamah, 2012
Chatt, Bob & Elaine, Tekamah, 1993
Christensen, Donald & Melva, Lyons, 1968
Christensen, Mrs. Marian O., Craig, 1981
Connealy, Neil & Joan, Tekamah, 1994
Cram, Gordan, Tekamah, 2001
Dallam Malone, Janet, Minnetonka, 2008
Dixon, Marie/ Kenneth, Lyons, 1987
Dyson, Donald, Lyons, 1999
Engdahl, Alan & Sonya, Oakland, 1981
Eriksen, Howard, Craig, 1989
Eriksen/ Sommerer, Mrs. Violet/ Mrs. Marjorie, Craig, 1979
Everett O'Connor, Margaret, Lyons, 1968
Follstad, Virginia, Whitewater WI, 2000
Gatzemeyer, Hazel, Bancroft, 1991
Gatzemeyer, Louis, Bancroft, 1980
Gilbert Hansen, Sandra, Fremont, 1995
Gill, Robert W., Tekamah, 2013
Going, Walter, Lyons, 1978
Goodsell, Ira C. Estate, Craig, 1986
Grothe, Willard T. & Tyrone E., Craig, 1986
Gustafson/ Anderson, Clifford/ Helen, Oakland, 1987
H. C. Larson & Associates, Decatur, 1989
Hancock, Paul, Herman, 1982
Healea Geary, Nelson & Mildred, Craig, 1969
Heitshusen, Dale, Oakland, 1980
Helms, Alby & Lula, Craig, 1981
James Cornish Farms, Inc., Tekamah, 2009
Jensen, A. Russell, Decatur, 1990
Johanson, Anthony Scott, Oakland, 2014
Johnson-Crump, Elizabeth, Franklin W. Johnson & Nile K. Johnson, Craig, 2004
Johnson, Elroy O., Oakland, 1983
Johnson, Eugene & Carole, West Point, 2013
Johnson, H. Bayard & Gladys, Tekamah, 1960
Johnson, Richard (Rick) & Loraine (Kristine), Lyons, 2014
Johnson/ Hansen, Leonard/ Leona, Oakland, 1979
Kennedy, Margaret, Craig, 2009
Kohlmeier, Marian R. Kohlmeier, Craig, 2004
Lange-McGill Jr., Kenneth, Decatur, 2005
Lange, Oscar, Tekamah, 1957
Larsen, Kenneth, Decatur, 1998
Larsen, Mary, Decatur, 1999
Larsen, Mary, Decatur, 2011
Larson, Dorothy, Oakland, 2005
Larson, John S., Lyons, 1990
Lindquist, Lloyd, Oakland, 1984
Lofdahl, Mrs. John A., Craig, 1990
Loftis, Robert, Craig, 1985
Lydick, Samuel, Rose, Tekamah, 1957
Lydick, Wally & Beverly J., Tekamah, 1994
Magnusson, Dorothy Magnusson, Oakland, 2004
Major, Barry & Mindy, Omaha, 1994
Major, Ethelyn & William, Omaha, 1966
Major, Leslie, Frances Major Clements, Decatur, 2007
Marr, Vere, Tekamah, 1968
McMullin, Asa & Adda, Tekamah, 1961
McNear, LaDonna & Lew, Tekamah, 2003
Moberg, Hervert & Edna, Craig, 1983
Monson, John S., Craig, 1975
Moseman, Vernolt H., Oakland, 1984
Myers, Charles, Lyons, 2009
Myers, Charles and Gloria, Lyons, 2016
Neal Rogers Farms, Inc., Lyons, 1968
Nelson, C. Gordon, Oakland, 1980
Nelson, Harlan, Oakland, 1976
Neuhaus, Dorothy, Lyons, 2010
Neuhaus, Larry & Helene, Lyons, 2010
Newill, Gerald, Lyons, 2016
O'Connor, Dr. Edward, Omaha, 1968
Olinger, Oliver, Tekamah, 1958
Olson, C. A., Craig, 1989
Oman Sundquist, Anna, North Platte, 1990
Palmer, Keith, Tekamah, 1995
Paulsen, Gerald & Mary, Oakland, 2014
Peck, John L., Tekamah, 1994
Penke, Jason, & Kelly Abrahams, Craig, 2012
Peters, Roland, Pender, 1998
Petersen, Mr. & Mrs. Lynn, Craig, 1967
Petersen, Norman & Sons Inc., Tekamah, 2003
Peterson, Maruice, Norma, Oakland, 2001
Peterson, Arnold, Oakland, 1983
Pipher, Bernard T., Tekamah, 1974
Rabe, Muriel, Oakland, 2008
Reyman Jack, Mrs. Betty Bell, Tekamah, 1961
Robertson, Anna T., Omaha, 1994
Schlichting, Guido, Oakland, 1995
Sears, W. Kurtz, Norfolk, 2006
Seeman, Susan, Stanon, 2016
Shafer, Lee & Ethel, Tekamah, 1963
Skinner, Bruce & Prudence, Herman, 2012
Sklenar, Joseph, Tekamah, 2011
Sklenar, Mrs. Lillian & Richard, Omaha, 1978
Sklenar, Robert T., Tekamah, 1995
Snow, Lyle and Judy, Tekamah, 2008
Stock & Family, John H., Tekamah, 1967
Stone, S. Kenneth, Lyons, 1973
Stoner Jr., William C., Stockton, 2008
Strelo, Carl & Esther, Craig, 1978
Tippery, Roy G. & Myrtle, Tekamah, 1967
Tranmer, Frances, Tekamah, 2009
Tranmer, John, Craig, 1975
Valder, Dee, Tekamah, 1961
Von Essen Trust, Oakland, 1995
Wallace, Kenneth, Oakland, 1988
Wallerstedt, Norman, Oakland, 2010
Weeces, Gerald F., Tekamah, 2008
White, Paul A., Tekamah, 1969
Young, Andrew Family, Oakland, 1967

Butler County

Adamy, Leo, Bellwood, 1982
Adamy, Leonard, Bellwood, 1982
Adamy, Robert, Bellwood, 1982
Allen, Donna R., David City, 2009
Ball, Herman & Doris, David City, 1971
Barcal, Richard, Linwood, 1972
Barlean-Hoeft, Sharon, David City, 2014
Barlean, Mrs. Anna, David City, 1973
Bartunek, Mrs. V.A., Bruno, 1978
Becker Rogers, Kathryn, Columbus, 1988
Bell, Robert, Bellwood, 1973
Bemis, Charles, Columbus, 1986
Beringer, Keith, Bellwood, 1984
Beringer, Peter, Bellwood, 1984
Birkel, Clarence J., David City, 1974
Birkel, Olivia, David City, 2002
Blatny, Adrian, Brainard, 1985
Blatny, Vitus, Linwood, 1985
Bohuslausky, Victor, Brainard, 2006
Bougger, Leo, David City, 1989
Buresh, Ernest & Denise, Dwight, 2007
Byam, Dean, Ulysses, 1984
Callaghan, Greg, Rising City, 1979
Carskadon Jr, Clay, Genoa, 1991
Cathcart, Mabel, Ulysses, 1973
Creech, Minnie M., Lincoln, 1972
Dallegge, Delmar & Wanda, Garrison, 1985
Davenport, Mr. & Mrs. Frank, David City, 1972
Davis, Leonard, Rising City, 1989
DeWispelare, Gerald, Octavia, 1974
Dodendorf, John Lee Estate, Bellwood, 1988
Duren, Grace, Rising City, 1975
Dvorak, Edwin, Brainard, 1985
Edwards, Floyd, Dwight, 1983
Family, Besch, Bellwood, 2010
Fenstemaker, Burch, Fullerton, 1980
Flansburg, Frank, David City, 1980
Fricke, Eilert & Margaret, Ulysses, 2001
Fuxa-Cuba, Lucille, David City, 2010
Gannon, Margaret, David City, 1993
Glock, Clarence & Iris, Rising City, 1977
Glock, Edwin & Anna, Rising City, 1977
Glock, Frank, Rising City, 1987
Glock, Fredrick W., Rising City, 1977
Glock, Lydia M., David City, 1980
Glock, Miller D. & Jessie, Rising City, 1977
Glock, Mrs. John, Rising City, 1976
Glock, Phillip, Rising City, 1977
Glock, Ralph, Springfield, 1995
Grau, Margaret, Hebron, 1992
Greenslit, R.A., Stanton, 1978
Gulbrandson, George, Waco, 1985
Haight, Ernest, David City, 1971
Helgoth, Catherine, St. Libory, 1982
Hermsen, Harold, David City, 1987
Hiller, Anthony, Bellwood, 1995
Hinze, Merle, Rising City, 1984
Homan, Marie Estate, David City, 1977

Hookstra, Mildred, David City, 1972
Horacek, Don, Brainard, 1996
Hotovy, John & Sylvia, Dwight, 1998
Hotovy, Ludvick, Dwight, 1985
Hottovy Fischer, Delores, Loveland Co, 1996
Hottovy, Adolph, Dwight, 1983
Hromas, Eugene, Ulysses, 2011
Hudson, Artie, David City, 1973
Jakub, Pete, Brainard, 1993
Janecek, Virgil & Elizabeth, Omaha, 2006
Janicek, Gerald, Bellwood, 1978
Jisa, David, Brainard, 2000
Johnson, L. Palmer & Lucile, Lincoln, 1974
Kabousek, Curt & Judy, Brainard, 2010
Kallenbach, Louis, Rising City, 1993
Kallerbach, Mrs. B.A., Shelby, 1973
Kilgore, Verdayne, Davd, 2000
King, Woodrow, Kathryn King, Linwood, 2011
Kirby, Dennis & Carole, David City, 2010
Klaus, Elizabeth, Columbus, 2001
Klingemann, Kenneth, Ulysses, 1987
Kobza, Marion & Betty Lou, Andrew Kobza, Dwight, 2011
Koester, Bernell, David City, 1982
Koranda, Rosemary, Wahoo, 1995
Kosch, Arthur, David City, 1981
Kosch, Elizabeth, David City, 1977
Kosch, Tom, Gerald, David City, 1977
Kovar, Marcel, David City, 1985
Koza, Louis, Bruno, 1974
Kozisek, Wilfred & Leola, Bruno, 1988
Krenk, Clarence, Linwood, 1997
Kriz, Clair, Brainard, 1995
Kriz, Dean & Lynelle, Brainard, 2009
Kriz, Margaret A., David City, 2004
Kriz, Phillip, Linwood, 1976
Kroft, Robert, Richard, & Lyle, David City, 1990
Krov, Joseph, Abbie, 1989
Kucera, Edward T., Linwood, 1976
Kudlacek, Cory & Kay, Eureka, MO, 2011
Lawson, Floy F., Lincoln, 1978
Lepert, Frank Family, Plattsmouth, 2005
Lettellier, Anthony, David City, 1980
Lichliter, Wendel & Virgene, David City, 1978
Loomis Nekl, Eileen, Bellwood, 1972
Lux, David, Shelby, 1983
Mahlin Farms Inc., David City, 1996
Matthews, Gene & Roseela, Grand Island, 2002
Meyer, Martin, Bellwood, 1975
Meysenburg, Joseph, Bellwood, 1989
Mick, Lorene, Columbus, 1988
Miratsky, Estella, Bruno, 1976
Morbach, Marie, Bellwood, 1983
Neujahr, Glen, Gresham, 1991
Neujahr, Mrs. Rudolph, Seward, 1973
Nicolas, Joseph & Phyllis, Columbus, 2011
Novacek, Phyllis & Marland, Linwood, 1977
Novcek Family, Bruno, 2012
Oehme, Mrs. Adolf, Brainard, 1982
Palensky, Edwin, Dodge, 1992
Papa, Ralph & Margaret, Bellwood, 2011
Pavel, Jerry, Linwood, 1978
Piitz, Henry C., Brainard, 2004
Piitz, Leon & Nancy, Brainard, 2014
Polacek, Gary & Geraldine, Bruno, 2004
Polacek, Louis & Marilee, Bruno, 2001
Polivka, Agnes, Brainard, 1984
Polivka, James & Jean, Bruno, 1995
Potter, Robert, Valparaiso, 2007
Prochaska, Anthony V., Abie, 1978
Prochaska, Brad, Dwight, 1995
Proskovec Aerts, Marilyn, Malcolm, 1994
Proskovec, Gary, Lincoln, 1994
Raitt, Dean, Rising City, 1976
Rech, Ernest & Darlene, David City, 1983
Reisdorff, Harry, David City, 1973
Rerucha, Nadine, Lincoln, 2006
Reyda, Roger, Brainard, 2007
Rezac, Kenneth R & Peggy L, Lincoln, 2015
Rezac, Rudolph, Norfolk, 1991
Ruth, M. Dean, Rising City, 1978
Ruth, Wayne, Rising City, 1984
Ruth, Wendell, Rising City, 1978
Sabata, Kenneth E., David City, 2006
Sauser, Mark, David City, 1981
Schmid, Bernard, Bellwood, 1990
Schmit, Dwaine & Selma, Bellwood, 1998
Schmit, Loran & Irene, Bellwood, 2010
Schroeder, Jonas & Edna, Ulysses, 1990
Selden, Lester, Rising City, 1983
Semin, Alvin J, Ulysses, 2009
Semin, Fred & Evelyn, Brainard, 1993
Semrad, Emil & Elsie, David City, 1989
Shcmid, Robert, Bellwood, 1989
Sims, Olga, Mayme Viglicky, Fremont, 2007
Slama, Victor, Columbus, 1983
Smith, Galen L, Allen, 2007
Spatz, James, Bruno, 1979
Speicher, Joseph, David City, 1989
Spitzli, Pauline, Virginia Beach, VA, 1999
Stanislav, Bernard, Linwood, 1993
Sternberg, Ronda, Fairfax, VA, 2005
Stouffer, Ray G., Cheyenne, 2013
Struebing, Richard & Vera, David City, 2004
Stryker, Harold, Rising City, 1984
Syring, Ben & Jean, Schuyler, 1977
Tejral, Helen, David City, 1992
Tesina, Emil, Bee, 2002
Timoney, Larry & Janet, Ulysses, 1995
Towner, Abraham Homestead, Surprise, 1971
Vandenberg, Arnold & Mary Ann (Potter), Brainard, 2013
Vanek, Rose, Lincoln, 1999
Vrbka, Denis, Brainard, 1992
Webster 1991 Trust, Las Vegas NV, 1995
Weverka, Henry & Mary, Brainard, 1992
Woodruff, Charles Robert & Gloria, Ulysses, 1980
Wynegar, Stanley, Ulysses, 1984
Yindrick, Gary & Charlene, David City, 1983
Zavodny, Arlene, Lincoln, 2006
Zeilinger, Margaret, David City, 2000
Zeilinger, Sylvia, Lincoln, 2000
Zeller, Mr. & Mrs. Gordon, Bellwood, 1977

Cass County

A & D Farm Inc, Milford, 2011
Ahl, Harry N & Alverta, Louisville, 1985
Albert, John & Irene, Louisville, 1987
Albert, Ray & Donna, Louisville, 2006
Althouse, Steven & Tamara, Waverly, 2015
Ayers, Katherine & Richard, Plattsmouth, 1994
Backemeyer, Gene, Murdock, 1998
Backemeyer, Lavina, Elmwood, 1984
Banning, Minnie, Union, 1975
Barkhurst, Mr. & Mrs. John, Nehawka, 1973
Baxter, Buster & Dixie, Murray, 1981
Behrns, Mrs. Lena, Avoca, 1966
Beil, Forest & LaVonne, Plattsmouth, 1984
Benoit, Janice L., New Iberia, LA, 2003
Bierl, Leona, Plattsmouth, 1989
Blaikie, Alvina, Lincoln, 2013
Boedeker, Jr., Charles, Murray, 1976
Bornemeier, Chester & Valda Trust, Elmwood, 2005
Bornemeier, Roger M., Laurel A. McReynolds, Elmwood, 2012
Bornemeier, Vernon & Roberta, Elmwood, 1984
Bornman, George & Mary, Lincoln, 1992
Bose, Lynn, Omaha, 1984
Brockhoff, Ralph and Linda, Avoca, 2003
Bruns, Maynard & Geraldine, Avoca, 1984
Buell, Fred, Murdock, 1969
Buell, Mr. & Mrs. Merle, Murdock, 1969
Campbell, Robert, Plattsmouth, 1960
Chriswisser, Roy & Ana Marie, Nehawka, 1997
Clements, Richard, Elmwood, 2002
Cole, Mr. & Mrs. Roy, Plattsmouth, 1972
Comer, Glen B., Gladys Comer Eacrt Trust, Union, 2006
Conrad Acres Inc., Plattsmouth, 1984
Cross, Carl, Union, 1956
Cuthrell, Clarence & Margaret, Plattsmouth, 1983
Daniel, Oniebell Hoenshell, Omaha, 1979
Dankleff, Lyle & Carol, Avoca, 1985
Davis, Stephen & Mary Lou, Plattsmouth, 1970
Day, Eugene, Weeping Water, 1969
Dill, Oscar, South Bend, 1964
Dorr, Russell, Falls City, 1984
Drake, Kirby & Cindy, Avoca, 1998
Eaton, David, Union, 1967
Eaton, Patricia, Alvo, 2014
Ebert, Ella, Syracuse, 1992
Eddy, Dennis & Ann, York, 2000
Engelkemier, Roy & Doris, Nehawka, 1980
Engelking, Dora Ellen, Elmwood, 1985
Farm Partnership, Ramge, Plattsmouth, 2010
Farmer, Robert, William Farmer, Ashland, 2009
Fischer, Robert, Elmwood, 1978
Foster, John & Rose Ann, Union, 1995
Fox, Robert, Pasadena CA, 2001
Frohlich Bird, Minnie, Eagle, 1977
Frohlich, Doris, Lincoln, 1989
Gaebel, Mr. & Mrs. LeeRau, Murdock, 1982
Gaines Steube, Sally, Plattsmouth, 1996
Gainse, 1st Sgt. Robert L., USMC, Las Cruces NM, 1996
Gonzales Family Trust, Charles, Margradelle Gonzales & Scott Gonzales, Guthrie Center, 2006
Greene, William, Ashland, 1994
Group, Maurice & Jean, Louisville, 2011

Guelker, Edna, Murdock, 1992
Haith, Golda, Elmwood, 1986
Hall, Dale & Norma, Elmwood, 1995
Hall, Gary, Glenn Hall, Shirley Wenzel and Marion Wenzel, George Buell, Maxine Buell, Dale Hall, c/o Shirley Wenzel, Elmwood, 2014
Hall, Murl, Elmwood, 1973
Halmes, Pete, Plattsmouth, 1967
Hansen, Anna, Fred & Kathleen, Nehawka, 1971
Hansen, John, Nehawka, 1956
Hansen, Mr. & Mrs. Ivan J., Nehawka, 1968
Hardnock, Forrest/ Helen, Waverly, 1996
Harshman, Wilmer & Marietta, Nehawka, 1973
Heier, Ivan & Ruby, Lincoln, 1985
Heier, Mr. & Mrs. Walter, Louisville, 1964
Heil, Herbert M. & Nancy G., Trustees Heil Family Trust, Sonoma, 2014
Heim, Jerald & Joyce, Plattsmouth, 2011
Hess, Marvin E., Bennet, 1987
Hild, Elsie, Plattsmouth, 1992
Hill, Joann, Descendents of John Ruhga, Elmwood, 2006
Hobscheidt, Katherine, Murray, 2008
Horn Goodchild, Grace, Plattsmouth, 1996
Horn, Denise, Weeping Water, 2003
Horn, Marie, Plattsmouth, 1996
Hostetter, Cecil Keith, Glenwood, IA, 2004
James, Robert, Lincoln, 1996
John, R. Theodore & Joan, Elmwood, 1969
Johnson Kintner, Lois, Lincoln, 1975
Johnson, Daisy I., Weeping Water, 1985
Johnson, Pauline, Alvo, 1999
K-Valley Farms Inc., Plattsmouth, 1993
Kaffenberger, Isola, Plattsmouth, 1999
Keetle, Doris & Roland, Ralston, 1991
Kellogg, Harold, Ashland, 1979
Kendall, Mrs. Myrtle, Union, 1957
Kiser, Robert & Rita, Plattsmouth, 2003
Klemme, Henry Estate, Murdock, 1982
Knabe, Harry, Nehawka, 1956
Kuehner, Elinor, Janet Caldwell, Susan Eisenberg, Red Cloud, 2006
Kunz, Dale & Sharon, Elmwood, 1984
Kupke, Paul & Malinde, Murdock, 1981
Laughlin, Wesley C., Weeping Water, 1992
Lauritzen, Kenneth & Alice, Doris & Bill Duff, Evelyn Richards, Alice & Lyle Sittler, David & Anne Lauritzen, Weeping Water, 2008
Leddy, William & Sophia, Ashland, 1966
Lemon, Cortis & Geraldine, Greenwood, 1985
Lepert, Stanley & Helen, Plattsmouth, 1998
Lindsey, Delbert & Fred, Nebraska City, 1989
Ludwig Kirchhoff, Caroline, Weeping Water, 1982
Luetchens, Ezra, Murdock, 1984
Magorian, Dorothy, Lincoln, 1995
Mathis, Marilyn & Raymond, Lincoln, 1995
Mathis, Raymond & Clarence, Lincoln, 1980
Meisinger, H. Fred, Plattsmouth, 1978
Meisinger, Mr. & Mrs. Stuart, Plattsmouth, 1970
Mendenhall, Eldon & Lucy, Elmwood, 1980
Metzger, William, Louisville, 1967
Mickle, Mrs. Hazel, Waverly, 1976
Miller, Betty, LaVista, 2000
Miller, Eleanor & Darold, Ceresco, 1992
Mills, George & Hallie, Murdock, 1988
Mills, Harvey & Cindy, Murdock, 1988
Moore, Joann L., Plattsmouth, 2013
Moren, Florence, Louisville, 2002
Mougey, Nettie, Union, 1971
Mueller, Henry, Louisville, 2002
Murdock, Guy, Nehawka, 1961
Murphy, Ann, Weeping Water, 1974
Murphy, Ruth, Louisville, 1974
Neitzel, Grant, Murdock, 1990
Neumann Lands LTD, Louisville, 1998
Nickel Bornemeier, Bernice, Murdock, 1984
Nickel, LaVern & Norma, Murdock, 1984
Nickels, Charles, Elmwood, 2001
Nickels, Dale & Marcia, Murray, 1997
Nickels, Robert, Murray, 1973
Nielsen, Margie, Elmwood, 2009
Oehlerking, Marvin & Ramona, Murdock, 1990
Oehlerking/ Schwaninger, Eldon/ Lavina, Elmeoow, 1992
Ost, Clara, Nehawka, 1984
Petereit, Mrs. Marvin, Plattsmouth, 1991
Philpot, Betty Anne, Plattsmouth, 1972
Philpot, Fride, Nehawka, 2002
Pollard, Larry, and Nolma Pollard Trust, Nehawka, 2009
Pollard, Merritt, Nehawka, 1980
Pollard, Raymond, Nehawka, 1958
Pollard, Verna, Nehawka, 1969
Puls, Delbert & Diana, Murray, 2009
Puls, Mrs. Earl, Louisville, 1993
Puls, Wilma, Louisville, 1981
Ragoss, Arlie & Lenora, Louisville, 1980
Reinke, Merlin & Ruth, Ashland, 1999
Resso, Mr. & Mrs. Donald, Nehawka, 1977
Rice, Harold & Paul, Murray, 1976
Rikli, Donald "Dutch", Murdock, 2014
Robb, Dorothy, LaVista, 2001
Rockenback, Mrs. Harry, Eagle, 1987
Roeber, Lynn, Ashland, 1984
Rosenow, Lucille, Elmwood, 1994
Ross, Clarence & Margaret, Nebraska City, 1988
Ruffner, Mr. & Mrs. E. G., Plattsmouth, 1978
Sack, Louis & Nellie, Murray, 1967
Schafer, Lyle, Nehawka, 1967
Schick, Mrs. Nora, Elmwood, 1988
Schlanker Family Farm c/o Shirley Wenzel, Elmwood, 2004
Schlichtemeier, Mr. & Mrs. Robert, Murray, 1966
Schlichtemeier, Stuart & LaVera, Murray, 1989
Schliefert, Mrs. Arnold, Murdock, 1985
Schomaker, Erma & Margie Nielsen, Elmwood, 2004
Schroder, Arland and Bobbie, Eagle, 2008
Schroder, Arnold and Marlene, Palmyra, 2008
Schweppe, Marie & Martha, Murdock, 1990
Sheldon/ Zertuche, Vilas/ Elizabeth, Otoe, 1970
Shrader, Mr. & Mrs. Gerald, Nehawka, 1968
Snoke, Donald & Georgia, Eagle, 1978
Sowards, Orie, Ashland, 1980
Splitt, Dorothy, Murray, 1994
Splitt, Roy & Ella, Nehawka, 1982
Sprieck, Mr. & Mrs. Arthur, Plattsmouth, 1974
Sprieck, Mr. & Mrs. Elmer J., Louisville, 1968
St. John, Ethel May, Plattsmouth, 2003
Stall, Frances, Eagle, 1995
Stander, David, Ashland, 2005
Stander, Herbert, Louisville, 1958
Stander, Mr. & Mrs. Frank, Murdock, 1977
Stander, Mrs. Andrew, Louisville, 1958
Stander, Mrs. Herbert, Louisville, 2010
Stander, Robert, Weeping Water, 1996
Stander, Virgil, Plattsmouth, 1991
Stege, Mrs Myrtle, Elmwood, 1979
Steinhoff, Arthur & Clara, Avoca, 1988
Stine, George, Union, 1966
Stock, Mr. & Mrs. Lyle, Murdock, 1975
Stock, Wayne, Murdock, 2001
Stohlmann, August, Louisville, 1958
Stohlmann, Christian C. & LaVone, Louisville, 1982
Stoll, Elmer, Nehawka, 1960
Story, Elbert, Murdock, 1980
Strabel, Ethel, Elmwood, 1977
Strabel, Ethel, Elmwood, 1982
Sullivan, Ralph, Plattsmouth, 1960
Swanson, Carl, Waverly, 1988
Switzer, Gerald and Carolyn, Nehawka, 2009
Terryberry, Mr. & Mrs. Glen, Louisville, 1968
Thimgahn, Eldred, Ashland, 1974
Todd, Melvin, Union, 1957
Todd/ Rogers, William / Mary, Union, 1992
Toplif, Ann, Murray, 1979
Tritsch Hennings, Wilma, Lincoln, 1989
Tritsch Meisinger, Esther, Louisville, 1983
Tritsch, Donna, Murray, 1995
Umland, Arthur & Clara, Hartford SD, 1999
Umland, Patricia, Alvo, 1989
Vallery, Mrs. John, Plattsmouth, 1966
Vielbig Trust et al, c/o W. LeRoy Cook, Green Valley, AZ, 2004
Vogler, Arnold, Louisville, 1973
Vogler, Verle & Betty Jean, Herman, 2007
Vogt, Howard & Eldean, Elmwood, 1981
Von Spreckelsen, Wililam & Elsie, Murdock, 1982
Watson, Lyndall & Willajean, Ashland, 1991
Wegener, William & Ruth, Louisville, 1980
Wehrbein, Flora Belle, Plattsmouth, 2009
Wehrbein, Fred, Plattsmouth, 1973
Wehrmann, Mrs. Erna, Louisville, 1967
Weichel, John V & Irene, Alvo, 1989
Wendt Ziegenbein, Myrtle, Murdock, 1991
Wendt, Lloyd, Murdock, 1991
Wenzel, John, Elmwood, 2004
Wenzel, Marion & Shirley, Elmwood, 1987
Wessel, Kenneth Richard, Avoca, 1993
Westlake, Clarence & Lucile, Elmwood, 1992
Weston, Mrs. Mildred, Ashland, 1972

Wetenkamp, Glenn, Plattsmouth, 1982
Wetenkamp, Mr. & Mrs. C. Arthur, Plattsmouth, 1971
Wiles, Cash, Plattsmouth, 1963
Wiles, Guy, Plattsmouth, 1965
Wiles, Howard, Plattsmouth, 1965
Wiles, Mrs. Della, Weeping Water, 1982
Williams, Mrs. Clara, Murdock, 1978
Wipf, Hazel & Roy Roger, Nebraska City, 1997
Wohlers, Wayne and Paula, Syracuse, 2016
Wolph, Evelyn, Nehawka, 1966
Wolski, Stanley & Rosie, Plattsmouth, 1985
Wulf, George, Eagle, 1975
Yearsley, Josephine, William Johnson, Julia Shelby, Betty Ann Fisser, Murray, 2011

Cedar County

Abts, Mr. & Mrs. Herbert, Belden, 1972
Albrecht, Arlan, Yankton SD, 2003
Albrecht, Floyd & Joyce, Fordyce, 2002
Anderson, Mr. & Mrs. Alfred A., Hartington, 1969
Anderson, Scot, Laurel, 1998
Ankeny, James B., Hartington, 1992
Arduser, Bernard, Coleridge, 1985
Arens Kleinschmit, Mr. & Mrs. Gerhard J., Hartington, 1974
Arens, Kenneth L., Hartington, 1969
Arens, Mr. & Mrs. Francis J., Fordyce, 1976
Arens, Roman W., Hartington, 1969
Asbra Steffen, Mr. & Mrs. George, Fordyce, 1973
Bauman, Leonard & Lorraine, Hartington, 1973
Becker, Loretta T., Hartington, 1987
Beste, LeRoy A., Wynot, 1980
Bobenmoyer, Gerald & Betsy, Hartington, 1999
Boehmer, Ignatius A., St. Helena, 1964
Bollich, Bruce & Donna, Fordyce, 1995
Bottolfsen, Martin B., Hartington, 1976
Brand, John, Fordyce, 1984
Brewer, Mildred, Lincoln, 2002
Brodersen, Kathleen, Yankton SD, 2000
Burbach, Joseph J., Wynot, 1986
Burney, D.W. & Mrs. W. W., Hartington, 1986
Burns, Hannah C., Laurel, 1992
Casal, Anita, Belden, 2007
Demdinger, Terry & LouAnn, Coleridge, 2015
Dendinger, Anthony & Jane, Hartington, 2001
Dendinger, Jack, Coleridge, 1992
Dirks, Joyce Anderson & Donovan, Herrin, IL, 2003
Dirks, Theodore H., Coleridge, 1971
Ebmeier, Arnold & Donna, Laurel, 1993
Eickhoff, Suellen, Cecilia Tideman, Margaret Doyle, Barbara Morris, Nebraska City, 2006
Falke, Joan M., Scribner, 2003
Feilmeier, Sheila & Neal, Fordyce, 2015
Ferber, Mrs. Margaret C., Wynot, 1968
Forsberg, Joseph E., Laurel, 1984
Frerichs, Arnold & Dorothy, Coleridge, 1985
Frerichs, Mrs. George B., Coleridge, 1984
Gillilan, Dalton, Wausa, 2000
Gillilan, Dorothy, Wausa, 1991
Gould, Helen, Laurel, 1995
Graham, Gary & Cindy, Hartington, 2013
Gubbels, Glenn J., Souix City, 1994
Guy, Willard L., Wynot, 1969
Haase, Shirley, Omaha, 2009
Haberman, Anthony, Fordyce, 1989
Hahn, Mrs. Margaret, Hartington, 1974
Hahne, Mr. & Mrs. Irvin L., Coleridge, 1971
Hall, Rod/Charlotte, Laurel, 1996
Hansen, Cleone B., Hartington, 2010
Heese, Don Heese, Laurel, 2004
Hefner, Avis S., Sioux City, IA, 2003
Heimes, Robert, Hartington, 2006
Heimes, Victor F., Wynot, 1968
Heine, Steve & Janet and Ron & Ann Heine, Fordyce, 2005
Henry, Darrel and Doris, Colerdige, 2008
Hesse Lauer, Mr. & Mrs. Donald, Hartington, 1972
Hines, Mrs. W. G., Coleridge, 1977
Hinrichs, Arthur G., Coleridge, 1976
Hinrichs, Mrs. Don Hinrichs &Ms. Judy Bertoloth Hinrichs, Laurel, 2014
Hirschman, William P. & Margaret M., Coleridge, 1969
Hochstein, Donald & Elaine, Hartington, 1994
Hochstein, Merlin, Fordyce, 1967
Hochstein, Neal & Sharlee, Wynot, 2006
Holm, Clarice, Hartington, 2013
Jansen, Arthur H and Twila Jansen, Fordyce, 2005
Janssen, Claire & Norma, Coleridge, 2000
Jones, Mrs. Mattie C., Wynot, 1967
Kaiser, Bernard & Shirley, St. Helena, 2007
Kaiser, Dolores, Wynot, 1996
Keiser, Steven and Elizabeth, Fordyce, 2016
Keiter, Anthony, Hartington, 1998
Kleinschmit, Duane & Dorothy, St. Helena, 1990
Kleinschmit, Kevin & Debra, Fordyce, 1993
Klug, Joseph F., St. Helena, 1969
Klug, Steven, Wynot, 2016
Koch, Donald & Mavis, Wynot, 1970
Koch, Jerry & Colette, Fordyce, 1989
Koch, Mr. & Mrs. Cletus, Crofton, 1976
Kollars, Henry C. & Frieda, Crofton, 1970
Konken, Kelly & Kelsey, Hartington, 2012
Kramer, Darlene, Omaha, 2009
Kremer, William & Melba, Crofton, 1974
Kruse, Marvin, Hartington, 1975
Kurtzhals, James D. Kurtzhals, Bristow, IN, 2004
Kvols, Morris C or Clara, Laurel, 1970
Lammers, Anton, Hartington, 1968
Lammers, Dean R. & Teresa M., Hartington, 2010
Lammers, Gary J. & Teresa, St. Helena, 2003
Lammers, Hildegarde T., Hartington, 1975
Lammers, I. A., St. Helena, 1965
Lammers, Steve & Janet, Hartington, 2006
Lange, Mrs. Edward J., Fordyce, 1972
Larson, LaVaughn, Holdredge, 2003
Lauer, Bill & Donna, Hartington, 2010
Leapley, Ronald L., Colerdige, 2008
Leapley, William C., Colerdige, 2008
Leise, John & Norbert, Wynot, 2000
Leise, Regis, Hartington, 2003
Linn, James D., Laurel, 2003
Magnuson, Wallace & Evonne, Laurel, 1992
Marsh, Dean & Laura Lou, Hartington, 1986
McGregor, David J/ Carol A, Hartington, 1996
Miller, Don G. & Darlene A., Hartington, 1986
Morrison, Lisle H., Coleridge, 1979
Morten, Monte and Diane, Hartington, 2008
Morton, Mrs. Ed, Hartington, 1966
Mueller, Allan & John Mueller, Fordyce, 2004
Nelson, John L., Crofton, 1968
Nelson Lewis, Olive, Crofton, 1968
Nelson, Lawrence, Hastings, 1973
Newton, Louisa, Obert, 1999
Olsen, Mr. & Mrs. William A., Hartington, 1971
Olson, Gordon & Latricia, Hartington, 2010
Peitz, Vera M., Hartington, 1995
Pflanz, Gaylord & Pearlanna, Belden, 1989
Pick, David & Marlene, Coleridge, 2000
Pick, Donovan & Nancy, Fordyce, 1998
Pick, Frank A., Hartington, 1975
Pick, George G. Pick, Hartington, 2004
Pick, Joeseph, Fordyce, 1968
Pinkelman, Ed H. Pinkelman, Hartington, 1974
Rathgeber, Larry & JoAnn, Coleridge, 2000
Reifenrath, Joseph F. & Donna M, Wynot, 1975
Schaefer Hoesing, Mr. & Mrs. Francis, Fordyce, 1973
Schaefer, Danny & Mary Schaefer, Fordyce, 2004
Schieffer, ValDean & Nancy, Crofton, 2001
Schindler, Leona, Hartington, 1997
Schroeder, Tim, Yankton, SD, 2011
Schulte, Donald & Laurie, Hartington, 1999
Schulte, Frank X., St. Helena, 1975
Specht, Mildred, Coleridge, 1977
Specht Hintz, Frances M., Coleridge, 1977
Stage, Marguerite, Laurel, 2009
Stapleman, Richard, Belden, 1985
Stappert, Hugo, Hartington, 1969
Steffen, Alfred & Marlys, Osage Beach, MO, 2003
Steffen, Mrs. Dorthea, Fordyce, 1973
Stevens, Mrs. Louise, Hartington, 1968
Stolpe, Vernon & Laura, Newcastle, 1981
Stratman, Kenneth J., St. Helena, 1975
Sudbeck Farms, Hartington, 1970
Sudbeck, Donavin W. & Henrietta M., Hartington, 1992
Sudbeck, Marvin J., Hartington, 1970
Sudbeck, Orville V. & Charlotte A., Hartington, 1970
Suhr, Dale, Wausa, 1991
Suing, Gerald & Joan, St. Helena, 1987
Suing, Mrs. Josephine, St. Helena, 1968

Theone, Alton G. & Delores, Hartington, 1970
Tolles, Louis C., Laurel, 1972
Tramp, Charles F., Crofton, 1976
Tramp, Joseph L., Crofton, 1976
Truby, LaVern & Ruth Ellen, Randolph, 1991
Uhing, Kenneth, Hartington, 1999
Vanderheiden, Anita, Laurel, 1988
Vogle, Norman, Wayne Vogle, Coleridge, 2009
Wickett, Marvin A. /Shirley M., Laurel, 1996
Wieseler, Gene, St. Helena, 1968
Wieseler, Mr. & Mrs. Leonard, St. Helena, 1972
Wilbeck, Warren & Corlynn, Hartington, 2002
Willenbring Schieffer, Joe W. & Irma, Crofton, 1974
Wohlman, Merle & Loretta, Hartington, 1998
Wortman, Mrs. Ferdinand, Hartington, 1990
Wortmann, Randy, Hartington, 2013
Wubben, Lawrence, St. Helena, 1963

Chase County

Appleyard, Audra G., Imperial, 1987
Beard, Burton, Champion, 1993
Beard, Don, Champion, 1986
Burtis, Dean, Imperial, 1988
Dart, Ronald, Enders, 2010
Davidson Hollyfield, Pauline, Champion, 1986
Davidson, Glen, Lamar, 1986
Dudek, Dorothy, Wauneta, 2000
Fanning, Frank & LaVerna, Wauneta, 1987
Hayes, Charlotte, El Sequndo CA, 2001
Heim, Joseph Earl Estate, Champion, 1986
Hill Marvin, Vincent L & Thyra Mae, Imperial, 1991
Hurd, Ralph/ Mava Jane, Imperial, 1986
Johnson Trust, Charles R & Betty J, Lincoln, 2009
Kelly Enterprise, L.L.C., Champion, 2007
Kimberling, Vinton & Fauniel, Champion, 1986
Kunnermann, Doug & Lee, Imperial, 2001
Lee, Betty, Haigler, 1986
Lee, Mr. & Mrs. Wayne C., Lamar, 1988
Luhrs & Rose FMS Inc., Enders, 2011
Lunkel, Levi Max, Lamar, 1986
Maddux Cattle Company, Wauneta, 1986
Mendenhall, Bob & Judy, Champion, 1999
Miller, Fred, Imperial, 1991
Moody, Vonna Jo, Wauneta, 2008
Ough, Doris, Wuneta, 2008
Silvester, Arthur & Oneta, Champion, 2000
Silvester, Arthur & Oneta, Champion, 2002
Statz, Louis C., Champion, 1988
Stenger-Pribbeno, Ltd., Imperial, 1988
Yaw, Gerald M., Champion, 1987
Yost Dinnel, Basil & Helen, Imperial, 1987

Cherry County

Adamson Trust, Nenzel, 2013
Airth, Dorothy, Dewinton, Alberta, 2009
Anderson, Evahee& Andy, Gorden, 2012
Andrews, Harley, Nenzel, 1988
Cole, Robert H. & LaVeta I., Gordon, 2004
Conner, Mr. & Mrs. Alfred, Sparks, 1985
Cox, AB, Mullen, 2006
Crawford Grummert, Carol, Inland, 2006
Eatinger Lands, Thedford, 1978
Fink, Eugene, Elsmere, 2004
Foster, Beatrice, Valentine, 2002
Foster, Charles & Evelyn, Sparks, 2002
Giles, Etta Mae & Bweta, Elsmere, 1986
Giles, George, Elsmere, 1985
Giles, Kenneth E., Elsmere, 1986
Gordan Valley Ranch, Ltd., Valentine, 2003
Grooms, Martin & Sylvia, Valentine, 1989
Gudgel Land Corporation, Valentine, 2011
Gudgel, William & Blanche, Ainsworth, 1999
Heelean, Phil J., Wood Lake, 1994
Hinton, Lois, Gordon, 2006
Hoyt, Gary, Whitman, 1984
Keller Losh, Vera, Valentine, 1984
Keys, Ted/ Keith, Mullen, 1996
Kime, Donald, Derby, 2013
Kime, Leslie & Betty, Valentine, 1997
Kime, Steve & Beth, Gordon, 2013
Krajeski, Paul & Ruth, Cody, 1988
Lee, Merrill E. & James R., Valentine, 1989
Lurz, Gilbert & Lucille, Valentine, 1987
Luther, Howard, Kilgore, 2002
McDuffee Bridges, James A. and Patricia, Mullen, 2008
Morris, Everett G., Wood Lake, 1983
Neumauer, Adah D., Purdum, 2004
O'Kief, Larry, Wood Lake, 2006
One Bar Inc., Cody, 2013
Osborn, Daniel & Shirley E., Mullen, 1997
Pearson Land & Cattle Co., Elsmere, 1986
Phillips, Joellen, Thedford, 2015
Reece, F.M., Valentine, 1993
Reimers, Calvin, Crookston, 1990
Reyman, Claude & Bruce, Wood Lake, 2000
Scheer, Kevin, Fink-Scheer Inc., Elsmere, 2015
Schlueter / Soles, Gene / Karen, Wood Lake, 1986
Scholtes, Merlin & Kathleen, Nenzel, 1991
Stolzenburg, "Bryon ""Bud"" & Jean", Crookston, 2014
Storer, Tanya & Eric, Whitman, 2012
Swendener, Mary Carr, Mullen, 2000
Van Metre, Earl L. & Olive R., Valentine, 1984
Weber Partnership, Crawford, 2006
Westover, Cal Estate, Ashby, 1985
Wolfenden & Sons, LeRoy & Kay, Valentine, 2015
Wolfenden, Alyce, Valentine, 2015
Wolfenden, Lawnie, Patty & Travis, Mullen, 2015

Cheyenne County

Anderson, Earl & Mary, Potter, 1999
Anderson, Earl & Mary, Potter, 1999
Borges, Helen, Dalton, 1986
Bowser Hillman, Alice, Lodgepole, 1987
Davis, Elizabeth, Primrose, 1986
Dickinson, Daniel, Crook, 1999
Dickinson, Billy & Clarabelle, Lodgepole, 1995
Easterly, Scott and Jane, Sidney, 2016
Eckert Sanders, Jessie, Dalton, 1997
Fecht, Herbert & Darlene, Dalton, 1989
Flessner, Lavern, Gurley, 1986
Florke, Vernon Herman, Lodgepole, 2000
Higgins, Myrtle, Sidney, 1986
Hinrichs III, John H., Sidney, 2015
Hornicek, Mrs. Gordon, Lodgepole, 2003
Horst, Virgil, Gurley, 2010
Jessen, B. Henry & Joyce Ingwerson, Lodgepole, 2003
Johnson, Glenn, Sidney, 2001
Jorgensen, Roger, Marjorie Jorgensen, Sidney, 2007
Krueger, Richard Earl, Sidney, 1990
Larson, Amy, Sidney, 1987
Lenzen, Jerome, Sidney, 2001
Lessman/ Snook, Marjorie / Ruby, Dalton, 1984
M.M. Johnson Farms Inc., Sidney, 1987
Mayer, Kim & Kathleen, Dalton, 2000
McRoberts Game Farm, Gurley, 1984
Miller, Kenneth & Frances, Dalton, 1989
Morgan Farms, Stuart and Linda Morgan, Dalton, 2010
Narjes Land and Livestock LLC, Sidney, 2016
Nelson, Robert & Alby, Bridgeport, 1988
Nelson, Leonard LeRoy, Potter, 1988
Niebaum, Glenna Belle, Sidney, 2007
Oberndorfer, Adella (Horst), Potter, 2010
Praire Park Farms Inc., Potter, 1986
Rasmussen & Sons LLC, Alliance, 2016
Rexroth, Keith & Nancy, Robert Rexroth, Roland Rexroth, Sidney, 2007
Rieken, Leonard & Mary Ann, Sidney, 1988
Rowan, Arthur, Gurley, 1990
Runge, Robert & Connie, Sidney, 1995
Runge, Robert & Connie, Sidney, 2010
Schuler, Chalire, Bridgeport, 1987
Sharman, Lillian, Sidney, 2014
Shaw, Hilma & Lloyd, Dix, 1988
Smith, Myrtle, Sidney, 1988
Stikker, Allen, Sidney, 1989
TECC Farms Limited, Potter, 1992
T.E.S.S.S. Farms, Inc., Esther Houser, President/Chairman, Sidney, 2014
Thurston, Clifton Palmer, Potter, 1988
Tremain, Gerald & Audrey (Zalesky), Sidney, 2004
Walsh, Stella, Dalton, 1989
Zeigler, Rog, Dalton, 1999
Darling, Edith, Lodgepole, 2005

Clay County

Ablott, George W., Inland, 1973
Alberding Family Limited Partnership Center, 2008
Alberding, Bernice M, Aurora, 1990
Anderson Snell, Ronalda, Sutton, 2002
Anderson, Bernice L., Ong, 2005
Anderson, Merle & Dorothy, Hastings, 1999
Barbee Hills Farm Corp., Sutton, 1993
Benson, Howard & Helen, Ong, 1979
Bermond, Laurence L., Harvard, 1995
Bieck, Kenneth & Wilma, Harvard, 1985
Boyd, Eunice, Ken & Ramona Spray, Center, 2006
Briggs, Harold G. Center, 1979
Brinkman Farms Inc., Shickley, 1984
Brodrick, Francis & Sharon, Hastings, 2005

Campbell, Mrs. Charles, Sutton, 1988
Carlson, Homer, Sutton, 1979
Crumbliss, Robert & Mary, Edgar, 2006
Dane, Robert & Joyce, Doniphan, 2014
DeHart, Shirley A., Nelson, 1987
Dieringer, Mary, Harvard, 1992
Eigenberg, Melvin A. & Marydae, Hastings, 1985
Eigenberg, Raymond R. & Lela W., Lexington, 1985
Englandton, Sutton, 1986
Erickson/ Brunken, Marvin A. / Bretta L, Lincoln, 1989
Fike, Marion & Sondra, Fairfield, 1983
Fisher, Frank C. & Donna R., Fairfield, 1986
Fitzke, Carlyle & Aubrey, Lincoln, 2006
Fitzke, Ward A., Elwyn R. Fitzke and Tyler D. Fitzke, Fairfield, 2014
Fitzke/ Dieringer, Mildred / Pauline, Harvard, 1984
Forberg, Edward D., Sutton, 1985
Frank, Flora L., Harvard, 1983
Gell, Wilson & Josephine, Shickley, 1986
Gemmer, Otto H. & Nadine, Sutton, 1989
Gerlach, Don, Akron, OH, 2005
Griess, Hilmer, Sutton, 1976
Hajny, Margaret J., Edgar, 1987
Hansen, Earl D. /Marie E., Hastings, 1996
Harmon, Anita & Mike, Trumball, 2011
Haseloh, Lloyd & Evelyn, Harvard, 1985
Housel, Lucille, Harvard, 2000
Hoyt, Jim, Fairfield, 1978
Hoyt, Melvin & Millie, Fairfield, 1986
Hultine, Leland, Saronville, 1981
Hultman, Lyle, Sutton, 1987
Iliff, Vernon & Cindy, Aurora, 2002
Jedounek, George, Fairfield, 1991
Jeffries Keller, Maxine, Harvard, 1987
Johnson, Aileen M. & E. Berlyn Center, 1987
Johnson, Clarence R. & Mildred L., Sutton, 1991
Johnson, Wesley & Myrtle, Sutton, 1979
Karnatz, Lyle & Deborah, Edgar, 2000
Keasling, Kenneth H. & Helen M., Harvard, 1988
Keasling, Z. Max, Harvard, 2008
Keller, Raymond Center, 1980
Kinyoun Family Trust, Shoreline, 2014
Kissinger, "Raymond, Stephen, & Craig", Glenvil, 1985
Koenig, Harolyn, Berthoud CO, 1999
Koenig, Paul & Shirley, Harvard, 1998
Koenig, William A., Harvard, 1986
Livgren Revocable Trust, Dale, Ruth Livgren Revocable Trust Center, 2011
Lobb, Harvey J. Center, 1991
Lovgren, Marvin & Doris, Harvard, 1992
McLaughilin Heirs Center, 1985
Meyer, JoAnn, Grand Island, 2011
Meyer, JoAnn, Grand Island, 2013
Nejezchleb, Mr. & Mrs. Ernest, Glenvil, 1987
Nelson, Lee, Sutton, 2001
Nelson, Marvin & Deanna, Harvard, 1981
Olson, Winston A & Dorothy, Hastings, 1986
Organ, Donald E., Edgar, 1980
Overturf, Abba, Hastings, 2002
Overturf, Martin L. Sutton, 1990
Overturf, Martin L., Sutton, 1998
Overturf, Pamella, Edgar, 1989
Pavelka, Edward, Diane Pavelka, Glenvil, 2011
Peshek, Orvis & Norma, Fairfield, 2004
Pope, Raymond & Dorothy, Aurora, 1991
Portenier, Ketih & Sharon, Harvard, 2001
Pratt, Edna, Firth, 1993
Rader, LaRue & Dolores, Trumbull, 1986
Rath, Mildred, Aurora, 2007
Ross, Elsie & Floyd, Harvard, 1985
Ross, Harold, Lincoln, 2002
Samuelson, Milan & Ethel Center, 1980
Samuelson, Sandra Kay, Harvard, 2009
Schliep, Clarence Arthur, Fairfield, 1985
Schmer, Floyd & Marjorie, Sutton, 1979
Schwenk, Bruce & Judy, Harvard, 1981
Sedersten, Albin, Sutton, 1983
Shalka, Rose Ann and Bryan, Deweese, 2016
Sheridan/ Hurst, Betty/ Gwen, Sutton, 1990
Sherwin Johnson, Fay, Seward, 1984
Shuck, Jerry, Edgar, 2002
Siefken, Kevin L, Trumbull, 2005
Skalka, Mr. & Mrs. James T., Deweese, 1986
Sowers/ Petersen, Mary Lou/ Shirley Marie, Hastings, 1985
Springer, Darrel & Jean, Edgar, 2006
Stimbert, George Heirs, Hastings, 1986
Storie Foster, Mary Alice, "Stillwater, OK", 2011
Strong, Earl L., Aurora, 1991
Taylor, Crit John Center, 1987
Taylor, John Center, 2002
Valentine, Lloyd & Donna, Glenvil, 2006
White, Alberta, Davenport, 2000
Woods, Royal, Beulah Woods, Fairfield, 2007
Yost, DeLores M., Harvard, 1993
Yost, LaVerne & Shirley, Harvard, 2012

Colfax County

Abraham, Lewis & Emma, Schuyler, 1983
Ahrens, Charles, Clarkson, 1993
Bachenberg, Esther, Norfolk, 2002
Bahns, Matt & Lisa, Clarkson, 2015
Bailey, Thomas W., Schuyler, 1985
Bergt, Ervin & Pearl, Schuyler, 1992
Blazek, Frank G., Dodge, 1983
Blazek, Marilyn, Robert Horak, Schuyler, 2007
Botsh, Eola Mae, Leigh, 1993
Box, Kenneth, Clarkson, 2002
Brabec, Mark, Sally & Renee, Clarkson, 1985
Brichacek, Gerald, Schuyler, 1984
Brichacek, Lillian D., Clarkson, 1992
Brodecky, Mr. & Mrs. Ernest, Howells, 1977
Cada, Marcelline, Clarkson, 1975
Cech Eggers, Lillian, Clarkson, 1975
Cech, Jerome, Clarkson, 1974
Cech, Russ & Janet, Rogers, 2015
Cerny, Mrs. Edward, Rogers, 1983
Chile Flats Inc., Howells, 2008
Clausen, Mr. & Mrs. LaVern D., Clarkson, 1976
Craig, Mrs. Norman J., Clarkson, 1981
DeBower, Elvin O, Schuyler, 1975
Dolezal, Jerry J., Schuyler, 1975
Dvorak, Joseph W., Howells, 1991
Dvorak, Robert C., Schuyler, 1976
Ernst, Allen & Kathy, Schuyler, 1988
Faltys, Robert D., Frances, Gary & Ricky, Clarkson, 2006
Figgner, Eugene & Jerane, Fremont, 2000
Fuhr, Helen M. Estate, Clarkson, 1987
Gerrard, Garrett, Schuyler, 1983
Gross, Mrs. Gerhard, Clarkson, 1984
Grotelueschen, Allen H., Richland, 1996
Grotelueschen, Martin, Leigh, 1998
Hajek, William F., Howells, 1978
Havrda, Emil R., Schuyler, 1995
Heerman, Charlotte, Lincoln, 2000
Hefti, Henry, Clarkson, 1991
Held, Jerlad & Betty, Leigh, 2002
Hledick, Tillie D., Clarkson, 1992
Holoubek, Verne, Clarkson, 1993
Horak, Joseph & Vlasta, Schuyler, 1992
Horn, Duane, Schuyler, 2000
Hugo / Inselman, W. / Edna, Leigh, 1993
Imholte, Alice, Schuyler, 2015
Indra, Dale & Laurie, Clarkson, 1999
Janousek, Dean J. & Penny, Clarkson, 1997
Jedlieka, Mr. & Mrs. John E., Schuyler, 1987
Johnson, Oliver & Marjorie, Clarkson, 2007
Jungbluth Payzant, Mrs. Lenora, Schuyler, 1977
Kasik, John F., Richland, 2015
Kassmeier-Lanphier, Kath, Omaha, 2016
Kehrli & Family, Galen & Diana, Carolyn R. Conner & Family, Schuyler, 2006
Klitz, Anita, Dodge, 2012
Klitz, Leonard, Dodge, 1978
Knust, Neil J. & Joan, Howells, 1992
Konicek, Stanley J., Schuyler, 1973
Krejci, Allan, Schuyler, 1984
Krejci, Robert, Lincoln, 1997
Krula, Lad H., Schuyler, 1975
Krupka, Alan Marvin, Howells, 2007
Kudera, Allen, Clarkson
Kudera, Kevin and Tamra, Clarkson, 2008
Kudera, Lumir and Geraldine, Schuyler, 2008
Kuhr, Russell & Beth, Clarkson, 2015
Kuzel, Hilarion, Rogers, 2002
Labens, Mrs. Bonnie M., Columbus, 1988
Lerch Family LLC, Clarkson, 2011
Littelman, Tillie, Leigh, 2001
Lodl, Lambert & Ebelyn, Schuyler, 1992
Lopour, Leonard and Lavonne, Clarkson, 2008
Lopour, Ronald, Clarkson, 1994
Loseke, Walter & Vera, Columbus, 1988
Lueke, Alfred G., Schuyler, 1975
Lueke, Edward W., Schuyler, 1975
Lukl, Jerry J., Clarkson, 1976
Lutjelusche, Ernest A., Columbus, 1973
Makousky and Makousky, Frances and Daniel, Howells, 2013
Makousky, Frances, Howells, 2013
Mares, Gale, Schuyler, 2011
Mares, Joseph R., Schuyler, 2008

Mares, Seth, Schuyler, 2011
Mares, Wylma, Schuyler, 2008
Marohn, Martin, Schuyler, 1962
Marohn, Michael, Cheryl & Christian, Schuyler, 1968
Martyn Kemp, Henry, Schuyler, 1967
Mastny, Milo & Libbie, Schuyler, 2004
McNally, Harry Heirs, Fremont, 1984
Moeller, Allen, Lincoln, 1999
Motycka, Carol, Leigh, 2016
Motycka, Delmar, Lincoln, 2016
Mueller, Merle, Richland, 2015
Muhle, Arnold & Pat, Leigh, 1989
Musil, Clarence P., Clarkson, 1985
Nemec, John E., Dodge, 1984
Nielson, Robert & Vivienne, Dodge, 1979
Novak, Gene and Florence, Schuyler, 2003
Novotny CahaNovotny Musil, Helen/Eleanor, Wahoo, 1985
Novotny, Joe W., Clarkson, 1973
O'Brien, John M., Schuyler, 1975
Oehlrich, Helen M. Fuhr Estate, Columbus, 1997
Perrin Heavican, Adelle, Schuyler, 1988
Pieper, James J., Dodge, 1988
Pojar, Emil & Helen, Schuyler, 1975
Prokupek, Marcella, Columbus, 1977
Prusa, Leonard R., Howells, 1984
Pruss, Richard J & Debra, Dodge, 2015
Reha, Russ & Carol, Reha Family Farms, Schuyler, 2014
Richards, Mr. & Mrs. Proctor, Schuyler, 1972
Richtig / Urbanek, Libbie / Darlene, Schuyler, 1988
Rickert, Alvin, Columbus, 2003
Robertson, Eugene, Rogers, 1970
Rocheford, Mary Ann, Schuyler, 2002
Ruskamp, Stephen F. & Joan M., Dodge, 2014
Ruzicka, Melvin, O'Neil, 1992
Ruzicka, Rodney & Kathryn, Rogers, 1989
Sayers Family Farm, Schuyler, 1990
Sayers, Alvin R. & Mary Lou, Clarkson, 1995
Sayers, Leonard & Beverly, Clarkson, 1975
Schmid, Melvin M., Columbus, 1991
Schmid, Melvin M., Richland, 1979
Sedlak, James, Schuyler, 1994
Severin, Ethel M., Leigh, 2004
Severin, John W., Leigh, 1989
Shonka, Allan, Schuyler, 2002
Shonka, Leonard, Schuyler, 2000
Sindelar, Melvin F., Clarkson, 1988
Sobota, Gene & Irene, Schuyler, 1985
Sobota, John F., Schuyler, 1970
Sobota, Joseph J. & Julia, Schuyler, 1985
Stibal, Louise A., Schuyler, 1973
Stuthman, Arnold, Platte Center, 2011
Svec, Rolland D. & Magdaline M. Svec Trusts, Rogers, 2012
Svoboda, Shirley A., Rogers, 1986
Telecky, Donald & Ann, Howells, 2002
Teply, Dean & Alice, Clarkson, 1998
Thalken, Laverne, Leigh, 2002
Tomka, Jerome, Dodge, 1973
Trojan, Thomas R., Murray, Utah, 2003
Tuma, Arnold A. & Loretta A., Schuyler, 2012
Urban, Calvin & Laurie, Clarkson, 2013
Vitek, Adolph & Lois, Clarkson, 1998
Vitek, Helen, Rochester MN, 2001
Vlcek, Armella, Omaha, 2012
Vogel, Clarence & Dorothy, Howells, 2012
Vogel, Thomas W. & Renae J., Dodge, 2012
Vogel, Wilbert, Howells, 2006
Votava, Donald A., York, 1986
Wachal, Dan, Schuyler, 2001
Wain, Antonia, Clarkson, 1976
Watchorn, Arlo, Arlene Gnirk, Ardith Schueth, Humphrey, 2005
Welch, Charles, Richland, 1973
Welch, Louis C & Barbara, Richland, 1976
Wendt, Delaine & Dianne, Leigh, 1995
Wendt, Wallace & Dorethy, Leigh, 2002
Wieneke, Kenneth, Dodge, 1983
Wisnieski, Brian & Karleen, Dodge, 2011
Wullschleger, John R., Deer Park TX, 1995
Wurdeman, Dennis & Linda, Leigh, 2010
Zastera/ Kudera, Alden/ Roma, Columbus, 1984

Cuming County

Abendroth, Fred & Delilah, West Point, 1963
Abendroth, Martin, Bancroft, 1985
Anderson, Carolyn J., Ames, 1981
Anderson, Delwin C. & Norma J, Oakland, 1989
Anderson, Merlin I., Oakland, 1981
Aschoff, Anthony, West Point, 1995
Aschoff, Donna, Omaha, 2015
Baumert, James H & Marianne, Howells, 1992
Beerbohm, Henry J., West Point, 1979
Beerbohm, John F., Wisner, 1994
Behlers, Albert & Edna, Wisner, 1974
Behlers, Ella V., Wisner, 1976
Behling, Julius and Wynona Behling, Beemer, 2004
Behling, Victor, Beemer, 1990
Benne/ Essman, Marvin / Arlene, West Point, 1992
Bettenhausen, Harvey K, Beemer, 1991
Bettenhausen, Henry & Augusta, West Point, 1972
Bettenhausen, Mrs. Dora, Oakland, 1979
Beutle, Mylon & Carol, Bandr, 1998
Beyer, Mrs. John A., West Point, 1981
Biles Farm Inc., Pender, 1996
Birchem, Adella, West Point, 1984
Blanknagel, Mrs. Gustave/ Willard & Alice, West Point, 1969
Bleyhl, Ardis Karlen, Beemer, 1979
Borgelt Inc., John, Wisner, 2010
Bracht, Edward Bracht, West Point, 2004
Brahmer, Burnis, Wisner, 1963
Brandt, Gary C., Lincoln, 2012
Brasch, Vernon & Shirley, Bancroft, 2001
Breitbarth, Arthur W, Bancroft, 1991
Brockemeier, Aaron, West Point, 1982
Brockemeier, Garylord & Grand, West Point, 2001
Brockemeier, Luann, Seward, 1993
Brockman, Mr. & Mrs. Walter, West Point, 1969
Brockman, Robert & Stella, West Point, 1967
Brockmann, Chris & Kim, West Point, 1999
Brundieck, Mr. & Mrs. Ernest, Wisner, 1970
Buchholz, Melva, Omaha, 2012
Bunjes, Edna, Wisner, 1994
Buse, Arthur, West Point, 1983
Crouter, Ada, West Point, 1962
Daberkow, Arnold & Mildred, West Point, 1980
Daberkow, West Point, 2006
Danielson Cull, Norma, Oakland, 1995
Danielson, Kenneth E., West Point, 1995
Dewitz, Douglas R., Ponca City, OK, 2003
Dinklage, John F. & Melvina, Wisner, 1994
Dirkschneider, Robert & Val, Howells, 1993
Doescher, John F. & Walinda, West Point, 1971
Drueke, Henry W & Bernadette M, West Point, 1997
Drueke, Joseph, West Point, 1969
Duhsmann, Willis J., West Point, 1985
Dunn, Randall F, Pender, 2003
Dvorak, Adella, Clarkson, 1999
Eggers, Randy, Wisner, 2014
Erickson, Janice, Craig, 1999
Ernesti, Janice, West Point, 2001
Ernesti, Ronnie & Mark, West Point, 2001
Eschenburg, Helen, West Point, 1974
Eucks Bendixen, Margaret A., Norfolk, 1995
Evert, Frank & Wilhelmina, Dodge, 1985
Fast, Roxann, & Joyce Weight, West Point, 2010
Fehliman Paulsen, Mrs. Wilma, Beemer, 1986
Feyerherm, Orville, Beemer, 1974
Fischer, Brian & Doreen, West Point, 2000
Frevert, Darleen, Norfolk, 1995
Fullner, Luetta, Stanley S., R.L., Daniel A., & Terry H, Beemer, 2015
Fullner, Arthur & Anna/ Harland & Luetta, Beemer, 1966
Fullner, Lydia /Alwin, Beemer, 1968
Galbraith-Palmer-Schmitz, Beulah F., Buena Vista CO, 1995
Gardes, Jack & Earlene, Bancroft, 1977
Gatsemeyer Hale, Odelia, Ft. Collins CO, 1962
Gentrup, Hilary, West Point, 1984
Gentrup, Terry/Tina, West Point, 1996
Goeden, West Point, 2006
Goranson, Martina, Oakland, 1978
Granatowicz, Kathryn D., Omaha, 1995
Greve, Bernhardt, Pender, 2007
Groskurth, Alice, Beemer, 1992
Groskurth, Earl, Bancroft, 1992
Guenther, Albert A., West Point, 1973
Gunn, Delores Jean Wegner & Darleen Faye Wegner Beals, Sonoma, CA, 2004
Hagedorn, Bernard W. & Roberta, West Point, 1967
Harstick, John, West Point, 1961
Harstick, Mark, West Point, 2000
Haskell, Mary Lynn (Aschoff) & Gordon, Madison, 2015
Hass, Ottlin D. & Vondell, Wisner, 1977
Hatterman, Eldon J., Dodge, 1986
Heimann, August, West Point, 1999
Heimann, Francis, West Point, 1983
Heinemann, Ardyce, Beemer, 2012

Heinemann, George, Wisner, 1995
Heller, Alvin & Kathleen, Wisner, 1998
Hermelbracht, Richard & Marlene, Bancroft, 2002
Herzinger, Rudolph, Beemer, 1995
Holland, Marjorie J. Qualified Annuity Trust, Wisner, 2004
Holtz, Mr. & Mrs. Norman, Beemer, 1975
Horst Sill, Lynn, Wisner, 1998
Horst, Arly & Anita, Wisner, 1994
Jacobsen, John, Wisner, 1987
Jahnka, Mrs. William, West Point, 1970
Jahnke, Dennis L., Bancroft, 1992
Jahnke, Loren & Dodie, West Point, 2013
Jindra, Milo M, Pilger, 1991
Johnson, Clifford C., West Point, 1983
Johnson, LaMae, Oakland, 2011
Johnson, Robert, Wisner, 1997
Kampschneider, Hubert & Johanna, West Point, 1999
Karlen, Gerald F., Pender, 1983
Kaser, Kevin & Audrey, Pender, 1985
Kaup, JoAnn (Aschoff) & Todd, West Point, 2015
Kaup, Leonard & Verona, West Point, 1969
Kaup, Richard & Bernice, Dodge, 2002
Kersten, Loha Ann, Beemer, 1976
King, Elton, Omaha, 1990
Kloth, Mrs. Ella, Bem, 1983
Kloth, Oscar, Wisner, 1992
Knobbee, Clem, West Point, 1981
Koch, Fred L. & Marie, West Point, 1971
Konicek, Janice M, Bancroft, 1990
Kreikemeier, Al, West Point, 1981
Kreikemeier, Anton J., West Point, 2007
Kreikemeier, Don, West Point, 1987
Kreikemeier, John B., West Point, 1962
Kreikemeier, Paul B. and Ladell M., West Point, 2008
Kucera, Betty /Roland/Gar, Howells, 1989
Kuester, Emil F. & Clara H, West Point, 1969
Kuester, Melvin & Dorothy, West Point, 1968
Kuhlman, Mrs. Louise A., Fremont, 1976
L & M Rief Inc, West Point, 1986
Laase, Leroy T., Lincoln, 1976
Lambrecht, Artwin & Lydia, Beemer, 1972
Landholm, Beverly, West Point, 2014
Leisy Properties L.T.D., Wisner, 1980
Lemm, William, Beemer, 1984
Lindsay, Twyla, Wayne, 1995
Linnemann, Bernice, West Point, 1992
Linnenbrink, Roseann, West Point, 2007
Lofgren, Mr. & Mrs. Clayton, West Point, 1980
Lofgren, Mrs. Marie, West Point, 1976
Long, Melvin & Helen, Beemer, 1981
Luebbert, Elaine, West Point, 2011
Luebbert, Fred & Marcella, West Point, 1987
Luebbert, Lester & Jean, West Point, 1999
Luedke, Georgia, Bancroft, 1984
Lueshen, Kevin, Beemer, 1999
Lutz, Keith A., Oakland, 1980
Lynn, Kevin & Kyle Borgelt & Kristi Plagman, Wisner, 2005
Maack, Donald & Audrey, West Point, 1967
Mackey, Leland & Lois, Bancroft, 1985
Maly, Patrick & Donna, West Point, 1988
Maly, William J. & Elizabeth C, West Point, 1967
Maresh, Richard, Dodge, 1982
Marik, Randall, Howells, 1983
Maurer, Don & Bernice, Fremont, 1977
McGuire, Michael & Rosemary, Omaha, 2012
McNamara, Fred & Rita, Beemer, 1969
McNamara, Lucill, Beemer, 1985
Meier, Eugene, West Point, 2001
Meier, Helen, Wakefield, 2001
Meier, Hilbert J Springfield, 1981
Meiergerd, "LeRoy & Dianne
Meiergerd, Leo & Ann Marie, West Point, 1983
Meister, Melvin, West Point, 2008
Meister, Mrs. Ben, West Point, 1984
Metzger, Kenneth & Marian Metzger, Bancroft, 2004
Meyer, Fred A & Gertrude, West Point, 1986
Miller, Donald F. & Katherine M., Dodge, 1985
Miller, Timothy P. Miller, Bancroft, 2004
Milligan, Irma, Wisner, 2009
Munderloh, Reinold H & Kathie G, Bancroft, 2015
Munderloh, Vernon F, Lyons, 1989
Murphy, Phil and Flo, Lincoln, 1984
Nebuda, Wally & Dorothy, West Point, 1984
Nelson Tolman, Suzzane, Rochester, 2008
Nelson, Mrs. Myron, Oakland, 1989
Ortmann, Alvin & Florence, West Point, 1993
Ortmeier, David & Aggie, West Point, 1989
Ortmeier, Henry & Helen, West Point, 1989
Pageles (Wegner), Wanda Anna Bertha, Norfolk, 2003
Peters, Carolyn A., West Point, 1983
Peters, Glen N., Morril, 1974
Peterson, Gary D., Oakland, 1981
Peterson, Nancy A, West Point, 1981
Pojar, Bonnie, Terry Haber & Haber Grandchildren, Elkhorn, 2004
Priebnow, Richard, West Point, 1976
Prinz, Elizabeth & Meister Trust, West Point, 1994
Raabe, Mr. & Mrs. Rudolph, Wisner, 1964
Raabe, Mr. & Mrs. Warren, Beemer, 1964
Rabe, Elton & Alice, Beemer, 1964
Rasmus, John, "Kristen Kusik, Sandra Rasmus", West Point, 2010
Rathke, Walden, Beemer, 1983
Reese, Irma M., Grand Island, 1978
Reeson, Charles F., West Point, 1985
Reiman, Gerald & Janet, West Point, 2011
Reinhardt, Mrs. Lavonne, Wayne, 1983
Rolf, Art & Arlene, West Point, 1986
Rueter, Marlin & Darlene Reuter, Norfolk, 2004
Ruskamp, Mary Ann, Dodge, 2014
Schademann, Lloyd & Anna Marie, Pender, 2002
Schademann, Otto, West Point, 1968
Schlautman, Bill D, Howells, 1991
Schlecht, Marcella, West Point, 2000
Schlecht, William H., Beemer, 1983
Schleusener, Arthur W, West Point, 2001
Schlueter, Edmund H, Wisner, 1988
Schlueter, Mr. & Mrs. Arnold, Beemer, 1969
Schlueter, Mr. & Mrs. Eldon, Beemer, 1969
Schlueter, Muriel, Beemer, 1983
Schlueter, Verona, Wisner, 1999
Schluter, Albert, Pender, 1985
Schmidt, Ervin & Pearl, Pender, 1997
Schmuecker, Leonard, West Point, 1984
Schoch, Harry J., Bancroft, 1991
Schold, Donald W., Oakland, 1979
Schultz, Arlie, Beemer, 1979
Schulz, Willard, Fremont, 2001
Schulzkump, Amos & Laverne, West Point, 2000
Schulzkump, Delmer & Mary, West Point, 1977
Schutte, Donald & Dorothy, Beemer, 1987
Schwartz, Milton and JoAnn, Howells, 2016
Schwedhelm, Mr. & Mrs. Albert J., Norfolk, 1968
Siebrandt, Glenn/Margaret, Pender, 1996
Simonsen, Duane, Pender, 1984
Smith, Hayes & Elsie, West Point, 1997
Smith, John H, Pender, 1986
Smith, Leslie, West Point, 2009
Smith, Merlyn, Pender, 2013
Smith, Mrs. Trexton, West Point, 1992
Sokol Jindra, Agnes, Stanton, 1991
Spenner, Mr. & Mrs. Paul J, West Point, 1984
Stalp, Mrs. Edward, West Point, 1982
Stark, Henry & Hildegard, West Point, 1968
Stark, Reynold & Helen, West Point, 1968
Steenken, Roger, Beemer, 2015
Steinmeyer, Beatrice, Wisner, 1994
Steuter, Paul, West Point, 2010
Stieren, Art, West Point, 1984
Stoltzman, Henry & Matie, West Point, 1963
Throener, Martn, West Point, 1998
Throener, Wilbert & Elizabeth, West Point, 1998
Toelle, Gene H., Beemer, 2013
Toelle, Leonard & Naoma, Beemer, 1977
Toelle, Morelle C. & Wiltrude M, Beemer, 1969
Uhing, Kenneth, West Point, 2011
Ulrich, Larry & Jan, West Point, 2009
Urban, Calvin & Laurie, Clarkson, 2013
Vincent & Vicki", Beemer, 2015
Vogt, Lester, Pender, 2001
Vollmer, Carl & VerGene, Wisner, 1980
Weber, Gayleen, Wisner, 2008
Wegner, Dean & Janice, West Point, 1995
Wegner, Harold & Berdina, West Point, 1990
Wegner, Lauren & Delores, Bancroft, 2002
Weiler, Dennis & Betty, West Point, 1987
Wendt, Mr. & Mrs. Harold F., West Point, 1969
Wiegert, Irma J., Beemer, 1983
Wiegert, Lester & Gertrue, Beemer, 1991
Wieneke, Edward, West Point, 1962

Wieneke, Gene and Deanne, Beemer, 2016
Wieneke, Marilyn, Dodge, 1982
Wostoupal, Robert L, Beemer, 2008
Wostoupal, Robert L., Beemer, 1991
Zimmerman, Mrs. Olive E., West Point, 1968
Zuhlke, Laura E., Bancroft, 1980
Zuhlke, Willis J., Bancroft, 1980

Custer County

Amsberry, Marilyn, Ansley, 2002
Anderson, Alan, Cozad, 2005
Anderson, Lawrence & Harriett, Gothenburg, 1988
Anderson, Leona Marie, Comstock, 2014
Anderson, Ted & Jennifer, Broken Bow, 1985
Anderson, Wesley, Gothenburg, 2011
Babbit, Steve & Sally, Broken Bow, 2000
Beavers, Ronald, Broken Bow, 2015
Bek, Mrs. Paul, Seward, 1978
Bentley, Gaylord & Donna, Sargent, 1992
Beshaler, Charles F., Arnold, 1986
Blakeman, Hazel L., Merna, 1984
Brand, Myrtle G., Broken Bow, 1986
Brand, Myrtle G., Broken Bow, 1993
Brown, Evelyn, Kearney, 1997
Burnett, Ruth G. & Clifford D., Broken Bow, 1983
Chartraw, Eugene & Michael, Mason City, 1997
Chesley, Elgin & Doris, Callaway, 1980
Christensen, Thomas P., Merna, 1991
Christensen, Thomas, Merna, 1984
Clarke, Hugh Jr. & Joyce, Berwyn, 2004
Cole, Cecil, Merna, 1991
Coleman, Harold & Mildred Coleman family trust, Merna, 2005
Coleman, Harold W & Mildred, Merna, 1994
Collier, Margaret M., Broken Bow, 1992
Connely, Claris J., Broken Bow, 2004
Conroy Land Co, Oconto, 2009
Cooksley, Leo C., Berwyn, 1980
Cool, Boyd & Mary, Callaway, 1991
Cool, Grayson & Family, Callaway, 1999
Coslor, Gerald E., Sargent, 1978
Dainton, Lois P, Anselmo, 1997
Desmond, Winnie, Merna, 1996
Dewey, Mrs. Alan S., Broken Bow, 1983
Dietz, Walter & Roberta, Broken Bow, 2001
Dittmar, Calvin F., Oconto, 2012
Dittmar, Ronald, Onconto, 1987
Downey, James & Joan, Merna, 2000
Downey, James & Joan, Merna, 2005
Downey, Paul & Mabel, Merna, 1980
Dowse, Phillip, Comstock, 1974
Eggleston, Allen, Broken Bow, 2001
Engelsgjerd, Severt, Broken Bow, 1982
Fagen, F.C. Family Trust, Anselmo, 1999
Finch, Faylin S., Arnold, 1980
Fleishman, Monica, Broken Bow, 1998
Fletcher, Floy C., Taylor, 1982
G'Schwind, W.A., Callaway, 1984
Glendy, Gary & Donna, Broken Bow, 1994
Govaerts, Lester, Broken Bow, 1997
Griebel, Leo, Milburn, 1981
Griffith, Clark T., Onconto, 1985
Grint, Norman & Merlene, Sargent, 1990
Gunnarson, C. & M., Sargent, 1983
Hall, Nellie Mae, Broken Bow, 1987
Halliwill, Edwin, Merna, 1988
Halouska, Helen, Arnold, 1991
Hammans, Earl, Kearney, 1984
Hansen, Jeanette Hansen & Cindy, Cozad, 2003
Hanson, William, Onconto, 1985
Hardin, George, Donald, & John, Arnold, 1990
Haumont Ranch Inc., Broken Bow, 1983
Haumont, Carol, Thomas Richter, Kearney, 2010
Haumont, Mrs. Frank J., Weissert, 1979
Haumont, Thomas E. Trust, Broken Bow, 1997
Hillside Ranch LLC, Lincoln, 2006
Hoblyn, Ray & Delpha, Ansley, 1995
Hoesel, Thomas & Jodi, Callaway, 2006
Holcomb, Carl, Broken Bow, 1984
Holm, Robert R. & Sarah Ann, Mason City, 1988
Horn, Dean, Broken Bow, 2004
Horn, Gilbert & Eugene Freeze, Johnson Lake, 2005
Hostick, Robert & Steven, Merna, 2000
Hunt, James & Anna, Anselmo, 1999
Hyde Family Farm LLC, Comstock, 2007
Jacquot, Cecil & Meta, Anselmo, 1984
Jenkins, Wayne & Frances, Onconto, 1978
Johns, Glenn V., Callaway, 1991
Johnson, James & Agnes, Callaway, 2000
Johnson, Robert Dale, Broken Bow, 1982
Johnston Klein, Carolyn M, Boise ID, 1996
Jorgenson, Neil, Callaway, 2010
Kastens, Keith & Marie, Merna, 2006
Kearney Land & Cattle Co., Sumner, 1991
Kenning, Kenneth L & Donna R., Gotenburg, 2006
Kinish, Lily Barbara, Gothenburg, 2001
Kleeb Jr., Harvey, Broken Bow, 2006
Kleeb, Lloyd, Broken Bow, 1986
Koubek, Peter Thomas, Arnold, 2006
Kramer, Earl D. & Barry, Litchfield, 1986
Lamb, Howard & Jo, Anselmo, 1983
Larson, Gertrude, Broken Bow, 1983
Leibhart, Jeanne, Merna, 2006
Lewis, Cecil, Arnold, 1984
Lindly, Lawrence, Anselmo, 1980
Lindly, Leonard V., Anselmo, 1995
Lindly, Lloyd, Anselmo, 1980
Loughran, Calvin & Thomas, Anselmo, 1976
Lucas, Robert, Arnold, 2000
Mackley, Francis, Gothenburg, 1987
Massie, Harold, Cozad, 1989
Mattox, Russell W., Broken Bow, 2012
McBride-Roswell, Joyce, Mary Sanders- Broken Bow, Sandy Jones-Anselmo, Ricki Fournier, Shirley Crouch, Roswell, 2005
McCandless, Ray E. & Verla, North Platte, 1990
McCullough, Donald W, Berwyn, 2006
McDowell, Letha Mae, Arnold, 1984
McGinn, Michael C. & Charles F, Anselmo, 1991
McGraw, Emily P., Broken Bow, 1980
Mendyk, Alberta, Arcadia, 2007
Miller, Karron, Scotia, 1998
Miller, Robert, Bill Miller, Ron Miller, Deb Golff, Broken Bow, 2004
Mills, Gordon & LouWana, Ansley, 1999
Mills, Paul & Donna, Ansley, 2005
Moroney, Vera M., Merna, 1985
Mougey, Burton, Kearney, 1991
Myers, Morris, Eddyville, 1977
Nelson, Cleo Trust, Oconto, 2005
Nelson, Mr. & Mrs. Larry, Callaway, 1979
Neth, Raymond J., Broken Bow, 1982
Nordin, Don, Gothenburg, 1984
Opal Cool Trust, Callaway, 1991
Ostergard, Jack & Mary, Gothenburg, 1997
Ostergard, Tonn, Lincoln, 1999
Ostrand, Benie Rex & Neil, Mason City, 1997
Ottun, Darrel, Sargent, 1983
Ottun, Gaylord & Anita, Sargent, 1985
Pandorf Land & Cattle Inc., Callaway, 1985
Patterson, Charles & Barbara, Callaway, 1999
Peltzer, Lyle & Karen Evans, James Eberle and Joe Eberle, Kristi Mason, Broken Bow, 2005
Petersen, John & Nelly Lou, Ansley, 2005
Peterson, Leo, Berwyn, 2004
Peterson, Mr. & Mrs. Ray T, Gothenburg, 2003
Pierce, Hazen V. & Emma, Onconto, 1982
Poland, Glenn & Douglas, Sargent, 1984
Popp, Phillip & Maxine, Broken Bow, 1997
Powell, Claude & Gail, Berwyn, 1981
R & M Acres, Gibbon, 2001
Read, Lilah Owen Life Estate, Oconto, 1997
Reinhard, Bertha Lee, Broken Bow, 1979
Rejda Bros. Inc., Burwell, 1984
Richard, Dennis & Mary, Oconto, 2006
Ridder, Marianne, Callaway, 2007
Roberts/ Knispel, Jean/ Shirlee, Kearney, 1997
Robertson, Norman & Judy, Tekamah, 2009
Sargent, Mr. & Mrs. Charles, Broken Bow, 1975
Sasek, Daniel O. & Marlene, Surgent, 2012
Sasek, Otto & Irene, Ord, 1995
Schmidt, Ron & Darrell, Anselmo, 1990
Schneider, Alex W. & Marjorie M, Callaway, 1986
Schneringer, Homer & Wanda, Callaway, 1982
Schultz, Sandie, Sargent, 2000
Scott, Esther, Broken Bow, 1984
Scott, Randy & Melea, Westerville, 1995
Scristmier, Alphne R., Broken Bow, 1984
Shoemaker, Ross, Mason City, 1991
Slagle, John H., Sargent, 2014
Slingsby, Kenneth N., Arcadia, 1984
Slingsby, Wayne & Debra, Arcadia, 2013
Smith, Robert & Margaret, Broken Bow, 2007
Snyder, Mary, Sigmund Kalinowski, Thaddeus Kalinowski, & Joan Beebe, Broken Bow, 2007

Kavanaugh, Wilma Jean, Dixon, 2011
Keller, Elman & Evelyn, Newcastle, 2003
Kinslow, Berlene (Echtenkamp) & Maydelle (Echtenkamp) Messerschmidt, Wayne, 1997
Kirchner, Edward W., Concord, 1983
Kneifl, Peter & Elizabeth, Newcastle, 1975
Koch, Martin C., Newcastle, 1974
Koch, Rodney, Omaha, 2009
Kramer, Ervin F. & Frances L. Family Trust, Laurel, 2012
Lockwood, Percy & Sophie, Allen, 1971
Loetscher, Eliene, Kendall & Cheryl Loetscher, Azala Jo & Thomas Lewis, Wayne, 2008
Lois Hall/Joyce Carpentier/ Lavonne Johnson, Echtenkamp Estate, Wayne, 1996
Lueth, Janice, Emerson, 2004
Martinson Farm, Maskell, 1965
McCabe, Michael D., Newcastle, 1975
McLaughlin, Mrs. Randal, Emerson, 1982
Meyer, Reuben & Alta, Wayne, 1989
Morgan, Vern, Dakota City, 1982
Moseman, Densel, Gerald Moseman, Steve Moseman, Emerson, 2003
Mrs. Clarence Nelson, Thyra Nelson Life Estate, Laurel, 1993
Muller, Vivian Estate, Wakefield, 2010
Munson, Conrey, Wakefield, 2015
Munson, Mr. & Mrs. Willard, Souix City, 1971
Nelson, Bonnie A., Wayne, 1991
Nelson, Raymond O., Newcastle, 1972
Nelson, Rick & Kellee, Ponca, 2013
O'Neill, Mrs. Sherlock & family, So. Souix City, 1979
Olson, Lucille& Arden & Ave, Concord, 1993
Osbahr, Myron Sr. or Julie, Allen, 1981
Packer, Kenneth W. & Ethel E., Wakefield, 1985
Panning, Alberta L., Kearney, 2013
Patten, C. Jean, Thomas Patten, Reno, 2009
Penlerick, Delwin T. & Lorna Y., Alliance, 2013
Reuter, Marvin & Mae, South Sioux City, 2013
Rewinkel's Incorporated, Wakefield, 2006
Rice, Derald N., Concord, 2003
Riter, Margaret, Richardson, 2007
Rockwell, Mr. & Mrs. William G., Ponca, 1973
Roeber, Merle/Donna, Wayne, 1996
Ruwe, Harlan, Wayne, 2009
Schrieber, Elmer & Bonnie, Wakefield, 1985
Schultz, Archie D., Allen, 1989
Sherlock, Thomas & Erma, Emerson, 1994
Sherman, Mr. & Mrs. Merl, Dixon, 1972
Stark, Alberta E., Ponca, 1984
Steecker, William E., Emerson, 2004
Stingley, Duane C. & Susan, & Randall Stingley, Laurel, 2010
Sullivan, Nicholas D & Patricia A, Ponca, 2015
Swanson, Ernest & Lyla Swanson, Wayne, 2005
Van Lent, William, Emerson, 1997
Von Minden, Harold & Darlene, Ponca, 1979
Warner, Mrs. Jimmie R., Allen, 1992
Weiershauser, Alma, Wayne, 1989
Wendt, Mrs. Oscar (Effie), Newcastle, 1982
Wendt, Robert & Susan, Newcastle, 1982
White, Ray & Florence, Ponca, 1971
Woodward, Iram I., Concord, 1983
Woodward, Thaine & Doris, Concord, 1989
Wriedt, Derwood H. and Ellen K Wriedt, Wakefield, 2003

Dodge County

Alberts, Donald, Scribner, 1980
Allan, Alfred & Marguerite A., North Bend, 1993
Anderson, Dennis, Hooper NE, 2001
Arneal, Maxine, North Bend, NE, 1999
Becker Family, Scribner, 1970
Beebe, Mr. & Mrs. Henry A., Fremont, 1956
Belardinelli, A.A. & Dorothy, Fremont, 1983
Bergquist, Henning Orville, Hooper, 1970
Bergquist, Keith & Mary, Hooper, 1990
Blomendahl, Herbert, Hooper, 2006
Bowman, Mr. & Mrs. Melvin, Hooper, 1969
Brainard, David & Elizabeth Brainard, Tekamah, 2005
Brugh, Mrs. C.S., Fremont, 1966
Bunn, Rex, Fremont, 1984
Busselman, Mr. & Mrs. Raymond H., Hooper, 1969
Carlson, Mrs. Lois, Hooper, 1991
Catherwood, James J. & Grace, Valley, 1975
Christiansen, Byron & Elaine, Fremont, 1987
Dahl, LaVonne, Scribner, 2016
Dames, Mr. & Mrs. Arlan, Scribner, 1969
Dames, Ronald, Scribner NE 68057, 2000
Diers Heller, Ida L., West Point, 1983
Diesterhaupt, Henry P., Dodge, 1982
Eason, Thomas K., North Bend, 1990
Ebel, Gene H. & Bernita M., Omaha, 1978
Egbers, Alfred, Hooper, 1985
Eikmeier, Anna Mae, Dodge NE 68633, 2000
Ellermeier, Evelyn, Scribner, 1993
Engelbrecht Family, Scribner, 1970
Erickson, Charles & Eunice, Hooper, 1993
Erickson, Douglas N., Hooper, 1977
Faltin, Geo G. & Agnes E., Dodge, 1967
Feala, Larry G., North Bend, 1996
Foy, Mr. & Mrs. Charles, Scribner, 1972
Francis, Irene, Fremont, 1984
Geisler, Lowell, Hooper, 1976
Groeteke, Mr. & Mrs. Arthur, Fremont, 1969
Grunke, Mr. & Mrs. Clinton, Scribner, 1969
Hamilton, David T., North Bend, 1992
Hansen, C. Edward, Fremont, 1967
Hansen, Mr. & Mrs. Murriel, Ames, 1978
Hansen, Mrs. Clara, Scribner, 1981
Harms, Dale G. & Elaine A., Fremont, 1981
Harms, Paul H., Fremont, 1982
Harnisch, Mr. & Mrs. Fritz, Scribner, 1980
Hasebroock, LaVerne, Fremont, 1985
Hasebroock, Meta, Scribner, 1996
Hasemann, Fred Jr. & Gladys, Scribner, 1975
Hasemann, Harvey, Richard &Helen Langemeier, Scribner, 1984
Hasemann, Vernon H., Scribner, 2004
Haun, Ellwood C., North Bend, 1970
Havecost, Leonard G., Hooper, 1968
Havekost, Arden & Georgean, Scribner, 2006
Havekost, John & Marilyn, Scribner, NE, 1998
Havekost, LaVerna, Scribner, 2004
Heese, Norman J., Scribner, 2012
Heimann, Marvin, Dodge, 1991
Helgenberger, Mr. & Mrs. Hilbert, Scribner, 1969
Helgenberger, Mr.& Mrs. Dean, Scribner, 1979
Heller, Howard J. E., Hooper, 1967
Heyne, Adolph G., Hooper, 1974
Hilbers Family, Hooper, 1971
Hilbers, Lois, Omaha, 2002
Hoegemeyer Family, Hooper, 1970
Hoge, Harold & Vera, Scribner, 1975
Howe-Larson, Myra, NorthBend, 1966
Johnson + Nancy Coetzee, Mark & Ken, North Bend, 2015
Johnson, Art W., Hooper, 1977
Johnson, E. Thome, Fremont, 1959
Johnson, Marguerite, Lavone & Robert, Scribner, 1997
Jorgensen, Ernest & Viola, Fremont, 1983
Keller, Elmer & Ethel & Lois Timm, Hooper, 1982
Keller, Norman L., Hooper, 1982
Klamt, Jeanne L., Scribner, 1974
Kluthe, Daniel J., Dodge, 1982
Knoell, Frank, Fremont, 1959
Knoell, Laurie C. & Mary Ellen, Fremont, 1977
Knoell, Theodore C., Fremont, 1967
Kreikemeier, Patrick, West Point, 1976
Kroenke, Lou, Dodge NE 68633, 2000
Kupfer, Mrs. Gertrude J., Fremont, 1982
Langewisch Cuda, Darlene J., Hooper, 1980
Larsen, Eugene L., Fremont, 1982
Larsen, Robert & Louella, Fremont NE 68025, 2000
Leadhoff, Rueben & Ella, Dodge, 1976
Lechtenberer, Gertrude, Fremont, 2011
Legband, Sandra, Scribner, 2004
Lennemann, Theodore, Scribner, 1983
Liston, William J. & LaVerle, Hooper, 1995
Low, Kenneth & Gladys, Scribner, 1974
Luebker, Leonard L./Anna May, Fremont, 1996
Luecking, Mr. & Mrs. Henry, Scribner, 1969
Lueninghoener, Irvin, Hooper, 1965
Malloy, Mrs. John/May Luther Heirs/c/o Jack Luther, Nickerson, 1989
Maresh, Mr. &Mrs. Joseph E., Dodge, 1971
Marxsen, Betty, Rogers NE 68659, 2001
Mehaffey, Bob & Shirley, North Bend, 2005
Meyer-Moeller, Merlin E. & Joan, Hooper, 1968

Meyer, Harry, Scribner, 2002
Meyer, Layne H., Hooper, 1986
Meyer, Mr. & Mrs. Henry M., Scribner, 1968
Milligan, Harland S., Hooper, 1972
Mintken, Gary, Cynthia Schaulis, Richard Mintken, Donna Reeson, Judy Zimmerman, Omaha, 2007
Mitties, Mrs. Anna Mary, Fremont, 1976
Moeller, Donald W., Hooper, 1996
Moeller, Mrs. Frances, Scribner, 1969
Mohr, Gilbert, Scribner, 1985
Mohr, John F., Scribner, 1976
Muller, Duane, Scribner, 2005
Mulliken Inc., Alan, Nellie, Jerry, Bestsy, Jack & Pamela Mulliken, Nickerson, 2008
Munderloh, Shawn, Michelle, & Karleen, Scribner, 1992
Nelson, Kenneth & Doris, Hooper, 1985
Nelson, Leslie J., Fremont, 1967
O'Hara, Mildred, North Bend, 1997
Olson, Martin, Mrs. Hannah E. Olson, Hooper, 1970
Olson, Mr. & Mrs. Martin S., Hooper, 1971
Osterloh, Elmer, Hooper, 1968
Otteman, Edith, Leslie Parsons, Hooper, 2007
Paasch, Norman, Scribner, 1984
Pacas, Jane L., North Bend, 1993
Palm, James & Bennie Sue, Scribner, 1983
Panning, Irvin, Hooper, 1983
Pathfinder Co., Alycemae Kastle, Fremont, 1982
Pedersen, Duane, Ames, 2002
Peters, William E. & Myrtle C., Scribner, 1971
Peters, William H., Hooper, 1976
Pieper, Mr.& Mrs. Raymond, Dodge, 1979
Popken, Mr. & Mrs. Norman, Hooper, 1969
Popken, Stephen, Scribner NE 68057, 2001
Poppe, Donald, Scribner, 1980
Powers, Greg, Lorena D. Powers, North Bend, 1981
Pribnow, Mr. & Mrs. Otto, Scribner, 1979
Quick, V. Margaret, Schuyler, 1974
Rastded Low, LaVaughn, Scribner, 1970
Rebbe, Miss Katherine W., Fremont, 1973
Rebbe, Mr. & Mrs. Carson A., Hooper, 1972
Rebbe, Mr. & Mrs. Louis F., Fremont, 1972
Richards, Mrs. Proctor, Quin H. Scott Family, Schuyler, 1968
Ries, Mr. & Mrs. Nicholas, Scribner, 1969
Ristau, Nadine, Fremont, 2007
Romberg Family c/o Mrs. O.F. Hoye, Grant, IA, 1969
Roubal, Mr. and Mrs. Phillip G., North Bend, 2002
Ruwe, Edward Trust, Lincoln, 2002
Ruzicka, Gene A. and Joyce E. Ruzicka, North Bend, 2004
Ruzicka, Marie, Emil R. Ruzicka, Gene A. Ruzicka, Billy D. Ruzicka & Larry L. Ruzicka, North Bend, 2004
Sampaon, Luella, Hooper, 1969
Sass Family c/o Stanley Sass, Fremont, 1969
Schafer, Howard J., Scribner, 1983
Schueth, Mrs. Henrietta, Scribner, 1978
Sloss-Morgan, Ruth & Janet, North Bend, 1959
Smeal Family, Snyder, 1971
Snover, Mark, Stuart, 2016
Snover, Terry, North Bend, 2016
Soukup, John, Fremont, 2011
Spalding, Denny, North Bend, 1983
Spath, Mrs. Phyllis, Scribner, 1983
Sprick, Steven & Ann, Hooper, 2006
Stecher, Joe & Denny, Hooper, 1979
Stockfleth, Ronald, Hooper, 2003
Stoltenberg, Kenneth, Schuyler, 1978
Strand, Loell & Shirley, Hooper, 2011
Stumpe, Dennis, Scribner, 2001
Tank, Gerald, North Bend, 2001
Tank, Mr. & Mrs. Richard E., Fremont, 1980
Thernes, Louis, Hooper, 1990
Thomas, Mrs. Lydia, Omaha, 1971
Thomsen, Elmer & Mona, Fremont, 1971
Thomsen, Josephine C., Fremont, 1982
Uehling, Elmer & Mildred, Hooper, 1968
Uehling, Homer, Oakland, 1983
Voegler, Mr. & Mrs. Vern W., Snyder, 1972
Von Seggern, Mr. & Mrs. Clinton, Scribner, 1969
Von Seggern, Shirley, Hooper, 2001
Wagner, Charles R & Betty J, Cindy Wagner, Fremont, 2009
Wagner, William, Hooper, 1980
Walla, Elaine, Marion Kern, Blue Bell, PA, 1994
Walla, Leona, Fremont, 2007
Walter, Joe A., Dodge, 1979
Wegner, Mr. & Mrs. Willis C., Scribner, 1971
Westphalen, Carrol, Scribner, 1981
Windhusen, Mr. & Mrs. Theo, Hooper, 1969
Wisnieski, Diana, Dodge, 2011
Witte, Mrs. Arland, Scribner, 1988
Wobken, Doris, Scribner, 1998
Wolcott, George W., Hooper, 1967

Douglas County

Andresen, Sophus & Hazel, Omaha, 1973
Andresen, William J, Elkhorn, 1974
Backhaus, Herman & Marguerite, Omaha, 1983
Backhaus, Murray D, Omaha, 1983
Backhuus, Mrs. Peter C (Pearl) II & Mrs. Frank W. (Olive) Backuus, Bennington, 1984
Batten, Tom J., Valley, 1987
Bunz, Peter Jr. Estate, Bennington, 1979
Carlson, Robert & Amy, Dennis Carlson, Valley, 2011
Clark, James L., Waterloo, 1964
Dickinson, Mr. & Mrs. James, Omaha, 1983
Dillon, Sr., Jay, Omaha, 1979
Donahoo, Dwayne & Carlene, Elkhorn, 1992
Dyer, John E., Waterloo, 1991
Fitch, E. Quay, Omaha, 1971
Gottsch, Otto, Bennington, 1967
Grau, Emma Lou/ Theodore E. Grau Jr., Bennington, 1992
Hadan, Mrs. August, Omaha, 1980
Heintze, Frank B., Omaha, 1965
Heise, Mrs. Emma, & Mr. & Mrs. Harold, Omaha, 1972
Heitmann, Frances E, Omaha, 1996
J.C. Robinson Seed Company, Waterloo, 2003
Jacobsen Farms, Inc., Omaha, 2013
Klinker, Mrs. John, Omaha, 1977
Kowal, Dawn Christensen, LaVista, 2016
Kruse, Fred C. & Martha E. (Heuhaus), Omaha, 1978
Kruse, Hugo & Helen, Omaha, 1972
Lamprecht, Betty, Donald Glasshoff, LaVada Armbrust Ferguson, Waterloo, 1992
Logemann, Dale W., Omaha, 1978
Logemann, Russell, Bennington, 1978
Magee-Knapp, Alice Magee &/ Or Wayland G. Magee, Jr. Trustee, Bennington, 1982
McCleneghan, Sam, Valley, 1973
Melotz, Paul & Julia, Omaha, 1972
Mohr Family, Glen Mohr, Lyle Mohr, Dale Mohr, and Lois Musel, Omaha, 2006
Mohrview Farm, Inc., Omaha, 2012
Noyes, Clark B, Elkhorn, 1967
Ohrt (or Mrs. Tim), Ella R., Bennington, 1982
Paasch, Henry G., Omaha, 1978
Parisot, Ralph & Josephine, Omaha, 1982
Parson, Wendell, Valley, 1985
Phalen, William E, Omaha, 1995
Plambeck, Marvin, Millard, 1968
Plambeck, Mrs. Joe E. & Family, Omaha, 1984
Ritner, George & John, Douglas, 2007
Safford, Helen M., Waterloo, 1990
Smith, Mr. & Mrs. William H, Irvington, 1959
Stolley, Mr. & Mrs. William, Omaha, 1964
Stoltenberg-Fowles, Arline, Lincoln, 1978
Stoltenberg, Robert L. & Ruth Stoltenberg Bock, Omaha, 1978
Stoltenberg, William A., Omaha, 1978
Warren, Charles T., Waterloo, 1965
Warren, Frank, Elkhorn, 1966
Westphal, Edward A. & Wayne E., Omaha, 1983
Winterburn, Mrs. Fred, Elkhorn, 1978
Wyatt, Linda, Jane d FLEMING TRUST, Tullahoma, 2006
Zeis Gitt, Kathleen J., Valley, 1986

Duel County

Akeson, Berton & Donna, Chappell, 2001
Bailey, Clinton & Donna, Big Springs, 2009
Brownell, Gary & Joan, Chappell, 1994
Carlson, Gilbert & Doreen, Chappell, 1987
Kalb, Leo, Julesburg CO, 1994
Leef, Alvin, Julesburg CO, 1987
Ness, Kristi, Bob & Norma Wright, Julesburg, CO, 2011
Neumann, Robert & DeLoris, Chappell, 1994
Pidgeon, Harold, Julesburg CO, 1992
Smith, Haven & Virginia, Chappell, 1986
Timm, Robert & Wilma, Chappell, 1987
Zimmerman, Ralph, Big Springs, 1985

Dundy County

Balderson, Ray, Benkelman, 1986
Bernard, Mabel, Enders, 1987
Brynoff, Nels & Naomi, Benkelman, 1987
Crosson Majors, Bonnie Jean, Wauneta, 1985
Crow, Budd & Eunice, Wauneta, 1995
Dicke, Bill, Connie Dicke Linscott, Lincoln, 2008
Fish, Freda & Glenn, Benkelman, 1986
Foster Brots. Inc., Benkelman, 1989

Frasier, Kenneth & Elaine, Max, 1999
Gardner, Mr. & Mrs. Richard, Max, 1987
Hamm, Lawrence, Benkelman, 1992
Hansen, Thomas J. & Mabel J., Benkelman, 1986
Harford, Don & Virgina, Haigler, 2007
Hofer, David/ Sally, Benkelman, 1996
Jessee, Doran & Dorothy, Benkelman, 2000
Kitt, Donnaline, Wauneta, 1997
Lazy T V Ranch, Inc., Benkelman, 1986
Lemon, Glen & Shirley, Max, 1987
Lutz, Carl & Jean, Parks, 2011
Majors, Donald L & Bonnie Jean Crosson, Wauneta, 2015
Melba & Guy Fries, Benkelman, 2005
Morehead, Marvin, Shubert, 1988
Osborn, Jim, Benkelman, 2016
Roesener, Ruby, Imperial, 2008
Schrader, E. Dale & H. Beth, Benkelman, 1985
Schrader, Howard J. & Fairy P, Benkelman, 1986
Schwartz, Edwin C. & Betty J., North Platte, 2004
Shillington/ Manning, Roger/ Patricia, Max, 1989
Stock Farms Inc., Wauneta, 1986
Stroup, Boyd F., Benkelman, 1990
Tecker Ranch Inc., Parks, 2004
Thomas, Robert Paul & Norma May, Stratton, 2012
Tracy, Frank V., Max, 1990
Tracy, Frank, Ardith Tracy and Irene L. Gooder, Max, 2012
Warning, Boyce, Benkelman, 2000
Williams, Royce & Lula, Haigler, 2007

Fillmore County

Albro, Helen First Trust, Carol Sieber Trustee, Geneva, 2009
Allgood, Karen, Strang, 1994
Anderson, Rosa May, Geneva, 1989
Badger Halsey, Mary, Fairmont, 1971
Barnell, Mrs. Merrill, Sutton, 2001
Baumann, William H. Estate, Sutton, 1979
Becker, Cyril & Margaret, Exeter, 1985
Becker, Mark and Meg, Exeter, 2016
Becker, Jr., William, Exeter, 2004
Bernasek/ Suda, Mrs. Matilda/ Mrs. Anna, Milligan, 1981
Biba Marek, Leo & Beverly, Exeter, 1987
Biba, Anton, Exeter, 1982
Biegert, Virgil, Shickley, 1993
Biester, Alvin, Geneva, 1990
Black, Ida & Ruth, Fairmont, 1973
Black, Ruth I, Fairmont, 1978
Bode, Bill and Virginia, Exeter, 2016
Bohlen, Mrs. Will, Geneva, 1992
Bornschlegl, Weldon J., Ohiowa, 2012
Bridgman, Mr. & Mrs. Federal, Fairmont, 1976
Brunkow, Claremont & Angeline, Tobias, 1984
Bures, Erwin, Exeter, 1981
Bures, Mildred, Geneva, 1985
Burns, Evelyn Medlar, Falls City, 1995
Burt, John, Geneva, 1974
Burt, Keith, Grafton, 1974
Capek, Arnold, Milligan, 2007
Capek, Raymond & Janet, Milligan, 1985
Carlson, Gary, Shickley, 1991
Christiancy, Clair & Marilyn, Ohiowa, 1979
Christiancy, Rodger, Geneva, 1983
Dawson, Lloyd & Magdalene, Fairmont, 1977
Diederich, Philip Joseph, Geneva, 1978
Dinneen, John J. & Judith, Exeter, 1978
Dunse, Herman & Emma, Ohiowa, 1988
Eberhardt Jorgenson, Gloria, Red Lodge MT, 1995
Edzards, Larry & Elizabeth Edzards, Bruning, 2005
Eich, William E, Geneva, 1978
Elznic, Evelyn, Sheila Crays, Tobias, 2007
Fisher, Sam & Doris, Shickley, 1978
Fitzgerald, John, Grafton, 1980
Flory, Donald L., Shickley, 1976
Gewacke, Clyde, Ohiowa, 1986
Gewacke, John, Geneva, 1986
Green, Melva, Aurora, 2000
Guthrie, Kenneth/ Mrs. Aaron, Exeter, 1989
Hanson, Monty & Dianne, Exeter, 1991
Hassler, Paul, Lincoln, 1995
Heckman/ Valda, Norman/ Loris, Grafton, 1991
Heinrichs, Arnold & Mary, Bruning, 2007
Hendrickson, Lillie Estate, Shickley, 1983
Hendrickson, Robert, Shickley, 2011
Hermsen/ Flickinger, Karen Ann/ Joan Helen, Omaha, 1980
Hoarty, Gene, Fairmont, 1983
Hoarty, Richard, Grafton, 1983
Hofmann, William & Pamela, Sutton, 2000
Holsten, Bob/Mary, Carmel IN, 1996
Horne, James, Exeter, 1972
Horner, Donald L., Davenport, 1975
Houchin, Wayne Jr. & Joy, Strang, 2014
Houck, Scott & Jody, Strang, 2014
Hromadka, Elsie & Fred, Friend, 1993
Hromadka, John, Milligan, 1983
Janing, Richard & Darlene, Geneva, 2004
Jansen, Ralph and Barbara, Exeter, 2016
Johnson, Darrel and Deb, Shickley, 2016
Kamler, Linus & Frances, Shickley, 1988
Kamler, Mike, Geneva, 2016
Kassik, Ralph & Brenda, Milligan, 1992
Kassik, Richard & Judy, Milligan, 1992
Keenan, Francis & Jerry, Grafton, 1980
Kimbrough, Dennis & Viola, Geneva, 1984
Klink, Amelia/ Doris, Fremont, 1976
Koahler, Earl Benjamin, Milligan, 1993
Koch, Mr. & Mrs. Kenneth, Shickley, 1980
Kohler, Norma, Geneva, 1976
Kohler, Robert and Karen, Milligan, 2008
Korbelik, Joseph & Lillian, Ohiowa, 1993
Kotas, Jim & Don, Milligan, 2000
Kotas, Joseph, Milligan, 1985
Kotas, Kenry, Milligan, 1985
Kottas, Frederick & Angeline, Milligan, 1991
Krause, Clarence & Joyce, Shickley, 1972
Krejdl, Anton, Exeter, 1989
Krupicka, Richard, Milligan, 1988
Kubicek, Mike & Helen, Milligan, 1979
Kucera, Mr. & Mrs. Charles, Exeter, 1984
Kuska Farms Inc, Paul Kuska, Lois Kuska, Carol Leipzig, Fairmont, 2007
Kuska, Delmar, Exeter, 2003
Kuska, Harry, Emma, & Katherina, Exeter, 1982
Landgren, Norman, Lincoln, 2011
Lang, Norma Jean, Geneva, 1981
Lefever, Ken & Sandra, Florence Lefever, Nathan & Chelsey Lefever, Geneva, 2007
Lovegrove, Jean, Fairmont, 2011
Luksik, Mr. & Mrs. Ernest, Milligan, 1977
Luzum, Dwayne & Cherly, Milligan, 1993
Marble, Ronald, Norton, 2007
Marek, Theodore, Geneva, 1988
Maresh, Ellsworth & Bonnie, Omaha, 1992
Maresh, Richard, Geneva, 1992
Maresh, Ruth, Geneva, 1992
Martin, Teresa, Jason & Michelle Tatro, Geneva, 2011
Maser, Helen, Sutton, 1989
Matejka, Alfred, Milligan, 1984
Matejka, Opal, Strang, 2014
Matejka, Wayne, Strang, 1991
McHale, Max & Joann, Lincoln, 1989
Medlar Trust, Jacqueline D., Columbus, 1995
Meistrell, Lillian, Fremont, 1998
Mengshol, Marjorie, Lincoln, 1984
Menke, Bernard/ Dorothy, Las Vegas NV, 1996
Merryman, Derald & Douglas, Geneva, 1994
Messman, Mrs. Lowell, Geneva, 1995
Meyer, Dorothy, Ohiowa, 1992
Most, Morley & Darlene, Tobias, 1985
Moul, Mrs. Ninette, Fairmont, 1976
Mussman, Dallas, Ohiowa, 2002
Mussman, Harry, Emma, & Katherina, Tobias, 1984
Mussman, Kermit & Alice, Ohiowa, 1985
Nedza, Mrs. Mabel, Ohiowa, 1975
Nelson-Anderson, John Nelson & Christine Nelson Andersen, Omaha, 2005
Nelson, Erwin, Ong, 1985
Nichols, Florence, Fairmont, 1978
Nichols, W. Robert, Geneva, 1989
Nun, George, Ohiowa, 1977
Nunns, Edna Life Estate, Geneva, 1990
O'Brien, Michael R, Grafton, 1981
Ogden, Norman & Gwen, Fairmont, 2000
Olson, Melvin, North Platte, 1990
Ozenbaugh, Ruth, Ohiowa, 1980
Pangle Burke, Berniece, Omaha, 1971
Pangle Cook, Bethene, Omaha, 1971
Perkins, Mrs. Kenneth, Fairmont, 1986
Peter, Marvin H., Sutton, 2012
PO Box 65, Ronnie Hamouz, Milligan, 2005
Quist, John & Marjorie, Omaha, 1995
Rath, Michael H., Sutton, 2014
Rath, Thomas, Grafton, 2001
Reinsch, Dean, Bruning, 2010
Reinsch, Merle, Shickley, 1976
Rhoda, Fred W., York, 1978
Rippe, Willard, Ohiowa, 1979
Rolfes Farm, Sutton, 1982
Rothrock, Quintin, Davenport, 1977
Row Trust, Paul D. & JoAnn W., Bruning, 2015
Satlzman, Gary, Shickley, 1998
Satmann, Fredrick & Kathryn, Ohiowa, 1985
Schaldecker, Fred, Geneva, 1975

Schelbitzki, James A, Geneva, 1976
Schelbitzki, Lloyd & Ione, Geneva, 2011
Schelkopf, Robert, Geneva, 2007
Schwab, Bert A, Exeter, 1972
Serr, Marie, Sutton, 1984
Sheridan, Roger, Sutton, 1992
Sieber, Richard J., Ohiowa, 1976
Sladek, Joseph, Exeter, 1989
Slezak, Ruth, Milligan, 1988
Sluka, August Heirs, Milligan, 1983
Sluka, Donald, Milligan, 1999
Sluka, Irvin, Geneva, 1990
Songster, Carl E., Exeter, 1977
Springer, Hobart, Exeter, 1999
Stephenson Leefers, Barbara, Lowell Stephenson, James Stephenson, Fairmont, 2011
Stevens, Harold Everett, Grafton, 1978
Stone, Lindsey Dawn, Hastings, 2014
Stuckey Farms Inc., Grafton, 1984
Svec, Bernard, Milligan, 1983
Svec, Leslie Svec and Louis Dale Svec, Milligan, 2005
Swartz, George, Grand Island, 1983
Taylor Family Farm, Geneva, 2014
Taylor, John, Geneva, 1991
Thomas, Georgia, and Sharolyn Frey, Sandra Miller and Randra Hansen, Lincoln, 2016
Tobiassen, Emma, Geneva, 1990
Tobiassen, John, Geneva, 1979
Trauger, Charley, Exeter, 1979
Trautman, Lawrence, Sutton, 1983
Uldrich, Myron, Fairmont, 2000
Uldrich, Ronald & Mary, Milligan, 2007
Venell Zetterman, Verlie & Maurine, Shickley, 1984
Votipka, James, Geneva, 2015
Votipka/ White, Anna/ James, Lincoln, 1995
Wagner, Richard & Norma, Geneva, 2000
Watmore, William & Ramona, Geneva, 1997
Wernimont, Mrs. Mabel A, Geneva, 1979
Wulf, Bessie, Ohiowa, 2001
Wythers, Dave & Marcy, Fairmont, 1987
Yates, Norman, Geneva, 1982
Zelenka, John, Friend, 1985

Franklin County

Ackerman Trust, Raymond, Kearney, 2009
Adam, Robert & Delores, Wilcox, 1991
Albers, Ralph & Randall, Hildreth, 1989
Bauer, Carolyn, Hildreth, 2015
Betz, Grace, Campbell, 1984
Bislow, Peggy & Harry, Franklin, 1989
Brunke, Henry, Campbell, 1984
Bunger, Orval O., Upland, 2008
Bunger, William & Margaret, Upland, 1993
Bunger, William, Upland, 1991
Bush, Harold, Naponee, 1978
Choquette, Phil & Angela, Campbell, 1991
Copley-Hossler, JoAnn and Ruth Copley Hostetler, Riverton, 2004
Copley, Paul, Riverton, 2013
Dallman, Leola, Columbus, 2002
DeJonge, Gary & Patti, Hildreth, 1983
Dow Bush, Clara, Naponee, 1978
E.J.Bunger Corp., Franklin, 1993
Etta, June, Campbell, 1993
Fouts Potter, Lois, Wilcox, 2015
Frecks, Alfred, Hildreth, 1993
Freeman, Duane, Hastings, 1993
Frerichs, Clyde, Franklin, 1998
Fries, Vernon & Joan, Upland, 1994
Fuerst, Dick & Bonnie, Hildreth, 2006
Garrett, George Walter, Bloomington, 1978
Gramke, Merrill, Upland, 2011
Greening, Howard Charles, Alma, 2010
Greisen, Robert and Bonnie, Minden, 2013
Groothuis, Sophia, Hildreth, 2000
Hendricks, Mr. & Mrs. Clarence O., Upland, 1979
Hendricks, Vaughn, Hildreth, 1992
Hinrichs, Brent, Hildreth, 2002
Hinrichs, Elmer H, Hildreth, 2006
Hinrichs, Marvin, Hildreth, 2010
Jacobsen Nielsen, Paula Nikki, Franklin, 2013
Jameson, Bernard, Campbell, 1987
Jameson, Burdette and Gwen, Upland, 2016
Jenson, John, and Mary Lou Wissmann and Debra Smidt, Minden, 2016
Johnson, John & Euella, Upland, 2006
Johnson, Lyle & Delores, Bloomington, 1985
Johnson, Lyle & Delores, Bloomington, 2004
Johnson, Marjorie, Juniata, 1998
Jones, Norma & Orville, Fremont, 1993
Knapp, LaDonna G. & Shirley Ann (Knapp) Nelson, Hildreth, 2003
Koch, Grady, Upland, 2012
L'Heureux, Douglas, Campbell, 1991
Lutkemeier, Ryan & Mary, Campell, 2015
Marcelle, Elizabeth, Campbell, 1996
Marcum, Jerry & Esther, Franklin, 2011
Maska, Sophia, Franklin, 1996
Melton, Clifford & Betty, Naponee, 2007
Melton, Don, Naponee, 1978
Mengers, Anna, Naponee, 1985
Meyer, Gerhart & DeLores, Campbell, 1993
Meyers, Kenneth & Gary, Wilcox, 1993
Mietzner, Dennis, Hildreth, 2004
Miller, LaVern, Hildreth, 1982
Nebraska Land Trust, Howard C Smith, Gulfport, 2010
Neubauer, Stacy, Tobiassen Farms, LLC, Alma, 2015
Osterbuhr, Mrs. Lawrence, Upland, 1983
Overleese, Robert & Darlene, Riverton, 1987
Parker, Marvin & Connie, Campbell, 1989
Price/ Croley, H./ Nona, Naponee, 1978
Quadhamer, Edwin, Hildreth, 1980
Quadhamer, Florence, Hildreth, 1980
Quadhamer, Gerhard & Arnold, Hildreth, 1991
Quinn, Mr. & Mrs. Charles, Naponee, 1972
Reiber, Ivrin & Dorothy, Campbell, 2013
Rhodig, Phyllis Seybold & Eldon, Omaha, 2014
Rutt, Mrs. Margaret, Hastings, 1986
Saathoft, Bill & Jim, Naponee, 2006
Scheibel, Norman, Campbell, 1985
Schepler, Mrs. Ernest, Wilcox, 2003
Schepler, Richard, Wilcox, 1999
Schnuerle, Neil, Bloomington, 2001
Schoone, Ivan Family, Upland, 2005
Schubert, Clayton & Grace, Wilcox, 1983
Sindt Farms, Inc., Naponee, 2004
Sindt, Paul/ Irene, Lincoln, 1996
Smith, Richard, Upland, 1999
Stecker, James, Grand Island, 2002
Stecker, James, Grand Island, 2002
Steinkruger, Bruce, Upland, 2016
Steinkruger, Ethel, Roger Steinkruger, Rogene McPherson, Rodney Steinkruger, Rayma Delaney, Franklin, 2003
Stover, Jr., Lawrence, Naponee, 1978
Taylor Estate, Bloomington, 1978
Tobiassen, LaVon J, Hildreth, 2015
Tupper, Clinton, Mesa AZ, 1994
Valley View Farms, Franklin, 2002
Verbeck, Mrs. Robert, Minden, 1983
Warriner, Eugene, Franklin, 1978
Wentworth, L.N. & Gaylene, Franklin, 1992
Wilken, Herman, Hildreth, 1981
Wilken, Rena, Hildreth, 2003
Wischmeier, Bernie & Ellen Trust, Wilcox, 1995
Wohleb, Lawrence & Viola, Naponee, 1979
Wohleb, Vera, Hastings, 2015
Wohleb, Viola, Naponee, 1980
Wolf, Mr. & Mrs. Olen, Franklin, 1972
Ziebarth, Wayne, Wilcox, 2000
Zuellner, Joel, Campbell, 2006
Zuellner, Tobias & Edna Trust, Franklin, 1993
Zulauf, Russel, Wilcox, 1995

Frontier County

Abbors, Stephe & Carlene, Walnut Creek, 2008
Allen, Herbert, Cambridge, 1988
Andrijeski, George & Dorothy, McCook, 1987
Balcom, Clara & Shirley, Cambridge, 1986
Banzhaf, Ronald, Cambridge, 1991
Barrett, H. Truman & Ruth, Eustis, 1984
Bartruff Wahl Anderson, Lucile, Cozad, 1996
Bartruff, Max & Suzanne, Eustis, 1998
Berke, Glenn & Elizabeth, Eustis, 2006
Bortner, Heath J. & Jessica L., McCook, 2010
Brockmeier, Bruce, Eustis, 1988
Brown, Nick & Amy, Curtis, 2014
Bryant, Bill & Norma, Curtis, 2010
Buehner, Robert, Eustis, 2016
Calkin, Everett & Ila, North Platte, 2013
Cole, Harvey, Curtis, 2003
Crawford, Harley, Stockville, 1988
Dueland, Duane & Joanne, McCook, 1990
Elley, William & Doris Jean, Farman, 1995
Elson, Thomas & Merna, Curtis, 1993
Erickson, Carla Hill & Karen, Valley Mills, TX, 2004
Farr, Velma, Eustis, 1998
Flying E Farms, Inc, Rex & Dona Ealy, Moorefield, 2011
Fritsche, Paul, McCook, 2005
Gilliland, E. Howard, Curtis, 1994
Grabenstein, Gene & Lyle Grabenstein, Eustis, 2005
Hansen, Elwin & Dorothy, McCook, 1998
Hicks/ Edson, H. Gerald/ Mildred, North Platte, 1985
Hueftle, Mr. & Mrs. Edward, Eustis, 1983
Hueftle, Vernon & Janet, Eustis, 2010
Hueftle/ Rupe, Dorsie/ Neil, Eustis, 1995

Jack, Willard & Eva, Eustis, 1984
Johnson, Raymond D. & Beverly E., North Platte, 2004
Jurjens, Verl & Monika, Eustis, 2009
Kester, Edward L. & Marilyn J., Cambridge, 2004
Klinkebiel, Paul & Marilyn, Cambridge, 1986
Klooz, Marvin, McCook, 1993
Koch, Herb & Rosemary, Eustis, 1998
Larsen, Verna Dee, Lorne Wilson, Owen Wilson, Moorefield, 2007
Lashley, Mary, McCook, 1996
Lehmann, Sherman & Nora, Eustis, 1985
Max, Alva & Carol, Cambridge, 1999
MBJL, Limited, Arapahoe, 2008
Messersmith, Keith & Leora, Farman, 2007
Minnick, W. Tiffany Trust, Cambridge, 1985
Mousel, George, Cambridge, 1984
Mousel, Mrs. R. D., Cambridge, 1984
Neiman Burtch Inc., Curtis, 1985
Nelson, H.J., Curtis, 1988
Newcomb, W.D., Cambridge, 1985
Owens, Larry & Marcia, Curtis, 2011
Perks, Marlin S., Stockville, 1993
Pfeil, Robert & Ruth, Alma, 1998
Proud, Ruby, Cambridge, 1984
Puls, Gilbert & Ethel, Eustis, 1982
Rieker, Raymond & Virginia, Eustis, 1992
Ruppeert, C. Otis, Bonnie Ruppert, McCook, 2007
Schmeeckle, Roy & Mary Catherine, Eustis, 1986
Schroeder, Carroll & Althea, Eustis, 1993
Shelley, W.M & Margaret L., Stockville, 1976
Shifflet, Mr. & Mrs. Burnell, Cambridge, 2002
Smith, Kenneth & Janet, Eustis, 1986
Snider, W./O. Elizabeth, Maywood, 1996
Stencel, Everett & Sandra, Curtis, 1982
Swanson, C.K., McCook, 1994
Teel Snyder, Delma, McCook, 2006
Tiller, Lois & George, McCook, 2005
Tiller, Robert & Vonnie Family Trust, McCook, 2014
Tillotson Jr, LeRoy, Farman, 1985
Trista Jesch, Steve Hall, Brad Hall, Curtis, 2014
Welch, Leland & Teresa, Curtis, 1999
Wilson, Owen & Deanna, Moorfield, 2006
Wilson, Vernon, Moorfield, 2006
Young, Alice M., Fairfax VA, 1995

Furnas County

Arganbright, Sharon, Lincoln, 2001
Babcock Cass, Alma Rose, Beaver City, 1984
Baily, Robert, Oxford, 1993
Bates, LaVerne, Beaver City, 1993
Breinig, R. Scott, Steve and Wendy, Holdrege, 2010
Brown, Eula, Beaver City, 2016
Brown, Mr. & Mrs. Joseph, Cambridge, 1984
Coffey/ Patsios, Karen/ Arlene, Beaver City, 1981
Courtright, Clarence & Mary, Beaver City, 2005
Ellis, Bernard, Alma, 1985
Fewell, Frank, Edison, 1985
Franklin, Helen, Beaver City, 1983
Fults, Dale & Twila, Beaver City, 1999
Garey, Douglas & Betty, Beaver City, 1978
Garey, Douglas & Betty, Beaver City, 2014
Garey, Douglas & Betty, Beaver City, 2014
Gibson Baily, Francis, Oxford, 2014
Gibson, Verna Clarine, Oxford, 1980
Haley, J. Leo & Elizabeth, Holbrook, 1979
Hardin Farms Inc., Beaver City, 1993
Harper, James & Julietta, Beaver City, 1987
Haussler Fau, Jeanene, Holbrook, 1992
Haussler, Mr. & Mrs. Robert, Arapahoe, 1980
Haussler, Terry & Vicki, Arapahoe, 1982
Hays, William & JoAnn, Edison, 1998
Heinen, Mr. & Mrs. Norbert, Arapahoe, 1987
Hockemeier, Wayne Edward, Arapahoe, 1996
Hunt, Deniis & Cynthia, Beaver City, 1985
Huxoll, Herman, Arapahoe, 1979
Inman, Harold C., Stamford, 1993
Johnson, Ervin, Stamford, 1984
Johnson, Roland, Arapahoe, 1986
Kinder, Norman & Arliss, Holbrook, 1993
Knappert, Gilbert/ Irene, Hendley, 1996
Knuth, Raymond & Robert, Oxford, 1983
Leising, Gilbert & Agnes, Arapahoe, 1997
Marble, Verlyn, Vivian Schluntz, Rick Schluntz, Oxford, 2011
Mart, Norval, Hendley, 1980
Martin, Mrs. George, Beaver City, 1985
Miller, R.V., Wisonville, 1975
Mock, Henry & Martha, Arapahoe, 1976
Niemeier, Walter, Arapahoe, 1983
Ordelheide, Dorothy, Oakdale, CA, 1999
Rankin, George D., Oxford, 1980
Riepe, Leon & Margaret, Arapahoe, 1982
Robinson, Don & Pat, Arapahoe, 1993
Robinson, Don, Arapahoe, 2007
Schoen, Franklin, Oxford, 1993
Schoen, Kenneth Schoen, Oxford, 2004
Shafer, Don & Karol, Oxford, 1998
Smith, Janet, Milton WI, 2001
Smith, Mary Joyce, Arapahoe, 1977
Stratton, James, Beaver City, 1979
Sturtevant, Irene, Holdrege, 1979
Swanson, Mr. & Mrs. William, Cambridge, 1973
Taylor, Cecil & Helen (Perkins), Kearney, 2004
Taylor, Mike & Jodi, Stamford, 2004
Taylor, Ruby, Beaver City, 1979
tenBensel, Gary, Cambridge, 2007
Traupe, Kenneth & Ruth, Holbrook, 1983
Traupe, Leo & Helen, Cambridge, 1983
Volgamore, Ervin and Shirley, Beaver City, 2008
Warner, Jerry D. & Arvida L., Beaver City, 2012
Wilmot, Merlin, Wisonville, 1974
Wineland, Mary, Denver, CO, 1998
Wolfe, Avis & Roscoe, Wilsonville, 1988
Woodruff, Donald, Beaver City, 2005
Yeager, Leon, Hendley, 1990
Youngquist, Lynn & Sharon, Beaver City, 1991

Gage County

Acton, Frederick, Holmesville, 1999
Acton, Ted, Liberty, 1991
Adam Trust, Lorna, Odell, 2011
Aden, William, Firth, 1974
Alpers, Mrs. Emma, Clatonia, 1982
Baehr, Elizabeth, Beatrice, 1979
Balderson, Lloyd & Dorothy, Cortland, 1977
Barnhill, Kenneth & JoAnn, Wymore, 2004
Beck, Galen/ Gwen, Wilber, 1989
Beerenstrauch, Beryl & Ervin, DeWitt, 1979
Bergemeier, Maurice & Delma, DeWitt, 1988
Bergmeier, Dean, Wilber, 1979
Bernard, James, Holmesville, 2016
Bever, Joe & Virginia, Beatrice, 2002
Blome, Arlin, Clatonia, 1983
Braun, Walter & Louise, Beatrice, 1988
Breunsbach, Willard, Liberty, 2002
Brewer, Laurel, Gretna, 1994
Brownawell, Lloyd & Thelma, Wymore, 1982
Brownawell, Rhonda, Wymore, 2015
Bruce, Tom & Jon, Half Moon Bay, CA, 2015
Buckridge, Mr. & Mrs. Wes, Holmesville, 1967
Burger, Edna, DeWitt, 1978
Burger, Edward, Clatonia, 1986
Burrows Hereford Farm Inc., Adams, 1978
Busboom, Henry, Beatrice, 2000
Busboom, Jerry & Dorothy, Beatrice, 1983
Buss, Iral, Beatrice, 1966
Buss, Ralph, Beatrice, 1985
Cacek, Marion, Odell, 1983
Carpenter, Fred & Sadie Pefferman, Beatrice, 1986
Carsten, Reed & Gertrude, Clatonia, 1980
Carstens, Gerhard & Louise, Pickrell, 1990
Claassen, Donald, Beatrice, 1977
Clifford Gibson, Cynthia, Beatrice, 2008
Closs, Ellsworth & Hazel, Wymore, 1978
Colgrove, Eugene, Odell, 2000
Cooper, Mrs. John, Adams, 2006
Cornelius, Lovelle, Beatrice, 2002
Cornelius, Lowell, Beatrice, 2002
Creevan, Gerald V. & Joan (Goes) Creevan, Wymore, 2004
Damkroger, Edwin, Beatrice, 1984
Daubendiek, Duane & Irene, Beatrice, 1983
Daubendiek, Fred, Pickrell, 1983
Debuhr, Darvin, Bricknell, 2000
Del-Tu Farm Inc., Beatrice, 2006
Disney, Dan & Charlene, Blue Springs, 1999
Dorn, Leonard & Doris, Adams, 2007
Dorn, Leonard, Adams, 1983
Dunn, Mrs. Alice, DeWitt, 1962
Epard, Betty & Elva, Beatrice, 1968
Erdmann, Stielf & Dorothy, Wymore, 2007
Esau, Robert & Ruth, Beatrice, 1997
Evers, Todd, Brian & Kent Evers, Lincoln, 2005
Fast, John & Esther, Beatrice, 1991
Faulder, William & Myrna, Beatrice, 2007

Feit, Nickolas, Blue Springs, 1983
Fisher, Reno, Beatrice, 1977
Forden, Amanda, Filley, 1980
Frerichs, Harm, Beatrice, 1983
Fritschel, William, Paul Fritschel, Barbara Holtermann, Sun City Center, 2006
Garrels, Billey & Jeanne, Filley, 2003
Gerdes/ Jurgens, J Delmar/ Betty, Wymore, 1982
Gerlach, Robert, Hallam, 1981
Goin, George, Wymore, 2002
Graff, Edwin, Beatrice, 1961
Gramann Jr., Henry, Adams, 1993
Grell, Kenneth & Dorothy, Blue Springs, 1984
Grell, Richard & Jacque, Beatrice, 2004
Grone, Laura Mae, Independence MO, 1995
Gronewold, Lawrence & Martha, Beatrice, 2006
Haecker, Bertha, Beatrice, 2001
Hagemeier, Harlan & Ardyth, De Witt, 2007
Hagerman, Mildred, Lincoln, 1992
Hahn, Joseph, DeWitt, 1992
Hajek, Theodore & Donnetta, Odell, 1991
Hajek, William, Charles, John, Odell, 1986
Hale, Kenneth, Cortland, 1999
Hamm, Mr. & Mrs. William, Beatrice, 1977
Hartwig, Calvin & Sharon, Cortland, 1985
Hawkins, Daniel & Barbara, Wymore, 1983
Heckman, Leland L., Lincoln, 2008
Heller, Gladys & Lee, Clatonia, 1966
Hellmer, Harvey/ Frieda/ Irene, Wymore, 1983
Helmrichs, John, Wymore, 1990
Hemminger, Verlene & Ralph, Wymore, 1978
Hess, Janice, Omaha, 2002
Hiatt, Robert, Huntsville, 2009
Hibbert, Dean, Adams, 1973
Higgins, Florence, Beatrice, 1980
Hill, Lori Ann Miller and Kathy Lynne Miller Kroeger, Gretna, 2003
Holsing, Kenneth & Karen, DeWitt, 1993
Hubka, Ernest Estate, Beatrice, 1982
Husa, Norman & Linda, Barneston, 1985
Hylo Farms Inc., DeWitt, 1983
Ideus, Charlene, LaVon Trauernicht, Hickman, 2008
Ideus, Menn & Emma, Cortland, 1992
Ikerd, Harold & Ethel, Firth, 1986
Isley, Eugene & Darleen, Firth, 1983
Jobman, LeRoy & Irene, Beatrice, 1994
Johnson, Mr. & Mrs. WIlliam, Filley, 1972
Jones, Alfred & Jacqueline, Wymore`, 2006
Jurgens, Elmer, Pickrell, 1980
Jurgens, Reinder, Filley, 1988
Kees, Miss Cara, Beatrice, 1967
Kimmerling, Max & Viola, Beatrice, 2003
Kimmerling, Muriel & Charles, Beatrice, 1973
Klaus, Ernest, Odell, 1992
Koch, Arlin and Marlys, Wymore, 2016
Kracke, Fred, DeWitt, 1979
Kracke, Robert & Irene, DeWitt, 1983
Krauter, Meredith & Larry, Clatonia, 1995
Krauter, Mrs. Orris, Lincoln, 1993
Kreuscher, Jean/ Paul, Omaha, 1992
Krueger, Gerald & Dolores, Odell, 1997
Lane, Dorothy & Lloyd, Beatrice, 2009
Lee, Oren, Kathryn Bednar, Wymore, 2007
Lenners, Donald & Elnora, Filley, 1984
Lenners, Julia & Michael/Mark Jurgens, Beatrice, 2003
Lenners, Michael & Julia/Evalena Lenners, Beatrice, 2003
Lenners, Stan/ Mary, Filley, 1996
LePoiderin, Mr. & Mrs. Ezra, Beatrice, 1965
Lidolph, Dorothy & Tracy, Cortland, 1995
Lieneman, Leslie & Cindy, Filley, 1995
Linsenmeyer, Don & Joan, Blue Springs, 2011
Linsenmeyer, Mr. & Mrs. Henry, Beatrice, 1980
Loemker, Leulla, Odell, 1990
Maschmeier, Carl, Beatrice, 1989
Masek Children's Trust, Beatrice, 2013
McKeever, Robert & Evelyn, Wymore, 1983
McPherson, Grace & Chester, Lincoln, 1993
Menke, Mrs. Florence, Beatrice, 1985
Meyer, Mrs. Clifford, Beatrice, 1977
Meyer, Vernon, Wymore, 1992
Miller, Melvin & Rosetta, DeWitt, 2000
Miller, Oren & Karoldeane, Clatonia, 1985
Mott, James & Joan, Beatrice, 1985
Mulder, Bill, Firth, 1979
Mulder, Marian, Firth, 2000
Mulder, Ray & Marian, Firth, 2002
Mulder, Roy & Sharon, Firth, 2006
Muller, George & Bonnie, Clatonia, 1995
Mumford, O.Dean & Ida, Beatrice, 1965
Nelson, Mrs. Stella Elizabeth, Beatrice, 1986
Neuman, Tony, Wymore, 2001
Nieveen (Living Trust), Wilma, Sterling, 2013
Novotny, Wilma, Odell, 1994
Oltman, Mr. & Mrs. Delmar, Cortland, 1982
Oltmans, Harlen & Elda, DeWitt, 1988
Papke, Dale & Phyllis, Cortland, 1999
Parde, Doug & Cindy, Adams, 2012
Parde, Duane, Pickrell, 1992
Penner Farms, DeWitt, 1991
Penner, Fred & Dora, Beatrice, 1989
Penner, Harold, Beatrice, 1978
Pethoud, Percy & Mildred, Beatrice, 1971
Pfeiffer, Nellie, Corland, 1974
Pieper, Arlen, DeWitt, 1991
Plucknett Waldo, Willard & Beulah, DeWitt, 1993
Plucknett, George, DeWitt, 1962
Ramsey, Donald E. and Shirley M., Wymore, 2008
Reiner, Lawrence, Beatrice, 2002
Remmers, Elvin, Firth, 1978
Remmers, Minnie, Sterling, 1981
Remmers, Mr. & Mrs. Orville, Pickrell, 1974
Remmers, Paul, Pickrell, 2016
Rhine, Boby and Tracy, Blue Springs, 2008
Rhodes, John, Beatrice, 2000
Riemann, Ralph J., Lincoln, 2007
Riemann, Ralph, Lincoln, 2001
Riesen, Daryl A. & Denise D., Beatrice, 2013
Rocking T Farm Inc, Beatrice, 2004
Russell, Mr. & Mrs. Robert, DeWitt, 1969
Scheiding, Raymond, Beatrice, 2011
Scherling, Rodney & Mary, Beatrice, 2014
Scheve, Mr. & Mrs. William, Beatrice, 1968
Schlake, Tady, Beatrice, 1983
Schneider, Floyd, Pickrell, 1975
Schuster, Albert & Hilda, Beatrice, 2000
Schwartzkopf, Mrs. Ed, Lincoln, 1981
Shaffer, Berwin R. & Roselyn A., Beatrice, 2013
Shaffer, Mrs. Alma, Liberty, 1967
Shalla, Mrs. Rosa, Wymore, 1982
Shanek, Walter & Betty, Odell, 1984
Smith, Judy, Lincoln, 1986
Smith, Wyman & Evelyn, Blue Springs, 1969
Smith/ Weibel, Velma/ Zelma, DeWitt, 1979
Spilker, David & Connie, DeWitt, 1984
Spitsnogle, Mildred, Odell, 1984
Steinmeyer/ Else, William/ Virginia, Clatonia, 1975
Stokebrand, Larry, Bellek Vista, AR, 2005
Synovec, Mr. & Mrs. Stanley, Wilber, 1981
Trahan, et al, Norma, Kansas City, 2010
Trauernicht-Lenners, Hildamae C, Grand Island, 2008
Trauernicht-Schuster, Elmer and Vivian, Fillen, 2008
Trauernicht, Donald & Luetta, Wymore, 1985
Trauernicht, Lorenz and LaVon, Pickrell, 2008
Trauernicht, Lorenz, Pickrell, 1991
Vlasak, Herbert, Ellen Vlasak, Wilber, 2009
Vrtiska, Virginia, Beatrice, 1970
Wardlaw, Kenneth & Maxine, Pickrell, 1971
Webb, Renetta, Lois Webb, Crystal Lake, 2009
Weiss, Robert, Virginia, 2015
Whittemore, Ernest, Beatrice, 1970
Wieters, Harold & Velora, O'Dell, 2004
Wilhelm Hurtz Farms Inc., Wymore, 1985
Wilson, Neola, Beatrice, 1962
Wollenburg, Carl, Roger Wollenburg, Marilyn (Wollenburg) Greenlee, Carol (Wollenburg) Meyer, De Witt, 2007
Wollenburg, Lorenz & Delores, Beatrice, 2004
Wymore, Virgil & Doris, Beatrice, 1993
Yoke, John & Beth, Filley, 1995
Ziegler, Donald & Glennys, Kirk & Kathleen Ziegler, Beatrice, 2009
Zimmerman Brownawell, Lloyd & Thelma, Wymore, 1994
Zimmerman Farms Ltd., Beatrice, 1984
Zimmerman, Bernadette, Wymore, 1982
Zimmerman, Elmer & Dianne, Beatrice, 1997
Zimmerman, Elmer C. Jr. & Diane K. Zimmerman, Beatrice, 2004
Zimmerman, Garold and Margaret, Beatrice, 2008
Zimmerman, Meryl, DeWitt, 1979
Zimmerman, Ranke & Friedericke, Pickrell, 1965
Zimmerman, William, Beatrice, 1967

Garden County

Berglund, Jennie, Lewellen, 1989
Brown, Albert Louis, Lewellen, 2015
Brunt, Clayton, Lewellen, 2001
Burke Seibert, Janice, Oskosh, 2011
Clary, Mrs. Naomi, Lewellen, 1997
DeLaet, David D. & DeLoris M, Lewellen, 2014
Delatour Land & Livestock Co, Lewellen, 1984
Dymak, David, Lewellen, 2009
Fiesterman, Theodore W., Lewellen, 2014
Guenin, Mr. & Mrs. Walter, Oshkosh, 1984
Guenin, Phyllis, BlackHawk, 2006
Hartman, Herbert & Shirley Lou, Oshkosh, 1987
Hisel, Martin, Ashby, 2009
Johnson Lehn, Robert & Marilyn, Julesburg CO, 1994
Lobner Land & Livestock Inc., Lewellen, 2014
McCarty-Davis Ranch Inc., Bingham, 2014
McCormick, Ronald & Darlene, Lewellen, 2007
Myers, Marlene & Lynn, Lewellen, 2014
Niebaum, Glenna Belle, Sidney, 2010
PGR Limited Partnership, Don Powles Family, Bingham, 2011
Rittenhouse, Dallas & Ruby, Lewellen, 1999
Rittenhouse, Merle & Edith, Lewellen, 1999
Robinson, Paul, Oshkosh, 1990
Suit, Donald, Oshkosh, 1995
Thelander, LeRoy & Arlene, Chappell, 1993
Walsh, Dennis & Irma, Lodgepole, 1990
Wilson, Marcia (Buske) & Marlis (Buske) Shearer, Lisco, 2003

Garfield County

Ballagh, Gordon A., Burwell, 1987
Brownell, Robert W., Burwell, 1987
Burson, Kevin J., Burwell, 2015
Draver, Mrs. Irma, Burwell, 1974
Helgeland, Orrin & Shirley, Burwell, 2005
Lowery, Belva Family (sons James, Melvin, Vernon), Burwell, 1991
Mary L. Scott, Albert J Mohr, Burwell, 1991
Maxson, Howard D., Burwell, 1974
Mohr, Arthur A., Burwell, 1993
Nelson-Bowers, Geneva, Burwell, 1997
Rowse, Freeman & Betty, Burwell, 1987
Sartain, David W., Burwell, 1991
Struckman, Rollin & Mary Ann, Burwell, 2011
White, Roland & Roger, Burwell, 2006

Gasper County

Anders Farms Inc. (Reuben L. Anders), Bertrand, 1984
Atkinson, Dale A., Wade E., Mark E. (Nutzman), Anne L. & Mary L. Lindsey, Nehawka, 1990
Bachmann, Charles Stuart/Janice Faye, Bertrand, 1996
Bell-Maxfeldt, Mrs. Susan, Holdrege, 1988
Benson, Charles L., Floyd A Benson, & Doris M. Phillips, Bertrand, 1981
Bloch, Mrs. Emil, Elwood, 1984
Bowen, William H., Lincoln, 2004
Brell, Robert & Eunice, Elwood, 1979
Brell, Wilber & Renata, Elwood, 1992
Briegel, Marvin & Esther Frese, Arapahoe, 2004
Bunsen, Walter & Edith, Smithfield, 1986
Burke, Larry L., Elwood, 1988
Chambers, Allen & Meta, Arapahoe, 1983
Cross, Leo & Hazel, Arapahoe, 1986
Cross, Ray, Arapahoe, 1989
Dawson, Lowell, Edison, 1982
Dawson, Myron F & Marilyn, Edison, 1982
Ebermeier, Delmar, Bertrand, 2004
Ebmeier, Shirley M., Bertrand, 2014
Evans, Berdina, Smithfield, 1988
Fastenau, Mable & Keith, Bertrand, 1993
Forster, Kevin, Smithfield, 2010
Foss, Eva & Gary, Elwood, 1984
Gardner, Mr. & Mrs. George, Arapahoe, 1982
Gennrich, Ricky L & Emma Jeanne, Bertrand, 1997
Gruber, George, Elwood, 2008
Halker, Howard N., Elwood, 1987
Hambidge, Donald C., Arapahoe, 1986
Hanson, Donald E. & Phyllis M., Smithfield, 2003
Hilker, Kevin/Tammy, Arapahoe, 1996
Houlden, Eldon & Delores, Elwood, 1983
Junkin, Eldred & Ernie, Smithfield, 1994
Krog et al, Eileen, Nebraska City, 1993
Langenberg, Caroline, Bertrand, 2006
Lerdall, Dorothy, Elwood, 2014
Maaske, Keith & Barbara, Arapahoe, 1981
Mahar, William E., Lexington, 1985
Meyerle, Gerald C., Holbrook, 1979
Meyers, Oscar E., Arapahoe, 1983
Mollring Hays, Janet, Hays Center, 2007
Monter, Glen & DeLilah & Iva Fritz, Elwood, 2005
Monter, Mrs. Reinhold, Elwood, 1987
Nott, Raleigh D. & Eleanor C, Arapahoe, 1992
O'Donnell, James G., Lexington, 1987
Oberhauser Kliewer, Loretta Fay, Aurora, 2014
Peterman, Ervin & Beata, Arapahoe, 1983
Petermann, Ervin and Beata, Arapahoe, 2003
Rightmire, Neita, Smithfield, 1994
Ross, Henry G, Arapahoe, 1976
Schmidt, Robert F., Eustis, 1990
Schutz, Ted & Janet, Elwood, 2013
Schwarz, Tom, Bertrand, 2006
Schwennienger Family Farm, Elwood, 2006
Scott, Lorena H., Lexington, 1984
Simon, Eldon M., Cambridge, 1975
Stehl, Vernon, Smithfield, 1987
Steinbrink, Leo & Ona, Elwood, 1980
Stillman, Ernest W., Arapahoe, 1978
Thompson, Mary & Helen I., Elwood, 1993
Tiedje, Richard V., Arapahoe, 1982
Tilson, Paul J., Elwood, 1992
Tull, Betty Gehrig & Carol Gehrig Albers, Omaha, 1993
Turner, Mrs. Erma, Eustis, 1985
Umberger, William G., Elwood, 1985
Vinzant, Alvina, Bertrand, 1991
W. Keller LTD (Wayne/Nancy Keller), Eustis, 1996
Wallace, W. Clyde, Lexington, 1985
Wilken, Joseph Douglas, Smithfield, 2011
Wilson, Sharlyn, The Villages, 2007

Grant County

Applegarth, Andrew & Binda, Whitman, 2015
Dubs, Arthur D., Ashby, 2014
Dubs, Paul & Alice, Ashby, 2015
Edelman Enterprises Inc., Hyannis, 2014
Hyde, Rhonda, Ashby, 2010
Lazy Bones Ranch Inc., Ashby, 2014
Lynch, Frank M. /Anne Lunch Falotico, Hyannis, 1987
Merrihew, Harry, Ashby, 2014
Rothwell, Robert L., Hyannis, 1988
Schufeldt, Larry, North Platte, 1986
Shepherd, M.D. & Janet, Hyannis, 1986
Simpson, Ben & Donna, Hyannis, 1986
Thurston, William E. & Twyla M., Hyannis, 2012
Wright, Lyle & Berneice, Hyannis, 1986
Wright, Opal / Rex Sikes, Hyannis, 1986

Greeley County

Babcock, Allen, Scotia, 1982
Ballweg, Aloysius & Louise, Lincoln, 1995
Ballweg, Frank & Mary, Spalding, 1978
Barnes, Louis & Loreen, Greeley, 1999
Barnes, Virgil & Thelma, Scotia, 2000
Bjorklund, Lorraine, Greeley, 2007
Blacklock, Robert, Greeley, 1981
Burke, LeMoine & Maybelle, Greeley, 2004
Christensen, James & Joan, Wolbach, 2000
Christensen, James & Joan, Wolbach, 2002
Christensen, Randall, Wolbach, 2001
Donovan, Mary Alice, Greeley, 2002
Esch, Dennis, Spalding, 1978
Gebhardt, Doug & Jeanette, Scotia, 2000
Gydesen, Harlan, Scotia, 1980
Harvey, Anna, Greeley, 1977
Hooker, Ted & Joan, Primrose, 1984
Ketchmark, John Trust, Scotia, 2015
Krick, Joseph, Spalding, 1977
Larson Rosander, Larry & Sharon, Wolbach, 1993
McCarthy, Alice Eileen, Spalding, 1993
McCarthy, Donald & Jean Anne, Spalding, 1993
McCray, M. Darlene, Fullerton, 2014
McGowen, Allen E. & Nancy A., Spalding, 2004
McGowen, Lawrence A. & Linda J., Spalding, 2004
McIntyre, Jack & Louise, Wolbach, 2010
Meyer, Richard & Vondell, Scotia, 1981
Mulcahy, Mary Ellen, Omaha, 2004
Noonan, Geroge, Spalding, 1983
Pfeifer, Florian "Bill", Spalding, 2004
Pfeifer, Leonard, Spalding, 1987
Poss, Melvin and Elizabeth, Scotia, 2016
Pritchard, Carl, Spalding, 1986
Pritchard, Mark, Spalding, 1986
Reilly, Angela, Omaha, 2006

Rother, Paul, Wolbach, 1987
Semper, Sr. Roberta, Annette Lammers, Marguerite Dixon, Frederick Semper, Spalding, 2005
Stam, Roy & Anita, "Stam Farms, Inc.", Scotia, 2015
Striker, Nancy, McCook, 2007
Sullivan, Joe & James Family, Spalding, 1977
Thayer, Mr. & Mrs. Louis, Wolbach, 1977
Thomas, Vernon Estate, North Loup, 1980
Towey, Gerald J., Philame Towey, Cedar Rapids, 2004
Winter, Dennis & Teresa, Scotia, 2001
Wood, Mary, Greeley, 1978
Zwiener, Elizabeth & Janet, Spalding, 2003

Hall County

Adams, Marion, Doniphan, 1988
Alberts, Darrel, Wood River, 1978
Barrows, Delbert, Hastings, 1979
Beukenhorst, Gail & Carol, Doniphan, 2011
Bilslend, Harold E. & Patricia L., Wood River, 2004
Boltz, Gene, Grand Island, 2009
Bonsack, Dennis, Wood River, 2009
Brittin, Nelson, Wood River, 1975
Brundage, Lenorad Heirs, Cairo, 1986
Buettner, Jerald & Mary Ann, Grand Island, 1985
Buettner, William Joseph & Nancy Jane, Grand Island, 2013
Costello/ Estate of Shirley Costello, Bernard, Omaha, 2013
Crosser, David, Marquette, 2009
Denman, Carline, Grand Island, 2015
Denman, Robert, Alda, 1979
Dority, Verle & G. Wayne, Shelton, 1974
Eggers-Nester, Opal, Cairo, 1999
Ellis, Addren & Patsy, Hastings, 2011
Elstermeier, H. E., St. Paul, 1998
Engel, Mrs. Mildred, Wood River, 1973
Ewoldt, Gaylon, Art, Bob, Delbert, Velma, Grand Island, 1986
Ewoldt, Edmund, Grand Island, 1967
Ewoldt, Herman, Grand Island, 1959
Frei, Helen, Grand Island, 1980
Gannon, Wilber & Eleanor, Grand Island, 1986
Gloe, Mark and Dian, Wood River, 2008
Goettsch, Richard Dale, Grand Island, 1989
Graf, August, Doniphan, 1982
Guenther, Edward, Grand Island, 1977
Gulzow, Lester & Mildred, Grand Island, 2000
Hagge, Clara, Grand Island, 1959
Haldeman, Charles William, Wood River, 2003
Hannon, Michael Patrick, Shelton, 2000
Hanssen, Harold & Bonnie, Doniphan, 1987
Hanssen, Norman & Thelma, Grand Island, 1987
Hanssen, Willis & Sharon, Grand Island, 1987
Harders, Hugo, Wood River, 1986
Harders, Lyle, Wood River, 1994
Harrell, Bryan, Doniphan, 1966
Heitz, Edison, Doniphan, 1979
Hendrickson, Mrs. Robert, Cairo, 1984
Hinkson, Keith, Cairo, 1991
Hoffman, Rex, Doniphan, 1988
How, Hazel, Hastings, 1982
Huebner, Ed, Alda, 1981
Hughes, Karl & Shirley, Wood River, 1984
Hulme, Orville, Cairo, 1979
Huxtable, Dennis, Wood River, 1980
Jensen, Mr. & Mrs. Arvon, Grand Island, 1972
Johnson, Peter, Grand Island, 1978
Johnson, Robert & Harriett, Hastings, 1989
Johnston, Robert, Emmet, 1979
Jones, Gerald & Patricia, Doniphan, 1986
Karr, Lavern, Doniphan, 1974
Kieffe, Hugh/ Mrs. Don, Hastings, 1983
Kleeb, Kelvin & Susan (Wiese), Alda, 2010
Knuth, Larry & Karen, Grand Island, 2007
Koch, Herbert & Helen, Grand Island, 1998
Kranz, George, Grand Island, 1980
Kroeger, Lawrence, Doniphan, 1999
Krohn, Robery Henry, Grand Island, 2015
Langenheder, Addie & Harold, Grand Island, 1995
Lockwood/ Short, Virginia/ Burr, Kenesaw, 1979
Lorenzen, John & Louise, Grand Island, 1966
Lorenzen, Louie & Frances, Grand Island, 1969
Lorenzen, Louis, Grand Island, 1967
Loucks, Mrs. Raymond, Grand Island, 1980
Lowry, Mrs. Roland, Doniphan, 1973
Mader, Geraldine, Grand Island, 1990
Marsh, Lawerence, Doniphan, 1977
Mettenbrink, Elmer & Norma, Grand Island, 1994
Mettenbrink, Leo & Jeanne, Grand Island, 2006
Mullen, William, Grand Island, 1972
Muller, Darlene, Grand Island, 1993
Muller, Lloyd, Grand Island, 1993
Myers, Oscar, Shelton, 1985
Nelson, Harry & Leona, Grand Island, 1980
Niemoth Hendrickson, Elsie, Cairo, 1992
Niemoth, Walter, Grand Island, 1982
Nietfeldt, Myron, Grand Island, 1978
O'Nele, Lloyd & Mae Hein, Grand Island, 1978
Obermeier, Ronald, Cairo, 2000
Opp, John, Wood River, 1980
Ostermeier, Albert, Grand Island, 1999
Quisenberry, Gary & Ida, Wood River, 1995
Quisenberry, Paul & Minnie, Wood River, 1995
Quisenberry, Paul, Wood River, 1985
Rainforth, Steven & Janet, Doniphan, 2003
Rathman, Lawney L. & Jim C., Grand Island, 2014
Rauert, Donald, Grand Island, 1993
Rauert, Elmer, Grand Island, 1984
Rauert, Fred, Grand Island, 1980
Rauert, Paul & Carroll, Grand Island, 2009
Rewerts, Mrs. Elfrieda & Children, Doniphan, 1979
Rolofson, Lawrence & Iona, Cairo, 1993
Rosenkotter, Fritz, Grand Island, 1983
Rowley, Bernard, Wood River, 1984
Sass, Mildred, Grand Island, 1981
Sassen, James & Rose, Grand Island, 2009
Schaaf, Minnie & Elmer, Grand Island, 1959
Schaaf, Rosa & Pearl, Grand Island, 1960
Schimmer, Leslie, Grand Island, 2001
Schoel, Harry H., Grand Island, 1973
Schultz, Donald, Grand Island, 2007
Schultz, Douglas, Grand Island, 2009
Schultz, Marjorie, Cairo, 1992
Schultz, Rose, Larry Hadenfeldt, Cairo, 2006
Schultz, Wilbert & Joyce, Wood River, 2009
Senseney, Arthur, Wood River, 1985
Stelk, Arthur, Grand Island, 1962
Stelk, Mrs. Detlaf, Grand Island, 1959
Stoltenberg, Anita Thompson, Grand Island, 1964
Stoltenberg, Bartny, Grand Island, 1962
Stoltenberg, Donald & Marie, Grand Island, 2001
Stuhr, Leo, Grand Island, 1959
Sundermeier, Doris, Cairo, 2005
Thompson, Florence, Grand Island, 2002
Thompson, Maude, Wood River, 1980
Thomsen, Thomas, Grand Island, 1982
Thomssen, Mr. & Mrs. William, Grand Island, 1972
Thomssen, Mrs. John, Grand Island, 1959
Toben, Martha Porth, Doniphan, 1978
Voss, Lenora & Duane, Cairo, 1979
Waddington, Mabel R., Palisade, 2008
Wagoner, Thomas & John, Wood River, 1982
Walker, Donald, Shelton, 2005
Weaver, Howard, Doniphan, 1962
Weinrich, Dale Estate, Alda, 1982
Whitelowry, Robert, Cairo, 1999
Wickersham, Maurice, Shelton, 1991
Wiese, Glen, Grand Island, 1998
Wiese, Lawrence, Grand Island, 1983
Wilhelmi, Clara, Grand Island, 1998
Winter, Charles, Grand Island, 1985
Woodman, C. Raye & Carolyn, Kenesaw, 1999

Hamilton County

Ackerman, Albert & Darlene, Sutton, 1972
Andersen, Ardena & Annebell, Marquette, 1988
Andersen Bankson, Herbert & Doris, Hampton, 1986
Anderson, James & Sandra, Aurora, 2007
Bald Mecham, Gladys, Aurora, 1977
Bankson, Everett, Hordville, 1978
Bannister/ Butt, Lillian/ /Marilyn, Hordville, 1994
Bavinger III, Wm., Wm. F. Bavinger Jr. Trust, Fremont, 2015
Berggren, Frances, Aurora, 1984
Bish, Alvin & Eldon, Aurora, 1992
Bish, Donald & Leola, Giltner, 1998
Boesen, Helen, Phillips, 1992
Budnick, Mrs. Raymond, Hampton, 1985

Bute, Eva, Anne & Florence, Aurora, 1991
Cain, George & Louise, Aurora, 1985
Carlson, Roger, Owego, NY, 2011
Carnes, Cecelia & Richard, Marquette, 2001
Carnes, Cecelia, Marquette, 1993
Chaffee, Mrs. Gladys, Aurora, 1966
Clarke, Myrtle Randahl family, Central City, 1987
Crossroads Farms Inc. (Wilton & Eileen Forsman), Aurora, 1991
Cutts, Steven, Higgins Family Farms Co., Giltner, 2014
Dangler, Lucile, Giltner, 1988
Dangler Robertshaw, Charles & Eva, Trumbull, 1987
Dose, Gaylord & Linda, Hampton, 1994
Erb, Gary, Myra Erb Higgins, Kay Erb Miller, Giltner, 2011
Erickson, Marjorie, Hordville, 1997
Erickson, Robert, Hordville, 1983
Ernstmeier, John, Phillips, 1990
Friesen, Arlie & Elsie, Henderson, 1978
Friesen, Mr. & Mrs. Elmer, Henderson, 1977
Garrido, Alvena, Kearney, 1998
Gustafson, Keith, Phillips, 1984
Gustafson, Richard & Deloris, Hordville, 1989
Hanson, Patricia & Don, Aurora, 1989
Hardt, Ruth, Seward, 1992
Harrison, Dorothy & Robert, Harrison Inc., Lincoln, 2015
Harrison, Dorothy, Robert, Mark & Scott, Lincoln, 2015
Henriksen, Levi & Maren, Hampton, 1972
Hilligas, Ralph, Hampton, 1979
HLC Inc., Giltner, 2004
Holtorf, Marx & Maynard, Hondville, 2011
Horst, Merle & Imogene, Grand Island, 2000
Jacobs, Cecil Walther, Phillips, 1991
Jacobsen, Elmer, Marquette, 1992
Jensen, Larry, Marquette, 1999
Johansen, Arvid, Marquette, 2005
Joseph, Jerl/ Debbie, Hampton, 1996
Juett, Lester, Giltner, 1976
Kauffman Farms Inc., Grand Island, 1993
Keller/ Melton, Opal/ Fayle, Phillips, 1985
Kirk, Francis L., Omaha, 2005
Kirk, James & Eleanor, Aurora, 1972
Kirk, Jeanne A., Downey, CA, 2005
Klute, LaVern, Hampton, 1984
Knutson, Roy, Central City, 1987
Kohtz, Harlan, Albion, 1978
Kuehner Trust, Marie F., Doniphan, 2006
Kuehner, William, Carolyn, Frieda, Doniphan, 1989
Lagoni, Harold & Leona, Hampton, 1990
Larson, Margaret, Hordville, 1988
Larson, Ruth & Gerald, Hordville, 1983
Ljunggren, Lester & Laura, Harvard, 1986
Lock, Bernice, Central City, 1994
Lock, Donald, Central City, 1985
Luff, J. LeRoy & Bette, Marquette, 1995
Luthy, Dallas & Babe, Phillips, 2006
Matthes, Ernest & Irene, Hampton, 1982
Mihm/ Kirkpatrick, Hazel/ Judith, Aurora, 1987
Miller, Don, Polk, 1994
MIller, Dwayne, Polk, 1994
Necker, Loyal & RoJean, Hordville, 2006
Nelson, B. Marie, Kearney, 1986
Nelson, Leroy, Hordville, 1981
Nickolaus, Robert & Althea, Aurora, 1972
Niklous, Maggie Pauline, Aurora, 1989
Nilson, Arthur, Marquette, 2002
Nissen, Garu & Rex, Aurora, 2008
O'Donnell, Paul, Giltner, 1986
Obermeier, Ronald E. & Dorothy J., Cairo, 2003
Ormond, Patrick, Harvard, 2007
Ortegren, Harold & Donna, Marquette, 1991
Oswald, Dale, Aurora, 1993
Ott, Justin & Tammy, Aurora, 2009
Parpart, Raymond & Anna, Hampton, 1986
Perry-Holloway, Karen M., Aurora, 2014
Peters, Alvin & Nadine, Hampton, 1981
Peterson, Deryl & Linda, Hordville, 1982
Pickerel, Dorlene & Maurice, Lincoln, 2007
Purdy, Alvin & Dorothy, Phillips, 1982
Quiring, Gordon, Hampton, 1983
Roberts Family Trust, Leonard, Malcolm, 2011
Roberts, Kenneth & Jennie, Lincoln, 2002
Roberts, Rodney & Janet, Aurora, 1992
Roehrs, Orville & Dennis, Hampton, 1993
Ross, Gladys, Aurora, 1988
Sandberg, Geraldine M., Port Hueneme, CA, 2005
Sandell, Neal & Ruth, Polk, 2007
Schultz, Stephen & Russell, Doniphan, 1988
Scott, Mr. & Mrs. Lee, Aurora, 1984
Shafer, John, Giltner, 1996
Smith, Morris & Dorothy, Aurora, 1980
Splinter, Wilbur, Hampton, 1981
Springer, Eleanor M., Aurora, 2006
Springer, John D. & Donita M., Aurora, 2005
Springer, John D. & Eda Aspegren, Ralph Springer, Louis Oswald, Aurora, 2005
Stranberg, Alvin, Hordville, 1982
Strotman, Joe, Aurora, 2009
Stunkel, Raymond, Grand Island, 1985
Thomsen, Harlan, Marquette, 1984
Troester, Bertha, Hampton, 1991
Troester, Lyle, Hampton, 1990
Troester, Walter & Florence, Hampton, 1977
Veburg, Richard, Hordville, 1978
Vetter, Stephen & Doris, Aurora, 2009
Von Housen, Ardith, Omaha, 1992
Vosika, Carolyn, Lincoln, 1992
W. & M. Regier Farms Inc., Henderson, 1993
Wagner, Robert & Erma, Grand Island, 2006
Willcock, Dwight, Aurora, 1987
Williamsen, Lester, Hampton, 1984
Wilson, DeRoy, Giltner, 1982
Wochner, James & Lois, Bradshaw, 1988
Wolff, Harold & Gladys, Aurora, 1981
Ziegelbein, Kenneth & Shirley, Polk, 2006
Zierott, Ardeen, Hampton, 1996

Harlan County

Amboese Inc., Orleans, 2003
Anderson, Billie, Republican City, 1988
Arehart, Marvin & Mary, Wilcox, 2006
Artz, Bonita, Alma, 1993
Artz, Dean, Alma, 1993
Artz, Lloyd Or Anona, Alma, 1984
Ashley Farms, Inc, Wilcox, 2006
Bantam, Larry, Orleans, 1978
Boehler Trust, June Elaine, Orleans, 2007
Boehler, John, Orleans, 1994
Broeker, Wallace & Ortha, Orleans, 1997
Carlson, Ernest, Alma, 1983
Christensen, Greg, ", Agatha Walker, Jean Christensen", Alma, 2006
Christensen, Raymond, Orleans, 1993
Colgan, Vincent & Phyllis, Holdrege, 1989
Collins, Virginia & Neil, Orleans, 1972
Delimont, Virgil & Virginia, Stamford, 1978
Delimont, Wallace, Stamford, 1995
Delimont, Wayne, Stamford, 1978
Dixon, John, Alma, 1978
Dobberstein, Mr. & Mrs. Stephen, Naponee, 1971
Flodine, "James, David, Randy", Omaha, 1993
Flodine, David, Stamford, 1991
Flodine, Lloyd, Stamford, 1979
Fouts, Larry & Alice, Naponee, 1994
Garrison, Donald R. & Yvonne, Alma, 2004
Gehrig, Johnie & Ruth, Stamford, 1986
Gorman, Mrs. William, Holdrege, 1993
Graf, Jay & Susan, Alma, 2007
Haffey, Leo & Clementine, Oxford, 1985
Hardin, Harry D. and Ruby F., Alma, 2008
Hodge, W. James & Sally, Holdrege, 2003
Hollertz, Alan & Lana, Holdrege, 1988
Holste, Edward Holste & Robert, Oxford, 2003
Howsden, Lee & Margie, Alma, 1982
Hunt Goedeken, Otto & Dolly, Republican City, 1983
Hunter, Kathryn, Orleans, 1993
Johnson, Willie, Holdrege, 1978
Kammerer, Arlene, Republican City, 1986
Kauk, Richard, Alma, 2000
Lans, Roland, Stamford, 1986
Lenneman, Phyllis, McCook, 1992
Lindsay, Mary, Grand Island, 1989
Lubeck, Leo & Ilene, Orleans, 1992
Lucking, Marlene, Holdrege, 2011
Lueking, Bernard, Oxford, 1986
Lush, Verlie and Phyllis Lush, Wilcox, 2005
Lutjeharms, Helen family, Republican City, 1996
Massey, Floyd L. & Marietta E., Holdrege, 2003
McArthur, Mr. & Mrs. Kenneth, Holdrege, 1972
McNiel, George Wesley, Holdrege, 1987
Miller, Alvah & Eunice, Alma, 1986
Ott, Ervin & Mildred, Republican City, 1985
Peterson Livestock Inc., Orleans, 1985
Peterson, Virgil & Leah, Orleans, 1987
Pond Heirs, Orleans, 1983
Rebman, Eldin, Alma, 1988
Reiss, Oscar c/o Orletha (Reiss) Block, Grand Island, 2004
Reynolds, Sara, Orleans, 1979
Riebley, Agnes, Orleans, 1975
Rogers, Mrs. William, Orleans, 1984

Schippert, Richard & Julie, Republican City, 2000
Schlatz Trust, Patricia, Republican City, 2008
Schluntz Lands Ltd., Republican City, 1976
Schluntz, LeRoy, Republican City, 1993
Shaver, Barbara, Republican City, 2016
Snyder, Roy, Alma, 1976
Spaque, Charlote, Republican City, 1999
Stevens, Paul & Sharon, Orleans, 2012
Struve, Rollo, Oxford, 1972
Tams, Dale, Oxford, 1990
Titkemeier, Marvin & Sylvia, Wilcox, 2002
Ulrichson, John & Juanita Ulrichson, Alma, 2005
VonHollen, Marie, Grand Island, 1988
Walker, Agatha, Alma, 2010
Weinand, John & Lucile, Alma, 2010
Woollen, Melvin, Wilcox, 1986

Hayes County

Alberts, Elmer & Leola, Palisade, 2003
Bixler, Duane, Hayes Center, 1984
Bodeman, Gerold & Norma, Palisade, 1985
Broz, Vincent, Hayes Center, 1985
Broz, Steve & Cherri, Hays Center, 2007
Callander, Glenn, McCook, 1998
Christensen, Edna, Wauneta, 2001
Clark, Bill, Elkhorn, 2002
Evans, Lanny, Hayes Center, 1985
Fielding, Court, Hayes Center, 1985
Fitzgerald, Freda, North Platte, 1991
Foroff, Kevin & Ann, Hayes Center, 1995
Franklin, Lauren, Wallace, 1988
Gathman, Robert, Hayes Center, 1987
Gohl, Melva, Culbertson, 1995
Hanna, Forest, Wauneta, 1994
Hummel, Donald Rees, Elsie, 1987
Jeffers, Charles & Larry, Palisade, 1984
Kittle, George, Cambridge, 1986
Korell, Robert, Culbertson, 1983
Laird, Faye/ Scott & Patty, Hayes Center, 1993
Leu Clark, Grayce, North Platte, 1993
Maucher, Donna, Hayes Center, 1994
McDowell, Grace, Palisade, 1986
Mintling, Eldon, Hayes Center, 1985
Nitsch, Ivan & Zalia, Hayes Center, 1985
Novacek, Esther, Wauneta, 1993
Orman, Paul & Delores, Maywood, 1989
Orman, Paul & Delores, Maywood, 2006
Plautz, Michael & Susan, Ravenna, 1988
Reed Barker, Dr. Irvin, Hayes Center, 1985
Repass, Robert & Rosa Belle, Hayes Center, 1985
Schmelzer, Betty & Edward, Maywood, 1985
Schmelzer, Josephine, Maywood, 1985
Schuman, Maurice & Velma, Kearney, 2001
Shafer, Victor, North Platte, 1986
Singleton, Alvin & Letha, Wallace, 1986
Stevens Brown, Esse, Maywood, 1995
Unger, Garald and Beverly, Hayes Center, 2009

Hitchcock County

Baker, Harold & Margaret, Palisade, 1985
Bakers Acres Inc. (Bernard & Marjorie Baker), Trenton, 1991
Bar Reverse D (Jack Dahnke), Stratton, 1986
Barger, Don & Dorothy, Culbertson, 1995
Barth, Gary & Donna, Culbertson, 2006
Bobinmyer, Robert, McCook, 1985
Brown, Cermit & Marjorie, Culbertson, 2011
Brown, William R., Trenton, 2014
Burk, Robert & Ray, Stratton, 1990
Cook, Elmer J. Trust (Laura Cook), Culberston, 1986
Diehl, John & Julia, Stratton, 2011
Frank, Edith & Lorraine, Trenton, 1986
Frecks, Pauline, McCook, 2016
Gohl, Jason and Laura, Culbertson, 2016
Hagemann, Pauline, Culbertson, 2009
Hawkinson, Arthur & Willa Mae, Stratton, 1984
Hay, Virgil, Trenton, 2006
Hoyt, Frank & Hazel, Culbertson, 1985
Hoyt, James & Wyona, Culbertson, 1985
Ingraham, M.Colene, Loveland CO, 1998
M Bar Farms, Culbertson, 2013
McCormick Land Company Partnership, Culbertson, 1993
Minary Family, Dorothy, Culbertson, 2006
Minary Family, Dorothy, Culbertson, 2007
Moses Brothers (Evan & Stanley Moses), Culberston, 1987
Myers, Louis Arvene, McCook, 1995
Owens, Jerry & Rose Ann, McCook, 2004
Pollman, Louise, McCook, 2006
Pollman, Louise, McCook, 2007
Ragan, Vonna, Stratton, 2016
Rhodes, Mary & Marvin, Lincoln, 1993
Stock, William W., Culbertson, 1985
Sydow, Inez R., Trenton, 1985
Underhill, G. Fred & A Mabel, Riverdale, 1984

Holt County

Baumeister, Larry and Jean, Stuart, 2016
Beelaert, Robert and Jayce, Page, 2008
Bell, Lana, O'Neill, 1997
Blain, George, Page, 1980
Bollwitt Farms, Ewing, 2002
Bonenberger, Margaret, Atkinson, 1990
Boshart Partnership, O'Neill, 2004
Bouska, Ron & Carol, Stuart, 1983
Brothers, Dvorak, Stuart, 1980
Carson, Albert, Lynch, 1999
Carson, Dona, Chambers, 1984
Christopher and Ramold, Joyce and Edward, West Plain, MO, 2013
Clark, Dick Family, O'Neill, 1985
Cole, Claude, O'Neill, 1985
Cole, Donna Rae & Robert, Emmet, 1988
Conway Wahrman, Dorothy, Omaha, 2014
Coolidge, Kenneth, Lincoln, 2000
Coufal, Mrs. Edward John, O'Neill, 1980
Cronk, Fred Anser, Page, 2007
Davis, Robert & Donna, O'Neill, 1997
Devall, Carol, Atkinson, 1992
Dexter, Don & Marlene, Amelia, 2015
Dobias, Raymond & Ruby, Atkinson, 2011
Donlin Greene, Eileen, O'Neill, 1979
Durre, Ernest & Patricia, Chambers, 1984
Elsberry, Eugene & Trudy, Page, 1998
Elsberry, Norman, Orchard, 1996
Ermer, Darold, O'Neill, 1985
Ernesti, Gertrude, Clearwater, 1991
Estes, Raymond & Mary, Stuart, 1984
Farrand, Morna, Clarnece, Irwin, Lynch, 1982
Farrand, Clarence & Irvin, Lynch, 1982
Farrand, Morna, Lynch, 1982
Flannery, Richard, Atkinson, 1986
Friedel, James & Barbara, Stuart, 2011
Funk, Joseph & Georgine, Ewing, 1993
Gallagher, W. James Family, Page, 1981
Gathje Heirs, Osmond, 1985
Gilg, John C and DiAnne Gilg, NE, 2005
Gilman, June & Harold, Atkinson, 1991
Goldfuss, Harold & Betty, Atkinson, 1984
Grass Estate (Ronald Grass), Page, 1985
Greger, Daniel & Joan, Stuart, 1999
Grimes Ranch, Chambers, 2007
Grimes, Gerald, Chambers, 1996
Grothe, Herman & Lela, Emmet, 1985
Grutsch, Maurice, O'Neill, 1987
Hale, Everett Testementary Trust, Atkinson, 1995
Hall, Arnold/Sharon & Virginia, Orchard, 2015
Hanson, Gregory J. & Laurie, Amelia, 2007
Harkins, Albert & Lela, Ewing, 1983
Hawk, Leonard & Joan, Ewing, 2010
Hayes Ulrich, Nona, Atkinson, 1983
Heiss, Ivan & Rose, Page, 1982
Henning, Albert & Margaret, Atkinson, 1985
Hoffman, Donald & Delores, Chambers, 1984
Hoffman, Jeanette (Morgan), Cathi (Morgan) Irish, Sheryl Morgan, Mary (Morgan) Krysl, Jean (Morgan) Quinn, Stuart, 2003
Holcomb, LeRoy, Chambers, 1984
Holliday, Holly H. & Gretchen M., Orchard, 2014
Horton, Oliver, Stuart, 1983
Huston, Kenneth and Laurine, Emmet, 2008
Hynes, Loretta, O'Neill, 1982
Johnson, Andrew, O'Neill, 1999
Johnson, Gordon, O'Neill, 1981
Johring, Bennie & LaVone, O'Neill, 1982
Jonas, Robert, Stuart, 1985
Kaczor, Edward & Margretha, O'Neill, 2001
Kelley, Mrs. James & Ned, Inman, 1983
Kelly, Barry & Lenore, Page, 2011
Keyes, Donald L, Inman, 1981
Knox, Melvin & Patricia, Ewing, 2013
Koenig, Kenneth J. and Brittney, Ewing, 2012
Kohle, Eric & Brenna, Stuart, 2015
Kollman, Walter, Atkinson, 2001
Kramer, Paul & Noreen, Stuart, 1984
Krobot, Jim & Marilyn, Stuart, 1995
Kubart, Charles, Atkinson, 1989

Lambert, Ida, Ewing, 1983
Langan, Tom & Joanne, O'Neill, 2002
LaRue, Kenneth, Ewing, 1985
Lehmann, Richard, Chambers, 1993
Malloy, Tom, Atkinson, 1984
Marston, Howard & Harriet, Lynch, 1981
Martens, Helen, Atkinson, 2009
Melcher's Herefords Inc., Page, 2011
Miller, Roy, Chambers, 1984
Mitchell, Allan, O'Neill, 1992
Mitchell, Boyd, Atkinson, 1992
Mlinar, Harry & Jennie, Atkinson, 1977
MMK Ranch Inc. (Karen Kaup), Stuart, 1984
Morgan, Harold & Diane, Stuart, 2004
Morrow, Veldeen & Robert, O'Neill, 1995
Murphy, Dan, Stuart, 1983
Murphy, Frank & Edith, Stuart, 1983
Murphy, Henry & Helen, Stuart, 1989
Nelson, Paul, O'Neill, 1979
Olberding, Gary C. & Deborah A., Stuart, 2014
Olsen + Mardelle Dennis, Regina, Grand Island, 2015
Pacha, Arthur & Dolores, Atkinson, 2009
Pribil, Elsie, O'Neill, 1985
Pribil, Leonard and Ruth, O'Neill, 2008
Ramm, Helen, Stuart, 2010
Riffey, Wesley & Ruth, O'Neill, 1993
Risor, Raymond & Joyce, Saronville, 2011
Rohde, Francis & Maureen, O'Neill, 1984
Rosenkrans, Jerry, Lynch, 1984
Rouse, Lloyd & Delbert, O'Neill, 1979
Schaaf, Doris, Eldon Krugman, Russ Krugman, Jan Stampe, Matthew, Julie, Lonnie Krugman, joni Khan, Christie Ram, O'Neill, 2003
Schaff, Lois, Stuart, 2009
Slaymaker, Cathy, Atkinson, 1998
Slaymaker, Lyle & Ethel, Stuart, 1982
Slaymaker, Mrs. Ray, Stuart, 2008
Smith, Anna & Kenneth, Inman, 1981
Sorensen, James & Janet, Page, 1982
Spangler, Donald, Kearney, 2007
Stearns, Maynard & Fern, Clearwater, 1985
Steinhauser, Harvey & Edith, Stuart, 1999
Stevens, Catharine, O'Neill, 1985
Storjohann, Robert, O'Neill, 2009
Summerer, Duane & Myra, Ewing, 1998
Thoendel, Albert, Ewing, 2001
Thoendel, Dennis & Elaine, Ewing, 1988
Thoendel, Joseph, Ewing, 1988
Thramer Tagel, Ludwig & Loretta, Grand Island, 1984
Tompkins, Harvey, Inman, 1984
Troshynski, William J. & Elizabeth L., Atkinson, 2014
Trowbridge, Dona M., Page, 2014
Urban, Maud, Ewing, 1983
Vandersnick, Donald & Marie, Ewing, 1984
Vandersnick, Lewis, Ewing, 1984
Von Robertson, Charles, Chambers, 1983
Wallinger, Jerry, Stuart, 2002
Wedige, Angela, Stuart, 1998
Welch, Jack & Ellen, Verdigre, 1985
White, Glenn, Amelia, 1983
Widfeldt Trust, O'Neill, 2001
Widtfeldt Skjold, Norma, O'Neill, 1983
Williamson, Richard & Ruby, Orchard, 2011
Wilson, Garry/ Sally, Lynch, 1996
Wilson, Grace, O'Neill, 1985
Wondercheck, John, Atkinson, 1985
Wood, Edgar, Page, 1982
Young, Junior & Charlene, O'Neill, 2003
Young, Robert & Nancy, O'Neill, 1991
Zahradnicek/ Van Horn, Louis/ Arlene, Atkinson, 1980
Ziska, Lawrence & Alice, Atkinson, 1984
Ziska, Lucy, Atkinson, 1999

Hooker County

Cox, Marvin, Mullen, 2006
Erny, Beulah, Mullen, 2001
French, Larry & Georgann, Mullen, 2006
Harding, Pearl, Frank & Michael, Mullen, 1988
French, Roger & Beverly & Children, 2005
Tucker, Jack, Mullen, 1988

Howard County

Arterburn, Paul, St. Paul, 1978
Barr, Dale & Vonnie, Wolbach, 1991
Berggren Irrevocable Trust, Vivian, Grand Island, 2011
Berggren, Lois Berggren, NE, 2005
Boehle Land & Cattle Co., St. Libory, 1986
Brezina Sazama, Theodore, Dorothy Florence Sazama, St. Paul, 2006
Bryce & Geri, Pearson, Elba, 2002
Burkhardt, Rose Ann, Palmer, 2001
Burry, Roy, Jamston ND, 1996
Christensen Grand View Stock Farm, Wolbach, 1992
Christensen, Willard, Elba, 1978
Clement/ Sevcik, Henry/ Edwin, Casper WY, 1987
Cyboron, David & Marilyn, Boelus, 1984
Deertz, Carl & Pauline, Palmer, 2011
Dvoracek, John, Farwell, 1999
Dvorak, Evelyn, St. Paul, 2009
Dvorak, Lawrence Elva, Camas, 2009
Fisher, Mrs. Henry, Grand Island, 1976
Fisher, Mrs. Henry, Grand Island, 1977
Forbes, Doris, St. Paul, 2002
Greenwalt, Eugene & Helen, St. Libory, 1994
Hadenfeldt, Norman & Dylece, Dannebrog, 1999
Hansen, Mrs. Charles, Farwell, 1982
Hilmer, Walter & Cathryn, Boelus, 1978
Holly, Arlene, Omaha, 1996
Jensen, Ronald & Carol, Cotesfield, 1982
Jerabek, William & Bertha, Farwell, 1993
Jorgensen, Marlin, St. Paul, 2009
Kment, Charles, St. Paul, 1989
Knuze, Reinhardt, Palmer, 2001
Koperski, Henry, Farwell, 1989
Kosmicki, Louis & Evelyn, St. Paul, 2009
Krzycki, Roman, St. Paul, 2000
Lauritsen, Jr., George, Dannebrog, 2000
Lemburg, Earl, Boelus, 1992
Lemburg, Henry & Mollie, Boelus, 1982
Lemburg, Russel & Charlene, Dannebrog, 1982
Lukasiewicz, David, Farwell, 1979
Markvicka, James & Ema, Farwell, 1982
Meyer-Matsui-Erickson, Brad Meyer & Staci Matsui, Jeri Lou Erickson, Tom Meyer, Patti Meyer, NE, 2005
Meyer, Melvin Meyer & JoAnn Meyer, NE, 2005
Moraczewski Brothers, St. Paul, 1991
Moravec, Eleanor, Cotesfield, 2006
Mrkvicka, Thomas & Rosemarie, Farwell, 1998
Nelson, James Nelson, NE, 2005
Nelson, Kenneth, Boelus, 2002
Nelson, Lyle & Jayne, Boelus, 1978
Nesiba, Milton, Ravenna, 1994
Nielsen, Eleanor, Dannebrog, 2014
Nielson, Richard & Linda, Boelus, 1989
Nitsch, Victor, Rockville, 1977
O'Mara, Wm J.D. & M. Patricia, Albuquerque, NM, 2003
Obermiller, Ann, Boelus, 1988
Obermiller, Edward, St. Paul, 1988
Olsen, Hertha, St. Paul, 1981
Paul, Robert, St. Paul, 1978
Pedersen, Anita, Boelus, 1985
Petersen, Harland, St. Paul, 1994
Powers, Faye, Cairo, 1992
Puncochar, Judith, Marquette, 2006
Rasmussen, Charlotte & Gary, Elba, 2014
Reimers, Gerald, Ravenna, 1986
Roepker, Alfred & Betty, St. Libory, 1990
Sack Brothers, St. Paul, 1991
Safarik, Wilfred & Rick, Farwell, 1993
Samuelson, Judith, Kathleen Wall, Palmer, 2008
Sazama, Dorothy F., St. Paul, 2006
Schardt, Karen Nielsen, Lincoln, 1989
Sevcik, Jim, Omaha, 2012
Shanahan, John & Janet, Grand Island, 2002
Smith, Lola, Grand Island, 1992
Smydra, Don & Cecelia, Boelus, 1988
Sonderup, Stanley, St. Paul, 1985
Sosnoff Baird, Sandra, New Haven, CT, 2011
Spilinek, Alvin R & Ellen, Elba, 2009
Stevens, Raymond & Lois, St. Paul, 1991
Svoboda, Earl & Marilyn, St. Paul, 1993
Tekolste, Melvin & Eilene, CO Trustees TeKolste Living Trust, Beatrice, 2009
Vincik, Lonnie, Linda Obermiller, Kenneth Vincik, Grand Island, 2008
Vopat, Frank & Rogene, Timothy Vopat, Wolbach, 2011
Wagner, Charlie, Wolbach, 2007
Ward, Arthur & Joyce, St. Paul, 1999
Wulf, LaVern, St. Paul, 2007

Jefferson County

Achtemeier, Fred, Jansen, 2002
Bachle, Paul & Shari, Seward, 2002
Bakewell, Robert & Jane, Endicott, 2000
Bakewell, Robert & Jane, Endicott, 2004
Bartels, Iona M., Tobias, 2003
Bauer, Harold & Elden Brothers, Gladstone, 1974
Bishop, Darrel & Marvin, Hermosa SD, 2001
Blair, Grant & Vina, Fairbury, 1964
Block, Steve and Beth, Steele City, 2016
Boddye, Lawrence Edwin, Fairbury, 1973

Boeckner, Amos, Plymouth, 1981
Brandt, Donald, Plymouth, 2009
Buckow, Karen Sukovaty, Plymouth, 2005
Bugbee, Richard, Steele City, 1983
Cacak, Joe & Margaret, Fairbury, 1985
Camp, Clyde & Cleo, Plymouth, 1981
Cerveny, Roger A. and Jolene, Liberty, MO, 2004
Cunningham, James, Fairbury, 1998
Darrow, Charles & Faye, Fairbury, 1980
Dick, Wayne, Fairbury, 1975
Dux Henry, ReNee L., Sherri L. Dux Ideus, Michael T. Dux & Stephen D. Dux, Daykin, 2006
Dux, Richard & Cleo, Fairbury, 1994
Engelman, Dale & Phyllis, Jansen, 1990
Fairley, Charles & Donna, Geneva, 1984
Flatung Family Farm, Fairbury, 2012
Freese, Allen, Plymouth, 1991
Freese, Forrest, Plymouth, 2016
Frey, Wilbert, Fairbury, 2008
Friesen, Abraham, Jansen, 1975
G & K Holtmeier Inc., Tobias, 1992
Gerberding, Harold, Fairbury, 1996
Germer, Grace, Beatrice, 1983
Glenn, Wayne, Fairbury, 1975
Graff, Lyle, Endicott, 2007
Grefe, Alwin & Rosella, Plymouth, 1989
Gronemeyer, Eldred W. & Aleatha V., Plymouth, 2003
Grummert, Brad & Jeanette, Plymouth, 1986
Grummert, Norman & Frieda, Fairbury, 1990
Hammer, Leonard, Fairbury, 1985
Hansmire, Floyd, Fairbury, 1993
Hartwig, Elsie & Herbert, Diller, 1990
Heidemann, Philip, Daykin, 2016
Hein, Muriel, Fairbury, 1987
Helvey, Warren, Fairbury, 1977
Henkel, Ceres, Fairbury, 1988
Henrichs, Arnold & Eunice, Diller, 1984
Henrichs, Terry, Tucson, 2008
Henry, Wayne & Marilyn, Daykin, 1983
Hoeting, Walter & Janet, Hemet, 2009
Houser, Max, Fairbury, 1993
Huber, Joseph, Fairbury, 1984
Huber, Shirley, Hebron, 1995
Jantzen, Ernest, Plymouth, 1989
Jantzen, Henry & Gretl, Plymouth, 1991
Jarchow, Richard, Fairbury, 1985
Kapke, Darrell & Helen, Gladstone, 1984
Kenning, Herman R. & Dorothy Kenning, Endicott, 2004
Knobel, Ralph & Martha, Fairbury, 2007
Koch, Edna/ Verdell, DeWitt, 1987
Kugel, Marvin D. and Delores, Fairbury, 2003
Kujath, Fred Trust, Fairbury, 1982
Kyker, Dale & Ruth, Fairbury, 1985
Landkamer, Loren, Broken Bow, 1986
Maschmann, Bruce and Janet, Fairbury, 2003
McCord, Alvin & Ona, Fairbury, 1972
Meyer, Dale & Arlene, Beatrice, 1989
Meyer, Neil & Susan, Diller, 1993
Moeller, Herman, Beatrice, 1985
Molden, Lillian, Fairbury, 1993
Newell, Wallace, Fairbury, 1978
Nider, Leslie, Plymouth, 1968
Nuckolls Famrs Inc., Fairbury, 1984
Pohlmann, Gale & Lois, Plymouth, 1985
Pohlmann, Judy, Plymouth, 2001
Powell, Glenn & Verda, Powell, 1962
Priefert, Burdette & Virginia, Belvidere, 1979
Ressman, Heasty, Falls City, 1973
Rohlfing, Herman & Phyllis, Plymouth, 1983
Rothmeyer, Hilda, Gilead, 1983
Ruhnke, Wilmer & Doris, Fairbury, 1995
Sandman, Otto, Plymouth, 2000
Scheve-Hohbein, Della C, Cortland, 2006
Schipper, George & Lois, North Platte, 1998
Schmidt, Dell, Fairbury, 2016
Schwab, Bradley & Patrcia, Plymouth, 1985
Shea Farm Inc, Fairbury, 1984
Siemsen, Eugene & Clarice, Fairbury, 1999
Slater, William H., 1230 17th Rd, Washington, KS, 2013
Spilker, Robert & Vicki, 56525 722 Rd, Fairbury, 2013
Stainbrook Lewis, J. Sam & Dorothy, Daykin, 1972
Starr, Kenneth & Kathy, Fairbury, 2007
Steckelberg, Mathilde, Lincoln, 1979
Svajgr, Ed & Bessie, Fairbury, 2002
Swartz, Evelyn, Fairbury, 1980
Tate, Carolyn, Plyllis Kriesel and Kent Kriesel, Fairbury, 2016
Torrey, Robert, Fairbury, 1973
Traum, Rachel, Lincoln, 1978
Vorderstrasse, Carl, Plymouth, 1988
Warnke, William & Janette, Fairbury, 2010
Wehling, Neil D. and Pamela J., Diller, 2003
Westerhoff, Marlowe & John, Alexandria, 1973
Woods Howe, William, Phoenix AZ, 2001
Wrigley, William & Marilyn, Diller, 2001
Yantz, Robert, Fairbury, 1986
Zabel, Harold & Ruby, Daykin, 1999
Ziettlow, Betherl, Fairbury, 1995
Zulauf, Milton, Plymouth, 1990

Johnson County

Aerts-Smidt, Beverly Aerts & Roger Smidt, Omaha, 2014
Ahntoholz, Wilma Hahn, Cook, 1997
Armknecht, Robert & Ruth, Cook, 2006
Badertscher, Eldon & Aloma, Tecumseh, 1995
Bartels, Eugene, Steinauer, 1977
Bartels, Harold & Leona, Tecumseh, 1993
Bartels, Nathan & Jolene, Elk Creek, 1994
Baum, Glen/ Mary Lou, Tecumseh, 1996
Beethe, Rayman & Leona, Elk Creek, 1975
Bentzinger, Louis & Grace, Cook, 1984
Blythe, Harlan, Waverly, 1977
Boden, Roger A., Sterling, 2013
Boldt, Clinton & Helen, Sterling, 1969
Boldt, Stephen & Sandra, Sterling, 1984
Boone, Norman & Ramona, Elk Creek, 1980
Brehm, Willard & Gladys, Talmage, 1982
Brod, Dwight, Cook, 1992
Brown, Mr. & Mrs. Douglas, Cook, 1972
Buehler, Mr. & Mrs. Robert, Sterling, 1975
Buethe, Mr. & Mrs. Eldon, Elk Creek, 1969
Buethe, Willis & Joyce, Elk Creek, 1994
Burns, Patrick & Norma, Tecumseh, 2002
Cody, Lorain, Elk Creek, 1995
Dahl, Genevieve, North Bend, 1992
Davidson, Loren & Valda, Lincoln, 2009
Dierking, Victor & Esta, Tecumseh, 1980
Dirks, Norman & Gwendolyn, Sterling, 1997
Eckert, Virgil & Elma, Sterling, 1989
Eilers, Mr. & Mrs. Edward, Sterling, 1968
Feurer, Lester, Johnson, 1984
Fisher, Lyle, Cook, 2014
Fisher, William & Joyce, Cook, 2003
Fisher, William M & Joyce, Cook, 2015
Fritsch, Laverne & Shirley, Tecumseh, 1992
Geweke, Elmer, Tecumseh, 1969
Goracke, Richard & Betty, Tecumseh, 2001
Grove, Kenneth & Mary, Tecumseh, 1964
Hahn, Darold & Norma, Tecumseh, 1973
Hahn, James & Jeri, Tecumseh, 2011
Happell, Donald, Hooper, 2001
Harms, Mr. & Mrs. Lorin, Sterling, 1993
Hauptmeier, Henry, Sterling, 2007
Haynes, Ralph, Sterling, 1989
Hazen, Lloyd & Fay, Cook, 1976
Hervey Farms Inc., Cook, 2006
Hestermann, Mary Ann Trust, Sterling, 2003
Hoehne, Ellinor, Sterling, 1991
Holden, Dale, Tecumseh, 1965
Huls, John, Sterling, 2000
Hutt, Lila Mae & Keith, Tecumseh, 1995
Jeffery, Harold, Crab Orchard, 1989
Joekel, Keith & Betty, Sterling, 1991
Kavanagh Sharp, Randall/ Mary, Newton IA, 1996
Keebler, Archie, Sterling, 1968
Kelley, Gerald, Tecumseh, 1976
Kettelhake, Wilber & Mildred, Steinauer, 1976
Knoche, Donna, Beatrice, 2009
Knoche, Steve & Glenda, Virginia, 2009
Koehler, Lillie/ Alma, Tecumseh, 1996
Koehler, Robert, Tecumseh, 2005
Krueger/ Bergmeyer, Joan / Lois, Lincoln, 1994
Laflin Estate, Crab Orchard, 1985
Laue, Dorothy, Tecumseh, 1993
Lillich, George & Doris, Tecumseh, 2002
LIttle, Juanita Rose, Tecumseh, 1994
Little, Junior & Vicky, Crab Orchard, 1990
Mason, Dale & Ima Gene, Tecumseh, 1993
Mason, Eugene & Ella Mae, Tecumseh, 1993
McAuliffe, Terry, Sterling, 2012
McLaughlin, Elaine, Tecumseh, 1993
McLaughlin, John & Beverly, Lincoln, 1993
Mommens, Mr. & Mrs. Wilber, Tecumseh, 1965
Mommens, Steve, Tecumseh, 2014
Murphy, Arthur, Lorene, John, Tecumseh, 1972
Neemann, Mrs. John, Cook, 1978
Niemann, Eldred, Cook, 1975
Olmsted, Jerome & Lorraine, Tecumseh, 2000

Olmsted, Mr. & Mrs. Laverne, Sterling, 1970
Othmer, Wilber, Elk Creek, 1968
Peters, Raymond, Elk Creek, 1962
Putman, John, Tecumseh, 1969
Rathbone, Mr. & Mrs. Mort, Johnson, 1970
Roberts, Lyle, Tecumseh, 1973
Robinson, Thomas, Elk Creek, 1962
Schacht, Albert, Cook, 1983
Schmidt, Delmar & Wanda, Tecumseh, 1990
Seeba, Harold & Mardell, Cook, 2002
Straube, Theodore, Sterling, 1982
Stutheit, Herbert W., Johnson, 2004
Stutheit, Herbert, Johnson, 1996
Stutneit, H.T., Ann Stutneit & Family, Atchison, 2006
Tagart, Mrs. William, Vesta, 1959
Thiesfeld, Marie, David Thiesfeld, Carolyn Bartels, Beatrice, 2009
Truscott, Rawlong, Sterling, 1983
Watermann, Leona, Humboldt, 2016
Watermann, Jr., Leona & Henry, Elk Creek, 1981
Wilken, Paul & Diane, Cook, 2002
Wilkinson, Howard or Helen, Crab Orchard, 1988
Wilson, John & Suziane, Lincoln, 1975
Wolff, Marvin, Sterling, 1984
Woltemath, Laverne & Eileen, Elk Creek, 1985
Wright, Opal, Tecumseh, 2006
Wusk Thiesfeld, Marie, Carolyn Thiesfeld Bartels, Beatrice, 2009
Wusk, Marion & Magdalene, Sterling, 1991

Kearney County

Abrahamson Estate, Axtell, 2012
Abrahamson, Clark & LaVon, Axtell, 1986
Adam, Mildred, Hildreth, 2007
Albers, Henry, Minden, 1992
Anderson, Lloyd, Minden, 2013
Anderson, Mrs. Lawrence, Axtell, 1982
Anderson, Robert, Minden, 1979
Anderson, Roy, Minden, 1985
Ard, Bob & Linda, Burchell Family Partnership, Minden, 2015
Bayer, Joe, Minden, 1974
Binderup, Mark and Marian, Minden, 2003
Bloomfield, Willis, Minden, 1976
Boisen, Doug, Minden, 2009
Boisen, Warren, Minden, 1988
Boostrom, Willard, Axtell, 1985
Borgaard, Robert & Wanda, Minden, 1998
Boudreau, Michael & Jodi Boudreau,, 2005
Brooks, Gene, Minden, 1986
Bruce, Beulah & Glenn, Axtell, 1989
Burchell, Alvin, Minden, 1981
Burchell, James, Minden, 1988
Cameron, June, Hastings, 2001
Carlson, Norris, Axtell, 1984
Carlson, Weldon, Holdrege, 1985
Casper, Steven G., Minden, 2008
Cavanaugh, Mike & Janet, Minden, 2002
Cederburg, Reynold, Axtell, 1988
Chappell/ Saltzgaber, Mona / Pearl, Minden, 1980
Christensen, Elmer & Lydia, Minden, 1983
Christensen, Lois, Minden, 1994
Coon Jr, Glenn, Heartwell, 2001
Danielson, Harlan, Axtell, 1979
Davidson, Elmer, Minden, 1977
Detwiler, Thomas, Chadron, 1994
Dornhoff, Pansy, Minden, 1987
Dornhoff, Richard & Goldie, Heartwell, 1993
Dunmire, Donald, Minden, 1991
Ellis, Orval, Minden, 1981
Ericson Farms Inc, Axtell, 1992
Ericson, Mrs. Vera, Axtell, 1985
Falk, Donald, Holdrege, 1990
Favinger, Marvin & Wilma, Minden, 2001
Felzien, Winifred & Kermit, Minden, 1989
Filbin, Richard, Heartwell, 1978
Foster, Marilyn & Clifton Foster, 2005
Frisbie, Leonard & Jean, Campbell, 1986
Frost, Pearl M., Colubus, 1990
Gemmell, Allan, Eugene OR, 2001
Gustafson, Wayne, Kearney, 2002
Hagemann, Leona, Wayne, 2002
Hansen, Alfred, Minden, 1998
Hansen, Alma, Minden, 1984
Hansen, Dora, Minden, 1987
Harkness, Shirley, Cozad, 2009
Headberg, Raymond, Holdrege, 1985
Hoban, Evelyn, Heartwell, 1980
Hoban, James & Rosemary, Heartwell, 2000
Hoban, Roxie, Kenesaw, 1980
Holl, Julius, Kenesaw, 1990
Holl, Mrs. Lorenz, Shelton, 1994
Holstein, Verlene, Minden, 1999
Hubbard, Elihu Family, Holdrege, 1987
Jelden, Leland & Margaret, Hildreth, 2010
Johnson, Harold, Minden, 1986
Johnson, Oveline, Campbell, 1984
Johnson/ Mehrten, Rita/ Viola, Kearney, 1985
Jorgensen Jr., William and Meribelle, Minden, 2008
Jorgensen, Mr. & Mrs. Wilbur, Hastings, 1985
Joyce, Tad, Heartwell, 1999
Jurgensmier Farms, 1989
Kahle, Mary, Earl Kahle Trust, Minden, 2011
Kauffelt, Carl, Minden, 1993
Keen, Chester, Minden, 1999
Knaus, Ronald, Hildreth, 2009
Krick, Donald, Kearney, 1978
Larsen, Charles J and Nancy L, Minden, 2003
Larsen, Mr. & Mrs. David, Minden, 1985
Larson, Arthur, Axtell, 1982
Layton, Donald, Minden, 1999
Liebers, Max, Frederick, Jean, Minden, 1989
Lindgren, Tennis, Axtell, 1993
Lundstrom, Conrad, Minden, 1994
Lynn, C. W., Minden, 1979
Mankin, Dorothy C., Hastings, 2004
Meisenbach, James, Minden, 1978
Mitchell, Margaret, Cletish Mitchell, Nancy Petersen, Kearney, 2009
Muller, W & D Family Partnership, Kearney, 2014
Murray, Doris Irene Bleck, Axtell, 1993
Needham, Dale & Gwen, Campbell, 2015
Nelson, Delmar & Bernice, Kearney, 2001
Nelson Steen, Ruth Lee, Minden, 2007
Nelson, Amy, Minden, 1988
Nelson, Clarence, Minden, 1989
Nelson, Clarence & Lorraine, Minden, 1979
Nelson, Dennis, Axtell, 1978
Nelson, Gilbert, Upland, 1985
Nelson, Mrs. Winifred, Wilcox, 1983
Newbold, Duane & Minna, Minden, 1978
Nielsen, Orville & Selma, Aurora, 1987
Olson, Edward, Holdrege, 1990
Olson, Mary Ann, Axtell, 2003
Olson, Wayne, Minden, 1989
Osborn, Genevieve, Axtell, 2010
Petersen Land Inc., Minden, 2006
Petersen, Vernon S., Scottsbluff, 2006
Peterson, Anna, Minden, 1981
Peterson, Eldon, Axtell, 1983
Peterson, Ellery, Axtell, 1989
Peterson, Richard, Minden, 1985
Pinn, Louis Calvin, Minden, 2008
Raun, Chris, Minden, 1983
Reynolds Sager, Judith A., Belmont, 2006
Rost, J. Waldo & Marcia, Holdrege, 1984
Ryan, Michael & Cheryl, Heartwell, 2000
Ryan Farms Inc., Heartwell, 2007
Ryan Sr., Patrick, Axtell, 1985
Ryan, Eugene, Heartwell, 1988
Ryan, Jeff & Laura, Heartwell, 2013
Scheidies, Randy & Lavern, Minden, 2001
Schmidt, Martin & Marcele, Minden, 1979
Schmidt, Otto Max, Minden, 1978
Seven A Farms Inc., Axtell, 1993
Shearer, Robert, Minden, 1998
Sheffield, Edna Lundeen, Wilcox, 1993
Shurmann, Erma, Hildreth, 2002
Soderquist Farms LLC, Linda Soderquist, Shawn Soderquist, Michelle Soderquist Hartnett, Ryan Soderquist, Axtell, 2011
Sones, Georgene & Doyle, Omaha, 1990
South Winds, LLC, Steve & Julie Roeder, Kendra Roeder, Mitch & Jodi Bendix, Kearney, 2012
Space, Dale & Vivian, Minden, 1999
Space, Royce and Carolyn, Minden, 2003
Stade, Gary, Kenesaw, 1991
Stehlik, Theresa, Lincoln, 2013
Stratman, R. H. & Illene, Minden, 1993
Streff, James & Laurayne, Hastings, 1990
Strolberg, Marjorie, Axtell, 1992
Stuehm, Matilda, Kearney, 2007
Swanson, Robert, Vicki, Forrest & Trevor, Wilcox, 2006
Thomsen Family Corporation, Minden, 2010
Thomsen, Darlene, Hastings, 1989
Tomsen, Mrs. Nels & Family, Minden, 1988
Volentine, Wilbur, Kearney, 1985
Warp, Harold, Minden, 1976
Wendell, Ernest, Axtell, 1983
Wendell, Leona, Axtell, 1991
Werth, William, Minden, 1988
Wilkinson, James, Minden, 1979
Wommer, Virgil Ray & Lee Isla Mae, Minden, 1983

Keith County

Beam, Lanny, Mark, Rex, Holdrege, 2013
Beckius, Bernard and Cheryl, Keystone, 2008
Dresslar Join Venture, Brule, 1995
Geisert, Gerald Lee, Ogallala, 1995
Haldeen, Harold & Donald, Brule, 1985
Halligan, Inc., Ogallala, 2006
Hardessen, Carroll & Mary, Kearney, 2015
Harms, Inc, JoAnne Harms, Brule, 2009
Haythorn Land & Cattle Co, Arthur, 1984
Haythorn Ranch (Waldo & Criag Haythorn), Arthur, 1984
Johnson, Caleb W., Ogallala, 2014
Jorgensen McCracken, Francis & Karen, Clarkdake AZ, 1995
Keystone Ranch Co., Ogallala, 1984
Krab Inc., Paxton, 1991

Lawler Farm & Ranch, Paxton, 1984
Linquist Horner Co, Ogallala, 1986
Lute, Robert, Ogallala, 1984
McBride, Joe & Sheila, Brule, 2015
McConnell, Mark S. & Alice S. McConnell, Paxton, 2004
McFedden, Francis, North Platte, 1989
Most, Ronald & Kathleen, Ogallala, 1993
Peterson, David, Madrid, 1995
Peterson, Lloyd, Brule, 1993
Pounds, Minnie & Walter, Lemoyne, 1994

Keya Paha

3B Ranch Co., John Barmore, Omaha, 2012
Beattie, Gertrude Lear & Leo, Ainsworth, 1984
Billings-Linse, Merna Billings & Margaret Linse, 2005
Carr, Armond, Springview, 1983
Carr, Florence, Springview, 1999
Carr, Frank & Howard, Springview, 1983
Carr, James A. , Jan Epke & J.D. Carr, Springview, 2008
Clopton, David, Mount Prospect, IL, 2011
Connell, Robert & Donald, Newport, 1984
Cook, Tim, Springview, 1987
Cook, Zelda, Springview, 1987
Dixon, Arnold & Frances, Springview, 1985
Doyle, George W., Newport, 1984
Fullerton, Dennis & Carol, Valentine, 1994
Gierau, Paul & Rodney, Springview, 1984
Goodrich, Mickey & Ruth, Ainsworth, 1982
Hall Land Ltd., Bassett, 2001
Hallock, Glen & Ruth, Springview, 1983
Haugen, Glenn & Sadie, Newport, 1981
Kenaston, Robert, Springview, 1983
Kienke, Louis, Springview, 1984
Kienke, Mrs. Fern, Springview, 1983
Kirsch, Richard & Avis, Springview, 1984
Kurzenberger, Rolland, Springview, 1985
Libolt, Stanley & Janice, Springview, 2006
Lowe, Ruby, Springview, 1982
Munger, Gary & Laurene, Springview, 1985
Owens, Agnes, Sparks, 1987
Palmer, Denis, Springview, 2000
Ruther, Clarence & Agnes, Springview, 2001
Shaw, James, Bassett, 2007
Thiede, Frieda, Springview, 1984
Titus, Gerald & Lois, Springview, 2000
Wentworth, Roger & Edith, Springview, 1987
Weston, Brad, Boyne City MI, 2000
Zirnig, Frank & Orebelle, Sparks, 1983

Kimball County

Atkins, Howard and Peggy, Dix, 2008
Bertramson, Mrs. J. W., Potter, 1989
Brothers, Glen, Dix, 1990
Evertson, David M (Mick) & Mary M (Peggy), Kimball, 2008
Freeburg, Troy & Loretta, Bushnell, 2015
Gustafson, Mrs. Lois M., Dix, 2014
Halstead, Marilyn, Kimball, 1992
Johnson, Verner, Dix, 1985
Johnson, Norman & Betty J, Potter, 1987
Nelson, Harvey, Potter, 1986
Nelson, Leonard & Laura, Potter, 1995
Taylor, Kenneth & Shirley, Kimball, 2007

Knox County

Abbenhaus, Joseph/ Andrea, Bloomfield, 1996
Adams, Scott and Mary, Orchard, 2008
Anderson, Guy & Michelle, Wausa, 2003
Anderson & Sons, Inc, Anna, Grand Island, 2009
Arens, Clair & Diane, Crofton, 1988
Arens, Margaret, Crofton, 2014
Ausdemoe, Irene, Crofton, 1989
Ayers, Harold Willard, Winnetoon, 1979
Beckmann, Ronnie & Yvonne, Bloomfield, 2004
Bengston, Rob and Sybil, Wausa, 2008
Bengtson, Fridolph & Edith, Wausa, 1992
Bonnesen, Hans & LaVeve, Verdigre, 1979
Bruhn Schainost, Carly/ Ruby, Wausa, 1996
Buschelman Family, Crofton, 1983
Bushkamp, Harold & Alvina, Crofton, 1986
Cameron, Melvin & Jo, Niobrara, 1993
Carlson, Gertrude, Norfolk, 2007
Cernousek, Alfred & Bernice, Kearney, 1999
Cihlar, Wallace E., Verdigre, 2008
Condon Farms Inc., Creighton, 1987
Cook, Jr, William, Tamara Cook, Bloomfield, 2009
Dartman, Margaret, Mary Ellen Poppe, Creighton, 2008
Dobrichovsky, Frank, Verdigre, 1980
Doerr, Myron and Shirley, Plainview, 2016
Dysert, Lillie & Dora, Niobrara, 1992
Effle, Roger & Cindy, Orchard, 2000
Ehrenberg, Joe & Berniece, Creighton, 2006
Erbst, Kenneth, Bloomfield, 2008
Foner, Foner Farms LLC, Kenneth Foner, Niobrara, 2012
Franek, Raymond, Verdigre, 2011
Frank, Larry & Marilyn, Verdigre, 1999
Frevert, Adrian & Diana, Wausa, 1992
Fricke, Donald, Bloomfield, 1993
Gdowski, Lawrence & Martha, Fullerton, 2006
Goedeker, Sophie, Winnetoon, 1983
Greckel, Lonnie & Jolene, Bloomfield, 1991
Gregg, John, Crofton, 1979
Grimm, Wilbur & Lenarda, Bloomfield, 1985
Gustafson, Leona, Wausa, 2013
Haineraich, Leonard & Betty, Verdigre, 1986
Haug, Albert & Oscar, Bloomfield, 1981
Havlicek, Edward, Verdigre, 1983
Heggemeyer, Edward, Orchard, 2011
Hines, Lavaine & Dorothy, O'Neill, 1980
Hinzmann, August, Creighton, 1996
Horstman-Fischer, Lyle A. and Maryann, Creighton, 2008
Ibach, Ernest, Bloomfield, 2007
Jacot, Richard & An, Leigh, 1999
Johnson, Arlynd, Wausa, 2005
Johnson, Clayton, Verdigre, 1990
Johnson, Roy, Bloomfield, 1999
Jundt, Leo, Creighton, 1975
Kammer, Dennis & Kay, Bloomfield, 2011
Kienow, Loren, Winnetoon, 1982
Knigge, Lance & Lorie, Verdigre, 2011
Krepel, Bob & Esther, Creighton, 2006
Kreycik, Marvin, Verdigre, 1980
Kruse, William C. & Beulah F., Winnetoon, 2003
Kube, Kevin & Janita, Croften, 2012
Lamprecht Eggen, Beverly, Bloomfield, 1992
Larsen, Chris, Verdigre, 1979
LeMasters, Mrs. Jay, Orchard, 1975
Lemke, Bruce, Bloomfield, 2006
McElhaney, Hazel Marie, O'Neill, 1990
McGraw, Leland, Niobrara, 1973
Meyer, Alvin E., Albion, 2013
Milne, Merle, Creighton, 2000
Minarik, Terry, Sandra Beem, DeLynn Minarik, Omaha, 2007
Mlady, Duane, Bloomfield, 2016
Moles, Linda Lou, Bloomfield, 2012
Moline, Dale & Donna, Bloomfield, 2006
Montgomery, Bernard, Orchard, 1999
Moore, Mrs. Clara, Creighton, 1983
Morman Family Trust, Crofton, 2011
Mormann, Burl, Crofton, 1993
Morrill, Robert & Norma, Creighton, 2000
Motacek, Elvin J., Niobrara, 2008
Nelson, John L., Crofton, 1968
Nelson, Terry & Karla, Wausa, 2015
Neuhaus, Gary & Marinell, Omaha, 2015
Novacek, Mr. & Mrs. William, Verdigre, 1963
Novak, Brandon, Yankton SD, 2000
Otte, Arnold J & Marlene, Ft. Calhoun, 2007
Paesl, Kenneth, Creighton, 2003
Paustian, Raymond, Bloomfield, 1990
Pavlik, Joseph, Verdigre, 1979
Peed, Janice, Verdigre, 2000
Peters, Ralph & Esther, Osmond, 1995
Petersen, Thomas, Bloomfield, 2002
Pischel, William, Verdigre, 1975
Placek, Wilbur, Lorinda Hoferer, Creighton, 2011
Reetz, Carrie, Bloomfield, 1983
Risinger, Mr. & Mrs. Franklin, Brunswick, 1987
Ruzicka, Mr. & Mrs. Emil, Verdigre, 1979
Ruzicka, Robert and Delores, Verdigre, 2008
Ruzicka, Robert, Verdigre, 1977
Schellen, Melvin, Winnetoon, 1983
Schlote, Lisa & Duane, Bloomfield, 2010
Schneider Kershner, Joan, Creighton, 2011
Schroeder, Marian, Bloomfield, 2009
Schroeder, Philip, Bloomfield, 2009
Schroeder, Roger, Bloomfield, 2009
Schumacher, Gary, Crofton, 1979
Schwartz, Don & Mildred, Winnetoon, 1995
Seagren, Elden, Wausa, 1991
Sedivy, Edward, Verdigre, 1978
Shelton, Stanley, Lexington Park, 2006
Skokan, Frank, Verdigre, 1982
Smolek, Gordon & Violet, Verdigre, 2005

Soucek, Evelyn & George, Verdigre, 1991
Stenberg, Joel, Creighton, 2008
Stewart, Richard, Crofton, 2002
Stoural, Mr. & Mrs. Donald, Verdigre, 1983
Strope, Dora, Orchard, 1980
Sukup, Dale & Marilyn, Verdigre, 1998
Sukup, George L. Sukup, Winnetoon, 2004
Sukup, James & Rita, Winnetoon, 1999
Sukup, Kenny & Pam, Verdigre, 1998
Sukup, Lawrence & Helen, Verdigre, 1983
Sukup, Leonard, Verdigre, 1980
Sukup, Lyle, Verdigre, 1998
Sukup, Paul, Verdigre, 1981
Thorell, Floyd & Delores, Wausa, 1990
Thunker, Ethel, Crofton, 1997
Thunker, Verne & Lorrie, and Michael & Barb Jackson, Bloomfield, 2003
Tikasky, James & Karen, Niobrara, 1981
Tramp, Ronald and Bernice, Crofton, 2016
Trenhaile, Tom & Mary, Bloomfield, 2006
Tuch, Helen, Niobrara, 2010
Uhlir, Arden, Verdigre, 1976
Utecht, Dennis, Norfolk, 1991
Vesely, William, Verdigre, 1980
Von Seggern, Virginia & Burnell, Orchard, 2012Wagner, Herman, Creighton, 1980
Wagner, James & Rose, Winnetoon, 2009
Wagner, James, Winnetoon, 2002
Walter, Elvina, Crofton, 1999
Walter, Emma, Crofton, 1982
Walter, Frank, Crofton, 2002
Warriner, JoHanna, Center, 1989
Woodworth, Laura, Creighton, 1974
Young, Bradley & Jeanne, Wausa, 2005
Zoubek, Michael, Lincoln, 1995

Lancaster County

Aksamit, Edward & Gertrude, Hallam, 1976
Almy III, Horace, Greenwood, 1979
Anderson, Hilbert, Waverly, 1983
Anderson, Mr. & Mrs. Willard, Ceresco, 1977
Armstrong, Pauline, Lincoln, 1979
Beall, Elizabeth & Oma, Roca, 1986
Beck, Warren, Crete, 1979
Berg, Paul & Joan, Lincoln, 1984
Bingham, Lucile Trustee, Davey, 2004
Birt, Earl, Roca, 1978
Black, Robert, Diane Black, Ceresco, 2007
Block, Richard & Sondra, Roca, 1998
Blome, Aidean & Rosalee, Hallam, 1983
Boesiger, Mrs. Alice-Leone, Firth, 2014
Booth, Edwin, Walton, 1967
Booth, Joseph, Walton, 2000
Booth, Mark & Bonnetta, Walton, 1967
Borgman, Harry, Lincoln, 1966
Bosak/ Kasper, Mary/ Margaret, Omaha, 1981
Bouma, Mr. & Mrs. Arnold, Firth, 1979
Bouma, Mrs. Dillie, Panama, 1981
Brandt Farms Inc., Martell, 1996
Brockhoff, Rick & Randy, Bennet, 2015
Brown, Russell, Waverly, 1975
Buel, Phil & Josie, Hickman, 1977
Buffum, Dennis & Tammy, Ceresco, 1986
Burcham, Myrtle E Burcham,, 2005
Burgess, Clyde, Lincoln, 1999
Burgess, Joseph, Lincoln, 1983
Caha, Deborah and Mark, Ceresco, 2008
Chambers, Marvin & Jean, Bennet, 2001
Chrastil, Stanley & Mary, Hallam, 1984
Cropsey, Harry, Lincoln, 1987
Daugherty, Norma Jean, & Harvey Daugherty, Martell, 2010
Deinert, Harold & Verna, Lincoln, 2002
Deinert, Melvin & Linda, Lincoln, 2002
Deinert, Ronald & Ela, Malcolm, 1990
DeVries, LeRoy & Rosemary, Hickman, 1990
Doeschot Boettcher, William & Shirley Ann, Longmont, 2006
Egger, Dennis & Sharon, Hickman, 1999
Egger, Merlyn & Delores, Roca, 1981
Ehlers, Donald, Roca, 1981
Ehlers, Glen & Stella, Roca, 1999
Erskine, Jr., Horatio, Eagle, 1981
Erskine, Jr., Orest, Lincoln, 1985
Etmund, William, Lincoln, 1988
Faig, Willie, Martell, 1969
Finigan, Mr. & Mrs. Paul, Lincoln, 1977
Francke, Albert A. & Albert H, Lincoln, 1980
Francke, Elmer, Lincoln, 1979
French, Marty Joe & Lois, Lincoln, 1999
Furrer, John & Ferne, Lincoln, 1993
Gates/ Brown, Denise/ Diane, Martell, 1991
Genuchi, Albert/ Marie, Bennet, 1996
Giles, Raymond, Denton, 2000
Gordon, Norma Jean, Sedalia MO, 2001
Grage, Elmer & Vera, Crete, 1983
Gropp, Dale & Diane, Crete, 2002
Hall, Marvin & Doris, Lincoln, 1980
Hall, Vernon & Elizabeth, Greenwood, 1982
Halling, Bert & LaVerne, Waverly, 1983
Halling, Mr. & Mrs. Gerald, Waverly, 1982
Hansen, Arlie Dale & Ann, Ceresco, 1989
Hass, George, Malcolm, 1983
Heidtbrink, Lester N. & Cleora C., Malcolm, 2004
Henning, Benjamin & Cora, Hallam, 1984
Henning, Gerald & Donna, Crete, 1983
Hesser, Harold & Johanna, Adams, 1992
Hesser, Johanna, Adams, 2003
Hestermann, Margaret, Walton, 1990
Hietbrink, Marvin & Mabel, Firth, 1991
Hines, Gladys & James, Martell, 1970
Hockenbary, Robert, Lincoln, 1984
Hodgin, Clyde & Buanta, Ashland, 1978
Holland, Orpha, Lincoln, 2002
Hoppe, R. B., Denton, 1998
Hornung, Todd, Cindy Hornung, Lori Block, Wahoo, 2009
Hutchinson, Francis, Waverly, 1997
Jacoby, Henry & Mabel, Lincoln, 1970
Johnson, Cheryl R., Walton, 2014
Johnson/ Peterson, Clifford/ Doris, Lincoln, 1977
Juricek, John & MaryAnne, Denton, 2009
Kane, Mr. & Mrs. Henry, Valparaiso, 1979
Kats, Cornelius, Firth, 1983
Kats, John & Opal, Firth, 1983
Keel, Charles, Roca, 1989
Keel, Jr., Charles, Roca, 1987
Keim, Antoinette, Walton, 2007
Ketelhut, Orval & Duane, Walton, 1982
Ketelhut, Winona Francke, Lincoln, 1995
King, Harlan C. & Ruby R., Bennet, 2004
King, Keith & Doris, Bennet, 1999
Kratochvil, Stephen & Georgia, Raymond, 2002
Kreiner, Richard & Delores, Davey, 2000
Krueger, Della, Roca, 1984
Krueger, Marian, Firth, 1992
Krueger, Richard & Bonnie, Roca, 1984
Krull, Charles Family, Sprague, 1967
Kunz, Leona A. Family Trust, Denton, 2013
Kuster, Ervin & Marilyn, Roca, 1986
Kuster, Harlan & Phyllis, Hickman, 2003
Kuster, Ivan & Esther, Roca, 1975
Kuster, Wendell & Wilma, Roca, 1986
Lange, Virgil & Bernice, Malcolm, 1983
Larsen, Vernon & Lillie, Lincoln, 1977
Lemke, Milton, Walton, 1969
Lemke, Robert, Walton, 1981
Lesoing, Melvin & Lorraine, Hickman, 1990
Lewis, Mrs. Ruby, McCook, 1975
Lidolph, Mrs. L.C., Roca, 1975
Loder, John, Waverly, 1957
Lohmeier, Earl & Adeline, Lincoln, 1970
Lostroh, Melvin & Joanne, Lincoln, 1970
Maahs, Mr. & Mrs. Clarence, Walton, 1975
Madden, Donald & Cleo, Martell, 1981
Marolf, Darleen, Waverly, 2001
Merriman, Fred & Marie, Lincoln, 1985
Meyer, August, Lincoln, 1975
Meyer, Ronald & Linda, Hallam, 2002
Mills, Freda, Lincoln, 1973
Mitchell, Donald, Roca, 1976
Mocroft, Thomas & Mary Jane, Waverly, 1984
Molzer, Emil & Roberta, Hallam, 1984
Moser, Brad, Hallam, 2002
Mulder, Ray & Marian, Firth, 2002
Mundt, Leon & Ruth, Crete, 1984
Nelson, J. William, Bennet, 1972
Nelson, Lynette, Davey, 2011
Nelson, Zelda, Omaha, 1972
Nielsen, David, Lincoln, 2014
Nitzel, Howard, Donna Hillgren, Crete, 2008
Nordstrom, J. Steve, Waverly, 2002
Nordstrom, Paul & Kathleen, Waverly, 1993
Northrop, Viola, Martell, 1991
O'Donnell, Tom & Pauline, Davey, 1983
Oelling, Ronald & Judeen, Roca, 2006
Oelling, W. Jean, Lincoln, 1999
Oldemeyer, Virgil & Shirley, Firth, 1983
Olson, Ralph & Sara, Ceresco, 1998
Ossenkop, Carl, Lincoln, 1984
Ossenkop, Dale, Lincoln, 1984
Pallas, James, Ceresco, 1979
Parks, Don & Grace, Greenwood, 1970
Parrott, Harold & Lynnelle, Lincoln, 1989
Paschold, Harold & Aldyth, Firth, 1983

Payne, Mr. & Mrs. Murle, Lincoln, 1983
Peckham, C. Gladys & Harry, Lincoln, 1982
Peterson, Melvin & Violet, Martell, 1987
Pickel, Catherine & Robert, Lincoln, 1966
Pickel/ Nord, Grace/ Erma, Raymond, 1981
Piening, Burdette & Virginia, Lincoln, 2011
Polak, DeRoyce J., Valparaiso, 2013
Pomajzl, Rudolph, Crete, 1975
Prophet, Norma & Russell, Firth, 1978
Rapp, William, Denton, 1994
Reddish, Arthur & Lorine, Lincoln, 1970
Reihart, Josephine, Lincoln, 1998
Retzlaff, Ernest, Walton, 1957
Retzlaff, Robert, Walton, 2014
Ringland, LeRoy, Bennet, 1990
Rippe, Jack & Rebecca, Lincoln, 1981
Rohrs, Ruth & Albert, Martell, 1978
Rolofson, Mrs. Virginia S., Attn: Benjamin L. Rolofson, Lincoln, 2003
Schaal, Janice, Lincoln, 1993
Schlichtemeier, Mr. & Mrs. Melvin, Hallam, 1967
Schnieder, Orin & Ruth, Hallam, 1990
Schrader, Richard, Lincoln, 1973
Schultz, Wilma, Walton, 1996
Severin, Raymond & Eleanor, Martell, 1989
Sieck, Ardell & Fran, Crete, 2009
Sieck, Dale, Martell, 1979
Siekman, Fern Erskine, Lincoln, 1982
Sittler, Harvey & Vera, Martell, 1973
Sittler, Melvin, Martell, 1977
Skinner, M. Eleanor, Lincoln, 1982
Sklenar, Evelyn, Ceresco, 2008
Smith, Grace Morgan & Orville W. Smith, Davey, 2004
Smith, Julia, Wilber, 1984
Spellman, Sam & Berdina, Lincoln, 1967
Speth, Ora, Hallam, 1984
Stahly, Robert & Gertrude, Hickman, 1999
Sullivan, Eugene & Laura, Denton, 1979
Sullivan, Harold, Lincoln, 1987
Swanstrom Two, Wylene, Sandy UT, 1999
TeKolste, Ronald & Ardith, Firth, 2006
Templeton, Virgina, Roca, 2000
TenHulzen, Amy & Paul, Firth, 1980
TeSelle, Agnes & Lurence, Martell, 1975
Tiedeman, Melvin & Beverly, Hickman, 1985
VanDerslice, Stan, Lincoln, 1993
Vannoy, J. Warren, Lincoln, 1984
Warner Hereford Farms, Waverly, 1983
Warner, Mabel, Waverly, 1976
Wells, Beulah, Hallam, 1983
Wendelin, August & Phyllis, Hallam, 2004
Wetenkamp, Woodrow & Lola, Lincoln, 1966
Wiechert, Arthur, Lincoln, 1983
Williams, Harold & Myra, Hickman, 1970
Williams, Mrs. Hartley, Lincoln, 1969
Williss, Glen & Winifred, Lincoln, 1982
Wilson Moore, Julia May, Lincoln, 1966
Winch, Stanley & Leoma, Roca, 1993
Wismer, Emma, Hickman, 1998
Wissenburg, James & Theresa, Crete, 1987
Wittstruck, Craig/ Colleen, Denton, 1996
Wittstruck, Dwayne & Joan, Denton, 1995
Wittstruck, Harlan & Leona, Roca, 1967
Wright, Gladys, Phoenix AZ, 2002

Lincoln County

Atkinson, Mrs. Vernon, Gothenburg, 1980
Auble, Reta, Arnold, 1993
Bay, Riley & Barb, Arnold, 2004
Bergstrom, Vincent, North Platte, 1994
Bloom, Harold, Brady, 1992
Brittenham, Eldon & Shirley, Moorefield, 2001
Brogan Trust, Alice, Keystone, 1988
Brogan, Bernard, Keystone, 1988
Clymer, Jean, Brady, 1987
Cohen, Gerald M. - Cohen Land LLLP, North Platte, 2004
Dahl, Ruth M., Sun City West, AZ, 2003
Downs, Laura R., Hershey, 2003
Elsasser, Evelyn, Brady, 1993
Foran, Carolyn and Robert, North Platte, 2008
Fox, Benjamin & Mary Ann, Arnold, 2007
Fulk, Fern, Wallace, 1991
Gaudreautt, Donald & Janice, Farnam, 1993
Greenwood, Regis & Marguertie, Wellfleet, 1985
Haist, William, Hershey, 1992
Hammond, Gail & Richard, Curtis, 1985
Hansen Ranch Company, North Platte, 1993
Hansen, Lawrence & Gloria, Wellfleet, 1987
Hansen, Marguertie, Curtis, 1993
Hansen, Vernon & Cora Jane, Curtis, 1993
Hoatson, Claude & Blaine, Hershey, 1993
Huebner, George, Hershey, 1985
Isackson, Richard, Brady, 1985
Jensen, William, Curtis, 1979
Johnson + Pamela J Weekly, Tedd L & John C, North Platte, 2015
Johnson, Don & Rosemary, North Platte, 1998
Koch, Wayne & Mae, Hershey, 1993
Lemoin, Kenneth, Wallace, 1984
Long, Robert, North Platte, 1993
Lynn, Marjorie, Gothenburg, 1980
Maham Family Trust, North Platte, 1999
McCune, Myron W. & Kathleen M., Sutherland, 2014
McNickle, Dale, Farnam, 1992
Merrill, Phil & Marian, Hershey, 2007
Meyer, Virginia, Wellfleet, 2005
Miller, Lois Y., Brady, 2004
Mortensen, Ivan, Curtis, 1992
Murray, Lorraine Gail, Ozark, 2008
Peckham, Hubert, Gothenburg, 1980
Peckham, Rich, Brady, 1984
Peterson, Brent & Deanna, Brady, 2005
Raney, Douglas & Karol, Wallace, 2007
Rielly, Mrs. Mary Frances, Brady, 1982
Roberts, Ronald, Maxwell, 1982
Scherz Family, North Platte, 1971
Sellers, Harrison & Marion, Wellfleet, 1992
Steel, Edd & Janie, North Platte, 1991
Sullivan, Edward, Maxwell, 1982
Voss family, Lincoln, 2013
Votaw, Delmar, Wellfleet, 1984
Votaw, Eli, Wellfleet, 1984
Walz, Leo Carl, Stapleton, 1992

Logan County

Brosius, Vernon, Stapleton, 1984
Diamond Bar Ranch, Stapleton, 2001
Hall-Schloz Partnership, Lamoine Hall, 2005
Halstead, John Irwin, Arnold, 2013
Joedeman, Harold C., Stapleton, 2008
Loughery, Hornbacher & Diane, North Platte, 1984
Rogers, Anna, Dunning, 1993
Salisbury, Larry, Stapleton, 1984
Santo Land & Cattle Co, Stapleton, 2006
Trumbull, Jacqueline R., Douglas, 2008
Tunnell, Adair, Stapleton, 1992
Wells, Ann, Wells Land Limited, Stapleton, 2006
Winkler, Marilyn, Loomis, 1999

Loup County

Britton Ranch Inc., Taylor, 1987
Dilsaver, Lyle, Taylor, 2001
Dilsaver Owen, Irma, Milburn, 1984
Dunbar, Bob & Kristy, Milburn, 2003
Dunbar, L.D. & Inez, Milburn, 1988
Dunbar, Paul and Debra, Taylor, 2010
Dunbar, Willis & Enola, Taylor, 2003
Fales, Hannah, Burwell, 1990
Hall, Myrtle, Sargent, 1989
Hastert/Bromwich, Ava/Joy, Sargent, 1979
Hetfield, Beverly, Cairo, 2015
Homestead Inc., Burwell, 1993
Howard, Dwayne & Nelva, Sargent, 1991
Hyde, Melvin, Johnny, Delores, Taylor, 1991
Hyde, Gladys, Taylor, 2010
Hyde, Mable, Taylor, 1988
Janicek, George G., Burwell, 2003
Kraus, Robert & Arlene, Burwell, 2003
Lindsay, Dale, Burwell, 1984
Mann, Gaylord, Burwell, 2004
Moninger, Fred Heirs, Taylor, 1985
Moore, Thomas, Taylor, 1976
Murphy, Donald L., Neva J. & Bryan K., Milburn, 2014
Petersen, H. David & Ruth Ann, Burwell, 1987
Petersen, Russell J., Burwell, 2003
Scherzberg, Eula M. et al, Burwell, 2004
Sheldon, Ruby, Sargent, 2006
Stevens, Glen Family Trust, Taylor, 1991
Strohl , Donzel & Winafred, Taylor, 1992
Strohl Mikkelsen, Geneva, Taylor, 1980
Worth, Walter Family, Taylor, 1985

Madison County

Adelman, Albert, 1974
Aldag, Dennis & Janice, Battle Creek, 1979
Anderson Family Trust, Willis J., Newman Grove, 2008
Anderson, Layton, Newman Grove, 1973
Armstrong, William David, Russellville, AR, 2015
Avery, William J. & Chera Lee, Norfolk, 2012
Bickley, Lyle & Alfreda, 2006
Boschen, Dennis & Gloria, Newman Grove, 2002
Boyer, Eugene & Doris, Tilden, 1980
Boysen, Wiliam, 1980
Bradford, James, Norfolk, 1995
Brahmer, Mrs. Dave, Norfolk, 1984

Brand, Leo, 1979
Brand, Leo, 1982
Bredehoft, John, Battle Creek, 1976
Brown, Margaret & James, Tilden, 1982
Bruan, Lena Braun, Norfolk, 2005
Choutka, Alvin, 1992
Deckert, Archie & Ruth, Columbus, 1984
Deering, Arnold, Norfolk, 1972
Deuel, Cleo, Norfolk, 1973
Dittberner, Clarice, 1973
DittbernerMADISON, Jack & Hildagard, Humphrey, 1985
Dorr King, Martha, Newman Grove, 1992
Dorr, Ella Barney, 1971
Dover, Earl & Lisetta, 1973
Dover, Mr. & Mrs. Ralph, 1973
Duhachek, Raymond, Meadow Grove, 1993
Eucker, Bertha & Arthur, 1976
Feldmann, Eudell & Verna, Meadow Grove, 1990
Frederick Terry Estate, Tilden, 1985
Freudenburg, Glen, Tilden, 2010
Freudenburg, Harlan & DeLene, 1995
Freudenburg, Melvin, 1999
Frisch, Allan & Evelyn, 1992
Gambill, Dorothy, Newman Grove, 1996
Gambill, Richard, Johnstown, 1997
Gross, Charles & Laura, 1976
Hanke, Delores, 2008
Heath, Janet, Cedar Bluffs, 2000
Herbert, Lucille, Newman Grove, 2006
Hoefelman, Loreen & LaVern, Humphrey, 1986
Hoepfinger, Morris, Neligh, 1980
Hohenstein, Emma, Norfolk, 1979
Horrocks, Clyde, Meadow Grove, 1983
Horrocks, Dale & Doris, Meadow Grove, 1990
Hughes, Nancy, Lawrence KS, 1994
Hughes, Steven, Towson MD, 1994
Jansen, Herbert, Norfolk, 1992
Kleensang, Lonnie, Norfolk, 1999
Klein, Ernest & Adela, Battle Creek, 1994
Knapp Jr., Adolph, 1978
Knapp, Afred, Battle Creek, 1981
Knapp, Milton & Norma, Norfolk, 2006
Knight, Elizabeth, Norfolk, 1975
Koenig, Donald, Newman Grove, 2000
Kohl, Larry, 1988
Kohl, Ralph, Newman Grove, 1988
Korth, Vernell & Lugene, Lindsay, 2005
Kucera, Regan & Dawn, 2012
Kuchar, Edward, 1986
Lauver, Mr. & Mrs. Donald, Battle Creek, 1972
Lindsley Hughes, Nancy Steven, Lawrence, 2013
Loyd, Bonnie, Norfolk, 2007
Lulow, Mr. & Mrs. Walden, Meadow Grove, 1989
Makelin, Glenn & Betty, 1990
Makelin, Glenn & Betty, 1998
Markussen, Karen, Karrol Kamrath, & Kathy Valnes, Omaha, 2004
Marquardt, Erwin, Norfolk, 1980
Maurer, Roy & Verena, Humphrey, 1969
McCallum, Janice & Jerry, 1997
McLow, Rubie, Norfolk, 1975
Miller, Melvin & Annette, Battle Creek, 2001
Mintken, Mr. & Mrs. Elmer, Humphrey, 1974
Moenning and Roberts, Larry E. and Lucille A., Norfolk, 2013
Munger, Donald, Newman Grove, 1992
Nathan, Evelyn, Derald Nathan, Lindsay, 2007
Nathan, Fred & Robert, Humphrey, 1981
Nathan, Vertis, Humphrey, 2016
Nathan, Walter & Edna, 1984
Neidig, Edith & Bryce, 1992
Neidig Anderson, Florence, Elgin, 1981
Nelson, Mr. & Mrs. Craig, Newman Grove, 1992
Nelson, Mrs. Eva, Newman Grove, 1978
O'Brien, James & Jeanne, Tilden, 1982
Ochsner, Morris, 1992
Patzel, Danny & Peggy, Newman Grove, 2000
Pittack, Fredrick, Tilden, 1985
Pospisil Jenkins, Lee & Elsie, Newman Grove, 1995
Potmesil, John c/o Karen Etling, Newman Grove, 2004
Powell, Orlo, Wethersfield CN, 1994
Praeuner, Norman & Adelyn, Battle Creek, 1978
Prauner, Martin, Battle Creek, 1969
Preuss/ Gellerman, William/ Doris, Seward, 1979
Purchal, Matt & Darlene, Lindsay, 2002
Raasch, Darius, Norfolk, 1974
Raasch, Ferdinand, Norfolk, 1976
Radenz, Donald, Norfolk, 2005
Reeker, Gene & Ruby, Meadow Grove, 2010
Reeves Killion, Viola, Meadow Grove, 1980
Reeves, Richard, Meadow Grove, 1978
Roberg, Donald C. & Frances/ sons Dean & David, Newman Grove, 2004
Schmidt, Emogene, 1994
Schott, Victor & Adeline, Battle Creek, 1977
Schroeter, Paul & Helen, Lindsay, 2010
Schulz, Arnold & Elaine, 1994
Seegebarth, Delbert, Norfolk, 1976
Shabram, Lillie, Tilden, 1989
Spinar, Bonnie, Bruce Dittberner, 2008
Stolle, Orlin J. LeOra, Meadow Grove, 2003
Storck Novatny, Donald & Mary Ann, Lindsay, 2006
Sueper, Ferd, Humphrey, 1981
Sueper, Mark, Lindsay, 1981
Sunderman, Ella, 1979
Tannehill, George, Norfolk, 1985
Tiedgen, Fred & Dee, Meadow Grove, 1998
Tiedgen, Robert & Mary, Norfolk, 1974
Tiedtke, Irene, Norfolk, 1992
Uecker, David, Norfolk, 1976
Vaage, Delores, Newman Grove, 1991
Volk, Donald, Norfolk, 1993
Wachter, Paul, Norfolk, 1976
Wallin, Merlin/ La Vede, Newman Grove, 1996
Warrick, John & Robert, Meadow Grove, 1979
Wedekind, Allen & Janice, Lindsay, 1998
Wedekind, Mr. & Mrs. Glen, 1982
Wegener, Clifford, Norfolk, 1985
Weiand, Mervin, Newman Grove, 1992
Weinberger, Joseph, 1978
Werner, E. John, Battle Creek, 1990
Werner, Mildred, Meadow Grove, 1992
White, Gary, 1985
Willems, Melvin H. Revocable Trust, West Hills, 2014
Winter, Douglas, Olathe, 2016
Winter, James, Norfolk, 1978
Witzel, Dennis & Janie, 1983
Zessin, Robert & DeLoris, 2009

McPherson County

Compton, Carl, Derby, KS, 2003
Connell, Lloyd, Tryon, 1987
Neal, LaVerne & Doris, Tryon, 1988
Neal, Sanford, Tryon, 1990
Pinkerton, Marjorie J., North Platte, 2003
Rosalie, Ramona, Roger Tederman, Gothenburg, 2013
Ten Bar Ranch, Inc., James G. Snyder, Pres., Sutherland, 2003
Waits, Vernon, Tryon, 2007

Merrick County

Aurand, Bruce, Chapman, 1976
Bader, Harold & Evelyn, St. Libory, 1980
Baird, Robert C., Central City, 1973
Benner, Mr. & Mrs. Charles, Central City, 1976
Benson, Clarence & Louise, Chapman, 1993
Bock, Marie, Blair, 2005
Boelts, Frederick H., Archer, 1993
Bosselman, Charles & Norman, St. Libory, 1978
Burger, Jack & Dorothy, Silver Creek, 1998
Clark, Donald Y., Central City, 1974
Clayton, Gary K. and Diane, Mesa, 2008
Daberkow, Julius, Grand Island, 1984
Dammann, Arnold, St. Libory, 1975
Dammann, Cork & Sandra, St. Libory, 1983
Deichmann, David, Central City, 2006
Djernes, Marilyn & Edgar, Archer, 1980
Duennerman, Lonny/Wanda, Grand Island, 1996
Dugger, Glen & Etta Belle, Palmer, 1986
Fauquier Estate c/o Elizabeth Frauquier Mason, Chapman, 1976
Ferris, Donna, Central City, 1999
Ferris, Donna, Central City, 2002
Ferris, Lisle W., Archer, 1977
Fish, Ruth A., Clarks, 1971
Frauen, Mr. & Mrs. Gustav, Grand Island, 1975
Frauen, Otto P.A., Chapman, 1992
Frauen, Vern M. & Bernice C., Grand Island, 1988
Friederichsen-Scarborough, Mildred, Palmer, 2014
Glause, Mrs. Stella R., St. Libory, 1976
Glynn, James, Palmer, 2005
Gray, Donald & Georgie, Central City, 1975
Gress, LaVonne (Wymer), Palmer, 2004
Haddix, Loral A. & Dorothy F., Chapman, 1976
Hanke Farms Inc., (Roger Hanke)", Palmer, 1991
Heins Bollesen, Ella, Grand Island, 1985
Hildur Willhoft, Arnold Willhoft &, Central City, 1970
Holtorf, Hans, Central City, 1978
Hudnall, Warren, Chapman, 1976
Hummel, Raymond, Central City, 1983
Janky, Francis & Mary, Chapman, 1985
Jefferson, John & Lola Felske, Central City, 2004
Johnson, Helen W., Grand Island, 1976
Johnston, Carroll & Nadine, Central City, 1999
Kershaw, Eugene, Silver Creek, 2005

Kershaw, Ron & Janet and Mrs. Donna Kershaw, Clarks, 2005
Key, Estate of Mrs. J.W. c/o Mrs. Leonard Church, Elm Creek, 1981
Klingenberg, Aronold (Jack) A. & Gladys L., Chapman, 1971
Kohl, William H. & Barbara, Clarks, 1994
Kortum, Jack H., Chapman, 1986
Kruse, Louis C. & Mrs. Florence Niedfelt, Grand Island, 1981
Kuck, Harold & Rose, Palmer, 1983
Kuhlman Stelk, Elmer E. & Delma, Grand Island, 1994
Kuhlman, Ann H., Clarks, 1994
Kuhn, Joyce N., Central City, 2012
Kyes, Edward & Marilyn (Holtorf), Central City, 2014
Kyes, James & Maxine, Archer, 1973
Luebbe, Arthur, St. Libory, 1980
Marsh, Marion, Archer, 1973
Marsh, W. Warren, Central City, 1985
Mattson, Gerald & Eleanor, Central City, 1979
McClure, Thomas A. and Janet F., St. Charles, 2008
McHargue, Tom, Central City, 2002
McNeff, Ronald & Nelva, Palmer, 1984
Medinger, Michael, Archer, 2011
Merryman, Gladys Blevins, Chapman, 1994
Micek, Monica J. & James A. Micek, Silver Creek, 1997
Mohr, Peter, Grand Island, 1978
Morgan, Barbara, Clarks, 1975
Nelson, E. Vernon, Central City, 1978
Nelson, Ivan Farms Inc., Central City, 1978
Nelson, Zetha, Columbus, 1996
Newlon, Onilee Wymer (Mrs. Leroy), Palmer, 1976
Nicholas, Victor, Palmer, 1981
Niedfelt, Walter D., Grand Island, 1986
Nitzel, Mr. & Mrs. Lloyd I., Central City, 1976
Ogden, Barbara C., Chapman, 1975
Paczosa, Jerome & Bonita, Silver Creek, 2009
Phelps, Mrs. Harvey, Clarks, 1992
Pitman Stout McAmis, Becky Tamela Pamela, Hastings, 2013
Place, John D. & Olive G. Place, Palmer, 1976
Reeves, Barbara, Central City, 2011
Reeves, Portia, Central City, 1999
Rice, Jasper, Chapman, 1976
Rudolf, Willam A., Palmer, 1976
Ryan, Marion W. & Gwendolyn L., Central City, 1972
Santin, Joseph A., Palmer, 1986
Schaffert, Richard & Stella, Aurora, 1985
Scheer, Elton F. & Betty J., St. Libory, 1993
Schinkel Ruff, Mrs. Odella, Grand Island, 1978
Schipman, Mrs. Reinhard & Sons, St. Libory, 1980
Shelton, William Cheney, Central City, 1985
Shively, Thomas Klar, Clarks, 1980
Shoemaker, William N. & Sarah S. Martin, Grand Island, 1971
Simonson, Frieda & Rev. Armin Wagner, St. Libory, 1981
Simpson, Jerry and Barbara, St. Libory, 2016
Snider, Florence B., Central City, 1977
Soeth, Ella K. & Carol Larson, Chapman, 1983
Sullivan, Lena G., Central City, 1997
Sweet, Ethel, Clarks, 1998
Tague, Harry, Clarks, 1976
Thede, Harry & Rose, St. Paul, 1997
Thomas, Neal, Lincoln, 1998
Ueckert, Donald, St. Libory, 2000
Van Pelt, Charles L. & Esther Van Pelt, Hordville, 1979
von Aschwege, Arnold, Palmer, 1981
Von Ohlen, Fredie & Jetta, Chapman, 1994
Wait, Paul L., Fullerton, 1997
Wegner, Carol, Palmer, 2006
Wegner, Dale & Helen, Archer, 1982
Weller, Rex, Palmer, 2005
Werner, Mrs. Gladys, Grand Island, 1985
White, Mrs. Marion L., Chapman, 1976
Wieck, Alan & Carolyn, Omaha, 1994
Wieck, Gary & Nanna, Grand Island, 1994
Wiegert, Raymond, Grand Island, 1973
Williams, Darwin & Lorraine, Clarks, 1971
Willmes, Rose Marie and Rodney, Hastings, 2012
Willoughby, Darrell, Central City, 2007
Windfield, Edgar E., Chapman, 1978
Zamzow, Otto B, Archer, 1974

Morrill County

Coulter/ Schruben, Neta/ Edna, Scottsbluff, 1988
Covalt/C&L Corp., John, Angora, 2013
Dueker, Glenn, Scottsbluff, 2000
Gillespie Dykman, Betty, Dalton, 2007
Gilroy, James & Rose Ann, Dalton, 2011
Juelfs Ranch, Inc., Bridgeport, 2008
Kizzire, Doug, Bayard, 2014
Lease, Jennie, Bridgeport, 1986
Leytham, Ray, Bayard, 1986
McConnell, Rodney, Broadwater, 1989
Reimers Heizer, Lois, Dalton, 1996
Russell, Joyce & Lew, Bayard, 2008
Russell, Joyce F., and Jeannine Tyree, Bayard, 2010
Sauder, Roy W. and Anna May, Bridgeport, 2003
Sweem, Robert, Bayard, 2010
Waitman, John, Bridgeport, 1988

Nance County

Anderson, Cressa & Arthur E. Jr., Estate, Genoa, 2015
Bialas, Johnnie and Maxine, Fullerton, 2016
Brandenburger, Norman, Silver Creek, 2011
Cook, Homer, Belgrade, 2005
Crouse, Eddie, Fullerton, 1995
Czarnick, Emil P., Genoa, 2006
Doris M. Ekwall, Carol M. Nelsen, Valentine, 1987
Fehrs, John & Sharon, Norfolk, 2011
Ferris, Stanley D & Gail R, Central City, 2015
Forbes, Robert, Palmer, 1989
Francis, Wendell, St. Edward, 1991
Hellbusch, Roger, Belgrade, 2006
Hickerson, Melvin C., Fremont, 1992
Kaup, Lillian, Genoa, 2013
Kayton, JoAnn, Omaha, 2003
Kuhn Farms Inc. c/o Mrs. Bill Kuhn, Clarks, 1982
Kush Farms, Louis P. Kush, Alphonse A. Kush, Rita M. Tate, Leonard J. Kush, Raymond J. Kush, Elizabeth A. Zakrzewski and spouses, Genoa, 2012
McBride, Thelma P. Trust (Ruth A. Mendyk), Phillips, 1993
McCoig, Dennis & Lynda, Genoa, 2004
McIntyre, Thomas W., Wolbach, 1993
McKeeman, Ben, Belgrade, 1984
McNeff, Robert & Karen, Palmer, 2014
Nelson, Raymond & Jean, Genoa, 2005
Ober, Lucia Trust, Denver, 2013
Pelican, John and Shirley, Palmer, 2008
Peterson, Curtis & Anne, Genoa, 1990
Peterson, Phyllis E. (Janice R. Frankel & Jacqueline L. Peterson), Fullerton, 2003
Peterson, Richard, St. Edward, 1984
Rathje, Laverne, Fremont, 2014
Russell, Albert, Fullerton, 2005
Russell, John & Ethelene, Fullerton, 2007
Russell, Wilma, Fullerton, 2016
Shotkoski, Henry J. & Carolyn K., Genoa, 1989
Sock, Kendal & Nikky, Ray & Sharon, Genoa, 2011
Stevens, Rogalyn/Roberta Frazey/Ruth Anderson, Columbus
Tarnick, Stanley E. & Sandra C., Genoa, 1989
Taylor, Warren L. & Eugene E., Albion, 1983
Ulferts, Paul & Lois, Fullerton, 1994
Underwood, Claire, Harold, and Duane, Fullerton, 1985
Williams, James V. & Lavanna, Highland, CA, 2003
Wright, Jack & Connie, Genoa, 2007

Nemaha County

Grove, Sr., Francis, Elk Creek, 1980
Adams Farm Ltd., Peru, 1988
Adcock, Gerald L., Omaha, 2015
Aden, Alvin & Lareene, Humboldt, 1980
Alberts, Eiber & Marlene, Auburn, 2002
Alden, Richard & JuanaLee, Auburn, 1988
Aldrich, Hanson & Amber, Johnson, 1965
Allemand, Eugene, Stella, 1972
Allen, Edward & Lucille, Auburn, 1984
Andrew, Melvin & Donnabelle, Brownville, 1965
Andrew, Morris, Auburn, 1972
Antholz, Herman & Zola, Johnson, 1981
Armstrong, Gerald, Auburn, 1968
Aufenkamp, Del & Burtis, 1992
Badberg, Mark, Cook, 2015
Barnes, Barton, Auburn, 1977
Beckmann, Reinhardt & Emma, Auburn, 1971
Behrends, DeLynn & Hilda, Elk Creek, 1985
Behrends, Ivan & Anna, Johnson, 1994
Behrends, Willard & Cheryl, Elk Creek, 1985
Bergmann, Mrs. Anna, Auburn, 1984
Berlet, Victoria & Emma, Auburn, 1968
Bernard, Lawrence & Lenora, Auburn, 1972
Blinde, Edwin & Emma, Johnson, 1985
Blount, Loyal & Thelma, Brownville, 1973
Boatman, Lawrence & Marjorie, 2003
Boellstorff, Mr. & Mrs. Charles, Johnson, 1967
Bohling, Anna, Auburn, 1965
Bohling, Harlan, Auburn, 1999
Bohling, Harold, Auburn, 1997
Bohling, Ruth, Auburn, 1999
Bohling, Ryan & Denise, Auburn, 2015
Bohlken, Kathrena, Auburn, 1959

Bray, Gary & LuAnn, Johnson, 2012
Broady, Raymond & Mary, Tecumseh, 1983
Brody, Lydia, Johnson, 1965
Burow, Arnold, Table Rock, 1983
Burow, LaVern, Elk Creek, 1983
Cain, Ray & Connie, Glenwood IA, 1995
Caspers, Albert & Blanche, Auburn, 1974
Caspers, Mark, Auburn, 2013
Caspers, Marvin & Blanche, Auburn, 1993
Clarke Oak Acres LLC, Brock, 2009
Clarke, Rosa Trust, Brock, 1993
Coatney, Charles & Beverly, Peru, 1968
Corner, Ivan, Auburn, 1999
Coulter, Eldo & Agnes, Auburn, 1981
DeBuhr, Linn, Auburn, 2007
Demarest, Frank Clark, Talmage, 1979
Deuser, Joy, Lincoln, 1979
Deuser, Richard & Charles, Lincoln, 1978
Dryer, Mr. & Mrs Floyd, Cook, 1967
Duerfeldt, Delbert, 1978
Dunekacke, Mr. & Mrs. Fred, Johnson, 1966
Ehlers, Clarence & Emma, Auburn, 1990
Ehlers, Louise, Auburn, 1968
Ely, Eugene, Auburn, 1970
Engles, Frank & Marge, Auburn, 1977
Engles, Mary Alice, Omaha, 1978
Feistner, Grace Moody, Lincoln, 1980
Frerichs-Gerdes, Virginia, Humboldt, 2013
Frerichs-Gerdes, Virginia, Humboldt, 2013
Gauchet, Fred & Marcia, Brock, 1995
Gerdes, LaVerne & Shirley, Auburn, 1988
Gilbert, Franklin & Pearl, Johnson, 1971
Gilbert, Mr. & Mrs. Arthur, Johnson, 1967
Gobber, Arnold & Orva, Elk Creek, 2009
Gobber, Ervin & Marilyn, Table Rock, 1992
Grafton, JoAnn, Peru, 2002
Grube Bates, Phyllis, Le Mars, IA, 1995
Haack, Roger, Elk Creek, 1992
Hahn, Mrs. Bertha, Johnson, 1976
Hall, Mrs. Thelma, Falls City, 1974
Hannaford, Martain, Peru, 2002
Harms, Mrs. Lucille, Auburn, 1989
Hawley, John & Cerella, Brock, 1988
Hawley, Mr. & Mrs. Charles, Brock, 1962
Hawxby, Mr. & Mrs. Clarence, 1969
Haynes, Henry & Mary Lou, Auburn, 1995
Heidzig, Henrietta, Johnson, 1983
Heidzig, Robert, Auburn, 1983
Henning, Charles, Peru, 1979
Heskett, Maud, Auburn, 1964
Higgins, Joseph, Lincoln, 1970
Hitzemann, Elmer & Annetta, Table Rock, 1983
Hodges, Mary, Peru, 1989
Holman, Charles & Helen, Auburn, 1970
Holthus, Herman & Berneda, Auburn, 1968
Ihrig, William Estate, Johnson, 1977
Jodry, Dean & Marie, Auburn, 1998
Keister, Irene & Joe, Auburn, 1972
Kennedy, Clay & Barbara, Brownville, 1982
Kiechel, Don & Marietta, Johnson, 2002
Kinghorn, Mr. & Mrs. Wilbur, Auburn, 1980
Knipe, William & Doris, Auburn, 1994
Koehler, Glenn & Verna Mae, Humboldt, 1986
Kohrs Adams, Keith, Pueblo CO, 2001
Kohrs Adams, William & Luella, Brock, 1976
Kotouc-Hedberg, Jane, Denver CO, 2000
Lambert, Mrs. Mella, 1971
Lewis, Hubert, Omaha, 1980
Linn, Carl, Prairie Village KS, 1969
Lowrie, W. J. & Nona, Auburn, 1969
Majors, Mr. & Mrs. Same, Peru, 1968
Markland, Blanche, Auburn, 1979
Marquardsen, Adela, Humboldt, 1983
Marshall, Sena, Talmage, 1984
Mathews, Guy, Auburn, 1982
Mayfield, Bob & LaVerne, 1990
McIninch, Mr. & Mrs. W.M., Brownville, 1967
McKenney, James & Mary, Auburn, 1991
Merwin, Edwin & Dorothy, Humboldt, 1970
Meyer, Arthur, Johnson, 1982
Michel, Edwin & Elsa, Auburn, 1989
Moerer, John & Marie, Johnson, 1965
Moerer, Wayne & Willy, Johnson, 2000
Moody, Beverly J, Auburn, 2009
Moore, Gailen Farm, 1968
Nagle, Agnes, Brock, 1976
Neddenriep, John & Lydia, Johnson, 1972
Nichols, Lloyd, Brock, 1979
Niedermeyer L E, Dorothy C., Johnson, 2014
Norvell, Mr. & Mrs. Sterling, Auburn, 1979
Norvell, Ralph & Iona, Auburn, 1979
Nowack, Robert & Joyce, Brock, 1992
Oestmann, Norvel, Melvin & Luella, Johnson, 1979
Oestmann, Clarence & Ellen, Auburn, 1971
Oestmann, Harold, Auburn, 1979
Oestmann, Victor, Auburn, 1990
Ord Jones/ Ord Wells, Gladys/ Esther, Auburn, 1977
Osborn, Mr. & Mrs. Merlyn, Talmage, 1981
Parriott, Richard & Fern, Peru, 1968
Peery, Harriet & Mildred, Brownville, 1957
Peterson, Wilma, Auburn, 1965
Phillips, Paul/ Janice, Humboldt, 1996
Pickerel, Bessie, Brownville, 1964
Piper, Frederick A. and Velma, Brock, 2004
Piper, Margaret, Brock, 1993
Plager, Eli & Lydia, Humboldt, 1972
Quante, Lawrence & Veve, Brock, 1972
Rademacher, Heye, Johnson, 1984
Remmers, Allyn, Auburn, 1995
Remmers, Gene, Auburn, 1995
Remmers, R. Wiley & Maxine, Auburn, 1977
Ritchie, Arleene & Ruth, Auburn, 1972
Rockemann, Alfred, Johnson, 1974
Rogge, Adolph, Auburn, 1987
Rogge, Gary, Auburn, 1988
Russel, Lester & Esther, Peru, 1999
Sandfort, Lenore & Ross, Brock, 1991
Schlange trust, Elmer, Auburn, 2011
Schmucker, Madeleine, Brock, 2015
Seid, Francis E, Falls CIty, 2009
Shires, Helen Louise Matthews, Porterville, CA, 2003
Shires, Preston D. Shires & Sylvia A., 2003
Sierks, John, Brownville, 2010
Simpson, Mr. & Mrs. Albert, Auburn, 1994
Skeen, Carl, Auburn, 1957
Skeen, Dale, Brownville, 1959
Skeen, Emil, Brownville, 1960
Smith, Merle & Thelma, Auburn, 1976
Smith, Merton, Beatrice, 1978
Smith, Mr. & Mrs. Harold, Auburn, 1981
Snyder, Dorothy, Lincoln KS, 1994
Sodman, Arthur & Adela, Humboldt, 1965
Sodmann, Mr. & Mrs. Edward, Humboldt, 1961
Solie, Mary, Brownville, 2006
Steffens, Elmer & Pearl, Auburn, 1989
Stiers, Donald & Arlene, 1992
Stoddard, Harold, Auburn, 1980
Stoddard, Hugh, Auburn, 1983
Stutheit, Ella, Tabble Rock, 1980
Stutheit, Norman & Donna, Humboldt, 1993
Stutheit, Werner, Elk Creek, 1983
Thomas, Mr. & Mrs. Everage, 1970
Trogel, Arnold, Auburn, 1965
Tynon-Morris, Elizabeth, Peru, 1989
Volker Knippelmeyer, Erhardt & Maxine, Humboldt, 1964
Volker, Lawrence & lenora, Humboldt, 1964
Wagner, Harold, Auburn, 2001
Wheeler, Gale and Jacqueline, Auburn, 2003
Wilkie, Mr. & Mrs. Harvey, Nebraska City, 1967
Williams, Eimer & Meriel, Auburn, 1963
Woerlen, Allan, Brock, 1979

Nuckolls County

A & W Acres, Davenport, 2014
Becker, Steven, Nelson, 2006
Biltoft, Roy & Georgia, Edgar, 2002
Buescher, George & Margaret, Lawrence, 1988
Buresh, Mildred I., Edgar, 2008
Clabaugh, Lee & Charlotte, Nelson, 2014
Collins, Eleanor, Lawrence, 1985
Corman, Kenneth, Superior, 1981
Corman, Vernon, Edgar, 2009
Crawford, Celia, Madrid, 1986
Culbertson, Joseph & Alice, Oak, 1992
Danehey, Donald, Lawrence, 1985
Davis, Melroy, Larson, Jacqueline Wehrman Davis, Patricia Wehrman Whiteley Melroy & Debroah Wehrman Larson, Kenesaw, 2014
Delynn/ Hoelting, Terence/ Betty, Austin TX, 1996
Dolnicek, John, 2005
Ebsen, Leonard & Betty, Superior, 2006
Farver, Clarence & Daisy, Nelson, 1979
Ferebee, James, Nelson, 1974
Greer, Mary & Merlin, Edgar, 1984
Grummert, Eldrick H., Nelson, 2003
Grummert, Raymond & Mary Lou, Superior, 1993
Hansen, Harold, Ruskin, 1986
Hanson, Don L. Limited Partnership, Gretna, 2006
Hanson, James & Pauline, Oak, 1981
Harris, James, Nelson, 1973
Herbek, Donald, Deweese, 2008
Holst, Emma, Edgar, 1979
Horst, Jame, Jeff & Jerry, Edgar, 2014

Jacobitz, Bruce & Rebecca S, Superior, 2015
Jagels, Elliot & Eunice, Davenport, 1985
Jennings Fisher, Alice, Grand Island, 1994
Jensen, Barbara, Edgar, 2006
Jeppesen Farms Inc., Ruskin, 1984
Johnson, Gale, Davenport, 2006
Johnston, Aaron, Oak, 1981
Karmazin, Michael & Judy, Lawrence, 1991
Karmazin, Tim, Lawrence, 2007
Kathman, Donald & Lisa, Lawrence, 2006
Keifer, Jo, Bostwick, 1971
Keifer, Oswin, Nelson, 1973
Keim, Dean, Davenport, 1992
Keim, Larry, Davenport, 2013
Kotinek, Robert & Nancy, Lawrence, 1990
Kucera, Clement & Patricia, Deweese, 1988
Lang, John, Edgar, 1983
Lowery, Sr., Mrs. Ray, Oak, 1974
Lyne Trust, Blaine, York, 2009
Madson, Theodore, Nelson, 2010
Maschmann, Esther, Ruskin, 1993
Mazour, Craig, Lawrence, 2009
Mazour, Len & Jolene, Dave Mazour, Lawrence, 2009
McCutcheon, Willard & Glennis, Hardy, 2002
Menke, Ray & Carolyn, Lawrence, 1984
Montogmerty, Robert & Lois, Oak, 1971
Ostdiek, Clarence, Lawrence, 1980
Overturf, Ellen M., Oak, 2003
Peters Farm Partnership, Hardy, 1993
Petsersen, Donald & Margrethe, Ruskin, 1979
Pritts, Elvin, Terrell & Wayne Pritts, Guide Rock, 2005
Ragan, Loren, Scappoose, 2006
Schendt, Eleanor, Lawrence, 2004
Schendt, William & Lola, Lawrence, 2003
Schroeder, Ann Heirs, Davenport, 2014
Schroeder, Norbert Family, Michael & Lori Schroeder, Davenport, 2005
Schroer, Arnold, Lawrence, 1978
Souchek, Ben, Deweese, 2000
Statz, Paul and Mary, Nelson, 2016
Stockam, Albert, Ruskin, 1979
Sullivan Dairy Inc., Superior, 1984
Svoboda, Eugene, Deweese, 2005
Tordrup, Janice, Superior, 2012
Wehrman Jr, Henry, Nelson, 1977
Wenig, Sharon, Nancy Fuller, Dean Adamson, Beverly Sutherland, Janet Adamson, Omaha, 2015
Williams, Elmer & Gwen, Nora, 1985
Wilton, Lynn & Roger, Superior, 2007
Wortman, Leonard Henry, Ruskin, 1980

Otoe County

Albers, Leland & Kathy, 2003
Antes, Mr. & Mrs. Harold, Syracuse, 1972
Antes, Mr. & Mrs. Wesley, Lincoln, 1972
Anville, Frank & Harriette, Talmage, 1965
Arends, Richard, Talmage, 1962
Balfour, Clark & Dorothy, Nehawka, 1969
Baltensperger, Daniel & Leone, Nebraska City, 1979
Baltensperger, Mr. & Mrs. Carl, Nebraska City, 1970
Baltensperger, Mr. & Mrs. Howard, Nebraska City, 1969
Baltensperger, Steve, Brad Baltensperger, Nebraska City, 2008
Bando, Johnny, Nebraska City, 1981
Banning, Minnie, Union, 1976
Bartels, Emma & Larry, Syracuse, 1995
Beccard, Mrs. William, Nebraska City, 1958
Beecham Family Trust, Beecham Family Trust, 2005
Beers, Allen & Louise, Syracuse, 1974
Bentzinger, Neal & Joyce, Cook, 2009
Berner, Mr. & Mrs. John, Lincoln, 1977
Biere, Dean, Dunbar, 2001
Bischoff, Arnold, Talmage, 1958
Boldt, Mr. & Mrs. Harold, Douglas, 1974
Bose, Henry R., Shirley Cappen, Mary Schechinger, Papillion, 2012
Brandt, Mr. & Mrs. John, Cook, 1971
Brehm, Milan & Christie, Unadilla, 1992
Brehm, Richard, Unadilla, 1992
Bremer, Glenn & leta, Dunbar, 1981
Bro, Ernest & Imogene, Syracuse, 1992
Brownlee, Don & Inez, Douglas, 1974
Brugman, Elizabeth, Douglas, 1992
Buckman Campbell, Susie, Nebraska City, 1961
Buckman Glockel, Anna, Omaha, 1961
Burger, Harvey and Maxine, Nebraska City, 2008
Cameron, Carl & Genevieve, Burr, 1978
Cameron, Robert & Joan, Dunbar, 1999
Campbell, Dorothy, Lincoln, 1997
Carey, Mildred & Don, Syracuse, 1982
Damme, Clarence & John, Cook, 1972
Damme, Mr. & Mrs. Arnold, Cook, 1972
Damme, Mr. & Mrs. Victor, Syracuse, 1980
Davidson, Loren and Valda, Lincoln, 2008
Davis, George Wright, Nebraska City, 1980
Davis, LaVerna, Syracuse, 1990
Dermann, William & Clara, Nebraska City, 1977
Dexter, Mary, Derrick Griepenstroh & Gretchen Griepenstroh, San Diego, CA, 2004
Dively, Jeffrey, Omaha, 2002
Doeden, Phillip & Eulabell, Cook, 1985
Douglas, Arthur, Palmyra, 1979
Douglas, Mr. & Mrs. John, Nebraska City, 1967
Dowding, Byron L., Palmyra, 2004
Dowding, Harold, Palmyra, 2002
Dowding, Pauline, Palmyra, 1982
Dowding, Winifred & Verne, Eagle, 1992
Dunbar, Pauline A. Trust, Pauline A. Wirthele Dunbar, Trustee, Falls City, 2004
Easter, Tom & Arlene, Dunbar, 1982
Eden, Mr. & Mrs. Dale, Burr, 1971
Eisenhauer, Emil & Irene, Syracuse, 1991
Eisenhauer, Ernest & Edna, Burr, 1991
Elizabeth McCarthy, Dolly, Nebraska City, 1982
Elliott, Mr. & Mrs. Harold, Weeping Water, 1978
Emshoff, Mrs. Helen, Avoca, 1978
Engelbrecht, Howard, Syracuse, 1991
Fallers, Helen, Syracuse, 1988
Ferguson, Calvin & Barbara, Omaha, 1974
Fey, Robert & Barbara, Palmyra, 1965
Fields, Fern, Palmyra, 1966
Fisher, Evelyn & Raymond, Johnson, 1987
Fisk, Raymond & Brenda, Burr, 1984
Floerke, Henry & Elsie, Cook, 1972
Freguson, Charles Sumner, Nebraska City, 1972
Frerichs, Ernest, Talmage, 1966
Gardner, Evelyn, Nebraska City, 1984
Gartner, Anna, Unadilla, 1993
Gartner, Clark, Unadilla, 2007
Gartner, Ralph & Viola, Unadilla, 1967
Goering, William K., Syracuse, 2003
Gress, Melvin & Norman Jean, Nebraska City, 1997
Gress, Raymond T. and Carolyn J., Nebraska City, 2003
Griepenstroh, Arlen & Kathleen, Nebraska City, 1994
Griepenstroh, Arnold, Nebraska City, 1990
Griepenstroh, Delphine, Nebraska City, 2007
Griepenstroh, James & Erma, Dunbar, 1989
Griepenstroh, Mary Ann, Syracuse, 2014
Griepenstroh, Mary Ann, Syracuse, 2015
Gross, Ernest, Syracuse, 1988
Gross, Naomi, Syracuse, 1988
Grotrian, Mr. & Mrs. Carl, Cook, 1970
Grundman, Darold A. and Rosalyn J., Syracuse, 2003
Guenther Jr., Alvin, Dunbar, 2003
Haeffner, Raymond & Leta, Burr, 1983
Hallowell, Allen & Donita, Palmyra, 2010
Hallowell, Boyne, Palmyra, 2012
Halvorsen, Forrest & Doii, Syracuse, 1987
Halvorsen, Mr. & Mrs. Carol, Syracuse, 1967
Hanau, Eleanor, Cook, 1999
Hanssen, Mr. & Mrs. Walter, Syracuse, 1969
Harms, Donald & Janet, Syracuse, 2010
Hauptman, Evelin & Eldon, Peru, 1986
Hauptman, Theodore, Peru, 1959
Heather, Mr. & Mrs. Delbert, Syracuse, 1969
Heng, Frank & Mark, Nebraska City, 1962
Heng, Oscar, Dunbar, 1999
Henke, Harvey, Syracuse, 2006
Herzog, Mr. & Mrs. Roy, Nebraska City, 1967
Heye, Mr. & Mrs. Edward, Dunbar, 1962
Hillman, Alvin & Viola, 1979
Hillman, Harold & Irene, 1982
Hillman, Maynard & Reva, 1983
Hoagland, Mrs. John, Nebraska City, 1958
Hobbie, Mr. & Mrs. Arthur, Dunbar, 1976
Holscher, Mr. & Mrs. Harvey, Cook, 1972
Holscher, William & Elizabeth, Unadilla, 1967
Hopp, Donald & Glenis, Talmage, 1978
Hopp, Leona & Richard, Syracuse, 1980
Horstmann, Mr. & Mrs. Frank, Lorton, 1964

Horstmann, Mr. & Mrs. Paul, Talmage, 1964
Hunt, Dell & Madeline, Douglas, 1983
Jacobmeier, V.V., Wayne, 1979
Jacobson, Raymond & Elaine, Syracuse, 1982
James, Harold, Nebraska City, 1979
Janssen, Clara, Syracuse, 1983
Janssen, Lloyd & Berniece, Burr, 1983
Jensen, Ray & Leona, Nebraska City, 1960
Johns, Beverly, Nebraska City, 1990
Johnston, Dora, Syracuse, 1974
Johnston, Mr. & Mrs. Reuben, Dunbar, 1970
Kaiser, Anna Mae & Clarence M., Nebraska City, 2003
Kasbohm, Don & Nancy Farm, Dunbar, 2003
Kastens, William, Nebraska City, 1977
Kettelhut, Raymond & Wilma, Bennet, 2001
Kimmel, Richard, Nebraska City, 1977
Klingsick, Ervin, Talmage, 1979
Knickman, Goldine, Syracuse, 1991
Kreifels, Bernard, Nebraska City, 2012
Kreifels, Irma & Francis, Dunbar, 1982
Kuehn, Arthur, Nebraska City, 1975
Kunz, Myrl & Edna, Talmage, 1982
Landwehr Farms LLC, Nebraska City, 2010
Lanning, Jessie Bunten, Lincoln, 1973
Lanning, Orin, Eagle, 1974
Latrom, Vernon & Doris, Eagle, 1981
Liesemeyer, Arlene & Victor, Syracuse, 1992
Lindsey, Delbert & Fred, Nebraska City, 1989
Ludwig, Otto & Flora Mae, Avoca, 1982
Luff, Mr. & Mrs. Dorman, Unadilla, 1971
Luff, Mrs. Floyd, Unadilla, 1971
Lutjemeyer, Fred, Syracuse, 1993
Lyon, Keith and Susan Lyon,, 2005
MacQuaig Heirs, Nebraska City, 1973
Malone, L. Bourke & Bertha, Palmyra, 1990
Mannschreck, Ray & Edith, Cook, 1974
Masters, Carl, Syracuse, 1963
Miller, Charles & Esther, Lincoln, 1979
Miller, Howard & Dorothy, Unadilla, 1983
Moeller, Mr. & Mrs. John, Cook, 1972
Moyer, Marjorie, Nebraska City, 1993
Moyer, Mr. & Mrs. Mark, Nebraska City, 1967
Nash, Harvey & Elaine, Lincoln, 1981
Nash, Madge, Palmyra, 1969
Neemann, Norman & Hilma, Cook, 1983
Neubauer, Mr. & Mrs. Roy, Nebraska City, 1973
Neubauer, Mr. & Mrs. Edward, Nebraska City, 1973
Niebruegge, Harold & Helen, Nebraska City, 1992
Niebruegge, Helen & Neal, Nebraska City, 1996
Nutzman, Mr. & Mrs. Ned, Nehawka, 1966
Nutzman, Susan M., Dunbar, 2004
Osborn, Mr. & Mrs. Carrol, Talmage, 1969
Overton, Mr. & Mrs. Carl, Nebraska City, 1969
Overton, Mr. & Mrs. Ernest, Nebraska City, 1961
Paap, Frank, 1969
Patton, Mr. & Mrs. Dexter, Dunbar, 1972
Petersen et al, Glenn, Syracuse, 2007
Petersen, Dorothy & Clarence, Nebraska City, 1996
Petersen, Randall/ Joni, Syracuse, 1996
Phillips, Forrest, Douglas, 1976
Philpot, LaVon, Winter Park, FL, 2011
Post, Randy and Cindy, Syracuse, 2003
Powell, Mrs. Dena, Cook, 1974
Pursel, Mr. & Mrs. Harold, Nebraska City, 1970
Ramold, Donald & Patricia, Nebraska City, 2010
Reed Lenker, Mr. & Mrs. William, Dunbar, 1978
Remmenga, Anna May, Dunbar, 1989
Renken, H. E., Millard, 1969
Reuter, August, Syracuse, 1989
Reuter, Virgil & Maxine, LeAnn Fossler L.E., Nebraska City, 2009
Riensche, Gilbert & LuElla, Johnson, 1991
Ritter, Arnold, Talmage, 1968
Ritter, Leonard, Talmage, 1974
Roos, David & Darcy, Omaha, 2007
Roos, Lyman, Dunbar, 1982
Ropers, Mr. & Mrs. Willard, 1972
Ross Land Ltd., Nehawka, 1981
Rowen, Wilbur & Ivy, Nebraska City, 1967
Royal, Larry & Gayle, Palmyra, 2012
Ruenholl, Harold & Betty, Syracuse, 1992
Ruenholl, Mrs. Fred/ Arthur, Nebraska City, 1968
Ruenholl, Walter & Evelyn, Syracuse, 1984
Rulla, Lena & Landis, Sterling, 1973
Schacht, Marvin, Cook, 1983
Schact Iwakiri, Lorene, Cook, 1961
Scharp, Rosemary, Nebraska City, 2013
Schmitz, Bernard, Nebraska City, 1978
Schreiner, Allen & Donita, Nebraska City, 2010
Schreiner, David & Linda, Kevin Sheppard, Doug Sheppard, Matthew Sheppard, Jennifer Koolen, Phoenix, 2010
Schreiner, David & Linda, Kevin Sheppard, Matthew Sheppard, Douglas Sheppard, & Jennifer Koolen, Phoenix, 2006
Schreiner, Eugene, Nebraska City, 1984
Schreiner, Harry, Nebraska City, 1985
Schreiter, Robert & Mildred, Nebraska City, 1994
Schriener, Paul & Louise, Dunbar, 1989
Schroder, Joy & Verle, Palmyra, 2012
Schutz, Calvin, 2004
Schweppe, Daniel G, Syracuse, 2015
Shallenberger, Larsh, Nebraska City, 1958
Sigerson, Mrs. Ruth Ann, Nebraska City, 1973
Sim, Ellen, Eugene & Robert, Nebraska City, 1980
Smallfoot, Robert, Nebraska City, 1982
Snyder, Melvin & Betty, Talmage, 1994
Springman, Mr. & Mrs. Alton, Palmyra, 1980
Springman, Mr. & Mrs. Ralph, Palmyra, 1974
Sreinhoff, Ralph, Palmyra, 2000
St. John, Leonard, Nebraska City, 1980
Stahlhut, Ralph & Mildred, Nebraska City, 1990
Stall, Frances Stall, Gary Stall, 2005
Stecker, James, Grand Island, 2002
Steinhoff, Edwin, Syracuse, 1982
Stillahn Devers, Anna Marie, Lewis IA, 1997
Stillahn, Mrs. Leota, Dunbar, 1997
Stilwell, Emmaline, Unadilla, 1974
Stilwell, George & Mattie, Unadilla, 1974
Straube, Richard, Sterling, 1982
Stubbendick, Gene & Moreen, 1983
Stubbendick, Glen & Margaret, Avoca, 1983
Stukenholtz, Earl, Nebraska City, 1983
Stukenholtz, John Stukenholtz, 2005
Stukenholtz, Mr. & Mrs. Herbert, Nebraska City, 1959
Teten, Duane, Dunbar, 1992
Teten, Mr. & Mrs. Theodore, Talmage, 1974
Teten, Mrs. Reinhold, Talmage, 1967
Thomson, Bobby, Lincoln, 1998
Thomson, Mr. & Mrs. Alfred, Palmyra, 1970
Thomson, Mr. & Mrs. George, Palmyra, 1970
Thormahlen, Walter & Lillie, Avoca, 1983
Todsen, Marilyn, Eagle, 1993
Trail, Robert & Lorene, Nebraska City, 1978
Ullsperger, Mr. & Mrs. Franics, Nebraska City, 1974
Wallen, Glen F. & Beatrice M., Unadilla, 2003
Wallen, Glen F. & Beatrice, Unadilla, 2007
Wallen, Lucile, Lincoln, 1978
Warner, Richard and Bonnie, and Ron & Ruth Gibbs, Nebraska City, 2014
Watermeier, Steve & Janice, Burr, 1987
Weible, Eugene, Nebraska City, 1994
Wendeln, Harvey, Burr, 1976
Wendeln, Mr. & Mrs. Raymond, Syracuse, 1968
Werner, Harold, Syracuse, 1991
Wheatley, Donald & Mildred, Eagle, 1993
Wheatley, Rick & Ronda, Eagle, 2002
Wiebusch, Mrs. Eda, 1972
Wiebusch, Mrs. Emma, Dunbar, 1972
Wieckhorst, Herman, Nebraska City, 1959
Wieckhorst, Robert & Norma Jean, Nebraska City, 1989
Wilberger, Mrs. Roy, Julian, 1984
Wilhelm, Ruth & Vernon, Unadilla, 1982
Wilson, Robert & Lois, Lincoln, 1995
Winkelhake, Henry & Maxine, Nebraska City, 1965
Winton, Isabella, Nebraska City, 1997
Wirthele, Paul Arnold, Burr, 2007
Witt, Bryce, Syracuse, 1980
Wittler, LaVern & Marilyn, Talmage, 1991
Wohlers, Wayne & Paula, Syracuse, 2009
Wolf, Mr. & Mrs Willard, 1969
Yont, Etta, Lincoln, 1966
Young, Leona, Syracuse, 1994
Zahn Floerke, Verone, Tecumseh, 1968
Zink, Marjorie Hull, Sterling, 1996

Pawnee County

Albers, Donald, Julie Albers City, 2009

Anderson, Dr. A Byford City, 1981
Argo, Shirley & Arlan, Liberty, 1978
Baker, Barbara Ann, Marshfield, 2015
Barr, Raymond or Colette, Liberty, 1981
Barr, Robert & Edwin, Liberty, 1981
Bausch, Jerald D & Eileen, Burchard, 2009
Beethe, Blenn & Ellen, Elk Creek, 1968
Beethe, Ida, Elk Creek, 1979
Beethe, Verlyn, Elk Creek, 1985
Beethe, Gary & Cathy Jo, Steinauer, 1984
Bernadt, Joseph, Steinauer, 1960
Bernadt, Carl, Table Rock, 1960
Binder, Beth Ellen, Springfield, 2001
Blecha, Howard & Debbie, Table Rock, 2011
Bock, Leslie & Mabel City, 1976
Bowhay, Ivan & Viola, Liberty, 1967
Bredemeier, Arden, DuBois, 1983
Bredemeier, Myrtle City, 1983
Burgert, Leo & Dorothy Mae City, 1979
Chittick, Helen City, 1979
Clark, Dean & Harriet, Lewiston, 1982
Clark, Duane & Elizabeth City, 1974
Conradt, John, Steinauer, 1963
Daniel, Raymond F. City, 2006
Duder, Lawrence & Mary, Table Rock, 1967
Eichenberger, Nora, Steinauer, 1980
Farwell, Lyle/ Dorothy, DuBois, 1996
Finney, Myron & Dorothy City, 1993
Ford, Glenda, Souix City IA, 1998
Fritch Jr., Rudy and Dorothy, Table Rock, 2008
Gilbert, Dwain & Delores, Table Rock, 1972
Good, Mr. & Mrs. Kenneth, Crete, 1973
Gottula, Florence, Steinauer, 1977
Gottula, Albert & Esther, Table Rock, 1969
Gyhra, Joseph, Burchard, 1978
Hansen, Marilyn, Omaha, 2010
Hinrichsen, Douglas & Luella, DuBois, 2002
Hoffman, Randolph & Lucile, Summerfield KS, 1972
Hothan, John, Steinauer, 1969
Hunzeker, Mr. & Mrs. Ernest, DuBois, 1956
Hunzeker, Magadlea, DuBois, 1956
Hunzeker, Mrs. Magdalena, DuBois, 1971
Jasa, Paul & Lynda, Table Rock, 2015
Kalina, Libbie, Table Rock, 2000
Killen Carver, Marguerite, Burchard, 1971
Koester, Fred, DuBois, 1956
Kozak, Emil & Patricia, Lincoln, 2007
Leitschuck, Michael & Abbie, Burchard, 2011
Luedeke, Edgar City, 1973
McClintock, Earl City, 1956
McClintock, Harvey City, 1956
Mcclintock Monagon, Mrs. Mary City, 1956
McNair, Mrs. John City, 1983
Meier, Arthur & Mildred, DuBois, 1989
Melichar, Wilmer & Natalie, Steinauer, 1968
Metzger, Byford & Virginia City, 1987
Meyer, Minnie & George, Elk Creek, 1976
Miller, Myrle, Liberty, 1980
Miller, Albert, Table Rock, 1974
Muscheites, B. Fred, Table Rock, 1981
Nemec, Alta & Jack, DuBois, 1987
North, Mrs. John City, 1967
Obrist, Gerald & Evelyn, Steinauer, 1970
Parks, Alodyne City, 2000
Pepperl, Edward & Josephine, Burchard, 1974
Poteet, Gwendolyn City, 2000
Powell, Myrna, Burchard, 2001
Rakosnik, Joseph, DuBois, 1979
Rees, Harold & Emly, Liberty, 1970
Reuter, Arthur & Christine, Table Rock, 1968
Reynolds, Hazel & Glen City, 1970
Richard, Alan Martie City, 2002
Rinne, Vernon, Ronald Rinne, Lincoln, 2008
Rinne, Alfred & Marguertie, Nebraska City, 1971
Rinne, Leo & Pauline, Steinauer, 1968
Rinne, Mary, Steinauer, 1968
Rinne, Leonard & Belva, Steinauer, 1969
Rinne, Mrs. Edwin, Steinauer, 1976
Rinne, Doloma, Syracuse, 2002
Rinne Brueggemann, Rose, Tecumseh, 1971
Schaardt, Elda, Elk Creek, 1980
Sejkora, Bradford & Linda, Burchard, 2002
Shinn, Ray, Seneca KS, 2002
Shuss, Junius & Harold, Table Rock, 1981
Smith, Richard & June City, 1968
Sommerhalder, Willard & Mary, Steinauer, 1973
Sowle, Stephen, DuBois, 1974
Spier, Elmer, Steinauer, 1973
Spier, Marie, Steinauer, 1981
Stehlik, Duane & Marna Sue, Table Rock, 1987
Stehlik, Loren, Pawnee City, 2016
Steinauer, Norbert & Genevieve, Steinauer, 1956
Steinauer, Mr. & Mrs. John, Steinauer, 1966
Stepanek, Robert, DuBois, 2008
Taylor, Lester, Humboldt, 1977
Taylor, Virgil City, 1965
Tegtmeier, Clarence, Burchard, 2001
Thiemann, Mary Alice & Ross City, 2007
Thomas, Mary Mitchell, Curchard, 1963
Tomek, Elsie M., Table Rock, 2013
Turnbull, Leonard, Burchard, 1967
Vondraskek, Robert and Gladys, Lincoln, 2008
Vritska, Mrs. Lena City, 1968
Vrtiska, Floyd & Lloyd, Table Rock, 1975
Vrtiska, Floyd & Doris, Table Rock, 1995
Wenzl, Lawrence, Steinauer, 1978
Wenzl, Clarence, Steinauer, 1988
Wilcox, Goldwin & Rosabelle City, 1972
Woltemath, Mr. & Mrs. Victor, Elk Creek, 1968
Young, James & Vera City, 1969
Watkins, Sutfin, Brockman & spouses, & James R. McGuire, Lincoln, 2003

Perkins County

Chrisman, Beverly, Wallace, 1987
Cooper Farms Ltd., Grant, 2007
Culver, John, Elsie, 1987
Cutler, Francies, Elsie, 1987
Dittmer, John, Ogallala, 1985
Dodson, Myron & Bonnie, Venango, 2007
Engel Mahnken, Carl & Dorothy, Grant, 1987
Erlewine Farms Inc., Grant, 1987
Faught Potts, Jane, Elsie, 2012
Ford, Glenn, Madrid, 1987
Hanson, Steve, Elsie, 2000
Hilferty, Dave, Grant, 1999
Jackman, Charles, Grant, 1987
Law, Davis, Elsie, 1995
Lee, Doris, Madrid, 1988
Lingbloom, Robert, Elsie, 1987
McClintock, Warren, Elsie, 1988
McCormick Burge + Dale Burge, Sharon, Madrid, 2015
McCormick, Gary & Diane, North Platte, 2013
McGahan, William Conrad, Elsie, 1986
Olson, Henry, Elsie, 1995
Osler, John, Imperial, 1986
Pankonin, Louise, Grant, 1985
Peterson, Max & Darlene, Ogallala, 1986
Simpson, Katherine, Denver, CO, 2011
Stanek, M. Glen & Paula, Elsie, 1989
Steinke, Richard, Venango, 1985
Sunnyknoll Farms, Inc., Ogallala, 2010
Thompson, Susie Fae, Elsie, 1987
Tjaden, Roger, Madrid, 2009

Phelps County

Abrahamson, Paul, Axtell, 1978
Anderson, Byron, Holdrege, 1983
Anderson, Dale R, Holdrege, 2015
Anderson, Dr. Robert, Holdrege, 1984
Anderson, Evert family trust, Kearney, 2005
Anderson, Helen, Funk, 2007
Asp, Wayne, Holdrege, 1983
Bachman, Wm. Wayne, Rancho Cordova, 2008
Benson Kelly, Mrs. Arlene, Holdrege, 1988
Benson, John, Kearney, 1987
Bergman, Grant & Donna, Loomis, 1978
Bergquist, Lloyd G., Holdrege, 2003
Bergstrom Farms Inc., Axtell, 1981
Black, Claude, Hildreth, 1991
Blythe, Marlene E. & Kerry L., Overton, 2013
Bunker Hill Farms Ltd, Holdrege, 1980
Burgeson Farms Inc, Clinton, Holdrege, 1990
Burton & Berdeen - Karlson, Holdrege, 2004
C & R Hendrickson Farms Inc., Holdrege, 1985
Cain, Robert, Holdrege, 1991
Carlson, Gustave & Aleda, Loomis, 1982
Carlson, Royle, Holdrege, 1986
Carlson, Wayne, Holdrege, 1983
Crandall, George, Wilcox, 1993
Crandall, R. Brent & Stephanie, Overton, 1994
Cruise, Hazel, Loomis, 1995
Dahlgren, Lilie Ann, Milton Dahlgren, Minden, 2009
Dahlgren, Vernon, Bertrand, 1983
Dahlstrom, Evelyn, Holdrege, 1995
Danielson, Milton & Neva, Holdrege, 1983
Ecklun, Hazel, Holdrege, 1986
Englad, Lamont & Jean, Holdrege, 1989
Englund, Rosemary, Holdrege, 1985
Erickson, Charlotte, Holdrege, 1989
Erickson, Erick Ray, Funk, 1988
Erickson, Stanley & Janet, Holdrege, 1981
Erickson, Waldon, Funk, 1978
Freburg, Morris, Loomis, 1979
Freed Brothers Farms, Loomis, 2009
Gardine, Hildur, Helen Tagge, Bertrand, 2006

Goodman, Charles J. and Rhoda M., Loomis, 2008
Gray, Gary, Sandra Slater, Holdrege, 2006
Gustafson, Harold, Funk, 1985
Gustafson, Orville & Otto, Axtell, 1986
Hallgren, Richard, Funk, 1983
Hamilton, Mary, Bertrand, 2002
Hanson, Earl J., Holdrege, 2006
Harris, Kenneth, Loomis, 1977
Hasselquist Farm Inc., Omaha, 2006
Hatting Pinkerton, Grace, Loomis, 1993
Hazel J. Farms Inc., Bertrand, 1987
Headberg, Gladys, Holdrege, 1989
High, James & Doris, Bertrand, 1979
Holen, Betty Lou, Bertrand, 2007
Ingram-Warren-Swenson Farms, Kearney, 1992
Isaacson, Oliver, Loomis, 1979
Johnson, Alwyn & Jeanne, Overton, 1992
Johnson, Blake & Renee, Holdrege, 2006
Johnson, Doris J., Grace E. Jones Estate, Holdrege, 2006
Johnson, Josie, Elm Creek, 1993
Johnson/ Roberts, Joan/ Jean, Axtell, 1992
Jurgens, Dick & Lois, Holdrege, 1995
Just, Verna Trust c/o Jim Just, Funk, 2004
Kirkpatrick, Mr. & Mrs. F. Glenn, Lexington, 1978
L.C. Jacobson Farms Inc, Holdrege, 1986
Larson, Don, Funk, 1980
Larson, La Vaughn, Holdrege, 2010
Larson, Ruth & Clinton, Holdrege, 1984
Larson, Winston, Holdrege, 1976
LaVern Johnson Farms Ltd., Holdrege, 1980
Leff, Donald, Holdrege, 1980
Lind, Carl & Roxana, Holdrege, 1994
Lind, Hilma, Holdrege, 1994
Linder, Greg, Loomis, 2009
Linder, James, Loomis, 2009
Linder, Mrs. Esther, Loomis, 1981
Linder, Robert & Jean, Loomis, 2009
Lindquist, Carl, Funk, 1980
Lindstrom, Aline, Elm Creek, 1995
Lofquist, Wayne & Roma Jean Freburg, Bertrand, 2005
Lund, Loren, Holdrege, 1981
Lundy, Ricky and Karen, Bertrand, 2016
Lyon, LaMonte & Barbara, Loomis, 1987
Marshall, Basil, Elm Creek, 1991
Marshall, Harold 7 Dorty, Holdrege, 1995
Marshall, Moses, Elm Creek, 1980
Mattson, Mr. & Mrs. Paul, Loomis, 1979
Mattson, Victor, Loomis, 1979
McClymont, Dean & Phyllis, Loomis, 1986
McClymont, Mrs. Clyde, Holdrege, 1979
McClymont, Robert & Mae, Holdrege, 1986
Meyers, Amerlia, Lincoln, 1989
Morrison Estate, Richard, Loomis, 1983
Nelson, Evelyn, Bertrand, 2005
Nelson, LeRoy, Bertrand, 1981
Nelson, Mr. & Mrs. Fred, Loomis, 1981
Nelson, Ruth Estate, Funk, 1985
Norberg, Bernard, Funk, 2010
Patterson, Murl, Kearney, 2002
Peterson, Hazel & Albert, Holdrege, 1986
Peterson, Jennie & Clarence, Holdrege, 1978
Peterson, Paul, Bertrand, 1981
Peterson/ Anderson, Lawrence / Rosceal, Holdrege, 1979
Polhemus, Laura, Holdrege, 1988
Polhemus, Lynn & Evelyn, Holdrege, 1988
Redfern Jones, Leona, Holdrege, 1993
Reed, Doran, Holdrege, 2002
Rosier, Dorothy, Holdrege, 2011
Sall, Melvin, Holdrege, 1983
Samuelson, Frances, Bertrand, 1992
Sand, Harold, Loomis, 1987
Sand, Mr. & Mrs. Charles, Holdrege, 1976
Sand, Mrs. John/ Charlie, Loomis, 1985
Sandell, Harold W, Holdrege, 2006
Schrock Farms, Funk, 2008
Shanahan, James, Axtell, 1988
Sjogren, Derrel & Yvonne, Funk, 1982
Sjogren, Donald & Gerald, Funk, 1993
Skoog, Dan, Holdrege, 1999
Skoog, Kathleen, Holdrege, 1999
Stephens, Keith & Dorothy, Elm Creek, 2002
Sullivan, Carolyn & Orville, Loomis, 2008
Swanson, Lois, Holdrege, 2015
Swanson, Reuben, Omaha, 1989
Swanson, Thomas & Lois, Holdrege, 1987
Swanson, Verna, Greenville TX, 1993
Verbeck, Don & Phyllis, Funk, 1985
Von Loh, Vivian, Lincoln, 1993
Waller, Lloyd Waller Feedlot, Inc., Holdrege, 2004
Waller, Richard B. & Susan J., Holdrege, 2006
Wendell, Paul, Holdrege, 2011
Wilke, Mr. & Mrs. Kermit, Holdrege, 1983
Wycoff, Retha, Lexington, 1985
Young, Albert & Alvin, Holdrege, 1983

Pierce County

Ahlman, Delbert & Sandra Buckendahl, 2003
Alexander, Edna, Plainview, 1991
Alexander Hoops, Anna Jean, Plainview, 1990
Alexander, Gaylyn, Plainview, 2007
Alfs, Linda M., Norfolk, 2003
Backer, Dennis & Michelle, Randolph, 2013
Birch, Owen, 1979
Bloomquist, Scott & Constance, McLean, 2000
Blunck, Norman & Arleath, Osmond, 1997
Blunk, Dale, Osmond, 2006
Boaumann, Michael & Kimberly, 2009
Bockelmann, Dwaine, 2005
Buckendahl, Duane & Vada, 1993
Burger, James, Plainview, 2013
Carstens, Dean & Ruth, Randolph, 1999
Carstens, Gerald D. & Jane A., 2003
Cates, Sandra, Susan Talcott, Grand Island, 2006
Clausen Trust, Irma A., Schuyler, 2008
Dedlow, Frederick, Plainview, 1971
Farms, Weyhrich, Randolph, 1993
Fischer, Clayton & Cathy, 2006
Foster, Larry & Karen, Plainview, 2007
Frederick, Clifford A., Randolph, 2006
Frey, Trustee, Rosemary, Frey Zephyr Revocable Trust, 2014
Fuelberth, Alton & Leoma, Osmond, 1994
Fulton Bosilevac, Elaine, Omaha, 2009
Gansebom, Delmer & Joan, Osmond, 1991
Gansebom, Vernon & Donna, Osmond, 1993
Gansebom, Vernon, Osmond, 2008
Gehm, Ewald & Betty, 1991
Goff, Mrs. Donna Mae, Fremont, 2015
Grahm, R. Alan & Eleanor, Randolph, 1992
Gutz, Byron & Carleen, Osmond, 1993
Hahn, Mrs. Ted (Ardith), 2003
Henzler, Don & Beverly, 1992
Herbolsheimer, Ardith, 1999
Herian, Norman Trust, 2013
Hoffmann, Julius & Amelia, 1981
Holmes, Gerald and Janis, Osmond, 2008
Holmes, Kenneth, Plainview, 1988
Hunter, Delaine, 2008
Kesting, Gustav, 1983
Kesting, Harold & Roger, 1993
Kienow, Lester Trust, Plainview, 1994
Kierstead Family, Lincoln, 1981
Kilterman, Esther & Erwin, 1969
Knaak, Alfred & Eleanor, Plainview, 1992
Knaak, Rolland, Plainview, 1987
Koehler, Frank & Marion, 1971
Koehler, Roger & Virginia, 1999
Koehler, Ronald, Osmond, 1998
Korth, John & Lori, Randolph, 2001
Krause, Elmer & Gertrude, Plainview, 2001
Krohn Special Trust, Norfolk, 2008
Krohn, Harold E. & Renata L., Norfolk, 2003
Krueger, Delbert, 1999
Krueger, Henry, 1973
Kruger, Duane, Randolph, 1991
Kuether, Margaret & Gordon, 1990
Kuhl, William & Judy, 2003
Kumm, Gerald & Hazel, Osmond, 2009
Kumm, Helen, Osmond, 1999
Larson, Frances, 1985
Larson, Wiggo, Plainview, 1982
Lehnert Family Trust, Eileen Lehnert Trustee, Plainview, 2009
Leiting, Margie, Randolph, 2009
Lenhoff, Timothy & Susan, Randolph, 2005
Lerum, Keith & Donna, Plainview, 1993
Lorenz, Josephine, Plainview, 1988
Lubke, David & Jean, Osmond, 2006
Luebbers, Mathias, 1998
Manske, Tillie, 1976
Meyer, Eileen, Randolph, 2016
Meyer, Elmer, Randolph, 2011
Miller, Irvin & Leon, Foster, 1983
Mittelstaedt, Ralph & Lorene, Norfolk, 1975
Monson, Arvid & Sharon, Wausa, 2000
Muhs, Larry and Anne, Pierce, 2016
Munger, Ivan, Meadow Grove, 1993
Oestreich, Dallas, Norfolk, 1983
Peters, Craig Harlan, 1993
Peters, Ralph & Esther, Osmond, 1997
Petersen, Cathryn, Plainview, 1995
Pfanstiel, Gary, Wausa, 1995
Planer Lang, Verada, 1984
Pochop, Helen, Plainview, 1993
Rath, Kenneth & Priscilla, Randolph, 1992
Reikofski, Jerry Jay & Michelle, Foster, 2011
Retzlaff, Delbert, Norfolk, 1976
Retzlaff, Robert & Jeana, 2006
Rohde, Lloyd, 1979

Rohrke, Ewald, Norfolk, 1978
Schlote, Verna, Plainview, 2016
Schulze, Isabelle, Tilden, 1993
Silhacek, Carol, 2002
Stark, Henry, Plainview, 1982
Steinkraus, Kathleen, Plainview, 1976
Steinkraus, Minnie, Plainview, 1976
Streich, Alton, Osmond, 1986
Streich, Herman, 1978
Sukup, Louise, Osmond, 1986
Synonec, Alvin, 1993
Synovec, Carlin T., & Judith J. Synovec Grantor Trust, 2010
Tacey Farms Ltd., Osmond, 1992
Theisen, Casper J., Agri-Vest, Inc., Omaha, 2006
Thomsen, Gene & Velma, Plainview, 2007
Tomek Nave, Catherine, Omaha, 1997
Voss, Merna, 1986
Vyhidal, Joe B. & W.F. (Donna Mae), 2004
Vyhlidal, Joe & Donna, 1993
Wachholtz, Raymond C & Meldene, Plainview, 2009
Wacker, Dennis, Osmond, 2000
Watson, Mr. & Mrs. L.E., Humboldt, 1972
Wattier, Francis/Susan, Randolph, 2008
Weber, Keith & Donna, Plainview, 1993
Zautke, Marlin & Jean, 2011

Platte County

Adamy, Leo, Columbus, 1986
Alderson, Cordelia, Humphrey, 1979
Alms Family Trust, Paula, York, 2006
Anderson, Lowell, Palos Park IL, 1997
Asche, Lowell, Leigh, 1977
Bakenhus, Arthur, Luther, and Edward, Columbus, 2008
Bargmann, Irene & David, Columbus, 2003
Barjenbruch, Mr. & Mrs. Otto, Leigh, 1972
Behlen, Harvey & Willard, Columbus, 1985
Benning Kummer, Esther, Columbus, 1995
Beran, Eunice Jenny, Howells, 1979
Berlin, Reynold, St. Edward, 1981
Blaser, Lynn, Columbus, 1995
Blaser, Chris & Amy, Duncan, 2011
Blaser, Hector & Minnie, Monroe, 1974
Bower, Mrs. Harold, Monroe, 1970
Bredehoft, Mr. & Mrs. Ted, Columbus, 1977
Brockhaus, Tom, Humphrey, 1985
Brockhaus, George & Elizabeth, Humphrey, 1999
Buggi, Eugene, Denton, 2009
Buhl, Dale and Mary Jo, Lindsay, 2008
Chana, Donald & Mildred, St. Edward, 1984
Christman, William & Wallace, Monroe, 1982
Connelly, James D., Lindsay, 2015
Cumming, Lonnie, St. Edward, 1993
Dasenbrock, Reinold, Leigh, 1982
Dicke, Mr. & Mrs. Leonard, Columbus, 1978
Dineen, Helen, Columbus, 1972
Dirks, Milton & Esther, Columbus, 1983
Ditter, Rodney, Monroe, 2013
Dohmen, Ervin, Humphrey, 1984
Ebel, Everette & Carolyn, Columbus, 1988
Ebner, Carroll & Arlene, Columbus, 1995
Ernst, Edward, Duncan, 1973
Ernst, Mrs. Eugene, Duncan, 1973
Fehringer, Daniel & Virginia, Humphrey, 2002
Feye, Hildegard, Fremont, 1984
Finecy, Dorothy & Vance, Columbus, 1988
Fittje, Lyle & Dianne, Humphrey, 1992
Foltz, Mark & Marcelline, Humphrey, 2001
Franzen, Mrs. Hilda, Leigh, 1976
Galley, George, Columbus, 1969
Gehring, Ernest & Louise, Columbus, 1984
Gehring, Steven & Cindy, Creston, 1994
Gehring, Larry & Norma Center, 1988
Gerber, Albert, Columbus, 1976
Gerhart Sr, Mrs. H.L., Newman Grove, 1982
Gottschalk, Frederick, Columbus, 1956
Greisen, Brian & Kara Center, 2014
Groteluschen, Lonnie & Terri, Humphrey, 2012
Gustafson, Helen, Newman Grove, 1986
Guter Pittman, Mrs. Mary, Columbus, 1956
Hamling, Larry, Creston, 1981
Harms, Gary, Monroe, 2002
Hassebrook, Rod Center, 2009
Heesacker, Alvin & Nancy, Humphrey, 1995
Heibel, Arthur & Mildred, Columbus, 1977
Heinen, Ted, Humphrey, 1982
Heitz, Eugene C., Leigh, 2004
Heitz, Eugene, Leigh, 2006
Hellbusch, Sam & Vicki, Creston, 2010
Hemmer, Everett & Norma, Humphrey, 2012
Hennessy, Francis & Madeleine Center, 1977
Herchenbach, Jerry & Darlene, Humphrey, 2000
Hoessel, Donald, Leigh, 1974
Hunteman, Clifford, Leigh, 1975
Jacobsen, Eldon, St. Edward, 1988
Jansen, DeForest & Luella, Creston, 1995
Janssen, Darrell & Patricia, Columbus, 2012
Jarecki, Joseph, Columbus, 1978
Joarosz, Raymond & Mary Ann, Columbus, 2006
Johnston, Gary, Lindsay, 1988
Kallweit, Leroy & Brenda, Columbus, 2010
Kapels, Roger, Columbus, 1981
Kapels, Keith & Colleen, Creston, 2014
Koch, Theodore, Creston, 1982
Korte, Ernst, Columbus, 1986
Korte, David, Leigh, 2011
Kosch, Talmah, Columbus, 1988
Kosch, Joe, Humphrey, 1982
Kringe, James, Humphrey, 1982
Krings, Marvin, Columbus, 1986
Krings, Kenneth, Columbus, 1987
Krzycki, Mrs. Phillip, Columbus, 1985
Kuhr, Dennis & Norene, Creston, 2000
Kummer, Paul & Esther, Columbus, 1984
Kunneman, John Gustav Center, 1977
Labens, Gustav & Minnie, Columbus, 1979
Larsen, Astrid, Lindsay, 1981
Letgen, Larry Center, 1989
Liebig, Richard & Mary, Columbus, 1987
Lindauer, Vern & Monica, Genoa, 1985
Long, Lloyd, Newman Grove, 1999
Loseke, L Fritz, Columbus, 1976
Loseke, Wayne & Ardeth, Columbus, 1982
Louis, Louise & Karl, Columbus, 1956
Lubischer, Donald, Humphrey, 1990
Luckey, Jerome & Dorothy Mae, Columbus, 1977
Luckey, Emil, Columbus, 1983
Lueschen, Clara, Creston, 1984
Lusche, Gustav & Elsa, Columbus, 1956
Lutjens, Lloyd, Tom & Dan Center, 1983
Mackenstadt, Lydia, Leigh, 1985
Maguire, Mark, Geoerge, Dick, Jim, Humphrey, 1990
Mark, Arno & Ruth Center, 1987
Mark, Clifford Center, 1987
Mark, Jerome & Margaret Center, 1988
Matson, Mrs. Elton, Newman Grove, 1986
Maurer, Oliver & Crystal, Columbus, 1993
Mausbach, James, Humphrey, 1987
McPhillips Sr., John, Humphrey, 1978
Meedel, Milton & Lillian, Columbus, 1995
Miksch, Alice & Margaret Center, 1992
Mitera, Mrs. James, Columbus, 1989
Mohlman, LaVern, Columbus, 1981
Mohlman, Lyle Center, 2010
Mohrman, Melvin & Nadine, Columbus, 1986
Molcyk, Philip A., Lindsay, 2015
Mueller, Marvin & Shirley, Columbus, 1976
Mueller, Emil & Adele, Columbus, 1982
Mueller, Mrs. Albert, Columbus, 1985
Muhle, David & Sandy, Leigh, 2001
Myhre, Mrs. Mary, St. Edward, 1974
Nansel, Fern, Monroe, 2006
Nelson, Lester & Mary, Newman Grove, 1992
Nienaber, Donald, Lindsay, 1994
Nyffeler, Lee & Jean, Columbus, 1992
Nyffeler, John & Marjorie, Columbus, 2009
Nygren, Mr. & Mrs. Alden, Newman Grove, 1983
Olmer, Arthur, Humphrey, 1985
Olmer, Gregory & Madonna, Humphrey, 2015
Olson, Alonzo, Newman Grove, 1984
Oppliger, Walter & Grace, Columbus, 1993
Osten, Arlyn & Nadine, Leigh, 1978
Paprocki, Rosmund, Columbus, 2009
Parry, Robert, Marjorie Parry Nyffeler, Columbus, 2006
Pearson, Jerold, Genoa, 1985
Petersen, Melvin & Lucille Center, 1980
Pfeifer, Melvin & Mary, Humphrey, 1978
Pfeifer, Keith & Pam, Humphrey, 1997
Pfeifer Jr., Ben & Donna, Lindsay, 2008
Phillip/ Drinnin, Samuel/ Anita, Columbus, 1995
Rasmussen, Lanny, St. Edward, 2002
Rasmussen, Wayne, St. Edward, 2002
Reinke, Martin & Edna, Columbus, 1980
Rickert, Adolph & Ruth, Columbus, 1974
Rollman, Ellen, Humphrey, 1978
Rollman, Mr. & Mrs. Thomas, Humphrey, 1978
Rosno, Stanley, Columbus, 1983
Rudat, Carroll, Columbus, 1979

Runge, Arthur, Columbus, 1976
Runge, August, Columbus, 1979
Sander, Mrs. Melvin, Columbus, 1991
Scharff, David, Columbus, 2001
Schneider, Peggy, and Nancy Harders, Columbus, 2012
Schreiber, Robert & Marilyn, Columbus, 1988
Schupback, Donald, Columbus, 1993
Schure, Richard, Humphrey, 1975
Sjuts, Donald & Norma, Humphrey, 1993
Sjuts, Calvin & Barbara, Humphrey, 1993
Smoots, William, Monroe, 1971
Spuit, Virgil, Genoa, 1983
Stempek, Adam & Victoria, Columbus, 1989
Stone, Leil, Lindsay, 1977
Stone, Howard, Lindsay, 1988
Stone, Lowell, St. Edward, 1988
Streblow, Edward, Humphrey, 1985
Ternus, Richard, Humphrey, 2001
Tessendorf, Gary & Gale Center, 1986
Theilen Vrana, Marilyn, Aurora, 2000
Thorson, Vendela, Mead, 1986
Tiaden, Quentin & Elaine, Columbus, 1986
Treadway, Joan, Osceola, 2009
Turner Jacobsen, Frances, Columbus, 1974
Van Dyke, Edwin, Humphrey, 1979
Wahl, Walter, Leigh, 2003
Wallin, LaVede, Newman Grove, 1996
Wegener, Earl, Humphrey, 1980
Welin, Lewis, St. Edward, 1985
Wemhoff, George, Humphrey, 1978
Went, Arlyn, Leigh, 1969
Wessel, Arthur, Humphrey, 1996
Wessel, Arlene, Lindsay, 2011
Westmore, Ervin & Frances, St. Edward, 1977
Wetgen, LaVern, Columbus, 1989
Widhalm, Earl/Edna, Humphrey, 1996
Wilke, Robert & Carolyn, Columbus, 2010
Wilke, Julian, Janice Wilke, Columbus, 2010
Will, Clare Center, 1986
Wurdeman, Marvin, Columbus, 1974
Wurdeman, Dennis, Leigh, 1987
Zach, Laurence, Columbus, 1988
Zach, Maurice, Omaha, 1988
Zuerlein, Mrs. Bruno, Leigh, 1984
Gerber, Jacob Children, Columbus, 1982
Rodehorst, Donald Trust, Columbus, 1994
Becher, Charles Farm, Columbus, 2003
Borg Farms Inc, Lindsay, 1984
Christman, Dana & Judy, Larry & Jean Andreasen, Monroe, 2003

Polk County

Adelson, Willard & Dorothy, 1974
Alt, Dave & Janet Alt, NE, 2005
Anderson, Allen & Dora, Stromsburg, 1993
Anderson, Bruce & Velma, Stromsburg, 1979
Anderson, Carl, Stromsburg, 1974
Anderson, Jack, Durham, 2006
Baker, Donald & Clarice, Stromsburg, 1988
Baker, Ronald, Stromsburg, 1973
Beebe, Raymond, Osceola, 1992
Beltz, Alta, Roca, 1994
Blasé, Lena Mae, Hordville, 1993
Bolton, Willard & JoAnne, Benedict, 1990
Borden, Phyllis, Holdrege, 1997
Boruch, Lambert Trust, Oscelola, 2015
Braasch, Janice Alt, NE, 2005
Brigham, Bill, W. James Brigham Living Trust, Shelby, 2014
Brigham, Roberta J., Shelby, 2004
Buhl, Dion & Cindy, Osceola, 2009
Burke, Raymond & LaVerne, Stromsburg, 1983
Burritt, Irene, Osceola, 1984
Cadwell, Samuel & Minnie, York, 1971
Campbell, Effie Lee, York, 1984
Carlson, Anna, Stromsburg, 1990
Carlson, August, Stromsburg, 1973
Carlson, Devon & Peggy, Osceola, 1993
Carlson, Nettie, 1975
Carlson, Paul D., Osceola, 2003
Dake, Rex & Marilyn, Silver Creek, 1992
Dawes, Merle, Osceola, 1973
Donelson, Dale & Kathy, Stromsburg, 2010
Dunn, Lewis, Osceola, 1971
Eggert, Orville & Elinor, Grand Island, 1980
Ernst, Max & Janice, Osceola, 2006
Ett Benson, John & Mary, Clear Lake IA, 1972
Farley, Homer & Virginia, Osceola, 1972
Fozzard Augustin, Helen, Shelby, 1979
Fyfe, Mrs Mae, Minatare, 1978
Gabriel, Mr. & Mrs. Arthur, Osceola, 1990
Gestrine, Irvin, 1974
Greene, Helen, Ashland, 1994
Gulbrandson, George, Waco, 1986
Hanson, Alfred & Carolyn, Osceola, 1989
Hasselbalch, August, Shelby, 1973
Hendricks, Dennis & Shirley, 2006
Hennings, Alice, Osceola, 1984
Hennings, Florence, Milan OH, 1997
Hughes, Don, Stromsburg, 2015
Johansen, Arthur, Osceola, 1983
Johansen, Virgil, Osceola, 1986
Johnson, Carl & Merle, Osceola, 1988
Johnson, Grace, Stromsburg, 1987
Johnson, Harlan & Mable, Stromsburg, 1988
Johnson, Morris, Stromsburg, 1971
Johnson, Morris, Stromsburg, 1973
Johnson, Rea Mae, Stromsburg, 2009
Johnson, Richard & Nancy, Stromsburg, 1989
Jones, David, Osceola, 1978
Jones, Duane & Carol, Osceola, 2006
Jones, Walden & Elaine, Osceola, 1993
Kinnan, Owen, Shelby, 1976
Kinzli, Lindo, Columbus, 1979
Kopke, Lavina, York, 1993
Kreis, Edna, Stromsburg, 1974
Kresha, Elaine, Columbus, 2003
Krumbach, Louise, Shelby, 1973
Kumpf Liss, Mildred, Shelby, 2009
Larson, William, Osceola, 1990
Lees, Mrs. Ernest, Shelby, 1978
Lewis, Kay, Chandler AZ, 1994
Lind, Lennis & Berniece, 1984
Lindblade Estate, Joseph, Osceola, 1972
Lindburg, Annette, Stromsburg, 2016
Lindburg, Maurice, 1986
Lindquist, Ansel, Grand Island, 2002
Lux, Ruth, Osceola, 1993
McKeag, Howard, Osceola, 1990
Messing, Sylvester, Shelby, 1992
Miller, Arlan, Lincoln
Miller, Dwayne & Elaine, 1994
Monson, Hannah, Osceola, 1973
Monson, Mr. & Mrs. Joe, Clarks, 1974
Moore Farms Company, Ulysses, 1986
Nelson, Harold Robert, Gresham, 1974
Norton, William, Osceola, 1986
Nowak, James & Joan, Osceola, 1993
Nyberg, Merlin, Stromsburg, 1988
Oberg, Robert & Vera, Stromsburg, 1991
Ostberg, Doris, Stromsburg, 1988
Peterson, Allen, Stromsburg, 1993
Peterson, Donald, Benedict, 1988
Peterson, Gustav, Stromsburg, 1973
Peterson, Vernon & Mary, Stromsburg, 1987
Peterson, Wallace, Stromsburg, 1989
Prososki Trust, Michael J. & Cheri L., Myron J. & Marla J., Jerome C., Kenneth A., Osceola, 2013
Rafert, Jlee R., Shelby, 2006
Rasp, Mr. & Mrs. L, Shelby, 1973
Reckling, Hollis & Kathaline, Lincoln, 1993
Roberts, Jane, Osceola, 2012
Rystrom Enterprises Inc., Stromsburg, 1985
Rystrom, Delwyn, Stromsburg, 1990
Sandell, Levar, Osceola, 1986
Schmid, Dale, Shelby, 1979
Shockey, Fern, Osceola, 1971
Shostrom, Don & Shirley, 2001
Smith, Jacob & Martha, Stromsburg, 1971
Smith, Lloyd, Shelby, 1986
Stevens, Patricia, 2006
Steward, Lenore, Shelby, 1985
Strickland, Mrs. Henry, Silver Creek, 1980
Sundberg, Dean, Stromsburg, 1982
Sundberg, Reuben, Stromsburg, 1982
Swedenburg, Ralph & Marjorie, Clarks, 1986
Thelen, Bertha & Children, Shelby, 1980
Timm, Agnes, Osceola, 1975
Timm, William, Lincoln, 1984
Tonniges, Clara, Columbus, 1973
Tooker, Norman, Ralston, 1987
Tooker, Robert, Silver Creek, 1987
Triba, Richard, Osceola, 1985
Twarling Donelson, Ruth Frances, Stromsburg, 1984
Urban, Francis, Osceola, 1993
Walgreen, Irene, Osceola, 1983
Waller, Philip & Timm, Stromsburg, 1984
Wescoat, Audrey, Osceola, 1971
Whaley, Frances, Central City, 1972
Whitmore, Mr. & Mrs. Howard, Shelby, 1979
Widga, Lester, Stromsburg, 1974
Widga, Lloyd Edward, 1973
Wieseman, Carl, Osceola, 1976
Wieseman, Robert, Osceola, 1985
Williths, Ralph & Catherine, Clarks, 1972
Wlagren, Paul, Osceola, 1989
Wyman, Ted & Donna, Silver Creek, 1988
Younglund, Harry, Stromsburg, 1976
Yungdahl, Dale & Sharon, Osceola, 2012

Red Willow County

Adams, Eurilla, McCook, 1974
Ball, William E, McCook, 2009
Ballou, Jeanette, McCook, 1995
Berndt, Dean, Helen Berndt, Danbury, 2007
Burton, Merwyn, Bartley, 2000
Dolph, Mrs. Myrtle, Danbury, 1973
Dutton, Esther, Omaha, 1995
Fitch, William Carol, McCook, 1974
Fritz, J.R. Family, Bartley, 1987
Galusha, Jack, Danbury, 1974
Haag, Raymond, Bartley, 1994
Hinz, Keneth, McCook, 1999
Hotze, Clarence, Indinaola, 1973
Kelley, Raymond, Danbury, 1973
Kircher, Jerry, and Donna Kircher, Indianola, 2012
Kircher, Kenneth and Karen, Indianola, 2012
Koetter, William Carol, McCook, 1991
Koslowski, Kenny and Chrissy, Denver, 2010
Lomaster, Glenn & Lillian, Wilsonville, 1986
Longnecker, Mitchell, Maywod, 1973
Maurer, Betty, Novi MI, 2000
Miller, Roger & Wanda, Danbury, 1993
Morris, Lawrence Lester, Hastings, 2012
Mustion, Fred & June, Culbertson, 1993
Newberry, Donal, Burlington, 2005
O'Dea Family LP, Andrew & Lori O'Dea, McCook, 2009
Parker, Ramond & Edna, Danbury, 1974
Quaduor, Edward, Albuquerque NM, 1999
Quaduor, Faye, Indinaola, 1999
Quigley, Stanley, Indinaola, 1973
Redfern, Patsy Lee & James Alan Redfern, McCook, 2004
Reiners, James & Wendy, Indianola, 2011
Schmelzer, Josephine, Maywood, 1992
Schultz, Gleyone, Danbury, 1994
Shields, Robert Harley, McCook, 1990
Sides, Joan, Indinaola, 1997
Sughroue, Bill & Sue, Indinaola, 1995
Teter, Marion & Grace, Bartley, 1989
Walters, Sam, Lebanon, 1974
Wesch, Carl & Nila, McCook, 1999
Weyeneth, Norris & Joyce, Danbury, 1977
Wil-A-Dar Farms Inc., Culbertson, 1984

Richardson County

Ahren, Gisela, Shubert, 1986
Ahren, John & Beatrice, Shubert, 1966
Ahren, William & Ruth, Shubert, 1967
Aitken, Janet, Falls City, 1989
Allen, Mr. & Mrs. Robert, Dawson, 1966
Anderson, Mr. & Mrs. Carol, Humboldt, 1976
Ankrom, William & Aline, Shubert, 1992
Armbruster, Marlin, Falls City, 2004
Arnold, Florence, Falls City, 2004
Arnold, Nellie Pearl, Humboldt, 1969
Ast, Mr. & Mrs. Elmer, Humboldt, 1960
Auxier, Edward/ LeRoy & Helen, Dawson, 1987
Avery, Jesse & Jean, Humboldt, 1967
Bachman, Mr. & Mrs. Loyd, Rulo, 1969
Bauer, Ruth, Verdon, 1982
Becker, Anna, Falls City, 1978
Belden, Mrs. George, Dawson, 1968
Berg, Arthur, Shubert, 1957
Berg, Orpha & Elmer, Auburn, 1975
Beutler, Caroline Christina, Falls City, 1986
Bletscher, Philip, Falls City, 2011
Bright, Mr. & Mrs. Dnoald, Shubert, 1969
Bucher, Errold, Humboldt, 1981
Bucher, Frederick & Ethel, DuBois, 1993
Buchholz, Anna, Falls City, 1965
Carico, Mrs. Wesley, Falls City, 1975
Castner, Kay, Columbus, 2006
Chism, Ralph Family, Lincoln, 1970
Church, Mrs. Ella, Dawson, 1967
Crook, Mr. & Mrs. Leo, Salem, 1956
Cummings, Ruth, Falls City, 1998
Dorste, Eugene & Ann, Rulo, 1970
Duerfeldt, Arthur, Falls City, 1966
Duerfeldt, Mr. & Mrs. Clifford, Barada, 1961
Duerfeldt, Mr. & Mrs. Elmer, Falls City, 1992
Duerfeldt, Mr. & Mrs. Marvin, Barada, 1961
Duryea, Polly Porr, Humboldt, 2016
Edwards, C. Harold, Lincoln, 1979
Eickhoff, Duane & Phyllis, Falls City, 1984
Eiguren, Jean, Boise ID, 1994
Emigh, Mr. & Mrs. Laurence, Humboldt, 1967
Estate, Portrey, Falls City, 1974
Fairbanks, Harry & Cleo, Dawson, 1983
Faller, Gilbert, Falls City, 1969
Faller, Herman William, Falls City, 1969
Fankhauser, Chris & Elmer, Humboldt, 1965
Fankhauser, Mr. & Mrs. Jerry, Humboldt, 1961
Fankhauser, Mrs. Fred, Humboldt, 1963
Farwell, David L., Humboldt, 2008
Fergus, James & Wilma, Humboldt, 1970
Fergus, Mrs. May, Humboldt, 1970
Fiedler, Amelia, Falls City, 1967
Finck, Raymond & Ruth, Falls City, 2009
Findlay, Harvey, Falls City, 1993
Fischer, Allen, Dawson, 1995
Fischer, Mrs. Fred, Rulo, 1975
Frank, Loraine & LeAnna, Humboldt, 1989
Frederick, James & Marcia, Falls City, 2006
Frey, Mr. & Mrs. Alfred, Humboldt, 1956
Fritz, Elmer, Verdon, 1989
Fugazzi, Patricia Kelly, Falls City, 1981
Gebhard, Harold D. & Zerlene E., "Nixa, MO", 2004
Gentry, Fred, Stella, 1982
Georges, Carol & Opal, Shubert, 1968
Gerdes, Dorothea & Loren, Humboldt, 1981
Gerlt, Mr. & Mrs. John, Falls City, 1965
Gilbert, Kathleen, Falls City, 1993
Griffiths, John, Pawnee City, 1970
Hanika, Leo, Shubert, 1980
Harkendorff Jr, Fred, Falls City, 1959
Harshbarger, Ted, Humboldt, 1999
Haynes, Mr. & Mrs. Henry, Auburn, 1983
Hayward, Glenn & June, Humboldt, 1982
Heim, Arthur, Dawson, 1974
Heim, Dr. H.S. Heirs, Dawson, 1982
Heim, Lowell, Dawson, 1984
Heineman, Joe & Mary Alice, Pawnee City, 1986
Hilgenfeld, Carl, Falls City, 1977
Hitzeman, Donald, Humboldt, 1980
Hoffman, Nellie, Humboldt, 1983
Hunker, Gary, Rulo, 2000
Hunker, Walter, Nebraska City, 1978
Hunter, Carol, Austin, 2007
Hunzeker, Mr. & Mrs. Paul, Humboldt, 1956
James, Erma Jean, Daniel F, and Dale E., Humboldt, 2008
James, Richard, Verdon, 1975
Johnson, James & Ella, Verdon, 2005
Jorn/Dodderer, Harriett/ Marion, Falls City, 1976
Kadavy, Alice Atwood, Lincoln, 1993
Kean Estate, Robert Emmet, Auburn, 1978
Kennedy, Lola, Nemaha, 1990
King, Ardis, Humboldt, 2011
Koso, Darrel & JoAnn (Scheitel), Rulo, 2013
Kuker, Norman & Maxine, Falls City, 2007
Kuper, John, Humboldt, 1967
Lang, Mr. & Mrs. Arthur, DuBois, 1979
Last, Albert & Lena, Shubert, 1976
Lazy Farms Inc., Rulo, 1989
Leatherman/ Ulmer, Marian / Ernest, Dawson, 1984
Lewis, Barye & Harriett, Verdon, 1965
Lewis, Martha Finck, Falls City, 1975
Lewis, Wayne & Dorothy, Falls City, 1965
Lichty, Mr. & Mrs. John, Falls City, 1969
Lippold, Mr. & Mrs. Gustav, Shubert, 1967
Loennig, Gordon, Humboldt, 1973
Luginbill, Mr. & Mrs. William, Humboldt, 1956
Luthy, Glendon, Humboldt, 1956
Luthy, Mr. & Mrs. Fred, Humboldt, 1956
Martin, William & Janis, Dawson, 1975
Masonbrink, John, Humboldt, 1992
Matthes, Mrs. Paul, Falls City, 1975
Matthes, Bernice, Falls City, 2016
Maze, Mrs. Gladys Ratekin, Rulo, 1967
McCloud, Patricia, Seattle WA, 1994
McConnell, Mr. & Mrs. Robert, Humboldt, 1964
McConnell, Mrs. Edna, Humboldt, 1964
McGinnis, Joe, Dawson, 1979
Mehlin, George, Humboldt, 1983
Merz, Rodger / Douglas Merz / Nancy Deklotz, Falls City, 2004
Mez Fuggett, Alta, Nebraska City, 1978
Mitchell, Carol Cornelius, Humboldt, 1969
Monnette, Constance & Raymond, Humboldt, 1972
Mosiman, Mr. & Mrs. Wayne, Falls City, 1979
Mullen, Mr. & Mrs. Ben, Humboldt, 1960
Mullen, Mr. & Mrs. John, Humboldt, 1960
Murphy, Edward, Rulo, 1980
Murphy, Mrs. William, Dawson, 1975
Nelson, Lester/ Anna, Falls City, 1996
Nemechek, Mr. & Mrs Jesse, Humboldt, 1969
Neville, Rita, Falls City, 2005

Niedfeldt, Kenneth & Janice, Columbus, 1997
Nolte, Mrs. Everett, Falls City, 1978
Ogle, Richard & Marilyn, Dawson, 1989
Owens, Mr. & Mrs. Roland, Verdon, 1976
Palmer, Verl Dean, Falls City, 1982
Parsons Von Seggern, Mary Margaret, Verdon, 1972
Parsons, Neal, Humboldt, 1972
Parsons, Neal, Verdon, 1975
Peck Jr., Henry Clay, Falls City, 1982
Peck Wilson, Halcyon, Falls City, 2009
Peck, Families of George, Cleon, Glenn & Marlin, Falls City, 1969
Percival, Clark, Falls City, 1986
Percival, Elmer, Scottsbluff, 1979
Phillips, Mary, Humboldt, 1965
Planer, Marilyn Irene Planer (Stauffer) & Clarence F. Planer, Newton, 2014
Pohlman, Elza, Humboldt, 1986
Poppe Lortscher, Harold & Marcille, Denver Co, 1995
Porr, Frank, Humboldt, 1957
Porr, Mr. & Mrs. Ivan, Humboldt, 1958
Porr, Mr. & Mrs. William, Humboldt, 1958
Quinlan, William, Dawson, 1984
Rawley, Robert, Omaha, 2000
Rears, Donald & Rose Marie, Stella, 1983
Rieschick, Don, Carolyn Rieschick, Benicia, 2007
Rieschick, Edith, Falls City, 1993
Rist, Crissie & Marjorie, Humboldt, 1983
Rohlmeier, Kenneth, DuBois, 1993
Rowell, LaVern, Falls City, 1987
Rowell, Loren Wayne, Omaha, 1999
Ruegge, Mrs. Henry, Falls City, 1979
Rumbaugh, Melvin & Annabelle Eis, Humboldt, 1993
Rutz, John & Edna, Dawson, 1968
Schenk, Irene, Humboldt, 1983
Schmid, Elizabeth & John, Lansing, 2005
Schulenberg, Henry, Falls City, 1968
Shafer, Max, Shubert, 1974
Simmonds, Mr. & Mrs. Earl, Humboldt, 1971
Slagle, Lee Roy & Beulah, Shubert, 1969
Slayton, Wilfred & Martha, Salem, 1976
Snethen, Mary Katherine, Humboldt, 1998
Snoke, Jon & Lacey, Humboldt, 2014
Stalder, Gordon, Humboldt, 1983
Stalder, Mr. & Mrs. Frank, Humboldt, 1963
Stalder, Mr. & Mrs. Joe, Humboldt, 1961
Stalder, Mr. & Mrs. Louis, Humboldt, 1961
Standerford, Joe & Kim, Humboldt, 2006
Standerford, Mr. & Mrs. James, Humboldt, 1966
Stappenback, A.P. & Leah Ruth Cornelius, Humboldt, 1969
Stevens, Loretta Lee, Falls City, 2013
Storant, Edd, Joe, & Katherine, Humboldt, 1980
Stump, Mr. & Mrs. Robert, Dawson, 1980
Swisegood, Mr. & Mrs. Lee, Verdon, 1978
Swisegood, Thelma, Falls City, 1998
Thacker, Dennis, Dawson, 1989
Thiltges, William & Catherine, Falls City, 1968
Thiltges/ Fischer, Mildred/ Raymond, Rulo, 1975
Thompson, Donald W, Steven Thompson, Dwight Thompson, Salem, 2006
Tichy, Mrs. Joseph, Humboldt, 1976
Timmerman, Howard & Dottie, Stella, 1969
Tjaden, Mrs. Mae, Humboldt, 1966
Trimble, Mrs. Willard, Humboldt, 1994
Tupa, Loren, Humboldt, 2013
Turco, Mary Ann, Humboldt, 1993
Tynan, Mr. & Mrs. Robert, Stella, 1958
Van Vleck, Luella Fisher, Lincoln, 1970
Veigel, Marcille, Falls City, 1995
Voegtli, Neola, Stella, 1985
Vogele, Lawrence, Rulo, 1974
Vogele, Mr. & Mrs. Charles, Rulo, 1968
Von Bergen, Mr. & Mrs. Ralph, Dawson, 1971
Watkins, Mr. & Mrs. Homer, Verdon, 1972
Watson, L.E., Humboldt, 1961
Weaver, Mr. & Mrs. Larry, Falls City, 1977
Weinert, Harold, Falls City, 1971
Weinert, Mary Lee, Jean Metzger, Linda Perue, Falls City, 2010
Wells, Darrell Millard, Verdon, 1971
Whitfield, Verne & Helen, Humboldt, 1989
Williamson, Robert, Dawson, 1989
Wilson, Mr. & Mrs. Kenneth, Shubert, 1969
Withee, Joe, Stella, 1963
Witt, Duane & Phyllis, Falls City, 1994
Wittrock, Johanna, Falls City, 1980
Yoesel, Dale W., Rulo, 2009
Yoesel, George, Falls City, 1975
Yoesel, Richard R., Rulo, 2009
Young, Nevada Hays, Stella, 1975
Zimmerman, Carl, Lincoln, 1989
Zimmerman, Robert, Falls City, 1980

Rock County

Ammon, Harold, Bassett, 1984
Boerger, Ben & Judy, Rose, 1986
Bouy, Darrell & Shari, Burwell, 2015
Buell, Lawrence & Nickie, Rose, 1986
Buoy, Harry & Faye, John H. Buoy Family, Burwell, 2015
Buoy, Harry & Faye, Phyllis Buoy Family, Burwell, 2015
Cameron, Larry, J.D. Cameron, & Marie Kerl, Summerfield, KS, 2003
Cook Jamison, Bernice, Newport, 2001
Copeland, Patricia, Bassett, 1996
Cosgrove Family Trust, C. Thomas & Margie Cosgrove, Long Pine, 2007
DeGroff, Geraldine, Atkinson, 2001
Dugger, Margaret, Bassett, 1991
Gesiriech Family, Newport, 1987
Harlan, Mr. & Mrs. James, Bassett, 1979
Hutton, Harold, Bassett, 1981
Iverson, Lewis & LaVerne, Bassett, 1987
Jilg Ranch Inc, Newport, 1987
Johnson, Allene, Norfolk, 2011
Johnson, Jim, Bassett, 2014
Kaup, Steve & Mary, Newport, 2006
Keller, Edith, Bassett, 2005
Kops, Lyle & Eleanor, Bassett, 1984
Kramer, Mrs. Mary, Bassett, 1984
Kreitman, Phyllis Sybrant & Evera, Bassett, 2004
Linke, Karl, Bassett, 1984
Maurer, Marcella, Bassett, 2001
May, DeLane & Larry, Bassett, 1991
McGuire, Frank & Donna, Bassett, 2003
Sandall, Keith & Delores, Bassett, 1991
Sandall, Nellie, Bassett, 1994
Seberger Iverson, Ruby, Bassett, 1986
Seberger, Fred, Newport, 1986
Seier, Johnnie, Bassett, 1980
Stewart, Jerold and Roy, Newport, 2008
Stout, Royce, Bassett, 2014
Sttrlow Family, Bassett, 1992
Swinney, Leona M., Basset, 2008
Sybrant, Allen, Bassett, 2000
Sybrant, William & Carolyn, Bassett, 2013
Thomas, Orrin & Louise, Bassett, 1990

Saline County

Ahlschwede, Mr. & Mrs. Robert, Crete, 1976
Akasamit, Steve & Libbie, Crete, 1976
Andelt, Arnold, Crete, 1981
Andelt, Donald/ Alice June, Western, 1996
Aron, Carl & Angie, Crete, 1969
Barta, Dolores, Swanton, 2006
Bartels, Dennis & Janet, Milligan, 2012
Bartels, Donald, Tobias, 1986
Bartels, Gaylord & Patricia, Tobias, 2007
Bednar, Henry, Wilber, 1977
Belohlavy, Roger & Janice, Crete, 2014
Birkett, Ivene & Jack, Wilber, 1985
Blankenship, Kathy, Nancy Richards, Placerville, 2009
Bors, Gary & Carole, Crete, 2014
Brabec, Joseph, Crete, 1962
Broz, Eugene & Doris, Wilber, 1993
Bruha, Shelley, Dorchester, 2008
Burns, Mary Virginia & Catherine C., Friend, 2014
Busboom, Phyllis, David & Thomas, Crete, 1989
Busch, Eugene, DeWitt, 2002
Caldwell, Jesse & Lola, Swanton, 1965
Capek, Rosemary, Crete, 1992
Cejka, Anna M., Lincoln, 2008
Cerveny, Donald W., S. Wanton, 2004
Chab, Mrs. Otto, Denton, 1982
Chaloupka, Mr. & Mrs. Mil, Wilber, 1971
Chalupa, Charles & Betty, Wilber, 1995
Chalupa, Helen, Crete, 1993
Chase, Doris, Friend, 1982
Chmelir, Gerald, Wilbur, 2015
Chmelir, Gerald, Wilbur, 2015
Crays, Don R. and Sheila M., Tobias, 2008
Crays, Donald & Anna, Tobias, 1990
Dlouhy, Duane, Branson, 2011
Donald, Joseph, Wilber, 1998
Drake, Ralph & Katherine, Friend, 2015
Durflinger, Gail and Eugene Krajnik Jr., Alexandria, 2016
Dvorak, Fred, Wilber, 1983
E & E Bartels LTD, Tobias, 1987
Elliott, Ann Barney, Lincoln, 1990
Endorf, Harold, Tobias, 1979
Eret, Larry and Betty, Goehner, 2016
Essman, Ervin, Swanton, 1987
Farms L.P., Palky, Dorchester, 2010
Fictum, Dwight & Shirley, Wilber, 2002
Fictum, Ramon & Carol, Wilber, 1984

Filipi Brydi, Henry & Anna, Tobias, 1994
Fink, Mr. & Mrs. Don, Crete, 1976
Fisher, Dan, Grand Island, 1993
Fittl, Irma Ann, Wilber, 1984
Fleischman, Valoris, Palmyra, 1993
Formanek, Allen & Camille, Crete, 1985
Freeof Revocable Trust, L & F, Leonard & Florence (Trustee), Wilber, 2004
Freeof, Maynard & Norma, Crete, 1984
Frew/ Williams, John/ Katherine, Crete, 1991
Fritz, Mr. & Mrs. David, Western, 1971
Fritz, Mr. & Mrs. Gus, Western, 1971
Fritz, Norman, Wilber, 1993
Gausman, Clarence & Gladys, Dorchester, 1970
Gillespie, Jason, Tobias, 1979
Graves, William, Tobias, 1987
Gross Kumpost, Henry & Genevieve, Wilber, 2000
Hansen, Edith Life Estate, Crete, 1994
Hatcliff, Roland, Crete, 1982
Heeren, Jim, Dorchester, 2010
Hegwald, Darlene, Lincoln, 1984
Heitman, William & Esther, Crete, 1993
Hilton, Dr. HD & Family, Lincoln, 1977
Holly, Joe & Julie, Dorchester, 2006
Homolka, Byron & Dorothy, Wilber, 2005
Homolka, Donald, Wilber, 1990
Horak, Ronald & Norma, Western, 1984
Hynek, George, Wilber, 1984
Inderlied, Twila, Wilber, 1992
Janda, Christine & Joe, Wilber, 1979
Janda, Harold, Friend, 2009
Javorsky, Irvin, Wilber, 1984
Jelinek, Albina, Wilber, 2000
Jelinek, Dennis, Wilber, 1998
Jelinek, Edward & Stephen, Omaha, 1995
Jezl, Joseph, Tobias, 1987
Jiskra, Gary Dale, Swanton, 2003
Jiskra, Rex A., Ohiowa, 2013
Jordening, Wilma, Tobias, 1988
Justa, Bertha, Dorchester, 1989
Kahle, Ron & Joyce, Dorchester, 2009
Kasl, Aaron / Karoline, DeWitt, 1996
Kasl, Richard, Wilber, 1997
Kaspar, Bernard & Helen, Dorchester, 1973
Kastanek, William & Sheryl, Crete, 2000
Kastanek, William, Crete, 1984
Kastenek, Lad & Dorothy, Wilber, 1973
Kay, Olga Vosoba, Jacksonville FL, 1997
Keller, Leon, Crete, 1985
Kellough, Max & Roberta, Friend, 1984
Kelso, Azariah I., Friend, 1978
Kemper, Mr. & Mrs. Ben, Crete, 1969
Kemper, Mr. & Mrs. John, Crete, 1969
Klasek, Alan, Western, 1993
Klasek, Betty, Wilber, 1993
Kohel, James, Wilber, 1993
Kohel, Mrs Anton, Wilber, 1981
Kohout, Elmer, Wilber, 1993
Kohout, Stanley, Wilber, 1993
Korinek, Delphin & Shirley, Wilber, 1993
Kottas, William James, Crete, 1997
Kovar, Elsie, Wilber, 1998
Krebs, June, Friend, 1982
Kreuscher, Allen & Mary Ann, DeWitt, 1993
Kreuscher, Allen and Mary Ann, DeWitt, 2003
Kreuscher, Glenn & Marion, Lincoln, 2003
Kubicek, Mr. & Mrs. George, Wilber, 1970
Kubicek, Terrane & Aleandr, Lincoln, 2000
Kucera, Charles, North Platte, 1987
Lampe, Linda, Olathe KS, 2001
Lautenschlager, Cecil, Swanton, 1984
Leavitt, Robert, Crete, 1982
Linhart, Frank & Nancy, Wilber, 1993
Littlefield, Franklin & Virginia, Lincoln, 1992
Loos, Everine Andelt, Lincoln, 2003
Lothrop, Gary & Denise, Crete, 2015
Mariska, Donna, Dorchester, 2006
Mariska, Mrs. Mildred, Crete, 1976
Mariska, Robert & Leonore, Dorchester, 1990
McAdams, Harlan, Western, 1988
McCracken, Charles & Norma, Friend, 1980
McNillan, Malcolm, Crete, 1982
Meier, Earl Radford, DeWitt, 1985
Meyer, Harlan, Tobias, 2013
Miller, O.F., Western, 1966
Miller, Paul & Warren, Columbus, 1984
Mitchell, Shirley Ann, Crete, 1983
Moore, Clarence & Ella, Wilber, 1983
Musil, Frank, Crete, 1975
Niederklein, Dallas, Tobias, 1993
Niederklein, Jacquelyn, Tobias, 2005
Nielsen, Donald & Janice, Tobias, 1985
Niemeier, Gary, DeWitt, 2012
Niemeier, Rachel, Fairbury, 2012
Nerud, Dan and Joan, Dorchester, 2016
Nolte, Romaine, Fairbury, 2007
Novak, Adolph & Evelyn, Dorchester, 1993
Novak, LaVern and Dorothy, Milligan, 2016
Overman, Charles & Mary, Alexandria, 1972
Patak, Joe & Erma, Crete, 1967
Pecka, Adolph, Crete, 1985
Pomajzl, Diana, Crete, 2016
Pomajzl, Larry & Linda, Crete, 2013
Pomajzl, Lumir & Lillian, Wilber, 1994
Pomajzl, Steve, Lincoln, 2001
Prachar Aksamit, Hattie, Wilber, 1976
Prachell, Arnold, Crete, 1983
Prochaska, Glenn & Mattie, Crete, 1993
Prokop, Henry & Bessie, Crete, 1971
Rada, Roger & Judith, Tobias, 1992
Rafert, Carl & Carol, Tobias, 1992
Rezabek, Frank, Lincoln, 1980
Rezabek, Raymond C. & Connie L., Dorchester, 2013
Richtark, Helen, Wilber, 1988
Ridpath, William Dwight, Western, 1983
Ripa, Glenn & Gary, Wilber, 1992
Ripa, Warren & Irma, Wilber, 1983
Roesler, Leola, Tobias, 2004
Runty, Ellen M., Edward L. Runty, Charles E. Runty, Bonnie J. Zaroban, DeWitt, 2003
Rut, John & Donna, Daykin, 2011
Safranek, Elmer & Lucill, Omaha, 1974
Salda, Bertha, Wilber, 1984
Sampson, Frank, Lincoln, 1986
Schleis-Roesler, Evelyn, Lincoln, 2007
Schleis, Joseph & Evelyn, Wilber, 1972
Schleis, Milo & Elaine, Wilber, 1991
Schleis, Wencil, Wilber, 1991
Schmidt, Duane & Vicki, Western, 1984
Schoenbeck, Vernon, Tobias, 2002
Schuerman, Esther, DeWitt, 1983
Schuerman, Marvin, DeWitt, 1983
Schultz, Frank, Lincoln, 1983
Sebek, Vernon & Darlene, DeWitt, 1993
Shafer, Louis, Tobias, 1992
Shestak/Mead, Herman/Grace, Falls City, 1982
Siedhoff, John and Francis, Crete, 2016
Skleba, Rod & Marsha, Wilber, 1995
Slama, Edward and Darlene, Crete, 2016
Slepicka, Leland, Clay Center, 1993
Smejdir, William & Margaret, Wilber, 1985
Smith, Andrew, Alexandria, 1972
Smith, Mr. & Mrs. Leonard, DeWitt, 1970
Smith, Ms. Lynn, Tobias, 1981
Steinacher, Gladys, Milligan, 1984
Steinbrook, Gordon, Columbus, 1992
Stokebrand, Harold and Doug, Dewitt, 2013
Stokebrand, Otto, DeWitt, 1997
Sukovaty, Georgia, Swanton, 2000
Sukovaty, Jack & Shirley, Martell, 1984
Sysel, Norma, Dorchester, 1983
Tenopir, Irene, Wilber, 1975
Tenopir, Mrs. Gladys, Crete, 1981
Tichy, Harold, Wilber, 1976
Timmel, Leslie, Fairbury, 1972
Turner, Irving S., Wilber, 1976
Tuxhorm, Don & Arlene, Crete, 1994
Tyser, Kenneth, De Witt, 2011
Vales, Robert, Swanton, 1993
Vana, Emil, DeWitt, 1973
Vavra, David & Dana, Friend, 2015
Vocasek, Robert, Crete, 1990
Vosoba, Joe T., Wilber, 1977
Vostrez, Robert & Bertha, DeWitt, 1973
Waldo, Mrs. H. O., DeWitt, 1982
Warta, Anton, Wilber, 1994
Weichel, Lonnie & Cindy, Jansen, 1993
Wielage, Gladys, Crete, 1993
Wilkens, Mrs. Everett, DeWitt, 1983
Wilkins, Everett, DeWitt, 1981
Winter, Lewis, Western, 1972
Witt, George, Lincoln, 1966
Wollenburg, Duane, Dewitt, 2013
Wollenburg, Paul & Warren, Beatrice, 1992
Yeck, Leon, Tobias, 1986
Zabel, Roger & Eloise, Western, 1963
Zajicek, Eugene & Vivian, Wilber, 1993
Zalesky, William & Darlene, Wilber, 1984
Znamenacek, Karel, Crete, 1974
Zoubek, Ernest & Rose, Wilber, 2011
Zoubek, Frank J., Milligan, 2004

Sarpy County

Allbery Farms, Gretna, 2004
Beck, Lois, Rex & Randy, Springfield, 2007
Bundy, Floyd, Gretna, 2002
Burton, Albert, La Platt, 1999
Cain, Richard, Omaha, 1966
Clark, Charles, Papillion, 1993

Daniell, Lester & Doris, Omaha, 1968
Eggers, Mr. & Mrs Lawrence, Gretna, 1976
Fornoff, Walter & Gladys, Omaha, 1969
Friche, Emil, Papillion, 1967
Fricke, Milton, Papillion, 1956
Gates, Mr. & Mrs. Harry, Omaha, 1956
Glesmann, Irene, Springfield, 1969
Glesmann, Mr. & Mrs. Alvin, Papillion, 1970
Glesmann, Mr. & Mrs. Chris, Papillion, 1970
Graham, MR. & Mrs. Ad, Papillion, 1956
Gramlich, Alois, Ft. Crook, 1956
Hintz, Mrs. Margaret, Gretna, 1978
HRL Properties LLC, Omaha, 2014
Jahn Farms Inc, George Jahn, Delores Snodgrass, Gretna, 2009
Keefer Brothers Farm LLC, David Keefer, Harold Keefer, Papillion, 2010
Keefer, Herbert & Helen, Omaha, 1984
Keyes, Donald, Papillion, 1973
Keyes, Mrs. Mary, Papillion, 1973
Koch, Ane Marie, Gretna, 2002
Krebs Farm (Ellen Nemec, Barbara Roth and Marcia Matthies), John Krebs, Papillion, 2005
Latham, Mayme, Springfield, 1975
Lienemann, Arthur & Mary, Papillion, 1984
Lilley, Mr. & Mrs. James, Omaha, 1957
Lind, Janice, Omaha, 1987
Linemann, Herbert & Helen, Papillion, 1968
Mann, Mr. & Mrs. Bill, Papillion, 1986
Milks, John, Springfield, 1980
Miller, Russell, Gretna, 1986
Morrison, Mr. & Mrs. John, Omaha, 1957
Otte, Harlan & Millie, Omaha, 1983
Peters, Mr. & Mrs. Henry, Omaha, 1957
Picconi, W.E.A.D. Eleanor, ", Ann, Dean", Elkhorn, 2006
Robinson, Mrs. Frances, Bellevue, 1962
Rothermund, Mr. & Mrs. Lewis, Ft. Crook, 1956
Runge, Paul & Helen, Gretna, 1982
Sautter, Alma, Papillion, 1971
Sautter, George, Papillion, 1971
Sautter, Sophia, Papillion, 1971
Sautter, William, Papillion, 1971
Schnack, Marcus Estate, Tretna, 1987
Schobert, Mark, Papillion, 1987
Schram, George & Marilynn, Gretna, 2001
Schram, Kenneth/ Julie, Papillion, 1996
Schwer, Mrs. Catherine, Papillion, 1974
Seibold Farms, Papillion, 2000
Seibold, Morris & Verna, Papillion, 1981
Snider, A.H, Springfield, 1974
Stender, Minnie, Papillion, 1975
Stender, Rinehart, Fremont, 1975
Tex, Ron & Sharon, Papillion, 1994
Thompson, Mr. & Mrs. Ward, Papillion, 1956
Timm, Mr. & Mrs. Otto, Papillion, 1956
Trumble Frederick, Judith, Manhattan KS, 1977
Trumble, Charles, Papillion, 1977
Trumble, Gertrude & Floyd, Papillion, 1983
Trumble, Harland, Papillion, 1984
Upjohn, Anna, Omaha, 1967
Weiss, Mr. & Mrs. Elmer, Papillion, 1956
Weiss/ Kittoe, Evelyn/ LoRene, Papillion, 1981
Williams, Leona, Omaha, 1982
Woodward, Ella Bolling, Valley, 1967
Zwiebel, Chet & Christina, Ft. Crook, 1956

Saunders County

Adams, Mr. & Mrs. Ernest, Mead, 1972
Anderson, August Alfred, Wahoo, 1975
Anderson, Dwight & Dorothy, Weston, 2002
Anderson, Irene & Adeline, Fremont, 1973
Arp, Allen, Yutan, 2014
Barnhill, Maxine & Rodney, Fremont, 1985
Barry Farms Inc., Malmo, 1997
Barry, Clarence, Wahoo, 1971
Barry, Kenneth, Weston, 1980
Bartek, Anna, Wahoo, 1992
Bauer, Daniel, Morse Bluff, 1978
Bauer, Jeff B. & Louise, Joe & Lillian, Morse Bluff, 1986
Bavlovic, Edmund & Betty, Prague, 2004
Beaman, Francis & Arvella, Ceresco, 1980
Beck, Daisy M., Fremont, 1987
Beetison, Dwight, Ashland, 1974
Belik, Vernon & Marie, Prague, 2002
Benes, Lillian, Weston, 1990
Bengtson, Mr. & Mrs. Lyman, Wahoo, 1970
Benson, Dwain & Sylvia, Wahoo, 1991
Berggren, Elva, Wahoo, 1971
Berggren, Leland, Denver CO, 2000
Boettner, Dorothy E., Fremont, 2007
Booth, John & Theresa, Malmo, 1976
Bouc, Ed & Stanley, Colon, 1989
Brabec & Family, Anton & Helen, Wahoo, 2006
Brabec, Helen, John & Dennis, Colon, 2010
Bronson, Beulah, Wahoo, 1992
Brtek, Frank, Linwood, 1979
Buresh, Joe & Mary, North Bend, 1975
Carlson, Harry & Delphia, Mead, 1976
Cernik, Elsie M., Wahoo, 2008
Chapek, Leonard & Rita, Weston, 2006
Chmelka, John, Leonard Chmelka, Edward Chmelka, Dorothy Juranek, Linwood, 2005
Christensen, Lyle & Velva, Wahoo, 1970
Chromy, William F. & Teresa, Linwood, 2006
Chvatal, Francis & Agnes, Glenn Chvatal, Lori Rosenbach, Prague, 2005
Cippera, Anton & Dennis, Morse Bluff, 1985
Clapham, Charles, Cedar Bluffs, 1980
Clark, Mardell, Omaha, 1979
Coady, Leo, Mead, 1987
Coady, Michael, Richard, Patricia, Mead, 2008
Cody, Richard & Sharon, Linwood, 1985
Conrad, Francis M, Paula J. Conrad, Fremont, 2006
Conrad, Mr. & Mrs. Francis George, Fremont, 1972
Cook-Johnson/Johnson, Betty, Carl A., Jr., Colon, 1987
Dauel, Bob & Lori, Malmo, 2009
Dietrich, Frieda M., Fremont, 1989
Divine, Edward, Lincoln, 1991
Drews, Wayne/ Ardyce, Yutan, 1996
Dugan, William, Fremont, 1995
Eggers, Harvey & Ruth, Fremont, 2005
Eichmeier, Milton, Ithaca, 1991
Erickson Estate, Emil, Fremont, 1971
Erickson, Conrad, Manhattan KS, 2001
Erickson, Morris & Jean, Mead, 2014
Feist, Georgia, Leshara, 1999
Fleming, Curtis & Geraldine, Morse Bluff, 1991
Focher, Joseph & Agnes, Prague, 1999
Franta, James, Morse Bluff, 1971
Freeman, Rita, Colon, 2014
Freeman, Rita, Colon, 2014
Fujan, Charles, Malmo, 1977
Fujan, Dennis and Patricia, Prague, 2005
Gaeth, Bonita, Fremont, 1982
Gamble, Miss Eliza, Lincoln, 1971
Govaerts Maude, Elizabeth Ann, & Timothy Charles Govaerts, Lincoln, 2006
Groff, DeForest & Jean, Morse Bluff, 1992
Gustafson, Duane, Ithaca, 1976
Gustafson, Harold, Cedar Bluffs, 1976
Gutzwiller, Robert, Yutan, 1990
Hageman, Ewalt, Ithaca, 1981
Hall, Jerry & Cecillia, Morse Bluff, 1981
Hall, Robert & Dona, Wahoo, 1970
Hanis, Adolph, Prague, 1998
Hanke, Kenneth & Grace, Wahoo, 1998
Hannan Estate, Mead, 1977
Hanson, Dale & Edna, Mead, 1978
Hanson, Forrest & Lyle, Mead, 1985
Hanson, Jeff & Mary, Ithaca, 2013
Happel, Lois, Morse Bluff, 1999
Havelka, Curtis & Jeanne, Weston, 2006
Havlovic, Thomas, Columbus, 1983
Heald, Mrs. R.L, Mead, 1971
Henderson, Hattie Ruth, Cedar Bluffs, 1972
Hengen, Donald, Omaha, 1985
Hines, Marvin & Karen, Morse Bluff, 1999
Hockinson, Reuel & Violet, Wahoo, 1976
Hollst Baumann, Darlene, Omaha, 2000
Holtorf, Melvin C., Cedar Bluffs, 1988
Holubar, Joe & Charles, Weston, 1979
Houfek, Bessie, Wahoo, 1979
Humlicek, Gerald & Roxan, Morse Bluff, 2003
Huscher, Mrs. Andrew, Cedar Bluffs, 1970
Hynek, Lawrence, Indian Shores, 2007
Jackson, Adele, Wahoo, 1978
Jambor, Rose, Prague, 1980
Janecek, Joseph & Donna, Weston, 1986
Jansa, Ferm, Weston, 1983
Jasa, Charles, Wahoo, 1977
Johnson, Arnold & Arlette, Colon, 1990
Johnson, Daryl, Wahoo, 1993
Johnson, Henry & Alice, Mead, 2005
Johnson, Kathleen, Colon, 2003
Johnson, Mrs. Ruben, Wahoo, 1972
Johnson, Philip, Mead, 1979
Johnson, Winsor Carl, Ceresco, 2003
Jonas, Vincent & Christine, Wahoo, 1981
Jurgens, Edward, Wahoo, 1994
Jurgens, Mr. & Mrs. Bert, Cedar Bluffs, 1978
Kadavy, LaVerne, Weston, 2007

Karloff, Glenn, Wahoo, 1993
Kavan, Bernard, Linwood, 1977
Kavan, Eugene, Morse Bluff, 2000
Kavan, Joie, Morse Bluff, 1976
Kavan, Leonard, Morse Bluff, 1976
Kavan, Ruben, Morse Bluff, 1976
Kavan, Stanley, Valparaiso, 1980
Keiser, Stanley and Evelyn, Ashland, 2016
Kilment, Max & Cheryl, Weston, 1988
Knuth, Chester, Cedar Bluffs, 1976
Konecky, Bernard & Maxine, Wahoo, 1977
Konecky, Mr. & Mrs Tom, Wahoo, 1977
Kubalek, Lillie, Wahoo, 1975
Kubalek, Richard, Weston, 1975
Kubik, Darrell E., Lincoln, 2006
Kucera, Edwin J., Weston, 1989
Kuhr, Robert & Michelle, Mead, 2011
Ladenburger, Gary D., North Bend, 2004
Landgren, Mary I., Columbus, 1985
Landstrom Farms, c/o James R. Landstrom, Lincoln, 1988
Lanik, Frank & Beckie, Ceresco, 1991
Larson, Clarence, Wahoo, 1977
Lehman, Ruth, Lincoln, 1987
Lehmkuhl, Gladys M., Minden, 1987
Lichtenberg, Louis, Cedar Bluffs, 1976
Lindgren, Dean W., Wahoo, 2008
Lindgren, Duane & Marjorie, Malmo, 1983
Lindgren, Robert & Maretta, Wahoo, 1989
Lindquist, Lawrence & Colleen, Wahoo, 1986
Luber, Vernon, Cedar Bluffs, 1999
Mach, Anton & Helen, Weston, 1978
Machovec, Anna, Wahoo, 1978
Machovec, Patrick & Deborah, Wahoo, 1990
Madigan, Marian, Weston, 1984
Magnuson, Plato & Ethel, Colon, 1972
Mahrt, Henry & Pearl, Colon, 1998
Maly, Eugene & Marie, Weston, 2005
Maly, Lumir & Dolores, Weston, 2005
Manske, Ervin & Esther, Ceresco, 1993
Mara, James, Ceresco, 1979
Martin, Larry and Shirley, Valparaiso, 2016
Masek, Dorothy, Gary & Michael Masek, Valparaiso, 2003
Masek, Fred, David City, 1977
McClean, Monty, Omaha, 2001
McDermott, William, Colon, 1992
McKee, Frances, Mead, 1990
McKlem, Frank, Wahoo, 1979
McReynolds, Dave & Judy, Ashland, 1990
Meduna, Henry & Gertrude, Weston, 1985
Morrison, Mary, Omaha, 2010
Morrissey, William, Yutan, 1977
Mueller, Monte, Yutan, 1979
Mumm, Lloyd, Yutan, 1979
Murren, Andrew, Colon, 1984
Musilek, Mr. & Mrs. Anton, Prague, 1985
Nelson, Clyde M. & Janet M., Valparaiso, 2004
Nelson, Glenn & Margie, Mead, 2005
Nelson, Larry L., Ceresco, 2008
Nelson, LeRoy & Bertha, Mead, 1980
Nelson, Mrs. Esther, Wahoo, 1976
Nelson, Sylvia, Weston, 1983
Nemec, Adolph, Prague, 1985
Noerenberg, Dale, Ithaca, 1990
Nord, Ivan, Valparaiso, 2003
Norgard, Mr. & Mrs. Ronald, Colon, 1992
Norris, Elmer & Mary, Ceresco, 1983
Nygren, Raymond, Mead, 1980
Odvody, Emma, Prague, 1999
Odvody, Milton, Morse Bluff, 2011
Ohnoutka, LeRoy & Lillian, Weston, 1985
Olson, Mark A. & Joyce, Malmo, 2004
Olson, Merle & Janet, Ceresco, 1983
Olson, Mrs. Percy, Wahoo, 1986
Ostry, Jerome, Prague, 1979
Otto, Glen & Lucile, Ceresco, 1979
Otto, Jerry/Gary, Ceresco, 1996
Pageler, William & Rosella, Cedar Bluffs, 1995
Palm, Mr. & Mrs. James, Scribner, 1986
Parmenter Yocum, Edwin & Annabelle, Yutan, 1968
Pellan, Genevieve, Mead, 1997
Person, Clarence, Wahoo, 1980
Pestal, Elmer, Wahoo, 1989
Petersen, Mrs. Martha, Cedar Bluffs, 1969
Peterson, Alfred E., Wahoo, 1985
Peterson, Eugene, Colon, 1984
Peterson, Eula Jane, Wahoo, 1996
Peterson, John and Cindy, Ceresco, 2010
Peterson, Marvin & Yvonne, Ceresco, 1998
Pokorny, Benjamin & Elaine, Wahoo, 1977
Pokorny, Louis, Firth, 1980
Polacek, Clara, Prague, 2003
Polacek, Robert & Arline, Wahoo, 2015
Pruss, Marlene & Eldon, Redwood City, 2007
Racek, Keith E. & Gilberta A., Morse Bluff, 2006
Racek, Louis, Morse Bluff, 1970
Racek, Nadine & Larry Racek, Morse Bluff, 2005
RasmussenKruce, Ellen, Colon, 1981
Reid, Evelyn, Omaha, 1981
Reim Trust, Herman F. & Rutha, Ashland, 2006
Replogle, Lillian, Pender, 1985
Revocable Trust, Gary D. Kratky & Maxine Kratky, Omaha, 2014
Rezac, Steven J., Valpraiso, 2014
Rezac, Thomas & Barbara, Weston, 2006
Reznicek, Dennis & Gladys, Louisville, 1977
Roberts, Thomas G. & Bonnie J., Cedar Bluffs, 1985
Rohwer, Arnold & Martha, Colon, 1973
Rohwer, Elmer, Elkhorn, 1973
Roseberg, Harold, Wahoo, 1973
Rosengren, Orion & Glen, Wahoo, 1973
Sabatka, Jacqueline, Valparaiso, 1998
Sanderson Family, Colon, 1970
Schmidt, Edwin D. & Marilyn D. (Schmidt) Guenther, Wymore, 2003
Schoen, Ruth Anne & Kenneth, Weston, 1980
Schulz Hanson, Roselyn, Mead, 1985
Schulz, Theodore, Yutan, 1996
Shanahan, Donald & Lillian, Morse Bluff, 1985
Shavlik, Stanley, Linwood, 1982
Sherman, Merle & Esther, Ithaca, 1987
Shires, Thomas & Margar, Mead, 2000
Siebken, Ronald, James Siebken & Stephanie Siebken Weiss, Leshara, 2008
Simodynes, George, Wahoo, 1976
Smith, Leo H., Wahoo, 2004
Smith, Mrs. Donald, Ashland, 1981
Smith, Roma, Wahoo, 2007
Snitily, Leonard & Loretta, Prague, 2010
Sorenson, Mrs. Alfred, Malmo, 1983
Soukup, Jerry & Barb, Morse Bluff, 2002
Sousek, James & Charlotte, Malmo, 1977
Sousek, John & Karen, Malmo, 2014
Stambaugh, Randolph, Ashland, 1959
Steinbach, Les & Marjorie, Prague, 1991
Stewart, Clarence & Evelyn, Ceresco, 1979
Stoeber, Mary Kaye, Antony R. Berkland, and Justin R. Berkland, Fremont, 2014
Storm, Clarence, Yutan, 1976
Stuchlik, Dennis and Kathleen, Wahoo, 2008
Sudik, James & Denise, Colon, 1992
Swanson, Larry & Roger, Colon, 1989
Swanson, Laurence, Weston, 1977
Swanson, Loren & Naomi, Wahoo, 1991
Swanson, Russel, Wahoo, 1992
Swanson, Russell & Karen, Wahoo, 2014
Swanson, Terry W. & Judy A., Wahoo, 2006
Swartz, Amalie, Prague, 2009
Tepoel, Raymond, Malmo, 1977
Thomas, Michael, Yutan, 1996
Thomsen, E.Howard & Maxine, Cedar Bluffs, 1979
Thorson, Milton & Myrtle, Mead, 1969
Thorson, Wayne D., Mead, 1986
Thorston, Russell & Ruth, Wahoo, 1977
Tomsicek, Laddie & Regina, Weston, 1978
Treptow, Gerald, Ithaca, 1992
Tyrdy, Julius & Isabelle, Wahoo, 2000
Valparaiso, Lloyd Ang, 2005
Vanek, Anton & Lillian, Wahoo, 1985
Vanek, Gilbert, Prague, 1984
Vermeline, Sharon, Malmo, 1978
Virgl, Donald & Alice, Weston, 1974
Virka, Clara M., Prague, 1988
Vlcek Dunn, Bill, Ruth, Omaha, 1977
Vopalensky, James, Morse Bluff, 1994
Vrana, Margie L., Wahoo, 2014
Vrana, Mrs. Alice, Wahoo, 1988
Widman, Terry, Rosanne Fetty, Yutan, 2007
Wiegand, William, Cedar Bluffs, 1977
Williams, Eldon, Yutan, 1969
Williams, Frank & Mary Eileen, Fremont, 1991
Wirka, Gary, Morse Bluff, 2007
Wirka, James, Cedar Bluffs, 1991
Wirth, Mrs. Evelyn, Omaha, 1979
Zakovec, Louis, Morse Bluff, 1984
Zaugg, Lois, Florissant MO, 1997
Zaugg, Marvin & Floy, Yutan, 1995
Zaugg, Wesley, Yutan, 1997

Scottsbluff County

Adcock Farms Inc, McGrew, 1989
Barbour, W.O, 1989
Bartz, Helen, Gering, 2005
Brown, Harlan, Mitchell, 2011
Burton, Vernita Ray, Denver, CO, 2005

Copsey, Alice & Jack, Minatare, 1987
Gatch, Neil & Maxine, Gering, 1987
George, John Estate, Gering, 1992
GlennRetta Inc, Mitchell, 1990
Juergens Anderson, Everett & Joan, Minatare, 1988
Lowry, Eyvone Estate, Gering, 1989
McElwee, Jack & Josehine, 1989
Mendenhall, Nellie, 1989
Nichols, Willis & Anne, Minatare, 1988
Pieper, Timothy & Joan, Jeffery & Somer Pieper, Mitchell, 2010
Russell, Raymond & Ruth, Mitchell, 1988
Smith, Doris, 2005
Smith, Doris, 2005
Thomas, Warren & Carl, Morrill, 1991

Seward County

Anderson, Esther Clark, Beaver Crossing, 1967
Anderson, Otto, Milford, 1981
Baack, Theodore, 1978
Banzhaf, Marvin & Helen, 2000
Banzhaf, Marvin & Helen, 2003
Bartholomew, Richard & Delores Bartholomew, 2004
Beekley, Jack, Dorchester, 1981
Bender, Gerhard, 1987
Berg, Eugene & Myrrl, Lincoln, 1984
Blanchard, Robert, Friend, 1984
Bluhm, Theodore, 1984
Bohaty, Marilyn & Richard, Co-Trustees, 2014
Braunsroth, Krista, Ronald Hinrichs, Pleasant Dale, 2007
Briggs Land Company, 1993
Brigham LLC, 2011
Brigham LLC, 2011
Brock, William, Cornoa CA, 2002
Brong, Mrs. Roscoe, Milford, 1985
Brown, Mr. & Mrs. Willis, Utica, 1971
Buecker, Robert, Sidney, 1994
Buller/Slawnyk, Arnold / Jane, Utica, 1983
Burianek, Richard, Pleasant Dale, 1983
Carnicle, Wayne, Lincoln, 1992
Cathcart, Mabel & Rodney, Ulysses, 1973
Chriansen, Lois Westerhoff, 1992
Christjaener, Ervin, Pleasant Dale, 1992
Daehling, Fred, Staplehurst, 1983
Daehling, Lucille, 1981
Daehling, Mrs. William, Staplehurst, 1967
Daehling, Otto, Staplehurst, 1967
Danekas, Ruth, Lincoln, 1991
Dankers, Henry & Anna, 1969
Dargel, Julius & John, Pleasant Dale, 1982
Downing, Willard & Pearle, 1968
Duer, Ben & Minnie, Staplehurst, 1969
Duffek, Janet, Betty Fehlhafer, Utica, 2010
Dvorak + Patsy Ann Hafer, Ervin E, Parkville, MO, 2015
Ebers, Merle, 1993
Eberspocher, Darrel & Carol, 2013
Eddy, Dennis & Ann, York, 2002
Ehlers, Mr. & Mrs. Herschel, Utica, 1977
Fehlhafer, Martin, Utica, 1976
Ficke, Agatha Danekas, Milford, 1986
Ficke, Kenneth, Pleasant Dale, 1988
Ficke, Wesley, Milford, 1983
Fosler, Dale, Milford, 1979
Fosler, Glenn & Emma, Milford, 1992
Gade, Carl & Ruth, 1994
Ganshorn, Harry, Beaver Crossing, 1991
Gard, Ray, Beaver Crossing, 1986
Gard, Wilma, Beaver Crossing, 1986
Gausman, Eugene & Marlene Gausman Behm, Milford, 2005
Geis, Carl, 1996
Geis, Estella, Beaver Crossing, 1992
Geis, Paul & Betty, Beaver Crossing, 1991
Gerkensmeyer, Norm & Evelyn, 2013
Gierhan, Mr. & Mrs. Darrell, Utica, 1978
Hackbart, Harold & Gertrude, Milford, 1983
Hackbart, Terry, 2001
Hanseling, Mrs. Art, Milford, 1980
Hartman Farms Inc, 1977
Hartmann, William & Doris, 2013
Heald, Catherine, 1999
Hegeholz, Timothy & Vickie, Staplehurst, 1988
Hegrholz, Gerald & Imogene, Staplehurst, 1998
Hentzen, Dennis & Bonnie, 2014
Heser, Richard & Mary, Beaver Crossing, 1984
Heyen, Boyd & Mildred, Milford, 2007
Hiemann, Loren, Staplehurst, 1982
Hildebrand, Paul, 1983
Hillmann, Gerald E. Hillmann & Lynne Hillmann, 2004
Imig, Ella, 2001
Jurgensen, David & Nancy, Cordova, 2011
Jurgensen, David & Nancy, Cordova, 2016
Justice, Naomi Clark, Fremont, 1967
Kamprath, Martha, 1981
Keller, Harold Trust, 2001
Keller, Mr. & Mrs. Alvin, 1988
Kenney, Bernard Ray, 1984
Keslar, Travis, Garland, 2005
Klein, Wilhelm Carl, Milford, 1982
Klintworth, Mitch & Priscilla, 1998
Klug, Emma, Pleasant Dale, 1979
Knerr, Mrs. Dorcas, Beaver Crossing, 1973
Krasser, Dale, Beaver Crossing, 1993
Krasser, Virginia, Beaver Crossing, 1986
Lambe, James Estate Heirs, Grand Island, 1989
Langenheim, Roger, Fort Worth, 2007
Lila Garner Trust, Lila Garner Trst, Milford, 1995
Luebbe, Clarence, Pleasant Dale, 2002
Luebee, Norman & Naomi, 1984
Maier, Wilfred & Doris, 1986
Martens, Margaret & Dorothy, Gresham, 1989
Matzner, Carl, Gresham, 1987
McHale, Harold, 2001
Meyer Bredow, Eleanore, 1972
Mikkelson, Frank & Dencil, York, 1975
Miller Family, Lincoln, 1974
MRL Family Limited Partnership, 2010
Mundhenke, Helen, Milford, 1987
Mundhenke/Griffiths, Donald / Jeanette, Milford, 1984
Neujhar, Glen & Verna, Gresham, 1985
Niemann, Loren & Esther, Staplehurst, 1991
Nietenstein, Rudlph & Bertha, Pleasant Dale, 1969
Petri, Lee & Doris, 1982
Plautz, Betty, 2012
Radford, Harold and Lola, 2010
Rathje, Everett & Alice, Gresham, 1993
Robotham, Leonard, Pleasant Dale, 1990
Roebke, Norman, 1982
Schildt, Joan, 2010
Schildt, Larry, Pleasant Dale, 1980
Schildt, Minnie, 1983
Schluckebier, Merle, 1988
Schmieding, Carl, Malcolm, 1975
Schoepf, John & Verona, Columbus, 2011
Scott, Robert & Diane, Friend, 2005
Sieck, Harold, Pleasant Dale, 1975
Sieck, Thomas L. and JoAnn, Pleasant Dale, 2005
Sloup, Shirley, Staplehurst, 1994
Soren Martin Hansen Estate, Utica, 1993
Stahly, Clinton, Milford, 1985
Stahly, Frank & Dencil, Milford, 1983
Stauffer Family Trust, Norman C., Milford, 2006
Stauffer, John, Milford, 2001
Steffens, Gordon & Mildred, Bradshaw, 1980
Stephens, Hazel, Lincoln, 1983
Stephens, Wesley, 1985
Sternberg, Leroy, Cordova, 2013
Stillahn, Eric & Clara, 1987
Suhr, Alfred & Eunice, 2013
Swanson, Linda & James, Staplehurst, 2010
Swanson, Mrs. Marcus, Staplehurst, 1981
Tonniges, Dale & Harriet, 1986
Vrbka, Beth, 1990
Wake & Company, Aeward, 1985
Warner, William, Friend, 1985
Watts, Arnold & Neomi, Beaver Crossing, 1970
Wehrs, Maurine, 1984
Welch, John, Milford, 1995
Welsch, Ernest William & Connie Jo, Milford, 1978
Westerhoff, Clifford, 1976
Westfall, Fern Duling, Malcolm, 1991
Woebbecke, Gilbert, Pleasant Dale, 2003
Woebbecke, Rodney, 1984
Wolvin, Lorena, Utica, 1973
Wood, Alvin, Staplehurst, 1971
Wynegar, Steve & Karen, Gresham, 1984
Zillig, Mervin & Myrna, 1979
Zwick, Lyal, Staplehurst, 1993

Sheridan County

Asher, Bert, Hazard, 1975,
Barth, Lyle, Rushville, 1985
Blakey, Betty, Lakeside, 2000
Briggs & Family, David & Phyllis, Alliance, 2014
Britt, Lyle, Rushville, 2005
Burrows, Hazel, Hay Springs, 1985
Cuningham, Wayne, Gordon, 1993
Davis, Owen & Marlene, Rushville, 1994
Dorshorst, Gary, Hay Springs, 1988
Dramse, Joe Heirs, Hay Springs, 1985
Eichenberger, Albert, Chadron, 2007
Fisher, Albert, Rushville, 1985
Forney Ranch Inc., Lakeside, 2014
Frey, Cynthia, Hay Springs, 1999
Fuchser, Richard Donald, Gordon, 1985
Gealy, Ruth, Gordon, 2004
Gilliam, James & Lila, Rushville, 1999
Hardin, George & Clara, Gordon, 1985
Herian, Rick & Becky, Alliance, 2011
Higgins Post, Dale & Dorothy, Rushville, 1985

Hindman, Albert, Hay Springs, 1985
Hindman, Ivan M., Rushville, 2006
Hindman, William & Florence, Hay Springs, 1992
Hooper Brothers Ranch, Lakeside, 1987
Hunter, Dean & Evelyn, Hay Springs, 1984
Hunter, Janice & George Dale, Hay Springs, 1985
Kadlecek, Mabell, David & Philip, Hay Springs, 1990
King, Jessie, Hay Springs, 1986
Kleve, Nellie, Hay Springs, 1988
Krejci, George, Alliance, 1989
Krotz, Flora, Rushville, 2007
Kruml Hankins, Edgar & Amelia, Hay Springs, 1994
Leisy, Raymond A., Scottsbluff, 2008
Lundy, Mildred Reno, Hay Springs, 1990
Marcy, Charles & Barbara, Hay Springs, 1989
Marx, Arlan, Leta, Neal & Gale, Rushville, 2006
Michaelson, Richard, Rushville, 1986
Nielsen, Niel, Gordon, 1986
Oldenburg, John Estate, Gordon, 1986
Orr, John, Rushville, 2005
Perreten, Darlene, Hay Springs, 1985
Peters, Wayne & Rita, Hay Springs, 1992
Petersen, Adolph, Hay Springs, 1989
Rezek, Naomi, Hay Springs, 1985
Sandoz, Donald, Hay Springs, 1990
Schmidt, Pete, Rushville, 1988
Skoda, Clayton, Hay Springs, 1985
Stava, Gene, Hay Springs, 1985
Thorsen, Dwight & Geraldine, Gordon, 1991
Vybiral, Anna & William, Hay Springs, 1986
Woods, Jr., James C., Rushville, 2004

Sherman County

Baillie, Earle & Catherine, Loup City, 1976
Bandur, Tom & Becky, Loup City, 2006
Barent, Alvin and Mary Jo, Rockville, 2008
Beck, Barbara and Delwin, Litchfield, 2005
Burman, Lawrence, Rockville, 2005
Charlton, Jack, Arcadia, 1992
Chizek, Emil & Genevieve, Ravenna, 1999
Coulter, Donald & Rose, Valparaiso, 1997
Coulter, Wendell, Stromsburg, 1986
Dethlefs, Gary & Darlene, Rockville, 2011
Eckel, Velma, Ravenna, 1994
Eurek, Franklin & Maryann, Arcadia, 2011
Fowler, Bruce & Mary, Arcadia, 2002
Glinsmann, Melvin, Ashton, 1981
Glinsmann, Merle & Virginia, Loup City, 1984
Goc, James, Rockville, 2001
Habe, Henry & Barbara, Ravenna, 2003
Harrington, Robert & Margy, Loup City, 1979
Heapy, R. D., Litchfield, 1978
Hehnke, Clara, Grand Island, 2000
Heil, Ida, Loup City, 1992
Hurt, Robert & Marilyn, Ashton, 2003
Huryta, James A., Rockville, 2015
Johnson, Lynn L., Loup City, 2004
K & N Mac Farms Inc., Rockville, 1989
Kowalski, Kenneth S, Loup City, 2005
Kusek, Thomas, Loup City, 2009
Landon, Donald & Bette, Arcadia, 2002
Larson, Carolyn, Hazard, 2006
Lindell, Albert & Hazel, Cairo, 1985
Maciejewski, Dennis & Lillian, Loup City, 2005
McFadden, Harriet, Omaha, 1986
Moeller, Keith, Loup City, 1998
Pierson, Arthur, Arcadia, 1981
Pritschau, Leo, Ravenna, 1974
Psota, Raymond, Ravenna, 1974
Rasmusen, Greg & Cyndia, Hazard, 2000
Rasmusseen, Venus B, "Scotts Valley, CA", 2005
Reinertson, Reinert, Hazard, 1983
Schmaljohn, Albert, Ravenna, 1973
Slobaszewski, Alvin, Ashton, 1973
Snow, Rod & Jenelle, Litchfield, 2002
Spotanski, Eugene & Alozy, Loup City, 1992
Thomsen, Niels & Lillian, Rockville, 1986
Treffer, Edwin, Rockville, 1973
Wrehe, Ronald & Janice, Boelus, 1992
Young, Joseph, Rockville, 1973

Sioux County

Agate Springs Ranch, Harrison, 1979
Bannan, Ila Mae, & children, Harrison, 2008
Beatty, Myron & Kay, Marrill, 2005
Buckley, Jack F., Rowena Buckley & Emerson & Hazel Buckley, Harrison, 2008
Cawiezel, Bonnie, Morrill, 2014
Coffee, Bill, Harrison, 1973
Coffee, Vera, Harrison, 1981
Eberspecher, Anna, Harrison, 1991
Engebretsen, Nels Children, Harrison, 1987
Gillett, Terence, Harrison, 2009
Hanley, Martin & Lois, Crawford, 2011
Harold, Dick & Mary Ann, Jim & Mae Dunn, Harrison, 2003
Harrison Meng, Albert & Helena, Crawford, 1987
Howell Farms, LLC, Morrill, 2009
Kreman, John, Scottsbluff, 1985
Mack, Donald & Betty, Harrison, 1992
McChesney, Shirley, Tucson AZ, 1998
Molitor/ Corbin, Lois/ Virginia, Harrison, 1988
Nation, Thelma, Crawford, 2009
Nielsen, Mary, Gary Nielsen, Jay Warren, Ruth Warren, Minden, 2011
Norman, Bernard, Crawford, 1988
Norman, Earl, Crawford, 1988
Parsons Kaan, Cynthia, Harrison, 2011
Piper, Marie K., Harrison, 2012
Putnam, Lois/ Virginia, Harrison, 1993
Quintard Harkin, Gertrude Jordan, Harrison, 1991
Reece, Jim, Crawford, 2010
Ruffing, John, Harrison, 1985
Saxton, Milo, Crawford, 1990
Semroska, Raymond, Harrison, 2002
Semroska, Todd & Sharon, Harrison, 2002
Serres, George & Floyd, Crawford, 2001
Serres, Louise & George, Harrison, 1985
Surb's Farm, Inc., Gering, 2007
Thayer, Gary, Harrison, 1999
Thomas, Everett, Harrison, 1989
Wheeler Family Trust, Marsland, 2011
Wickersham, Dorothy, Harrison, 2002
Wilson, Tommy, Morrill, 2011
Wilson, William, Harrison, 1990
Zerbe, Loren, Harrison, 1986
Zimmerman, Keith, Harrison, 1985

Stanton County

Abels, Judene, 2015
Ahrens, Marilyn, Davenport, 1992
Asch, Mr. & Mrs. August, Pilger, 1970
Asplin, Viola, Barrington IL, 1999
Bachman Trust, Darlene, Creston, 2010
Barnas/ Nosal, Eleanor/ Irene, Wilber, 1984
Barr, Mr. & Mrs. Art, 1971
Bauermeister, Harold & Laura, Norfolk, 1978
Baumert, Ronald & Janice, Howells, 2001
Belina, Edwin & Lois, Pilger, 1990
Benson, George & Mary, 1982
Beutler, John & Irene, Norfolk, 1986
Bjorklund, Elaine, Moses Lake WA, 2002
Bourek, Jerry & Anna, Howells, 1995
Brabec, Dean and Karen, Clarkson, 2016
Bremer, Gus, Gene, & Anna, 1972
Brothers, Sprieck, Pilger, 1991
Brundieck, Ernest & Eleanor, Norfolk, 1992
Burtwistle, Oliver, 1980
Chance, Jack, Pilger, 1988
Chilcoat, Sr., Robert, 2009
Christ, Doran & Lola, Howells, 1987
Christensen, Emil & Olga, Pilger, 1994
Clinch, Gerald, Madison, 1994
Davidson, June, 2001
Denney, Dale and Lola, 2008
Denney, Joy, 1984
Engdahl, Norman, 1970
Ertzner, Arthur, Omaha, 1981
Eusterwiemann, Donald & Victoria, Howells, 2006
Falk, Harold, Norfolk, 1977
Frank, Donald E., 2012
Frank, Elvin & Gladys, 1997
Freiberg, Gladys, 1994
Freiberg, Robert, 1970
Frerichs, Virgie Burris, Pilger, 1993
Fuhrman, Marvin & Eleanor, Norfolk, 1978
Gnirk, Mardelle, Hoskins, 2000
Goeller, Dallas & Sharon, Wisner, 2002
Goeller, Gaylen G., Pilger, 2003
Goetsch, Paul, 1992
Goodell, Jerry & Geraldine, Norfolk, 1994
Greunke, Barbara Greunke, Mardell Kropp and Anthony Wemhoff, Fremont, 2014
Hansen, Howard & Barb, 1995
Hansen, Norman, Brenda Hansen-Styskal, Grant, 2009
Hansen, Raymond D., and Delores M. Hansen, Norfolk, 2010
Hartl Fischer, Mary, Omaha, 1994
Hartl Kratochvil, Elsei, Madison, 1994
Hartwig, Mr. & Mrs. Elven, Norfolk, 1980
Herscheid, LeRoy, 1982
Hradec, Larry, Norman Hradec, Norfolk, 2011
Hrouda, Darlene & Eldon, 2010
Hyde, Esther, Beemer, 1986
Jensen, David & Diane, 2007
Jindra, Milo, Pilger, 1991

Johnson, Mr. & Mrs. Ray, 1970
Kander, Melvin, Howells, 1985
Karmann, Louis, Winside, 1991
Kment, Wayne & Sue, 2002
Koch, Mr. & Mrs. Louie, 1969
Koehlmoos, Adolph & Lorraine, Pilger, 1987
Koehlmoos, Betty Lou, Pilger, 1997
Koehlmoos, Verona, Pilger, 2006
Kohlhof, Melvin & Gwen, Norfolk, 1993
Kuester, Harlan, Wisner, 1984
Lammli, Mark & Rosalie, 2011
Long, Gloria, Norfolk, 1998
Long, Mr. & Mrs. Alan, Madison, 1972
Luttman, Pauline L., 2003
Matthes, Don, 1991
Molacek, Donald & Ardyth, Howells, 2007
Moritz, R. Jerome, Lisetta A. Moritz, Rita Roenfeldt, 2008
Mortiz, Mr. & Mrs. Max, 1983
Nelson, Walter & Lillian, Pilger, 1969
Nykodym, Edna, Norfolk, 1999
Pekny, Joe, Clarkson, 1984
Pestel, Allan, Plattsmouth, 1993
Peterson, Eldon & Leanna, 2005
Peterson, Norman, Pilger, 1995
Pohlman, Beverley J. (Swoboda), Omaha, 2004
Pohlman, Neal & Rose, 1978
Prokopec, Richard & Robert, Clarkson, 2005
Puls, Mr. & Mr.s Emil, Norfolk, 1970
Reade, Carolyn Williams, Elkhorn, 2004
Renner, Galen & Janice, Howells, 1999
Renner, Orley & Norma, Howells, 2003
Roker Wantlin, Esther, 2000
Roker, Gerald, Norfolk, 1997
Rottler, Donald, 1977
Salmons, W. E., Pilger, 1983
Sasse, Edgar & Ethel, Pilger, 1969
Sateren, Eugene & Dennis, Pilger, 2004
Schellpeper, Fred & Margaret, 1981
Schellpeper, Margaret, 2000
Schellpeper, Mr. & Mrs. Dale, 1987
Schellpeper, Mr. & Mrs. Harold, 1971
Schellpeper, Walter, 1999
Schlautman, Alvin & Lorraine, Howells, 1993
Schultze Cedar Brook Farm Inc., 1987
Schwede Grantor Trust of 1987, Agatha, Norfolk, 2010
Sieck, Ervin Trust, Pilger, 1988
Sindelar, Stanley, Howells, 2015
Spence, Lynn & Mary Trust, 1994
Spence, Terry & Gloria, 2012
Sprieck, LeRoy, Pilger, 1987
Statler, Henry T., Pilger, 2008
Steffensmeier, Lawrence, Norfolk, 2001
Svenson, Mr. & Mrs. Iryl, 1970
Svitak, Edward and Mary, Howells, 2008
Svitak, Stanley & Betty Ann, Howells, 1991
Tejkl, Loren/ Marcelene, 1996
Tobias, David, Letha Shmigelsky, Lana Baldwin, Pilger, 2003
Trine, Eugene & Phyllis, Madison, 1989
Vogel, Gene, Panama City FL, 1992
Vogel, Lyle, Howells, 1992
Vogel, Morris, 1989
Vogel, Olga, 1989
Wattier, Betty Lou, Norfolk, 1992
Wegener, Clifford, Norfolk, 1989
Weiher, Frank & Linda, 1995
Weiher, Oscar & Ralph, 1983
Whisinnand, Marc and Nancy, Norfolk, 2008
Wiegert, Marjorie, Plainview, 1987
Wilken, Marvn & Elaine, Norfolk, 1995
Wittgow, Terral, Greenville TX, 1999
Wolff, Judith, Joyce Kennedy, Howells, 2007
Wolff, Vernon & Gladys, Norfolk, 1992
Zastrow, Otto, Norfolk, 1981

Thayer County

Achterberg, Galen, Hebron, 2002
Airy Knoll Farms Inc, Cozad, 2000
Bank, Joyce, Maxine Dusatko, Seward, 2006
Bates, Walter, Davenport, 1977
Bauer, Mrs. Edward, Hebron, 1979
Brettmann, Alvin, & Steven and Scott Brettmann, Fremont, 2010
Brettmann, Arthur & Twila, Deshler, 1993
Brettmann, Arthur, Deshler, 2002
Brettmann, Carl & Kathryn, Hebron, 1989
Bruning, Douglas, Bruning, 2012
Bruning, Leon, Bruning, 1979
Bruning, Victor & Leola, Bruning, 1982
Burbach, James, Denver CO, 2002
Coordsen, Walter & Marjure, Hebron, 1982
Crouse, Harold & Maxine, Chester, 1987
Culligan, Mr. & Mrs. Thomas, Alexandria, 1960
Currey, Mrs. Myron, Hebron, 1991
Degner, Richard & Viola, Hebron, 2000
Dill Ortman, Estella, Belvidere, 1970
Dill, Robert, Belvidere, 1970
Doll, Bruce and Delores, Geneva, 2008
Easley, Daryl, Alexandria, 1971
Easton, Harriet, Hebron, 2008
Elting, Dwayen & Beverly, Carleton, 1982
Fintel, Curtis B., Byron, 2008
Gausman, Erma, Duane Vorderstrasse, Hebron, 2011
Graham, William, Narka KS, 1999
Grugan Loch, Mildred, Fairbury, 2000
Harms, Donald O. and Lorna A., Byron, 2008
Harms, Joy, Davenport, 2009
Harms, Steven & Beverly, Davenport, 2015
Harrison, Charles & Roberta, Bruning, 1994
Hawks, Earl, Bruning, 1993
Hays, Cecil, Harold Hayes, Robert Hayes, Placentia, CA, 2003
Heitmann, Lavern & Verla, Hardy, 2004
Heller, Roger & Rolla, Hebron, 1991
Henrichs, Alic, Hebron, 2000
Hergott, Wilfrid, Belvidere, 1994
Hermann, Clifford & Grace, Bruning, 1982
Hinds, Russell, Davenport, 2012
Hinrichs, Fred, Bruning, 1992
Hinrichs, Max & Karen, Bruning, 2012
Hobbs, Ruth Hobbs, NE, 2005
Hoffmeyer Ryan, Jane, Los Angeles CA, 2000
Hoops, Michael & Yvonne, Byron, 2014
Hruska Holtzen, Genevieve, Reynolds, 1995
Husa, Bessie & Wilma, Hebron, 1992
Husa, Frank Joseph, Hebron, 1991
Irrevocable Trust, Dorothy E. Pohlman, c/o Dale Pohlman, Kearney, 2014
Kirchhoff Farms, Hardy, 2000
Kirchhoff, Eldon & Judy, Byron, 2013
Koester, Marvin & Barbara, & Adalia Frum, Deshler, 2004
Lautenschlager, Dean & Lucy, Kearney, 2004
Lichty, Inez, Carleton, 1990
Marquart, Irma Marquart, NE, 2005
Marquart, Irma, Byron, 1999
Marsh, Rob & Natalie, Belvidere, 2012
Maschmann, Dale, Deshler, 2006
McLaughlin, Leon, Larry McLaughlin, Carol & Victor Sorge, Deshler, 2006
Meyer, Dorothy, Ohiowa, 2000
Middendorf, Carl, Bruning, 1982
Miller, Hubert, Bruning, 2001
Miller, Neil & Berneta, Davenport, 1994
Mosier, Kenneth, Davenport, 1998
Murrell, John, Alexandria, 2000
Mussmann, Bernard & Edna, Ruskin, 1985
Onnen, Gerhard R. & Delores E., Hebron, 2013
Othling, Delores, Deshler, 1989
Pohlmann, Elmer, Deshler, 1992
Pohlmann, Orville/ Lois, Deshler, 1996
Reinke Diary In.c, Byron, 1989
Reinke, Arthur & Gleanye, Deshler, 1980
Reinsch, Darlene, Bruning, 2008
Rex/ Van Skiver, Wendell/ Deloris, Davenport, 1996
Row Trust, Paul D. & JoAnn W., Bruning, 2015
Schardt, Verland, Carleton, 1979
Schiermeyer, Stanley, Fairbury, 1998
Schmidt, Henry, Deshler, 1980
Schmidt, Joyce E. & Dennis G., Deshler, 2006
Sinn, Loren, Fairbury, 1993
Stables, John & Shelia, Carleton, 2000
Stairs, Mr. & Mrs. David, Belvidere, 1980
Stepp, Mary and Michael and Mary Isernhagen, Deshler, 2016
Swenson, Richard and others, Davenport, 2003
Tipton, B.D. & Mary, Chester, 1999
Van Dyke, Mary Lou, Lincoln, 2001
Van Westen, Donavon & Anna, Alexandria, 1992
Vieselmeyer, Louis & Madalyn, Lois (Vieselmeyer) & Daryl Nelson, Kaylene Vieselmeyer, Kearney, 2012
Vorderstrasse, James, Hebron, 2011
Vostrez, Ervin, Hebron, 1987
Wearin Humphries, Edna, Lincoln, 1971
Wenske, Luetta M., Byron, 2015
Werner, Elmer, Carleton, 1999

Thurston County

Burmester, Starr & Alice, Pender, 1995
Cooney Family Trust, Walthill, 2006
Frey, Woodrow A., 2013
Hennig, Marlon H., Rosalie, 2014
Jahde, Ardell & Arvaleen, Pender, 1994
Jorgensen, Barbara, Pender, 2003
Krusemark, Mrs. Fred, Pender, 1989
McQuistan, Geraldine, Pender, 1987
Meyer, Glenville & Hilda, Pender, 1993
Ostrand, Sandra & Lauren, Pender, 2010
Sunderman, Larry W., Pender, 2012
Wichman, Leon, Pender, 1992

Valley County

Avidano, Raymond & Mary, Ord, 1986
Beran, William, Elyria, 1984
Bredthauer, Eugene & Alma, Ord, 1999
Bresley, Dean & Heloise, Ord, 1999
Brickner, Bud, North Platte, 1994
Bruha, Eldon & Lucille, Ord, 1994
Bruha, Laddie & Darlene, Ord, 1989
Bruha, Laddie & Mary Lou, Comstock, 1992
Clement, George, Ord, 1974
Cook, Wilfred Daniel & Evelyn, Ord, 1983
Covert-Thomsen, Ruth, Ord, 1983
Cronk, Rolland & Marlene, Ord, 2006
Cruiskshank, Max, Ord, 1985
Czech 4 Lucky Stars, LLC, Burwell, 2012
Czech 4 Lucky Stars, LLC, Burwell, 2014
Danczak, Leon Danczak, NE, 2005
Dubas, Rene & Angie, Elyria, 2007
Foth, Will & Lillie, Ord, 1994
Hackel, Mae & Edward, Ord, 1974
Heirs, Novosad, Ord, 2006
Hill, John, Ross Hill, Daryl Hill, Linda Hill Boatright, Boulder, 2006
Hill, Willard & Doris, Arcadia, 1980
Hornickel, Kent & Twilla, Ord, 2000
Hosek, Joseph W. & Minnie A., Ord, 2003
Hruza, Dorothy, Ord, 1995
Jablonski, M.L. Unified Credit Trust, Omaha, 2012
John, Robert & Wilma, Thad & Roberta Meese, Ord, 2009
John, Rudolph & Mary, Ord, 1985
Jones, Wauneta, Ord, 1993
Koelling, Orel & Elinor, Ord, 2001
Kokes, Larry & Patricia, Elyria, 1999
Koll, John & Mary Lu, Ord, 1994
Kordik, Robert, Omaha, 1996
Kusek, Anthony & Carol, Albion, 2011
Lueck, William & Eleanor, Arcadia, 1984
Lutz, Jim, Arcadia, 2002
Maresh Land, LLC, York, 2014
Maresh, Wilma K., Ord, 2014
Mason, Archie, Ord, 1984
McDermott, Ardon, Ord, 1984
McIntyre, Mike & Ruth, Wolbach, 1994
Moudry, Leonard & Betty, Ord, 1999
Mulligan, Dale & Grace, North Loup, 1986
Murray, Fred & Gerald, Ord, 1990
Nelson, Alvin, Comstock, 1991
Nelson, Carl, Arcadia, 1987
Novak, Eugene & Ina, Ord, 1990
Osentowski, Myron & Sharon, Duane Lewandowski, Grand Island, 2008
Pokorney, Marilyn, Ord, 2004
Psota, Don, Sandra, and Ryan, North Loup, 2003
Psota, Michael E., North Loup, 2003
Remington, Adrian & Norma, Ord, 1974
Rose, Russel & Luella, Ord, 2001
Schaaf, Donald & Mary, Omaha, 2000
Scott, Clark S., Patricia L. Grudzinski, Grand Island, 2012
Vogeler, Wm. F. and Nancy, North Loup, 2009
Wadas, Jerome & Joyce, Ord, 2007
Walahoski, Ted & Margaret, Elyria, 1984
Walker, Shirley, Ord, 1982
Walkowiak, Simon and Garnett, Ord, 2008

Washington County

Allen, Mr. & Mrs. Pierce, Ft. Calhoun, 1977
Andersen, Alvin, Kennard, 1981
Andersen, Arthur, Herman, 1979
Anderson, Alvin J., & Lorene Andersen Revocable Trust, 2006
Andreasen, Kenneth, Fremont, 1985
Andreasen, Roger & Donna, Blair, 2007
Andreasen, Stanley, Fremont, 1985
Armstrong, Lynn, Blair, 1978
Auch, Londo & Lorna, Bennington, 1999
Barclay, Helen, Herman, 1993
Bartling, Alice, Fremont, 2000
Bartling, Arlan & Barbara, Herman, 1983
Bartling, Kenneth, Herman, 1983
Bartling, Marie Sohia, Fremont, 1983
Behrendt, Curtis & Frances, Herman, 1992
Bisbee, Mr. & Mrs. William, Arlington, 1970
Blum, Walter, Herman, 1999
Brown, Lawrence, Arlington, 1974
Cameron, Jack, Herman, 2007
Cameron, Mr. & Mrs. Joe, Herman, 1968
Carlson, Mrs. Grace, Omaha, 1966
Christensen, Clarence & Leora, Herman, 1987
Christensen, Russell, Blair, 2002
Christiansen, Mr. & Mrs. E.R., Fremont, 1970
Cook, Mrs. Curt, Arlington, 1977
Crist, Hilda, Blair, 1995
Cuda, Rita, Mark Cuda, Beth Cuda, Hooper, 2005
Dam, James & Alice Dam, Hooper, 2005
Dickmeyer, Albert, Arlington, 1976
Dunklau, Mr. & Mrs. Lorenz, Arlington, 1960
Fedde, Harry, Bennington, 1970
Flake, Martha, Herman, 1970
Franke, Mr. & Mrs. Herbert, Fremont, 1982
French, Gifford, Arlington, 1975
Fryman, Mr. & Mrs. Joe, Blair, 1979
Fuchs Farm Inc, Arlington, 2003
Fuchs, Robert & Cheryl, Arlington, 1991
Giesselman, Mr. & Mrs. Harry, Arlington, 1967
Goedker, Dale & Sue, Bennington, 2002
Grabbe, Mark, Blair, 1985
Greunke, Henry & Ellen, Omaha, 1978
Haggerbaumer, Mr. & Mrs. Harvey, Hooper, 1983
Hammang, Leonard, Fremont, 1973
Hansen, Clarence & Mae, Blair, 1987
Hansen, Walter & Lillian, Omaha, 1987
Happel, Howard & Doris, Hooper, 1971
Hartung, Henry, Fremont, 1991
Hastings, Irene, Blair, 1970
Heise, Marvin & Norma, Blair, 2013
Heybrock, Mr. & Mrs. Fred, Arlington, 1961
Hilgenkamp, Gary & JoAnn, Arlington, 2001
Hilgenkamp, Mr. & Mrs. Arthur, Arlington, 1960
Hilgenkamp, Mr. & Mrs. Floyd, Arlington, 1968
Hilltop Dairy Inc, Arlington, 1998
Hineline, Ronald & Ralph, Ft. Calhoun, 1985
Hovendick, Mr. & Mrs. Francis, Blair, 1969
Huber, Clark, Omaha, 1957
Iversen, Miss Goldie, Ft. Calhoun, 1956
Jahnel Henricksen, Bertha, Blair, 1981
Japp, Walter, Kennard, 1990
Jensen, Jack P. & Opal, Arlington, 2004
Jones, Elizabeth, Herman, 1999
Jungbluth, Lloyd & Margaret, Bennington, 1975
Kahnk, Louie & Grace, Omaha, 1969
Katt, Elmer & Norma, Herman, 2010
Kelly, John & Marie, Omaha, 1972
Kelly, Mr. & Mrs. Daniel, Omaha, 1983
Kerstetter, Earnest & Helen, Blair, 1971
Knecht, Adela, Arlington, 1997
Kruger Livestock Inc., Arlington, 1988
Kruse, Eugene, Herman, 1983
Kuhr, LeRoy & Jane, Blair, 1972
Laaker, Lawrence & Marian, Arlington, 1990
Lallman, Mr. & Mrs. Dietrich, Nickerson, 1972
Lallman, Mr. & Mrs. Gottlieb, Nickerson, 1972
Lang, Russel Clark, Craig, 1982
Langhorst, Esther, Hooper, 2000
Larsen, Alton, Blair, 1977
Larsen, Elaine, Blair, 2001
Larsen, Mrs. Ardeth, Omaha, 1980
Lauritsen, Herman, Blair, 1993
Liesche, Minnie Heirs, Bennington, 1985
Lippincott, Elna & William, Blair, 2002
Lippincott, Mr. & Mrs. Richard, Blair, 1961
Loosing, Mrs. Hazel, Arlington, 1963
Lorenzen Farms Inc., Blair, 1987
Love, J. Ross & Lela, Blair, 1975
Luebker, Leonard, Arlington, 1974
Luebker, Leonard, Fremont, 1987
Magill, Raymond, Herman, 1991
Maney, Francis, Omaha, 1972
Marquardt, Quentin, Nickerson, 1973
Matzen, Lorene, Bennington, 1970
McDonald, Mr. & Mrs. Allen, Blair, 1969
Meeves Bunz, Elizabeth, Omaha, 1993
Meeves, Wilfred, Bennington, 1994
Meierhenry, Harold & Irma, Fremont, 2000
Micheel, William Henry, Kennard, 1982
Misfeldt, Harold & Irma, Kennard, 1981
Moll, Howard, Hooper, 1973
Monke, Luther & Wilmeth, Nickerson, 1970
Mueller, Arthur, Kennard, 1973
Mueller, Mr. & Mrs. Orlando, Hooper, 1972
Nelson, Keith & Janet, Blair, 1980
Nelson, Roger & Donna, Blair, 2011
Newkirk Dimmitt, Mary, Fremont, 1993
Newkirk Miller, Fran, Blair, 1993
Nielsen, Mr. & Mrs. Gifford, Ft. Calhoun, 1966
Novak, Mr. & Mrs. Walter, Blair, 1975
Oerman, Barnard, Hooper, 2011
Ohrt, Alvin, Ft. Calhoun, 2000
Ohrt, Calvin Claus, Omaha, 2008
Otte, Marlene, Fort Calhoun, 2003
Penke, Donn & Ann, Kennard, 2001

Peters, Noel & Gina, Herman, 1990
Petersen, Alvis & Agnes, Hooper, 1972
Petersen, Howard & Betty, Herman, 1983
Petersen, Richard, Herman, 1977
Pfeiffer, Robert, Arlington, 2000
Quinlan, Patrick, Blair, 1974
Rabe, Gordon, Arlington, 1981
Rabe, Jerome, Arlington, 1983
Rasmussen, Robert & Charlotte, Herman, 1985
Rasmussen, Roland, Blair, 1987
Rathmann, John, Ft. Calhoun, 1974
Raver, Miss Anna, Blair, 1961
Rhea, Mr. & Mrs. William, Arlington, 1972
Rhoades, Mrs. Bessie, Blair, 1966
Rinas, Mr. & Mrs. Paul, Arlington, 1969
Rogers, Loran & Ina Deane, Arlington, 1991
Rohlff, Virgil L, Norfolk, 2015
Rohwer, Marv & Judy, Fort Calhoun, 2011
Rohwer, Mrs. George, Ft. Calhoun, 1957
Rolland, Gustav & LaVonne, Blair, 1978
Rood/ Dowler, Grace/ Genevieve, Omaha, 1980
Ruwe Johnson, Edna, Nickerson, 1985
Ruwe, David & Marsha, Blair, 1999
Ruwe, Mrs. August, Nickerson, 1987
Ruwe, Paul, Arlington, 1987
Salsbury, Glen & Carla, Blair, 1982
Scheer, Dalvin, Arlington, 2002
Scheer, Obert & Selma, Arlington, 1995
Schmidt, Mr. & Mrs. Donald, Arlington, 1961
Schmidt, Mr. & Mrs. Leo, Arlington, 1969
Schmidt/ Ubbelohde, Dean / Erma, Arlington, 1980
Shaner, Neale, Ft. Calhoun, 1997
Smith, Eva, Ft. Calhoun, 1976
Sprick, Wesley, Blair, 1994
Stork, Jefery & Sherone, Arlington, 2006
Stork, Martin, Arlington, 1956
Stork, Milo & Norma, Blair, 1986
Stork, Mr. & Mrs. Harlan, Blair, 1969
Stork, Robert, Arlington, 1967
Stork, Verdell & Deloris, Arlington, 1974
Stranghoener, Alfred, Arlington, 1974
Sully, Louis, Ft. Calhoun, 1966
Taylor, Howard & Nellie, Blair, 1975
Thielfoldt, Lyle & Mona, Herman, 1998
Thompson, Duane & Sharon, Arlington, 1968
Thompson, Earl, Blair, 1976
Thone, Mr. & Mrs. Clarence, Herman, 1969
Thunn, Pearl H, Omaha, 2015
Tonjes, Kenneth & Jean, Hooper, 1974
Triple P Dairy Farm, Herman, 2014
Tyson, Harry & Myrna, Blair, 1965
Voss, Louise, Elmer, Delores, Fremont, 1985
Wachter, Jack & Lorraine, Herman, 1988
Ward, Mr. & Mrs. Herle, Herman, 1967
Weaver, Ruth, Herman, 2001
Weberg, Frank A. & Patricia A. (Laughlin), Wausa, 2012
Weitkamp, Mrs. Martha, Fremont, 1973
Westermann, Delvin, Arlington, 1984
Wilcox, Harley, Omaha, 1966
Wilkening, Mr. & Mrs. Alfred, Fremont, 1965
Wilkening, Mr. & Mrs. Henry, Arlington, 1965
Wilkening, Mr. & Mrs. Raymond, Arlington, 1965
Wilson Riverview Farms Inc., Blair, 1989
Wilson, Mr. & Mrs. Frank, Blair, 1969
Wittman, Mr. & Mrs. Arthur, Arlington, 1960
Wolsmann, John & Betty, Blair, 2005
Wrich, William & Joyce, Kennard, 2012
Wright, Chauncey, Blair, 1966
Wulf, Lester, Blair, 1991

Wayne County

Allemann, Leslie & Arlene, Winside, 1993
Ambroz, Connie, Winside, 2002
Anderson, Leland & Adeline, Winside, 1995
Baier, Walter, 1992
Bard, Gordon & Coreen, Wakefield, 1992
Bargholz, Christain & Verona, 1993
Bargholz, Elmer & Mylet, Wakefield, 1995
Biermann, Homer, Wisner, 1980
Biermann, Walter O. & Saundra R., and Aaron D. & Theresa A. Biermann, Wisner, 2014
Bjorklund, Dwaine, Wakefield, 2002
Boeckenhauer family, Cedar, MN, 2013
Boeckenhauer, Maurice & Dee, Wakefield, 1990
Boeckenhauer, Ruth, Wakefield, 1990
Boldt Nykodym, Edna, Norfolk, 1994
Bottger, Erwin, Wakefield, 1992
Brumels, Randall, Norfolk, 2001
Bruning, Marcus J. & Julie M., Hartington, 2013
Carlson, Larry, Glenwood Springs Co, 2002
Carstens, George & Barbara, Hoskins, 1997
Christiansen, Louise, Pender, 1990
Clausen, Jerry Pete, Pender, 1995
Damme, Albert, 1988
Damme, Val, 1988
Daum, Leon, 1997
Deck, Myron & Luetta, Hoskins, 2012
Deck, Paul, Stanton, 1979
Dolph, Gerald / Mike & Lynnelle Sievers, Ft. Lupton, CO, 2004
Driskell, Dorothy, Wakefield, 1991
Dunklau, Randy, 2008
Erickson, Keiht & Fern, 1984
Evans, Dennis & Gloria, Winside, 2000
Finn Trust, Patrick, Carroll, 2011
Fredrickson, Clifford, Wakefield, 1984
Frevert, Kenneth & Darlene, 1991
Gelster, Helen & Melvin, Omaha, 1995
Greenwald, Robert, 1980
Greunke, Duane R. et al, 2004
Greve Farm, Inc., Howard & Barbara Greve and H. Harley & Vickie Greve, Wakefield, 2009
Grone, Edward H. & Dorothy H., 2003
Gubbels, Richard & Rita, Randolph, 2009
Gunnarson, Ted & Deanna, 2005
Haglund, Norman, Wakefield, 2000
Halsey, Renee, Norfolk, 2010
Hansen, Lynnett, Wakefield, 2001
Hansen, Robert & Elaine, Wakefield, 1995
Hattig, Tom & Joyce, LeRoy Hammer, 2006
Hausen, Marilyn, Wakefield, 2012
Henschke, Paul & Arlene, 1988
Herman, Leland Walter, 2008
Herscheid, James, Stanton, 2001
Hoffman, Erna, Hoskins, 2007
Holmes, Donna, Grand Island, 1984
Hugelman, Lloyd & Vivienne, Wakefield, 1991
James, Hazel, 1995
Janke, Dean, Winside, 1996
Jenkins, Zita, 2000
Johnson, Melvin & Dorothy, 1996
Johnson, Ruth L., Battle Creek, 2007
Kai, Rudolph, Pender, 1983
Kay, Willie C., 2006
Kenny, Bob, Carroll, 1991
King Harrington, Ella, 1971
Kniesche, Victor & Janet, 2013
Kniesche, Vince and Ranee May, 2013
Koch, August, Cody WY, 1996
Koepke, Lester, Norfolk, 1989
Koll, Thomas, Winside, 1995
Krusemark, Ron & Corliss, Wakefield, 2011
Lage, Anne, 1984
Lage, Willis & Lois, Carroll, 2004
Langenberg, Stanley, Hoskins, 1978
Loberg, Dan & Kris, Carroll, 2003
Loberg, William, Carroll, 1992
Lubberstedt, William & Jane, Gary Lubberstedt, 2007
Lundahl, LaVern & Mary, Wakefield, 2009
Lutt, Don, 1980
Mann, Victor/ Reba, Norfolk, 1996
Marotz, Lyle Kane, Hoskins, 2001
McGuire, Mrs. P.E., Pender, 1982
Meyer, Jerold D. & Connie, 2006
Meyer, Mrs. Don, 1996
Miller, Beverly, Sterling VA, 1997
Miller, Vernon, Hoskins, 1984
Morris, Edward & Corinne, Carroll, 2001
Nelson, Albert & Janelle, Wakefield, 2006
Nelson, Willis & Margaret, Walkefield, 1989
Noelle, Frank, Laurel, 1997
Nuernberger, Gordon, 1982
Nuernberger, Marvin Fred, Lincoln, 1984
Owens, Keith & JoAnn, Carroll, 2010
Owens, Milton, Carroll, 2005
Pearson, Lydia, Wisner, 2001
Puckett, Darrell or Pearl, West Point, 1977
Puls, Dallas & Ruth, Hoskins, 1991
Puls, Dennis & Ramona, Hoskins, 1984
Rees, John & Dorothy, Carroll, 1991
Rees, Ronald, Randolph, 1993
Roberts, Terry & Laurie, Carroll, 2015
Sandahl, Wayne, Wakefield, 1984
Schellenberg, Dallas & Marcela, Winside, 2000
Scheurich, James/ Sherry, Lincoln, 1996
Schroeder, Lyle & Lois, Anaheim CA, 1999
Schroeder, Marvin Fred, Norfolk, 1981
Sebade, Rhonda & Ron, 1991
Siefken, Arnold & Lois, 2004
Splittgerber, Alma, 1982
Splittgerber, Louise, Wisner, 1981
Splittgerber, Walter, Wisner, 1981
Stoltenberg, Loren & JoAnn, 1999
Stout, James & Barbara, Wakefield, 1992
Stout, James & Barbara, Wakefield, 1999
Strate, Ernest, Hoskins, 1971
Stringer, Mark & Margaret, 1980
Sundell, Alvin & Lawrence, Wakefield, 1995
Temme, Margaret, 1991
Thompson Cragt, Winifred, 1986
Thompson, Levi, 1991
Ulrich, Arthur, Hoskins, 1991
Ulrich, M Gene & Mary, Hoskins, 1986
Vernon Behmer, Ernest, Hoskins, 1983

Victor Duering Bruning, Marta & Mike, Marcia, Marlyce & Jerry, Wayne, 2013
Weich, Larry & Elizabeth, Hoskins, 1993
Wendt, William, Richard, & Dlae, Hoskins, 1978
Westerhold, Pauline, Pender, 2012
Williams, John & Judith, Carroll, 2007
Wittler, Charles, 1989
Youngmeyer, Leslie & Pearl, 1989

Webster County

Alber, David, Guide Rock, 1984
Alber, Mr. & Mrs. Adolph, Blue Hill, 1983
Alber, Victor A., Blue Hill, 2004
Albin, George, Guide Rock, 1989
Albin, Robert & Marlene, Guide Rock, 1979
Anderson, Darrell & Kathy, Blue Hill, 1999
Armstrong, William & Margaret, Blue Hill, 2007
Auten, LaVelle M. Family Trust (LaVelle M. Auten TTE), Phoenix, AZ, 2004
Bartels, Donald & Elaine, Campbell, 1999
Beavers/ Dominy, Beverly/ Karin, Superior, 1993
Berns, David & Sharon, Blue Hill, 2000
Berns, Gerald & Geraldine, Bladen, 2000
Brockman, Joseph & Frances, Caroline Brockman, Lawrence, 2009
Brooks, Wayne & Ardath, Bladen, 1987
Burg, Madie & Jessie, Campbell, 1987
Corner, James & Alice, Blue Hill, 1986
Crumley, Barbara, Red Cloud, 1993
Davis, Marvin, Bladen, 1988
Diederich, Elizabeth, Campbell, 1987
Fritz, Karl & Eunice, Bladen, 1984
Giger, Samuel, Guide Rock, 1971
Greenhalgh, Lloyd & Bonnie, Guide Rock, 1984
Guy, Claire, Guide Rock, 1985
Hansen, Lester, Red Cloud, 1976
Hansen, Richard & Irene, Red Cloud, 1989
Harman, Ron & Janie, Amy, Amber & Annette Hartman, Blue Hill, 2009
Harris, Verlin, Red Cloud, 1982
Hartman, Charles, Alamo, CA, 2011
Harvey, John & Ruth, Inavale, 1990
Heffelbower, Mr. & Mrs. George, Red Cloud, 1970
Heikes, Russell and Marilyn, Berryville, 2016
Hemberger, Roger and Diane, Hastings, 2016
Holcomb Arneson, Mary, Inavale, 1971
Howell, Ricky C., Nankato, 2014
Hubl, George J. c/o Julie Eckert, Aurora, CO, 2004
Hunter, Joe, Guide Rock, 1980
Janssen, Royland, Blue Hill, 2000
Johnson, Cecil, Bladen, 1983
Johnson, Howard & Donna, Hastings, 1986
Johnson, Thomas & Pam, Blue Hill, 2001
Karr Caughenour, Kathleen, Cheyenne WY, 2000
Karr, James (Jim) and Janet, Bladen, 2004
Keeney, Mr. & Mrs. John, Hardy, 1972
Knehans, Gerhard & lenora, Hastings, 1985
Kohmetscher, Curt, Larence, 2007
Konzack, Owen, Guide Rock, 2010
Kottwitz, Burnell & Cecil, Blue Hill, 1991
Kral, James & Virginia, Bladen, 1985
Krueger, Tim & Jody, Blue Hill, 2005
Kucera, Edmund and Dolores, Lawrence, 2003
Larrick, Richard & Jean, Bladen, 1977
Larrick, Ronald, Bladen, 1976
LaVon Rose Trust, Myron & Marlene Petr, Marilyn & Jose and Philip Petr, Blue Hill, 2012
Lewis, Vaughn, Bladen, 1973
Lindgren, Carol, Bladen, 1974
Lockhart, Marjorie, Bladen, 1982
Lutkemeier, Leon, Bladen, 2010
Lutz, Verna, Red Cloud, 1978
Magarin, Gene & Andrea, Blue Hill, 1983
McConkey, Ila Mae E., Blue Hill, 2014
McCoy, Dale, Loveland, CO, 2003
Meents, Wayne & JoAnne, Blue Hill, 1990
Muhleisen, LaVern & Madalynn, Blue Hill, 1988
Nemecek, Veldon, Blue Hill, 1992
Oak Valley Farms Inc., Guide Rock, 1984
Ocker, Glendoris, Glenvil, 1990
Ohmstede, Florence, Guide Rock, 1984
Pavelka, Darlene, Bladen, 2000
Peterson, Henry & Charlotte, Bladen, 1983
Richardson, Earl Dean & Patricia Ann, Guide Rock, 2004
Robb, Don & Susan, Guide Rock, 2001
Russel/ Latta, George/June, Red Cloud, 1986
Schroeder, Robert & Barbara, Blue Hill, 2005
Schroer, Irvin & Marie, Lawrence, 1982
Schutte, Allen and Ron, Guide Rock, 2003
Sidlo, Floyd, Red Cloud, 1979
Siebrass, Harlan & Mildred, Blue Hill, 1992
Siebrass, Harlan & Mmildred, Blue Hill, 2009
Siebrass, Harold, Blue Hill, 1982
Simpson Crews, Mrs. Lois, Guide Rock, 1980
Somerhalder, Robert, Clay Center, 2002
Stout, Edith, Hastings, 1981
Strobl, Rudolf, Red Cloud, 1979
Turner, Carrol, Bladen, 1974
Vance, Glenn & Elaine, Bladen, 2001
Vavricka, Eddie Roy, Red Cloud, 1990
Waechter, Keith & Linda, Bladen, 2001
Wagoner, Laban, Red Cloud, 1972
Wichmann, Carl, Guide Rock, 1974
Wickman, Mrs. Tressie, Guide Rock, 1978
Wilson, Jack & Marilyn, Clarks, 2006
Wilson, Johnny & Bonnie, Bladen, 1977
Wilson, Robert & Diana, Jayne & Bill Hogeland, Bladen, 2008
Wolfe, Mr. & Mrs. Laird, Red Cloud, 1972
Woodward, Allen and Linda, Guide Rock, 2016
Worley, Mae, Campbell, 1985
Zimmerman, William, Ayr, 1991

Wheeler County

Bartak, Charles, Ewing, 2001
Bollwitt, Victor & Mabel, O'Neill, 1992
Bumgardner Ranch, Louise Lane, Limon, 2009
Butler, Alice, Ray & Pauline, Greeley, 1987
Dahlsten Family, Ericson, 2001
Dell Thiele, Vera, Elgin, 1997
Dwyer, Michael, Bartlett, 2002
Esch, Richard & Betty, Spalding, 1989
Hoefener, Daryl & Elizabeth, NE, 2005
Hupp, Virgil, Ewing, 1993
Kusek, Mary, Omaha, 2016
Mignery, Heirs Of Erwin, Bartlett, 1978
Patrick, Paul, Ericson, 1984
Penne, Richard & Marie Penne, NE, 2005
Reimer, Henry & Eleanor, Ewing, 1981
Rossow, Eva & Earl, Elgin, 1985
Schrunk, Merlyn and Nila, Ewing, 2016
Triple L Farm, Walton, 1993

York County

Baker, Dr. Roy, Arlington, 1986
Barr, Robert & Mildred, 1979
Bartlett, Mary Alyce, Utica, 1979
Beck-Smith, Jane, Waco, 1999
Bedient, Owen, Bradshaw, 1992
Behm, Larry, Seward, 1988
Blum, Darrell, 1998
Blum, Gerhardt & Pearl, 1982
Blum, Mr. & Mrs. Herbert, 1972
Boehr, Stanley & Karren, Henderson, 1985
Boehr, Stanley Lou, Henderson, 2009
Bolton, Darrel, Benedict, 2008
Boyer, John E & Lora, Mary Boyer & Phil Waters, 2008
Brahmsteadt, Robert, 1979
Branz, Fred & Pat, Waco, 1984
Bredehoft, Gertrude, 2012
Bristol, Milton, Waco, 1978
Bruns, Anna, Waco, 1983
Buller, Dan & Anna, Bradshaw, 1983
Buller, Maria, Henderson, 1977
Christiansen Family LLLP, Larry & Julie Christiansen, Queen Creek, 2009
Clausen, Donna & Neal, Bloomington, 2012
Collett, William, Bradshaw, 1978
Cool, Harold & Judith Cool, Fairmont, 2005
Dinkelman, Ervin, 1992
Dinkelman, Paul, 1992
Due, Carla, Exeter, 1995
Due, Dean, Exeter, 2000
Ellis, Melvin or Reta, McCool Junction, 1979
Epp, Berdon, Henderson, 1987
Epp, Dale & Sharon, Henderson, 2002
Epp, Jacob, Henderson, 1987
Epp, Pete & Dale, Henderson, 1979
Epp, Peter & Esther, Henderson, 2002
Fassnacht Farms, Fairmont, 2007
Fay, Virginia, 1994
Flick, Mr. & Mrs. Maurice, Bradshaw, 1971
Franz, Gerhard, Henderson, 1976
Friesen, Arnold & Irene, Henderson, 1985
Friesen, Clarence & Gladys, Henderson, 1984
Friesen, Waldo, Henderson, 1978
Gotcher, Mabel, Henderson, 1981
Graves, Daniel, Bradshaw, 1976
Haws, Mrs. Eugene, 1978
Heiden, Arnold, 1972
Heiden, Ralph & Mrs. Lother, 1973
Hohlen, Lydia, 1987
Huber, Thomas, Sutton, 1986
Hurlbut Farm, 1985
Hyde, Virgil, Waco, 1986
Jackson, James K. & JoAnn, Fairmont, 2004
Jackson, Kenneth, McCool Junction, 1982
Jacobsen, James R. & Glenda K., McCool Junction, 2003
Jenkins/ Swartz, Clifford/ Mrs. Cecil, Fairmont, 1975
Jensen, Scott John, Beaver Crossign, 1990
Johnson, Dwight & Genevieve, 2002
Kaiser, Charles & Barbara (Brooke), 2014

Kealiher, Noel, Polk, 1995
Kirkpatrick, Frank, 2000
Kirkpatrick, Mabel Meeker, 1985
Klepper, Kenneth & Marilyn, Waco, 2015
Klone, Clara, 1974
Klute, Esthen, 2006
Kroeker, George, Henderson, 1975
Labart, Yvonne, McCool Junction, 1981
Landers, Hester Whitemore, McCool Junction, 1990
Lang, Claribell, Bradshaw, 1982
Liedtke, Mr. & Mrs. Robert, Stromsburg, 1972
Lloyd Morrison, Evelyn, Omaha, 1985
Lockridge Farms Inc., 1988
Lonsdale, Edward & Twila, 1986
Lutz, Gerald, Fairmont, 2012
Lytle, Don, Benedict, 1986
Maronde, Walter & Lydia, 1982
Merchant, Robert, Polk, 1973
Miltner, Arthur, Waco, 1983
Miltner, Donald & Darlene, Lincoln, 2002
Miltner, Donald, Darlene Miltner, Lincoln, 2006
Miltner, Donald, Lincoln, 2001
Miner, Howard & Verla, 1986
Monnier, Mr. & Mrs. Lawrence, 1973
Morgan, William, Exeter, 1986
Naber, Harlan, Waco, 1983
Neville, Dennis & Anita, 1985
Neville, Edmund, 1984
Neville, Eugene, 1984
Nienhueser, Ines, Waco, 1988
Norquest, Burton & Delbert, 1979
O'Brien, Helen, Lincoln, 1978
Obermier, Anna, Waco, 1983
Obermier, William, Waco, 1983
Osborne, Dr. Robert, 1998
Otte, Ted & Ardell, Lincoln, 1979
Overing, John, 2005
Peter Ellis Farms Inc., Hastings, 1981
Peters, Jacob & Erna, Henderson, 1975
Peters, Reinhart, Henderson, 1978
Peterson, Elaine & Willard, 1985
Plock Smith, Duane & Donna, Fairmont, 1988
Price, James, Trudy Dougherty, Robin Price, South Portland, 2011
Quintessential Acres, LLC, Waco, 2014
R & R Land Co., 1984
Ratzlaff, Marvin, Henderson, 1975
Reetz, George, Waco, 1972
Reetz, Mildred, Benedict, 2009
Reetz, Paul & Gayle, Waco, 1986
Regier, David & Elizabeth, Henderson, 1977
Regier, Fred & Betty, Lushton, 1977
Rempel, Marlin E. & Arlene, McCool Jct, 2012
Richardson & Quick Farms, Benedict, 1982
Romohr, Jennie E., Gresham, 2003
Romsdal Family, 1971
Roush, Thais Ann, Lincoln, 1995
Scheele, Fredrick, Waco, 1986
Sheets Trust, Clark, Grand Junction, 2010
Schlechte, Wilbur, Utica, 1984
Siebert, Peter Ellis Farms Inc., Henderson, 1981
Simmons, Ellen, Aurora, 2005
Smith, Alva, Exeter, 1972
Stephens, Edward, Waco, 1976
Stream/ Gill, Calvin/ Connie, 1979
Stuhr, Edgar & June, Waco, 1977
Stuhr, Reiny & Ella, Waco, 1999
Swan, David, Sheri Swan and Jordan Swan, Karol Swan, Frank Swan, Clay Swan, 2014
Swartz, Marvin & Delores, Fairmont, 2001
Teinert, Edward, 1992
Thiessen, Virgil & Nelda, Hampton, 1998
Trauger, Shirley B. Gillan, Exeter, 2003
Triple D Farms Co, Gresham, 1986
Trollope, Mr. & Mrs. Archie, Waco, 1972
Trollope, Wesley Conley, 2004
Uffelman, Mr. & Mrs. Marvin, Waco, 1972
Volzke, Marvin, Waco, 1985
Von Housen, Ardith, Omaha, 1983
W&M Regier Farms Inc., Henderson, 2013
Wagner, Frances Dreier, Morrow GA, 1998
Walkup, Mr. & Mrs. Claude, 1970
Watson, Gilbert, Waco, 2001
West, Phyllis, McCool Junction, 2011
Westwood, Lynn & Elinor, Waco, 1987
Whitemore, Isabelle Marie, McCool Junction, 2014
Wiemer, Mildred M., Seward, 2013
Wiswell, Frances, Lincoln, 1975
Wochner, George & Anita, Bradshaw, 2007
Wolstenholm, Alma Allen, 1985

Index

The following index includes family last names, farm names and county listings. Entries listed in the farm rosters are not included.